Science and Engineering of Casting Solidification

Doru Michael Stefanescu

Science and Engineering of Casting Solidification

Third Edition

Doru Michael Stefanescu
Metallurgical and Materials Engineering
The University of Alabama
Tuscaloosa, AL, USA

Materials Science and Engineering
The Ohio State University
Columbus, OH, USA

ISBN 978-3-319-33063-1   ISBN 978-3-319-15693-4 (eBook)
DOI 10.1007/978-3-319-15693-4

Springer Cham Heidelberg New York Dordrecht London
© Springer International Publishing Switzerland 2015
Softcover reprint of the hardcover 3rd edition 2015

Printed on acid-free paper

Springer International Publishing AG Switzerland is part of Springer Science+Business Media (www.springer.com)

The book of nature is written in mathematical language

Galileo

Contents

1   Length-Scale in Solidification Analysis 1
    References 5

2   Thermodynamic Concepts—Equilibrium
    and Nonequilibrium During Solidification 7
    2.1   Equilibrium 7
    2.2   The Undercooling Requirement............................. 9
          2.2.1   Curvature Undercooling 12
          2.2.2   Thermal Undercooling............................. 14
          2.2.3   Constitutional Undercooling 15
          2.2.4   Pressure Undercooling 18
          2.2.5   Kinetic Undercooling............................. 19
    2.3   Departure from Equilibrium............................... 21
          2.3.1   Local Interface Equilibrium 22
          2.3.2   Interface Nonequilibrium 23
    2.4   Applications... 26
    References 27

3   Nucleation and Growth Kinetics—Nanoscale Solidification 29
    3.1   Nucleation 29
          3.1.1   Steady-State Nucleation—Homogeneous Nucleation 30
          3.1.2   Steady-State Nucleation—Heterogeneous Nucleation 36
          3.1.3   Time-Dependent (Transient) Nucleation in Pure Metals 43
          3.1.4   Inoculation and Grain Refining....................... 43
          3.1.5   Dynamic Nucleation 45
    3.2   Growth Kinetics 48
          3.2.1   Types of Interfaces 48
          3.2.2   Continuous Growth 52
          3.2.3   Lateral Growth 53
    3.3   Applications... 55
    References 58

4   Fundamentals of Transport Phenomena as Applied to Solidification
      Processing 61
      4.1   General Conservation Transport Equations 61
      4.2   Flux Laws 64
      References 65

5   Diffusive Mass Transport at the Macroscale 67
      5.1   Solute Diffusion-Controlled Segregation 67
      5.2   Equilibrium Solidification 70
      5.3   No Diffusion in Solid, Complete Diffusion in Liquid
             (the Gulliver–Scheil Model) 72
      5.4   No Diffusion in Solid, Limited Diffusion in Liquid 73
      5.5   Limited Diffusion in Solid, Complete Diffusion in Liquid 75
      5.6   Limited Diffusion in Solid and Liquid 79
      5.7   Partial Mixing in Liquid, No Diffusion in Solid 79
      5.8   Summary of Diffusion-Controlled Macrosegregation 80
      5.9   Zone Melting 81
      5.10 Applications 83
      References 88

6   Diffusive Energy Transport at the Macroscale 89
      6.1   Governing Equation for Diffusive Energy Transport 89
      6.2   Boundary Conditions 91
      6.3   Analytical Solutions for Steady-State Solidification
             of Castings 93
      6.4   Analytical Solutions for Non-Steady-State Solidification
             of Castings 94
             6.4.1   Resistance in the Mold 97
             6.4.2   Resistance at the Mold/Solid Interface 100
             6.4.3   The Heat Transfer Coefficient 103
             6.4.4   Resistance in the Solid 104
      6.5   Thermal Analysis 105
             6.5.1   Direct Thermal Analysis 106
             6.5.2   Differential Thermal Analysis 106
      6.6   Applications 114
      References 117

7   Momentum Mass Transport at the Macroscale 119
      7.1   Shrinkage Flow 119
      7.2   Natural Convection 119
      7.3   Surface-Tension-Driven (Marangoni) Convection 122
      7.4   Flow Through the Mushy Zone 123
             7.4.1   The Hagen–Poiseuille Model 123
             7.4.2   The Blake–Kozeny Model 124
      7.5   Segregation Controlled by Fluid Flow 124

7.6   Segregation Controlled by Fluid Flow and Solute Diffusion 126
7.7   Macroshrinkage 128
     7.7.1   Metal Shrinkage and Feeding 128
     7.7.2   Shrinkage Defects 133
References 134

8   Diffusive Mass Transport at the Microscale; Microsolute Redistribution and Microsegregation 135
   8.1   Summary of Microsegregation Models 135
   8.2   Applications 143
   References 144

9   Solidification of Single-Phase Alloys; Cells and Dendrites 145
   9.1   Interface Stability 145
      9.1.1   Thermal Instability 145
      9.1.2   Solutal Instability 147
      9.1.3   Thermal, Solutal, and Surface Energy Driven Morphological Instability 150
      9.1.4   Influence of Convection on Interface Stability 153
   9.2   Morphology of Primary Phases 154
   9.3   Analytical Tip Velocity Models for Cells and Dendrites 157
      9.3.1   Solute Diffusion-Controlled Growth (Isothermal Growth) of Needle-Like Crystals and Dendrites Tip 157
      9.3.2   Thermal Diffusion-Controlled Growth 161
      9.3.3   Solutal, Thermal, and Capillary-Controlled Growth 162
      9.3.4   Interface Anisotropy and the Dendrite Tip Selection Parameter σ^* 169
      9.3.5   Effect of Fluid Flow on Dendrite Tip Velocity 170
      9.3.6   Multicomponent Alloys 171
   9.4   Dendritic Arm Spacing and Coarsening 173
      9.4.1   Primary Arm Spacing 173
      9.4.2   Secondary Arm Spacing 175
      9.4.3   Dendrite Coherency 181
   9.5   The Columnar-to-Equiaxed Transition 182
   9.6   Applications 189
   References 194

10 Solidification of Two-Phase Alloys—Micro-Scale Solidification 197
   10.1 Eutectic Solidification 197
      10.1.1 Classification of Eutectics 197
      10.1.2 Cooperative Eutectics 199
      10.1.3 Models for Regular Eutectic Growth 201
      10.1.4 Models for Irregular Eutectic Growth 207
      10.1.5 Divorced Eutectics 213
      10.1.6 Interface Stability of Eutectics 216

10.1.7 Equiaxed Eutectic Solidification 220
10.2 Peritectic Solidification 221
    10.2.1 Classification of Peritectics 221
    10.2.2 Peritectic Microstructures and Phase Selection 223
    10.2.3 Mechanism of Peritectic Solidification 228
10.3 Monotectic Solidification 234
    10.3.1 Classification of Monotectics 235
    10.3.2 Mechanism of Monotectic Solidification 235
10.4 Applications 240
References 248

11   **Solidification of Multicomponent Alloys** 251
11.1 Thermodynamics of Multicomponent Alloys 251
11.2 Thermophysical Properties 254
    11.2.1 Multicomponent Diffusion 254
    11.2.2 Interface Energy 255
    11.2.3 Microstructure 255
References 262

12   **Microshrinkage** 263
12.1 Defect Size and Shape 263
12.2 The Physics of Shrinkage Porosity Formation 267
    12.2.1 Pressure in the Mushy Zone 270
    12.2.2 Gas Pressure in Pore 272
    12.2.3 Gas Evolution in Liquid 273
    12.2.4 Pore Nucleation 274
    12.2.5 Pore Growth in the Mushy Zone 278
References 280

13   **Rapid Solidification and Amorphous Alloys** 283
13.1 Rapidly Solidified Crystalline Alloys 283
13.2 Metallic Glasses 288
References 294

14   **Semisolid Processing** 295
14.1 Phenomenology 295
14.2 Typical Process Routes 299
    14.2.1 Semisolid Slurry Processing 299
    14.2.2 Forming of the Semisolid Slurry 301
14.3 Material Models/Systems 301
References 303

15   Solidification of Metal Matrix Composites 305
   15.1 Solidification in the Presence of Freely Moving Particles 307
      15.1.1 Particle Interaction with a Planar Interface 308
      15.1.2 Material Properties Models 311
      15.1.3 Kinetic Models 312
      15.1.4 Microstructure Visualization Models 322
      15.1.5 Mechanism of Engulfment (Planar S/L Interface) 323
      15.1.6 Particle Interaction with a Cellular/Dendritic Interface 325
   15.2 Solidification in the Presence of Stationary Reinforcements; the
      Infiltration Pressure 326
      15.2.1 Surface Energy Considerations 327
      15.2.2 Transport Phenomena Considerations 329
      15.2.3 Microstructure Effects 331
   15.3 Processing of Ex-Situ MMCs by Solidification Techniques 332
      15.3.1 Stir Casting 332
      15.3.2 Infiltration of Reinforcements 335
      15.3.3 Spray Casting 336
      15.3.4 Ultrasonic Cavitation 336
   15.4 Processing of In-Situ Metal Matrix Composites 338
   References 339

16   Multiscale Modeling of Solidification 343
   References 344

17   Numerical Macroscale Modeling of Solidification 345
   17.1 Problem Formulation 345
      17.1.1 The Enthalpy Method 346
      17.1.2 The Specific Heat Method 347
      17.1.3 The Temperature Recovery Method 347
   17.2 Discretization of Governing Equations 348
      17.2.1 The Finite Difference Method: Explicit formulation 348
      17.2.2 The Finite Difference Method: Implicit Formulation 352
      17.2.3 The Finite Difference Method: General Implicit and
           Explicit Formulation 353
      17.2.4 Control-Volume Formulation 353
   17.3 Solution of the Discretized Equations 354
   17.4 Macrosegregation Modeling 355
      17.4.1 A Mixture-Theory Model 355
      17.4.2 Effect of Solid Deformation 359
   17.5 Macroshrinkage Modeling 360
      17.5.1 Thermal Models 360
      17.5.2 Thermal/Volume Calculation Models 362
      17.5.3 Thermal/Fluid Flow Models 363
   17.6 Impact of Macromodeling of Solidification on the Metal Casting
      Industry 366

17.7 Analysis of Shrinkage Porosity Models and Defect Prevention 369
17.8 Applications 371
References 375

18 Numerical Microscale Modeling of Solidification 379
18.1 Heterogeneous Nucleation Models 380
18.2 Continuum and Volume-Averaged Models 385
    18.2.1 Problem Formulation 385
    18.2.2 Coupling of Macro-transport and Transformation-Kinetics
            Codes 388
    18.2.3 Dendrite Growth Models 389
    18.2.4 Microporosity Models 399
18.3 Phase Field Models 407
18.4 Stochastic Models 410
    18.4.1 Monte-Carlo Models 412
    18.4.2 Cellular Automaton Models 416
    18.4.3 Lattice Boltzmann Models 426
18.5 Molecular Dynamics Models 428
18.6 Applications 430
References 431

19 Solidification of Some Casting Alloys of Commercial Significance 435
19.1 Steel 435
    19.1.1 Macrostructure 435
    19.1.2 Microstructure 437
    19.1.3 Dendrite Arm Spacing 439
    19.1.4 Nonmetallic Inclusions 440
    19.1.5 Simulation of the Solidification of Steel 441
19.2 Cast Iron 443
    19.2.1 The Structure of Liquid Cast Iron 445
    19.2.2 Graphite Shape 445
    19.2.3 Nucleation and Growth of Austenite Dendrites 447
    19.2.4 Nucleation of Graphite 451
    19.2.5 Growth of Graphite from the Liquid 454
    19.2.6 Eutectic Solidification of Cast Iron 469
    19.2.7 The Gray-to-White Structural Transition 480
    19.2.8 Thermal Analysis of Cast Iron 484
    19.2.9 Simulation of Solidification of Cast Iron 485
19.3 Aluminum–Silicon Alloys 494
    19.3.1 Nucleation and Growth of Primary Aluminum Dendrites . . . 494
    19.3.2 Eutectic Solidification of Al–Si Alloys 495
    19.3.3 Effect of Oxides 500
    19.3.4 Ultrasonic Processing 502
    19.3.5 Thermal Analysis of Aluminum Alloys 504
    19.3.6 Simulation of the Solidification of Aluminum-Based Alloys 505

19.4 Superalloys...509
    19.4.1 Microstructure of Superalloys...........................510
    19.4.2 Solidification Processing of Superalloys 514
    19.4.3 Simulation of the Solidification of Superalloys............522
19.5 Applications...527
References 528

Appendix...535
Appendix A: Some Solutions of the Diffusion Equations 535
Appendix B: Properties of Selected Materials.......................538
Appendix C: Selected Phase Diagrams 545
References 549

Index 551

NOMENCLATURE

C, C_o	alloy composition	p	probability
C_S^*	interface composition in the solid	q	diffusion flux
C_L^*	interface composition in the liquid	r	radius (m)
D	species diffusivity ($m^2 \cdot s^{-1}$)	t	time (s)
E	internal energy ($J \cdot mole^{-1}$ or $J \cdot m^{-3}$)	v	volume (m^3)
F	Helmholtz free energy ($J \cdot mole^{-1}$ or $J \cdot m^{-3}$)	v_a	atomic volume ($m^3 \cdot atom^{-1}$)
G	Gibbs free energy ($J \cdot mole^{-1}$ or $J \cdot m^{-3}$)	v_m	molar volume ($m^3 \cdot mole^{-1}$)
	gradient	ΔC_o	concentration difference between liquid and solid at the solidus temperature
H	enthalpy ($J \cdot mole^{-1}$, $J \cdot m^{-3}$, $J \cdot kg^{-1}$)	ΔG_v	change in volumetric free energy ($J \cdot m^{-3}$)
I	intensity of nucleation (m^{-3})	ΔH	change in volumetric enthalpy ($J \cdot m^{-3}$)
J	mass flux	ΔH_f	latent heat of fusion ($J \cdot mol^{-1}$, $J \cdot kg^{-1}$, $J \cdot m^{-3}$)
K	curvature (m^{-1})	ΔS_f	entropy of fusion ($J \cdot mol^{-1} \cdot K^{-1}$ or $J \cdot m^{-3} \cdot K^{-1}$)
	permeability of porous medium (m^2)	ΔT	undercooling (K)
	equilibrium constant (Sievert's law)	ΔT_c	constitutional undercooling (K)
P	pressure (Pa)	ΔT_k	kinetic undercooling (K)
	Péclet number	ΔT_o	liquidus-solidus interval (K)
Q	volumetric flow rate ($m^3 \cdot s$)	ΔT_r	curvature undercooling (K)
R	gas constant ($J \cdot mol^{-1} K^{-1}$)	Γ	general diffusion coefficient
T	temperature (K or °C)		Gibbs-Thomson coefficient ($m \cdot K$)

T_L	liquidus temperature (K)	Φ	phase quantity
T_S	solidus temperature (K)	α	thermal diffusivity ($m^2 \cdot s^{-1}$)
S	entropy ($J \cdot mol^{-1} \cdot K^{-1}$ or $J \cdot m^{-3} \cdot K^{-1}$)		dimensionless back-diffusion coefficient
V	velocity ($m \cdot s^{-1}$)	β_T	thermal expansion coefficient (K^{-1})
V_o	speed of sound ($m \cdot s^{-1}$)	β_c	solutal expansion coefficient ($wt\%^{-1}$)
c	specific heat ($J \cdot m^{-3} \cdot K^{-1}$)	γ	surface energy ($J \cdot m^{-2}$)
f	mass fraction of phase	δ	boundary layer, disregistry
g	volume fraction of phase	v	kinematic viscosity($m^2 \cdot s$)
$g, \boldsymbol{g}$	gravitational acceleration ($m \cdot s^{-2}$)		vibration frequency
h	heat transfer coefficient ($J \cdot m^{-2} \cdot K^{-1} \cdot s^{-1}$)	ρ	density ($kg \cdot m^{-3}$)
k	solute partition coefficient	λ	interphase spacing (m)
	thermal conductivity ($W \cdot m^{-1} \cdot K^{-1}$)	μ	growth constant
k_B	Boltzman constant		chemical potential ($J \cdot mole^{-1}$)
l	length (m)		dynamic viscosity ($N \cdot m^{-2} \cdot s$)
m	slope of the liquidus line ($K \cdot wt\%^{-1}$)	θ	contact angle
	mass (kg)	τ	momentum flux
n	number of atoms (moles)		

	superscripts		subscripts
het	heterogeneous	*cr*	critical
hom	homogeneous	*e*	equilibrium
m	molar	*eut*	eutectic
r	property related to radius of curvature	*f*	fusion
*	interface	*g*	glass
		het	heterogeneous
		hom	homogeneous
	subscripts		**subscripts**
E	equivalent, eutectic	*i*	component, interface
G	gas	*k*	kinetic
L	liquid	*met*	metastable
P	particle, pressure	*n*	atoms per unit volume
S	solid	*r*	property related to radius of curvature
T	thermal	*s*	surface, stability
c	constitutional, solutal	*st*	stable
		v	property related to volume

Chapter 1
Length-Scale in Solidification Analysis

The art of solidification is one of the oldest in human history, contemporary to Casting, which is the modern technology that most heavily depends on solidification science and engineering. Toward the end of 5000 BC, humans learned to smelt copper ores to yield almost pure copper. The refined copper was then poured into open molds to roughly shape the desired object. After solidification, the casting was forged to obtain the final shape. Clay crucibles and examples of such castings were found in Anatolia (*Anatolia, Cradle of Castings* 2004). By 3500 BC, the early metallurgists learned to produce the first cast alloy, arsenic bronze. Cast iron, the first man-made composite, is at least 2500 years old, but the word "solidification" still had to be invented, and it remained an art for many, many years. While modern physicists contributed to the understanding of atomic-level phenomena specific to solidification, it is not until 1956 that solidification became an engineering science with the development of the constitutional undercooling concept by Chalmers (1956).

As the human species gradually moves from the "iron age" to the age of "engineered materials," of all metal forming processes, the casting process remains the most direct and shortest route from component design to finished product. This makes casting one of the major manufacturing processes, while casting alloys are some of the most widely used materials. Between 2012 and 2013, while the world economy mostly stagnated, casting production increased by 2.4 and 3.4 %, respectively. The main reasons for the longevity of the casting process are the wide range of mechanical and physical properties covered by casting alloys, the near-net shape capability of the casting process, the versatility of the process (weight from grams to hundreds of tons, casting of any metal that can be melted, intricate shapes that cannot be produced by other manufacturing methods), and the competitive delivery price of the manufactured goods. While castings are "invisible" in many of their applications, since they may be part of complex equipment, they are used in 90 % of all manufactured goods.

© Springer International Publishing Switzerland 2015
D. M. Stefanescu, *Science and Engineering of Casting Solidification,*
DOI 10.1007/978-3-319-15693-4_1

Solidification is an inherent part of the casting and welding processes. Today, solidification science encompasses a vast body of knowledge pertaining to phenomena occurring over a large length scale of 10^{-9} (nanoscale phenomena) to 10m (macroscale phenomena).

During solidification, the as-cast structure of the casting is generated. Since many castings are used in the as-cast state (that is, without further thermal or mechanical processing), it follows that the structure which results from solidification, the as-cast structure, is often also the final structure of the casting. It also follows that the mechanical properties of the casting, which are directly dependent on the microstructure, are controlled through the solidification process.

About one billion tons of metal is solidified worldwide annually. The applications of solidified materials include shaped castings, but also a large variety of semifinished products such as atomized powders, continuously cast wires and sheets, continuously cast strands of various cross sections. It is obvious that the modern metal caster must be well educated in the science of solidification.

While solidification science evolved from the need to better understand and further develop casting and welding processes, today, solidification science is at the base of many new developments that fall out of the realm of traditional metal casting.

Castings are made with dimensions of a few millimeters up to tens of meters in length. It is natural, therefore, to assume that the important dimensions to use in describing castings are of that magnitude. However, as the microstructure of the casting (the structure which can be seen using an optical microscope) determines the properties of the casting, it, too, is important. Moreover, because solidification is the process of moving individual atoms from the liquid to a more stable position in the solid alloy lattice, the distances over which atoms must move during solidification are also important. For these reasons, the evolution of the solid/liquid (S/L) interface during solidification must be discussed at three different length scales, macro-, micro-, and nanoscale. However, solidification models often analyze solidification events at an intermediate scale, the mesoscale. These scales are described graphically in Fig. 1.1.

The Macroscale (macrostructure) This scale is of the order of 10^{0}–10^{-3} m. Elements of the macroscale include shrinkage cavity, cold shuts, misruns, macrosegregation, cracks, surface roughness (finish), and casting dimensions. These macrostructure features may sometimes dramatically influence casting and welding properties and consequently castings acceptance by the customer.

At this scale, only two phases are assumed to exist, the solid and the liquid, separated by a sharp S/L interface. The computational models that describe macroscale solidification are based on the solution of conservation equations for mass, energy, species, and momentum. The model output includes the temperature and composition (macrosegregation) fields when the energy and species diffusion equations are solved, and may include shrinkage prediction when the mass and momentum equations are also solved.

MACRO   MESO   MICRO   NANO

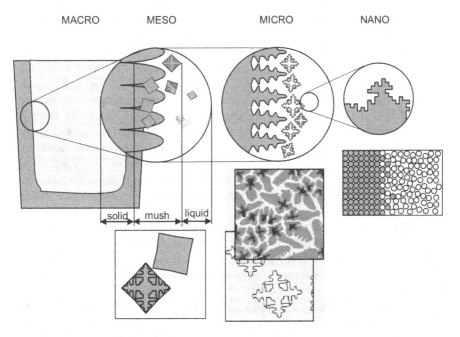

Fig. 1.1 Solidification length scale

The Mesoscale This scale allows description of the microstructure features at grain level, without resolving the grain boundary. Generally, it can be considered that the mesoscale is of the order of 10^{-4} m. It is seen on Fig. 1.1 that the S/L interface that appears as a line at the macroscale is more complex when examined at the mesoscale. There is no clear demarcation between the liquid and the solid. In fact, three regions can be observed: liquid, mushy (containing both liquid and solid), and solid.

The computational models that describe solidification at the mesoscale are typically based on the cellular automaton (CA) technique. The computer is transformed into a dynamic microscope as the evolution of grain morphology can be outputted during the run of the computer model. However, standard transport models of the type developed for the macroscale can be combined with transformation kinetics models to predict microstructure evolution. The older models can calculate volumetric grain density, they rely on average properties and cannot typically output grain morphology.

The Microscale (microstructure) This scale is of the order 10^{-6}–10^{-5} m. The microscale describes the complex morphology of the solidification grain. In a sound casting or weld, mechanical properties depend on the solidification structure at the microscale level. To evaluate the influence of solidification on the properties of the castings it is necessary to know the as-cast grain morphology (i.e., size and type, columnar, or equiaxed), the length scale of the microstructure (e.g., dendritic arm spacing), and the type and concentration of chemical microsegregation. The CA technique or the phase field methods that are used for modeling microstructure evolution at this scale generate all this information.

The Nanoscale (Atomic Scale) This scale is of the order of 10^{-9} m (nanometers) and describes the atomic morphology of the S/L interface. At this scale, solidification is discussed in terms of nucleation and growth kinetics, which proceed by transfer of individual atoms from the liquid to the solid. Currently, there is no database correlating elements of the nanoscale with the properties of castings. However, an accurate description of the S/L interface dynamics requires atomistic calculations. The present knowledge and hardware development does not allow utilization of the atomic scale in applied casting engineering. Nevertheless, accurate solidification modeling may require at least partial use of this scale during computation.

During the last decade, solidification modeling has exhibited a sustained development effort, supported by academic as well as industrial research. The driving force behind this undertaking was the promise of predictive capabilities that will allow process and material improvement, as well as shorter lead times. The most significant recent progress has been incorporation of transformation kinetics, for both the solid/liquid and the solid/solid transformation, in the macro-transport models. The results of these efforts have materialized in a proliferation of publications and commercial software, some of which have penetrated the industry. Numerous claims are made regarding the accuracy and capabilities of modeling methods. They include prediction of casting defects, of microstructure length scale and composition, and even of mechanical properties.

With the advent of faster computers paralleled by the rapid development of numerical methods, the metallurgical aspects of microstructure evolution have finally become a quantitative engineering science. Indeed, in the broadest definition of engineering science, we know only what we can predict through mathematical models. As late as 1975, the solution of the complete macro transport-transformation kinetics problem required for microstructure prediction was considered a "formidable problem" (Maxwell and Hellawell 1975). Microstructure prediction was strictly an empirical exercise, where elements of the microstructure were correlated with processing and material variables. However, today, as proven by the numerous papers that tackle various phenomena occurring during solidification through mathematical/numerical modeling, the task has lost its reputation for inaccessibility. Nevertheless, the problem is far from a final solution. The complete casting process, from initial mold filling to the final stressed component, including defect prediction, has not yet been modeled quantitatively. The main reason is the tremendous computational requirements. Indeed, full casting simulation requires the solution of highly nonlinear discretized equations that may involve a large number of continuum variables (10–20), a complex unstructured mesh (10^5–10^6 nodes), and substantial temporal resolution (10^3–10^4 time steps) (Cross et al. 1998). Certainly, this task can only be addressed using high performance parallel computers.

Table 1.1 Space scales and methods in materials simulation. (Adapted after Raabe 1998)

Scale (m)	Simulation method	Applications
10^{-10}–10^{-6}	Monte Carlo	Thermodynamics, diffusion, ordering
	Molecular dynamics	Structure and dynamics of lattice defects
10^{-10}–10^{0}	Cellular automata	Recrystallization, phase transformation, grain solidification and growth, fluid dynamics
10^{-5}–10^{0}	Large-scale finite element, finite difference, linear iteration, boundary element methods	Averaged solution of differential equations at the macroscopic scale (composition, temperature and electromagnetic fields, hydrodynamics)
10^{-6}–10^{0}	Finite elements or finite difference with constitutive laws considering microstructure	Solidification, microstructure evolution of alloys, fracture mechanics

A more general discussion of the time-space scales in computational materials science and the available simulation methods can be found in Raabe (1998) and Elliott (2011). A short summary is introduced in Table 1.1.

Computational modeling techniques are now widely employed in materials science, as they can enable rapid testing of theoretical predictions or understanding of complex experimental data at relatively low cost. However, many problems involve collective phenomena that occur over a wide range of time and length scales which are intrinsically difficult to capture in a single simulation. Thus, multiscale modeling has been the focus of researchers over the last decades. For a comprehensive review on the general subject, the reader is referred to the excellent review paper by Elliott (2011).

References

Ö Bilgi (ed) (2004) Anatolia, Cradle of Castings. Döktaş, Istanbul

Chalmers B (1956) Trans. AIME 200:519

Cross M, Bailey C, Pericleous KA, Bounds S M, Moran GJ, Taylor GA, McManus K (1998), in: Thomas BG, Beckermann C (eds) Modeling of Casting Welding and Advanced Solidification Processes VIII. The Minerals, Metals and Materials Soc., Warrendale, Pennsylvania, p 787

Elliott JA (2011) International Mater. Rev. 56 (4):207

Maxwell I, Hellawell A (1975) Acta Metall. 23:229

Raabe D (1998) Computational materials science. Wiley-VCH, Weinham, p 5–12

Chapter 2
Thermodynamic Concepts—Equilibrium and Nonequilibrium During Solidification

Thermodynamics is a useful tool for the analysis of solidification. It is used to evaluate alloy phase constitution, the solidification path, basic alloy properties such as partition coefficients, slopes of liquidus, and solidus phase boundaries.

2.1 Equilibrium

The free energy of any phase is a function of pressure, temperature, and composition. Equilibrium is attained when the Gibbs free energy is at a minimum (equivalent to mechanical systems for which equilibrium exists when the potential energy is at a minimum). Thus the condition is:

$$dG(P, T, n_i \ldots) = \left(\frac{\partial G}{\partial T}\right)_{P,n_i\ldots} dT + \left(\frac{\partial G}{\partial P}\right)_{T,n_i\ldots} dP + \left(\frac{\partial G}{\partial n_i}\right)_{T,P,n_j\ldots} dn_i + \ldots = 0$$

(2.1)

where n_i is the number of moles (or atoms) of component i. The partial derivatives of the free energy are called partial molar free energies, or *chemical potentials*:

$$\mu_i = \left(\frac{\partial G}{\partial n_i}\right)_{T,P,n_j,\ldots}$$

(2.2)

At equilibrium, and assuming $T, P = $ constant,

$$dG = \mu_i dn_i + \mu_j dn_j + \ldots = 0$$

(2.3)

For a multiphase system, a condition for equilibrium is that the chemical potential of each component must be the same in all phases (for derivation see inset):

$$\mu_i^\alpha = \mu_i^\beta$$

(2.4)

where the superscripts and β stand for the two phases.

© Springer International Publishing Switzerland 2015
D. M. Stefanescu, *Science and Engineering of Casting Solidification*,
DOI 10.1007/978-3-319-15693-4_2

Derivation of the Equilibrium Criterion

Consider two phases, α and β, within a system at equilibrium. If an amount of dn of component A is transferred from phase α to phase β at $T, P = ct.$, the change in free energy associated with each phase is $dG^\alpha = \mu_A^\alpha dn$ and $dG^\beta = -\mu_A^\beta dn$. The total change in free energy is $dG = dG^\alpha + dG^\beta = (\mu_A^\alpha - \mu_A^\beta)dn$.

Since at equilibrium $dG = 0$, it follows that $\mu_A^\alpha - \mu_A^\beta = 0$.

Although equilibrium conditions do not actually exist in real systems, under the assumption of *local thermodynamic equilibrium*, the liquid and solid composition of metallic alloys can be determined using *equilibrium phase diagrams*. Local equilibrium implies that reaction rates at the solid/liquid interface are rapid when compared to the rate of interface advance. This concept has been shown experimentally to be true up to the solidification velocities of 5 m/s.

Equilibrium phase diagrams describe the structure of a system as a function of composition and temperature, assuming transformation rate is extremely slow, or species diffusion rate is very fast. Two-component phase equilibrium in a binary system occurs when the chemical potentials of the two species are equal.

Phase diagrams were originally obtained from experimental cooling curves. The progress in thermodynamics and computational thermodynamics developed the method of constructing phase diagrams with the help of the Gibbs free energy curves. A simple example for the case of nonideal solution is given in Fig. 2.1. G_m is the Gibbs free energy of mixing, which for nonideal binary solutions is given by:

$$G_m = x_A G_A^o + x_B G_B^o + RT(x_A \ln x_A + x_B \ln x_B) + G_m^{Ex} \tag{2.5}$$

where

x Molar faction of components A or B
G^o Free energy of the pure component A or B
R Gas constant
      $G_m^{Ex} = G_m^{non-ideal} - G_m^{ideal} = H_m^{mix}(1 - AT)$ Excess free energy
A Constant to be evaluated through experiments

In Fig. 2.1, at temperature T_1, the energy of the liquid, G_m^L, is smaller than that of the solid, G_m^S, and the liquid phase is the stable phase at all compositions. At temperature $T_2 < T_1$, the free energy curves intersect. A tangent to the two curves gives the region where the two phases, L and α-solid, coexist. At temperature $T_3 < T_2$, the tangent construction produces two two-phase regions, L + α and L + β. At temperature T_4, the tangent is in contact with the three phases, L, α, and β, corresponding to a triple point, which is the eutectic point on the phase diagram. At temperature T_5, G_m^L is above the tangent, which means that there is no liquid. The central region is a mixture of the two phases α and β.

For an in-depth discussion of the thermodynamics of solidification the reader is referred to Fredriksson and Åkerlind (2012).

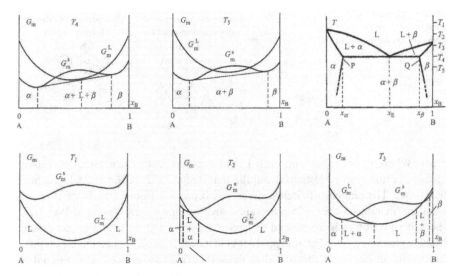

Fig. 2.1 Use of Gibbs free energy curves to calculate a binary phase diagram (Fredriksson and Åkerlind 2012). With permission from Wiley

2.2 The Undercooling Requirement

The driving force of any phase transformation including solidification, which is a liquid-to-solid phase transformation, is the change in free energy. The Helmholtz free energy per mole (molar free energy) or per unit volume (volumetric free energy) of a substance can be expressed as:

$$F = E + P \cdot v - T \cdot S \tag{2.6}$$

E Internal energy, i.e., the amount of work required to separate the atoms of the phase to infinity

P Pressure

v Volume

T Temperature

S Entropy

Thermodynamics stipulates that in a system without outside intervention, the free energy can only decrease.

The entropy is a measure of the amount of disorder in the arrangement of atoms in a phase. In the solid phase, the disorder results from the thermal vibrations of the atoms around their equilibrium position at lattice points. In the liquid phase, additional disorder comes from structural disorder, since the atoms do not occupy all the positions in the lattice as they do in solids. Indeed, the greater thermal energy at higher temperatures introduces not only greater thermal vibrations but also vacancies. Immediately below its melting point a metal may contain 0.1 % vacancies in its

Fig. 2.2 Schematic representation of long- and short-range
order regions (solid and liquid
metals, respectively)

long-range order in
crystalline metal
CN = 12

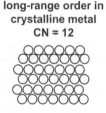

short-range order in
liquid metal
CN < 12

lattice. When the vacancies approach 1 % in a closed-packed structure, the regular 12-fold coordination is destroyed and the long-range order of the crystal structure disappears. The number of nearest neighbors decreases from 12 to 11 or even 10 (the coordination number (CN), becomes smaller than 12, as shown in Fig. 2.2). The pattern becomes irregular and the space per atom and the average interatomic distance are increased. Short-range order is instated. In other words, the liquid possesses a larger degree of disorder than the solid. Thus, the entropy of the liquid is higher than the entropy of the solid. The disorder resulting from melting increases the volume of most materials.

A certain amount of heat, the heat of fusion, is required to melt a specific material. Since the heat of fusion is the energy required to disorganize a mole of atoms, and the melting temperature is a measure of the atomic bond strength, there is a direct correlation between the two.

Let us start our analysis of solidification by introducing a number of simplifying assumptions:

a. Pure metal
b. Constant pressure
c. Flat solid/liquid interface, i.e., the radius of curvature of the interface is $r = \infty$
d. No thermal gradient in the liquid.

For constant pressure, Eq. 2.6 becomes the Gibbs free energy equation:

$$G = H - TS \qquad (2.7)$$

where $H = E + P \cdot v$ is the enthalpy.

Eq. 2.7 is plotted in Fig. 2.3. Since the slope of the line corresponding to the liquid free energy is higher (i.e., $S_L > S_S$), the two lines must intersect at a temperature $T_{e\alpha}$. This is the equilibrium temperature at which no transformation (melting on heating or solidification on cooling) can occur. Under normal nucleation conditions, when the temperature decreases under $T_{e\alpha}$, α-stable solid will form. If nucleation of α is suppressed, β-metastable solid will form at a lower temperature, under $T_{e\beta}$. If nucleation of both α and β is suppressed, then metastable glass forms. The metastable γ solid can only be produced by vapor deposition.

The equilibrium condition Eq. 2.4 can be written for the case of solidification as:

$$\mu_L - \mu_S = 0 \quad \text{or} \quad G_L - G_S = 0 \qquad (2.8)$$

Fig. 2.3 Variation of the free energy of the liquid and solid with temperature

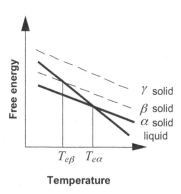

where the subscripts L and S stand for liquid and solid, respectively. This means that at equilibrium the change in chemical potential or in free energy is zero. At the equilibrium temperature, if the two phases coexist:

$$\Delta G_v = G_L - G_S = (H_L - H_S) - T_e(S_L - S_S) = 0$$

Thus, one can further write:

$$\Delta H_f = T_e \Delta S_f \quad \text{or} \quad \Delta S_f = \Delta H_f / T_e$$

Here, $\Delta H_f = H_L - H_S$ is the change in enthalpy during melting, or the volumetric latent heat. ΔS_f is the entropy of fusion (melting). At a temperature lower than T_e:

$$\Delta G_v = \Delta H_f - T\frac{\Delta H_f}{T_e} = \Delta H_f \frac{T_e - T}{T_e} = \Delta S_f \Delta T \qquad (2.9)$$

ΔT is the undercooling at which the liquid-to-solid transformation occurs. From this equation the undercooling is defined as:

$$\Delta T = \Delta G_v / \Delta S_f \qquad (2.10)$$

Note that if $\Delta T = 0$, $\Delta G_v = 0$. This means that, if there is no undercooling under the equilibrium temperature, the system is at equilibrium, and no transformation can occur.

Thermodynamics does not allow further clarification of the nature of undercooling. It simply demonstrates that undercooling is necessary for solidification to occur. Kinetics considerations must be introduced to further understand this phenomenon.

This analysis has been conducted under the four simplifying assumptions (a–d) previously listed. The analysis states that the only change in free energy upon solidification is because of the change of a volume of liquid into a solid, ΔG_v. However, when the four assumptions are relaxed the system will increase its free energy. This increase can be described by the sum of the increases resulting from the relaxation of each particular assumption:

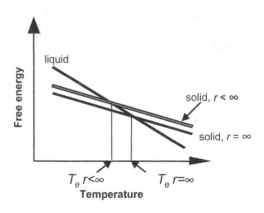

Fig. 2.4 Decrease of equilibrium temperature because of the curvature of the S/L interface

$$\Delta F = -\Delta G_v + \Delta G_r + \Delta G_T + \Delta G_c + \Delta F_P \tag{2.11}$$

The four positive right hand terms are the increase in free energy because of curvature, temperature, composition, and pressure variation, respectively. Let us now evaluate the terms in this equation.

2.2.1 Curvature Undercooling

In the evaluation of the equilibrium temperature presented so far, it has been assumed that the liquid–solid interface is planar (flat), i.e., of infinite radius (assumption c). This is seldom the case in real processes, and never the case at the beginning of solidification, because solidification is initiated at discrete points (nuclei) in the liquid, or at the walls of the mold that contains the liquid. As the volume of a solid particle in a liquid decreases, its surface/volume ratio increases and the contribution of the interface energy to the total free enthalpy of the particle increases. Thus, when the particle size decreases in a liquid–solid system, the total free enthalpy of the solid increases. The curve describing the free energy of the solid in Fig. 2.3 is moved upward by ΔG_r. This results in a decrease of the melting point (equilibrium temperature) as shown in Fig. 2.4.

If solidification begins at a point in the liquid, a spherical particle is assumed to grow in the liquid, and an additional free energy associated with the additional interface, different than ΔG_v, must be considered. This additional energy results from the formation of a new interface and is a function the curvature of the interface.

In two dimensions, the curvature of a function is the change in slope, $\delta\theta$, over a length of arc, δl, (Fig. 2.5):

$$K = \delta\theta/\delta l = \delta\theta/(r\delta\theta) = 1/r \tag{2.12}$$

Fig. 2.5 Definition of curvature

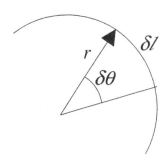

In three dimensions, the curvature is the variation in surface area divided by the corresponding variation in volume:

$$K = dA/dv = 1/r_1 + 1/r_2 \qquad (2.13)$$

where r_1 and r_2 are the principal radii of curvature (minimum and maximum value for a given surface).

For a sphere $r_1 = r_2$ and thus $K = 2/r$
For a cylinder $r_1 = \infty, r_2 = r$ and thus $K = 1/r$.

General Definition of Curvature

In general, if a curve is represented by $\mathbf{r}(t)$, where t is any parameter, the curvature of that curve is:

$K(t) = \dfrac{\sqrt{(\mathbf{r'} \cdot \mathbf{r'})(\mathbf{r''} \cdot \mathbf{r''}) - (\mathbf{r'} \cdot \mathbf{r''})^2}}{(\mathbf{r'} \cdot \mathbf{r'})^{3/2}}$ where $r = dr/dt$ and $r = d^2 r/dt^2$. In Cartesian coordinates, for a curve $y = y(x)$:

$K(x) = \dfrac{|y''|}{(1+y'^2)^{3/2}}$ where $y = dy/dx$, etc.

Assuming that the radius of the spherical particle is r, when the particle increases by dr, the work resulting from the formation of a new surface, $d(4\pi r^2 \gamma)/dr$, must be equal to that resulting from the decrease of the free volumetric energy, i.e., $\frac{d}{dr}\left(\frac{4}{3}\pi r^3 \Delta G_v\right)$. Equating the two, after differentiation, the increase in free energy is:

$$\Delta G_v = 2\gamma/r \quad \text{or, more general} \quad \Delta G_v = \gamma K \qquad (2.14)$$

where

γ Liquid–solid surface energy
K Curvature

Fig. 2.6 Bulk thermal undercooling

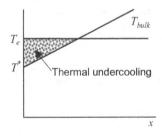

Then, from the definition of undercooling, Eq. (2.9), we obtain:

$$\Delta S_f \Delta T_r = \gamma K \text{ or } \Delta T_r = T_e - T_e^r = \left(\gamma / \Delta S_f \right) K = \Gamma K \qquad (2.15)$$

where ΔT_r is the curvature undercooling, T_e^r is the equilibrium (melting) temperature for a sphere of radius r, and Γ is the Gibbs–Thomson coefficient. The
Gibbs–Thomson coefficient is a measure of the energy required to form a new
surface (or expand an existing one). For most metals $\Gamma = 10^{-7}$ K m. In some calculations, molar ΔH_f and ΔS_f are used, for which the units are J mole^{-1} and J
mole^{-1}K^{-1}, respectively. Then the Gibbs–Thomson coefficient becomes:

$$\Gamma = v_m \gamma / \Delta S_f \qquad (2.16)$$

where v_m is the molar volume in m^3/mole.

For a spherical crystal $\Delta T_r = 2 \ \Gamma/r$. Using this equation it follows that for
$\Delta T_r = 2\,°C$, $r = 0.1 \ \mu m$, and for $\Delta T_r = 0.2\,°C$, $r = 1 \ \mu m$. Thus, the S/L interface
energy is important only for morphologies where $r < 10 \ \mu m$, i.e., nuclei, interface
perturbations, dendrites, and eutectic phases.

2.2.2 Thermal Undercooling

Let us now relax assumption (d), and allow a thermal gradient to exist in the liquid (Fig. 2.6). As long as nucleation of solid and subsequent growth of these nuclei
is rather fast, the only S/L interface undercoolings for the pure metal are kinetic
and curvature. However, if nucleation difficulties are encountered, or if growth of
the solid lags heat transport out of the liquid, an additional undercooling, *thermal
undercooling*, ΔT_T, occurs. When ignoring kinetic undercooling, this additional
undercooling is simply the amount the liquid is under the equilibrium temperature
of the pure metal solidifying with a planar interface (no curvature). Thus, the bulk
thermal undercooling is:

$$\Delta T_T^{bulk} = T_e - T_{bulk} \qquad (2.17)$$

Fig. 2.7 Interface thermal
undercooling

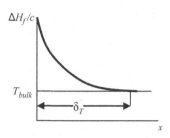

where T_{bulk} is the bulk liquid temperature (temperature far from the interface that can be measured through a thermocouple).

At the S/L interface the rejection of latent heat must also be considered. As shown in Fig. 2.7, a boundary layer of height $\Delta H_f/c$, and length δ_T will form at the interface (position $x = 0$), because of heat accumulation at the interface. The interface thermal undercooling can be calculated as:

$$\Delta T_T^* = T^* - T_{bulk} \tag{2.18}$$

The corresponding increase in free energy is $\Delta G_T = \Delta S_f(T^* - T_{bulk})$. Sometimes, metals can undercool considerably before solidifying. For example, pure iron can be undercooled under its melting (equilibrium) temperature by 300 °C, or even more, under certain controlled conditions.

2.2.3 Constitutional Undercooling

Up to this point, only pure metals have been considered (assumption a). For alloys, the solutal field introduces an additional change in the free energy, which corresponds to an additional undercooling. Fig. 2.8 shows the left corner of the phase diagram of a hypothetical alloy solidifying to form a single-phase solid solution. T_L is the liquidus temperature, T^* is the interface temperature at some arbitrary time during solidification, and T_S is the solidus temperature. Note that for alloys, T_L is the equilibrium temperature T_e. At temperature T^*, the composition of the solid at the interface is C_S^*, while the composition of the liquid is C_L^*. The bulk composition of the alloy, at the beginning of solidification, is C_o. The ratio between the solid composition and the liquid composition at the interface is called the *equilibrium partition coefficient, k*:

$$k = (C_S^*/C_L^*)_{T,P} \tag{2.19}$$

The indices T and P mean that calculations are made at constant temperature (isotherm) and at constant pressure (isobar). Note that at the end of solidification,

Fig. 2.8 Schematic region of a phase diagram for a solid solution alloy

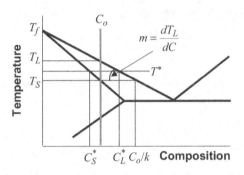

T_S, it can be calculated that the last liquid to solidify should be of composition C_o/k. For the particular case described in Fig. 2.8, there is more solute in the liquid than in the solid at the interface. This partition is the cause of the occurrence of macrosegregation and microsegregation in alloys, to be discussed later in more detail.

The partition coefficient is constant only when the liquidus slope, m, is constant. Since for most of the alloys m is variable, so is k. Nevertheless, for mathematical simplicity, in most analytical calculations m and k are assumed constant. Note that $k < 1$ when the left-hand corner of a phase diagram is considered. However, $k > 1$ when the slopes of the liquidus and solidus lines are positive.

The following relationships exist between various temperatures and compositions in Fig. 2.8:

$$\Delta T_o = T_L - T_S = -m \cdot \Delta C_o \text{ and } \Delta C_o = C_o(1 - k)/k \qquad (2.20)$$

where ΔT_o is the liquidus–solidus temperature interval at C_o, and ΔC_o is the concentration difference between liquid and solid at T_S.

For dilute solutions, the Van't Hoff equation for liquid–solid equilibrium, $d(\ln k_e)/dT = \Delta H_f/RT^2$, holds and can be used to calculate k. Integrating between the melting temperature of solute B, T_f^B, and solvent A, T_f^A, gives:

$$k = \exp\left[\left(\Delta H_f^B/R\right)\left(1/T_f^B - 1/T_f^A\right)\right] \qquad (2.21)$$

In addition, k and m relate as:

$$k = 1 - m\Delta H_f^A/\left[R\left(T_f^A\right)^2\right] \qquad (2.22)$$

Here, ΔH_f^i is the latent heat of phase i, T_f^i is the melting temperature of phase i, and R is the gas constant. The index i stands for the pure solvent, A, or the solute, B.

The difference between the solid and liquid solubility of the alloying element is responsible for the occurrence of an additional undercooling called as *constitutional, or compositional, or solutal, undercooling* (ΔT_c). The concept was first introduced by Chalmers (1956). Consider the diagrams in Fig. 2.9. The first diagram in the

Fig. 2.9 The thermal and
solutal field in front of the
solid/liquid interface

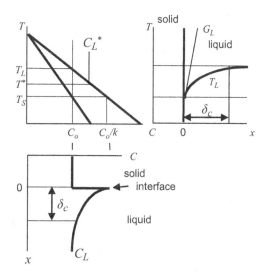

upper left corner is a temperature—composition plot, that is, a phase diagram. C_o is the composition of the solid at temperature T_S, while C_o/k is the composition of the liquid at the same temperature. These compositions have been translated onto the lower diagram, which is a composition—distance (x) diagram. A diffusion boundary layer, δ_c, is shown on the diagram. This layer occurs because at the interface the composition of the liquid is higher (C_o/k) than farther away in the bulk liquid (C_o), and consequently, the composition of the liquid, C_L, decreases from the interface toward the liquid.

The third diagram, on the upper right, is a temperature–distance diagram. It shows that the liquidus temperature in the boundary layer is not constant, but increases from T_S at the interface, to T_L in the bulk liquid. This is a consequence of the change in composition, which varies from C_o/k (at temperature T_S) at the interface, to C_o (at temperature T_L) in the bulk liquid. A liquidus (solutal) temperature gradient, G_L, can now be defined as the derivative of the $T_L(x)$ curve with respect to x at the temperature of the interface, T^* (Fig. 2.10).

Since heat is flowing out from the liquid through the solid, there is also a thermal gradient in the liquid, G_T, which is determined by the evolution of the thermal field. The two gradients are compared in Fig. 2.10. If $G_L < G_T$, the temperature of the liquid ahead of the interface is above the liquidus temperature of the alloy. If on the contrary, $G_L > G_T$, over a certain distance ahead of the interface, the liquid will be at a temperature lower than its liquidus. Thus, while the bulk liquid may be at a temperature above its liquidus, the liquid at the interface may be at a temperature below its liquidus, because of the solute concentration in the diffusion layer. This liquid is constitutionally undercooled. The undercooling associated with this liquid is called *constitutional, or compositional, or solutal, undercooling*, ΔT_c. Based on Fig. 2.9 it can be calculated as:

Fig. 2.10 Constitutional undercooling diagram comparing thermal (G_T) and liquidus (*compositional*) (G_L) gradients

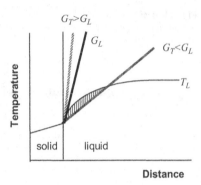

$$\Delta T_c = T_L - T^* = -m(C_L^* - C_o) \tag{2.23}$$

Note that the sign convention here is that m is negative. The corresponding increase in free energy is:

$$\Delta G_c = -\Delta S_f m(C_L^* - C_o) \tag{2.24}$$

2.2.4 Pressure Undercooling

Let us now relax assumption (b) and consider that local pressure is applied on the S/L interface, or that pressure is applied on the whole system. The change in free energy of the liquid and solid with small changes in pressure and temperature can be calculated from Eq. 2.6 as:

$$\Delta F_L = v_L \Delta P - S_L \Delta T \text{ and } \Delta F_S = v_S \Delta P - S_S \Delta T$$

This is true assuming that the internal energy, the volume, and the entropy of the condensed matters (liquid and solid) change little under the proposed conditions. Then, from the equilibrium condition, $\Delta F_L = \Delta F_S$, the change in equilibrium temperature because of the applied pressure is:

$$\Delta T_P = \Delta P \Delta v / \Delta S_f \tag{2.25}$$

This equation is known as the Clapeyron equation. During solidification, the change in volume Δv is positive. Thus, an increase in pressure ($\Delta P > 0$) will result in an increase in undercooling.

For metals, the pressure undercooling is rather small, of the order of 10^{-2} K/atm. Hence, pressure-changes typical for usual processes have little influence on the melting temperature. However, in certain applications, such as particle engulfment by the S/L interface, the local pressure can reach relatively high values, and ΔT_P may become significant. Furthermore, starting again with Eq. 2.6, and using

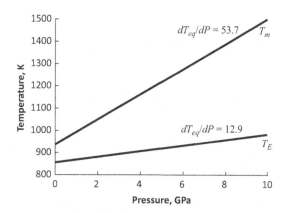

Fig. 2.11 Melting point of aluminum (T_m) and eutectic temperature of Al–Si system (T_E) as function of pressure (after Sobczak et al. 2012)

atoms in solid $\Rightarrow$ atoms in liquid	melting
atoms in liquid $\Rightarrow$ atoms in solid	solidification

differential notations, at constant temperature $dT = 0$, and the equation becomes $(\partial G/\partial P)_T = v$. This means that at constant temperature, the free energy of a phase increases with the increase in pressure, and a new phase with a smaller molar volume may form.

The Clapeyron equation also implies that a change in pressure will impose a change in the equilibrium temperature. Most metals and alloys expand upon melting so that in Eq. 2.25, $\Delta T_P/\Delta P > 0$. Consequently, a pressure increase will lead to an increase in melting temperature (see example in Fig. 2.11). Exceptions include Bi, Sb, Si, and graphitic cast iron, all of which expand upon solidification resulting in a decrease of the melting temperature.

For other effects of pressure on solidification phenomena the reader is referred to the review paper by Sobczak et al. (2012).

2.2.5 Kinetic Undercooling

The concept of undercooling can also be understood in terms of atom kinetics at the S/L interface. While this analysis is done at the atomic scale level, and a more in-depth discussion of this subject will be undertaken in Chapter 3, some concepts will be introduced here for clarity. When an S/L interface moves, the net transfer of atoms at the interface results from the difference between two atomic processes (Verhoeven 1975):

The rate of these two processes is:

$$\text{Rate of melting} (S \rightarrow L) = (dn/dt)_L = p_L n_S v_S \exp(-\Delta G_L/(k_B T)) \quad (2.26)$$

$$\text{Rate of solidification} (L \rightarrow S) = (dn/dt)_S = p_S n_L v_L \exp(-\Delta G_S/(k_B T)) \quad (2.27)$$

Fig. 2.12 Requirement of kinetic undercooling based on atomic kinetics considerations

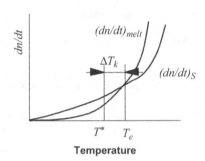

where n_S, n_L are the number of atoms per unit area of solid and liquid interface respectively, v_S, v_L are the vibration frequencies of solid and liquid atoms, respectively, ΔG_{melt}, ΔG_S are the activation energy for an atom jumping through the interface during melting and solidification, respectively, and p_M, p_S are probabilities given by:

$$p_{M,S} = f_{M,S} \cdot A_{M,S} \tag{2.28}$$

Here $f_{M,S}$ is the probability that an atom of sufficient energy is moving toward the interface, and $A_{M,S}$ is the probability that an atom is not kicked back by an elastic collision upon arrival.

At equilibrium, the flux of atoms toward and away from the interface must be equal, that is, $(dn/dt)_M = (dn/dt)_S$. Thus, the two curves must intersect at T_e (Fig. 2.12). For solidification to occur, more atoms must jump from L to S than from S to L. Consequently, the solidifying interface must be at lower temperature than T_e by an amount that is called as *kinetic undercooling*, ΔT_k.

Another approach to this problem (e.g., Biloni and Boettinger 1996) would be to consider that the overall solidification velocity is simply:

$$V = \text{Rate of solidification} - \text{Rate of melting} = V_c - V_c \exp(-\Delta G / RT_i)$$

where ΔG is expressed in J/mole. V_c corresponds to the hypothetical maximum growth velocity at infinite driving force. Then, using series expansion for the exponential term $(1 - e^{-x} \approx x)$, neglecting 2nd and higher order terms, and assuming that Eq. 2.10 is valid near equilibrium we obtain:

$$V = V_c \frac{\Delta H_f \Delta T_k}{RT_e^2} \quad \text{or} \quad \Delta T_k = \frac{RT_e^2}{\Delta H_f} \frac{V}{V_C} \tag{2.29}$$

Two hypotheses have been used to evaluate V_c. The first one (e.g., Turnbull 1962) assumes that the rate of forward movement (atoms incorporation in the solid) is the same as the rate at which atoms can diffuse in the melt. Thus, $V_c = D_L/a_o$, where a_o is the interatomic spacing. The second one, the so-called *collision limited growth model* (Turnbull and Bagley 1975), assumes that the solidification event may be

Table 2.1 Hierarchy of equilibrium. (Boettinger and Perepezko 1986)

Increasing undercooling or solidification velocity ⇓	I. Full diffusional (global) equilibrium
	A. No chemical potential gradients (composition of phases are uniform)
	B. No temperature gradients
	C. Lever rule applicable
	II. Local interfacial equilibrium
	A. Phase diagram gives compositions and temperatures only at liquid–solid interface
	B. Corrections made for interface curvature (Gibbs–Thomson effect)
	III. Metastable local interface equilibrium
	A. Stable phase cannot nucleate or grow sufficiently fast
	B. Metastable phase diagram (a true thermodynamic phase diagram missing the stable phase or phases) gives the interface conditions
	IV. Interface nonequilibrium
	A. Phase diagram fails to give temperature and compositions at the interface
	B. Chemical potentials are not equal at the interface
	C. Free energy functions of phases still lead to criteria for impossible reactions

limited only by the impingement rate of atoms with the crystal surface. Then $V_c = V_o$, where V_o is the speed of sound. Note that V_o is approximately three orders of magnitude higher than D_L/a_o. Experimental analysis of rapidly growing dendrites in pure melts (Coriell and Turnbull 1982) has confirmed the collision limited growth model. Typically, for metals the kinetic undercooling is of the order of 0.01–0.05 K.

2.3 Departure from Equilibrium

We have demonstrated that for solidification to occur a certain amount of undercooling is necessary. Solidification cannot occur at equilibrium. Depending of the amount of undercooling different degrees of departure from equilibrium may occur, following a well-defined hierarchy. As shown in Table 2.1, as the undercooling or the solidification velocity increases, the liquid-to-solid transformation changes from fully diffusional to nondiffusional.

Global equilibrium, (I), requires uniform chemical potentials and temperature across the system. Under such conditions, no changes occur with time. In solidification processing such conditions exist only when the solidification velocity is much smaller than the diffusion velocity. Such conditions truly exist only when solidification takes place over geological times (Biloni and Boettinger 1996), or after long time annealing (see Application 2.1). When global equilibrium exists, the fraction of phases can be calculated with the lever rule, and the phase diagram gives the uniform composition of the liquid and solid phases.

During solidification of most castings, both temperature and composition gradients exist across the casting. Nevertheless, in most cases, the overall kinetics can be described with sufficient accuracy by using the mass, energy, and species transport equations to express the temperature and composition variation within each phase, and equilibrium phase diagrams to evaluate the temperature and composition of phase boundaries, such as the solid/liquid interface. This is the local equilibrium condition, (II). Most phase transformations, with the exception of massive (partitionless) and martensitic transformations can be described with the conditions present under (II).

Metastable equilibrium, (III), can also be used locally at the interface. The most common case is the gray-to-white (metastable-to-stable) transition in cast iron that occurs as the cooling rate increases. The stable eutectic graphite-austenite is gradually substituted by the metastable iron carbide-austenite because of difficulties in the nucleation of graphite and the higher growth velocity of the metastable eutectic. Metastable transformation can occur at solidification velocities exceeding 0.01 m/s. Usually, solidification occurring at rate above this value is termed rapid solidification.

For both stable and metastable local equilibrium, the chemical potentials of the components across the interface must be equal for the liquid and for the solid. However, at large undercooling, achieved for example when using high-solidification velocities, this condition ceases to be obeyed. The solidification velocity exceeds the diffusive speed of solute atoms in the liquid phase. The solute is trapped into the solid at levels exceeding the equilibrium solubility. These conditions, (IV), correspond to rapid solidification. Typically, for solute trapping to occur, the solidification velocity must exceed 5 m/s (Boettinger and Coriell 1986).

The preceding analysis is useful in attempting to classify practical solidification processes based on the degree of equilibrium at which they occur as follows:

- Processes occurring with local interface equilibrium : shape casting, continuous casting, ingot casting, welding (arc, resistance), directional solidification.
- Processes occurring with interface nonequilibrium : welding (laser), melt spinning, atomization, surface remelting.

2.3.1 Local Interface Equilibrium

For the time scale (cooling rates) typical for solidification of castings, the assumption of local interface equilibrium holds very well. However, the interface temperature is not only a function of composition alone, as implied by the phase diagram. Interface curvature, as well as heat and solute diffusion, affects local undercooling. Accordingly, to express the condition for local equilibrium at the S/L interface all

Fig. 2.13 The various components of interface undercooling with respect to the bulk temperature under the condition of local interface equilibrium

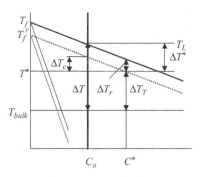

the contributions to the interface undercooling must be considered. The total undercooling at the interface with respect to the bulk temperature, T_{bulk}, is made of the algebraic sum of all the undercoolings previously derived (see Fig. 2.13):

$$\Delta T = \Delta T_k + \Delta T_r + \Delta T_c + \Delta T_T + \Delta T_P \qquad (2.30)$$

Ignoring the kinetic and pressure undercooling, and since $T_L = T_f + m\,C_o$, the interface undercooling under the condition of local equilibrium for castings solidification can be written as:

$$\Delta T = \Delta T_T + \Delta T_c + \Delta T_r = (T^* - T_{bulk}) + (T_L - T^*) + \Gamma K$$
$$= T_f + mC_o + \Gamma K - T_{bulk} \qquad (2.31)$$

where T_f is the melting point of the pure metal (see Application 2.2).

In practical metallurgy, the solidification velocity is increased by increasing the cooling rate. As the cooling rate increases the length scale of the microstructure (e.g., dendrite arm spacing (DAS)) decreases. For cooling rates up to 10^3 K/s, local equilibrium with compositional partitioning between the liquid and solid phases at the solidification interface is maintained. The interface undercooling is small. However, when the cooling rate increases above 10^3 K/s nonequilibrium solidification occurs.

Local equilibrium can occur even at significant undercooling under the equilibrium temperature if nucleation is avoided. In this case, the liquidus and solidus lines can be extended as metastable lines, as shown in Fig. 2.14.

2.3.2 Interface Nonequilibrium

It has been shown that for a multiphase system a condition for equilibrium is that the chemical potential of each component must be the same in all phases, as stated by Eq. 2.4. This is shown graphically in Fig. 2.15. It is noticed that, while the chemical potentials in the liquid and solid are equal, the compositions are not. The necessary

Fig. 2.14 The stable Pb–Sn phase diagram (*solid line*) with superimposed calculated metastable extensions (*dotted lines*) of the liquidus and solidus lines, and measured data. (Fecht and Perepezko 1989)

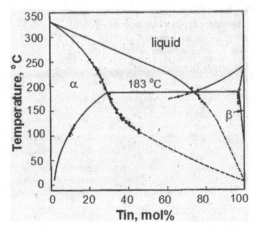

Fig. 2.15 Interface composition and chemical potential for equilibrium and diffusionless solidification (*solute trapping*)

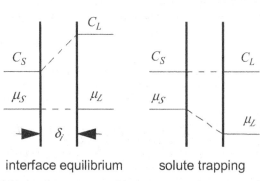

interface equilibrium   solute trapping

condition for interface equilibrium is $V << D_i/\delta_i$, where V is the solidification velocity, D_i is the interfacial diffusion coefficient, and δ_i is the atomic jump distance. Note that D_i is smaller than the bulk liquid diffusion coefficient, D_L. The equilibrium partition coefficient is calculated from the phase diagram with Eq. 2.19. If the ratio between the two velocities is reversed, that is $V >> D_i/\delta_i$, as shown in Fig. 2.15, the equality between the chemical potentials is lost, but the composition becomes uniform across the interface. The partition coefficient becomes one. Solute trapping occurs. Using the typical values of $D_i = 2.5 \ 10^{-9} \ \text{m}^2/\text{s}$ and $\delta_i = 0.5 \ 10^{-9}$ m, the critical velocity for solute trapping is calculated to be 5 m/s.

For solute trapping to occur, the interface temperature must be significantly undercooled with respect to T_L. During partitionless solidification ($C_S^* = C_L^*$), a thermodynamic temperature exists which is the highest interface temperature at which partitionless solidification can occur. This temperature is called the T_o temperature, and is the temperature at which the molar free energies of the solid and liquid phases are equal for the given composition. The locus of T_o over a range of compositions constitutes a T_o curve. The liquid and solid phase compositions are equal along the T_o curve.

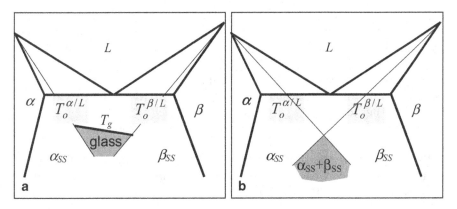

Fig. 2.16 Schematic representation of T_o curves for two different eutectic systems. (Perepezko and Boettinger 1983)

Some examples of such curves are given in Fig. 2.16. They can be used to evaluate the possibility of extension of solubility by rapid melt quenching. If the T_o curves are steep (Fig. 2.16a), single phase α or β crystals with compositions beyond their respective T_o cannot form from the melt. The solidification temperature in the vicinity of the eutectic composition can be depressed to the point where an increased liquid viscosity stops crystallization (glass temperature transition, T_g). If the T_o curves are shallow (Fig. 2.16b), for composition below both T_o curves, a mixture of α and β crystals could form, each phase having the same composition as the liquid.

Baker and Cahn (1971) formulated the general interface condition for solidification of binary alloys by using two response functions:

$$T^* = T\left(V, C_L^*\right) - \Gamma K \tag{2.32}$$

$$C_S^* = C_L^* k^*(V, C_L^*) \tag{2.33}$$

At zero-interface velocity (equilibrium), the functions T and k^* are directly related to the phase diagram. Indeed, $T(0, C_L^*)$ describes the liquidus temperature of the phase diagram and $k^*(0, C_L^*)$ is the equation for the equilibrium partition coefficient, Eq. 2.19. The dependence of k^* on interface curvature is ignored.

Several models have been proposed to describe the dependence of the partition coefficient on velocity. The most widely accepted is the one proposed by Aziz (1982). Ignoring the composition dependence of the partition coefficient, its functional dependence for continuous growth is:

$$k^*(V) = \frac{k_e + \delta_i \cdot V/D_i}{1 + \delta_i \cdot V/D_i} \tag{2.34}$$

where k_e is the equilibrium partition coefficient.

Note that for $V = 0$, $k^* = k_e$, and for very large V, $k^* = 1$. D_i is unknown. In some other models liquid diffusivity rather than interfacial diffusivity is used. The atomic diffusion speed $V_i = D_i/\delta_i$, is usually obtained by fitting Eq. 2.34 to experimental curves showing velocity dependence on partition coefficients. Some typical values for V_i are 17 m/s for Sn (Hoaglund et al. 1991), 33 m/s for Ni-0.6 at% C (Barth et al. 1999), and 5 m/s for Ag-5at% Cu (Boettinger and Coriell 1986). From this analysis it follows that for solute trapping to occur two conditions are necessary: $k^* = 1$ and $T^* < T_0$.

By evaluating the change in free energy and assuming a linear kinetic law for the interface velocity (from Eq. 2.29), Baker and Cahn (1971) calculated the two response functions for a flat interface to be:

$$T^* = T_f + m_L(V)C_L^* + \frac{m_L}{1 - k_e}\frac{V}{V_o} \text{with} m_L(V) = \frac{m_L}{1 - k_e}\left[1 - k^*\left(1 - \ln\frac{k^*}{k_e}\right)\right]$$

(2.35)

$$C_S^* = k^* C_L^* \tag{2.36}$$

Boettinger and Coriell (1986) have proposed a slightly different derivation, substituting the last term in Eq. 2.35 for interface temperature with the kinetic undercooling given by Eq. 2.29, to obtain:

$$T^* = T_f + m_L(V)C_L^* - \frac{RT_e^2}{\Delta H_f}\frac{V}{V_o} \tag{2.37}$$

Note that if $D_i/\delta_i = 0$ and $V_o = \infty$, then the conditions for local interface equilibrium revert to the equations previously introduced:

$$T^* = T_f + m_L C_L^* \text{and} C_S^* = k_e C_L^* \tag{2.38}$$

2.4 Applications

Application 2.1 Calculate the time required for the directional solidification of a rod having the length $l = 10$ cm, so that full diffusional equilibrium operates during solidification.

Answer Assume $D_L = 10^{-9}$ m/s. For equilibrium solidification to occur diffusion will have to go to completion; that is the solute should be able to diffuse over the entire length of the specimen. The diffusion velocity for complete diffusion over the sample of length l is $D_L/l = 10^{-9}/10^{-2} = 10^{-7}$ m/s. The solidification velocity must be much smaller than the diffusion velocity, i.e., $V_S << D_L/l$. Assume $V_S = 10^{-10}$ m/s. Then, the solidification time is $t = l/V_S = 10^{-2}/10^{-10} = 10^8$ s $= 3.17$ years.

Application 2.2 Consider a Cu-10%Sn bronze (phase diagram in Appendix C). Assume solidification with planar S/L interface under local equilibrium conditions. A thermocouple placed far from the interface reads 950 °C. What is the interface undercooling at the beginning of solidification? Calculate the change in interface undercooling when the average (bulk) composition has changed from 10 to 12 %.

Answer The interface undercooling is given by Eq. 2.31. The contribution of curvature is ignored as the interface is planar. From the phase diagram $T_f = 1085$ °C. The liquidus slope can be calculated using values at the temperature of 798 °C, as follows: $m = \Delta T/\Delta C = (1085 - 798)/(-26) = -11$. C_o is given as 10 %. Substituting in Eq. 2.31 we obtain the initial interface undercooling to be $\Delta T = 25$ °C.

The change in interface undercooling when the bulk composition increases to 12 % is simply $m(C_o - C_{bulk}) = -11(10 - 12) = 22$°C.

References

Aziz MJ (1982) J. Appl. Phys. 53:1158

Baker JC, Cahn JW (1971) in: Solidification. ASM Metals Park, OH, p 23

Barth M, Holland-Moritz D, Herlach DM, Matson DM, Flemings MC (1999) in: Hofmeister WH et al. (eds) Solidification 1999. The Minerals, Metals and Materials Soc., Warrendale PA, p 83

Biloni H, Boettinger WJ (1996) Solidification. In: Cahn RW, Haasen P (eds) Physical Metallurgy. Elsevier Science BV, p 670

Boettinger WJ, Perepezko JH (1985) in: Das SK, Kear BH, Adam CM (eds) Rapidly Solidified Crystalline Alloys. The Metallurgical Soc., Warrendale PA, p .21

Boettinger WJ, Coriell SR (1986) in: Sahm PR, Jones H, Adams CM (eds) Science and Technology of the Supercooled Melt. NATO ASI Series E-No. 114, Martinus Nijhoff, Dordrecht, p 81

Chalmers B (1956) Trans. AIME 200:519

Coriell SR, Turnbull D (1982) Acta metall. 30:2135

Fecht HC, Perepezko JH (1989) Metall. Trans. 20A:785

Fredriksson H, Åkerlind U (2012) Solidification and Crystallization in Metals and Alloys. Wiley

Hoaglund DE, Aziz MJ, Stiffer SR, Thomson MO, Tsao JY, Peercy PS (1991) J. Cryst. Growth 109:107

Perepezko J H, Boettinger WJ (1983) Mat. Res. Soc. Symp. Proc. 19:223

Sobczak JJ, Drenchev L, Asthana R (2012) Int. J. Cast Metals Res. 25(1):1

Turnbull D (1962) J. Phys. Chem. 66:609

Turnbull D, Bagley BG (1975) in: Hannay NB (ed) Treatise on Solid State Chemistry. Plenum, NY, 5:513

Verhoeven JD (1975) Fundamentals of Physical Metallurgy. John Wiley & Sons, New York, p 238

Chapter 3
Nucleation and Growth Kinetics—Nanoscale Solidification

As discussed in Sect. 2.2, liquids have short-range order, while solids have long-range order. However, liquids and amorphous solids possess a significant array of short- to medium-range order, which originates from chemical bonding and related interactions. Solidification is the result of the formation of stable clusters of long-range order atoms in the liquid (nucleation), followed by their growth. These are phenomena occurring at the atomic scale (nanometer) level. The present understanding of the beginning of formation of solid crystals from their liquid is based on the classical theory of homogeneous nucleation. This theory uses macroscopic concepts and classic thermodynamics to describe the appearance of the first microscopic crystals in the melt.

Nucleation and growth control the morphology and the fineness of the as-cast microstructure, and even the phases that it contains. In turn, the microstructure controls the mechanical properties of the casting. Thus, in our effort to quantify the evolution of microstructure during solidification, the ultimate goal is to be able to describe the structure as the result of movement of individual atoms. This task is yet out of reach for the castings of commercial size, but significant progress has been made in this direction and calculations can be made for thin films.

3.1 Nucleation

The classic nucleation theory is a phenomenological theory that assumes that clusters of atoms or molecules form spontaneously in the matter undergoing transformation. In other words, steady-state, or time-independent nucleation is assumed, resulting in a constant nucleation rate. While the classical theory is applicable to the study of a wide range of nucleation phenomena, in many cases the assumption of a constant nucleation rate is wrong. Time-dependent nucleation is relevant to many first-order phase transformations including condensation from vapor, crystallization of undercooled liquids, crystallization of glass (devitrification), and others (Kelton 1991).

© Springer International Publishing Switzerland 2015   29
D. M. Stefanescu, *Science and Engineering of Casting Solidification,*
DOI 10.1007/978-3-319-15693-4_3

Alternatively, nucleation can be classified based on the nature of the nucleating substrate as follows:

- homogeneous nucleation
- heterogeneous nucleation
- dynamic nucleation

3.1.1 Steady-State Nucleation—Homogeneous Nucleation

Above or below the equilibrium transformation temperature, fluctuations in density, atomic configurations, heat content, etc., occur in the liquid. They make the formation of minute particles of crystalline solid (long-range order) called *embryos* possible. Consequently, a liquid/solid interface is created, and associated with it is an interface energy. As a result, the free energy of the system increases, and, unless sufficient undercooling is available, the embryo will remelt. If the undercooling of the melt is sufficient, the embryo will survive, and will grow to form a *nucleus*. As the nucleus has the same composition as the liquid and solid, this is called homogeneous nucleation. It can be demonstrated that the embryo must grow to a certain *critical size* in order to become stable and form a nucleus.

Consider that an embryo of radius r is formed in the liquid. This will result in a change in free energy, first, because of the decrease in the free energy resulting from the change of the volume of radius r from liquid state to solid, and second, because of the increase in the free energy due to the newly created liquid/solid interface. The change in free energy is:

$$\Delta G = -v_S \Delta G_v + A_{LS}\gamma_{LS} = -\frac{4}{3}\pi r^3 \Delta G_v + 4\pi r^2 \gamma_{LS}, \qquad (3.1)$$

where v_S is the volume of solid formed, A_{LS} is the newly created liquid/solid interface, γ_{LS} is the surface energy associated with the newly created surface of the grain. The first right hand term (RHT) of this equation, is the driving force for nucleation. The second term represents an energy barrier. The whole equation is represented by the ΔG curve in Fig. 3.1. The maximum of this curve corresponds to a radius r_{cr}, which is called the *critical radius*. If the embryo has reached size r_{cr}, then it can be seen from the figure that further increase of the embryo will result in a decrease of the free energy. This means that spontaneous growth of the embryo, which is now a nucleus, is possible. On the contrary if, $r_{embryo} < r_{cr}$, the embryo will melt, unless the additional energy is removed from the system. To find the value of r_{cr}, one has to find the maximum of the curve described by Eq. 3.1, that is, to equate the first derivative of this equation with respect to r, to zero, $\partial \Delta G / \partial r = 0$. Differentiating and using Eq. 2.9, the size of the critical radius is:

$$r_{cr} = \frac{2\gamma_{LS}}{\Delta G_v} = \frac{2\gamma_{LS}T_e}{\Delta H_f \Delta T} = \frac{2\gamma_{LS}T_e}{\Delta H_f(T_e - T)} = \frac{2\gamma_{LS}}{\Delta S_f(T_f - T)}. \qquad (3.2)$$

Fig. 3.1 Variation of the free energy of the liquid—solid system with the radius of the embryo

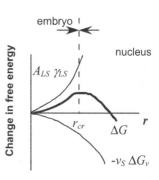

Note that r_{cr} depends inversely on ΔT. For negative values of both, a sphere of liquid in a solid crystal is implied (melting).

Experimental work summarized by Turnbull (1956) has demonstrated that the undercooling required for homogeneous nucleation, ΔT_N, is, for most liquids, larger than 0.15 T_f. For face centered cubic *(FCC)* and body center cubic *(BCC)* metals, $\Delta T_N/T_f \approx 0.18$. However, using the droplet emulsion technique, Perepezko et al. (1988) obtained undercooling almost twice as large as that obtained by previous investigators. According to Turnbull (1981), this may be because continuous coatings on small droplets that have different thermal contraction may generate sufficient stress to displace the thermodynamic equilibrium temperature. Using Turnbull's value for $\Delta T_{\max}$ in Eq. 3.2, and remembering that the size of the nucleus is related to the number of atoms in the nucleus, n_{cr}, by:

$$n_{cr} v_a = (4/3)\pi r_{cr}^3, \tag{3.3}$$

where v_a is the atomic volume (m^3/atom) and n_{cr} has no units, it can be calculated that a nucleus of critical radius includes several hundred atoms (see Application 3.1). Note that if the undercooling is zero, r_{cr} must be infinite, which means that solidification will not occur. In other words, the system is at equilibrium.

Equation 3.2 can be manipulated to obtain the undercooling due to the radius of curvature, ΔT_r, as follows:

$$\Delta T_r = T_f - T_e^* = \frac{\gamma_{LS}}{\Delta S_f} \frac{2}{r} = \Gamma \frac{2}{r}, \tag{3.4}$$

where T_e^* is the temperature of local equilibrium, and Γ is the Gibbs–Thompson coefficient. It can be calculated that for most metals, Γ is approximately $10^{-7}\,°C \cdot m$. Then, for example, for $r = 10$ µm the curvature undercooling is $\Delta T_r = 0.05\,°C$. Thus, the liquid/solid interface energy is important only for morphologies with $r < 10$ µm, i.e., nuclei, interface perturbations, dendrites, and eutectic phases.

The excess free energy of the critical nucleus can be calculated by substituting Eq. 3.2 into 3.1 to obtain:

$$\Delta G_{cr} = 16\pi\gamma^3 / \left(3\Delta G_v^2\right) = 16\pi\gamma^3 T_f^2 / \left(3\Delta H_f^2 \Delta T^2\right).\qquad(3.5)$$

Let us calculate the nucleation velocity. The thermodynamics of embryo forma-tion can be related to the rate of appearance of nuclei, I (intensity or velocity of homogeneous nucleation), through the description of the population distribution of embryos. Consider an undercooled liquid. The molecules of the liquid are collid-ing and continually forming clusters that redissolve because they are too small to be stable. Denoting α' as a molecule of liquid and β_i' as an embryo of solid con-taining i molecules, the sequence of the bimolecular reactions leading to nucleus formation is:

$$\alpha' + \alpha' \leftrightarrow \beta_2'$$

$$\beta_2' + \alpha' \leftrightarrow \beta_3'$$

$$\beta_3' + \alpha' \leftrightarrow \beta_4'$$

$$\dots\dots\dots\dots\dots\dots\dots\dots ,$$

$$\dots\dots\dots\dots\dots\dots\dots\dots$$

$$\beta_{i*-1}' + \alpha' \leftrightarrow \beta_{i*}'$$

$$\beta_{i*}' + \alpha' \leftrightarrow \beta_{i*+1}'$$

where a cluster containing $i*$ molecules is considered as a critical size nucleus which will continue to grow. The equilibrium number of embryos containing i atoms can be obtained by minimizing the free energy of the system with respect to the number n_i of such embryos of size i. The free energy change of the system upon introducing n_i embryos in a liquid containing n_L atoms per unit volume is:

$\Delta G =$ *enthalpy of system with embryos – enthalpy of system without embryos,*

or $\qquad\qquad \Delta G = n_i \Delta G_i - T \Delta S_i.$

The entropy of mixing is $\Delta S_i = -k_B\, n[C \ln C + (1 - C)\ln(1 - C)]$ with $n = n_i + n_L$ $C = n_i/(n_i + n_L)$ and $1 - C = n_L/(n_i + n_L)$. Then:

$$\Delta S_i = -(n_i + n_L)k_B \left(\frac{n_i}{n_i + n_L}\ln\frac{n_i}{n_i + n_L} + \frac{n_L}{n_i + n_L}\ln\frac{n_L}{n_i + n_L}\right),\quad \text{or}$$

$$\Delta S_i = -k_B[-(n_i + n_L)\ln(n_i + n_L) + n_i \ln n_i + n_L \ln n_L].$$

Minimizing the free energy: $\frac{\partial \Delta G}{\partial n_i} = \Delta G_i - T\frac{\partial \Delta S_i}{\partial n_i} = 0$

$$\frac{\partial S_i}{\partial n_i} = -k_B\left[(n_i + n_L)\frac{1}{n_i + n_L} + \ln(n_i + n_L) - n_i\frac{1}{n_i} - \ln n_i\right], \text{ and then}$$

$$\Delta G_i + kT\ln\frac{n_i}{n_i + n_L} = 0$$

Since $n_L \gg ni$ the number of embryos of critical size (nuclei) is:

$$n_i^{cr} = n_L \exp\left(\frac{\Delta G_i}{k_B T}\right). \tag{3.6}$$

Note that $\Delta G_i = \Delta G_{cr}$ is the excess free energy of the critical nucleus given by Eq. 3.5. Assuming that each critical nucleus grows into a crystal and is thereby removed from the distribution of cluster sizes, the velocity of homogeneous nucleation, I, is:

$$I = (\textit{no. of critical nuclei}) \times (\textit{rate of incorporation of new atoms in nuclei})$$
$$\text{or } I = n_i^{cr} \cdot dn/dt,$$

where n_i^{cr} is expressed in m^{-3}, and dn/dt in s^{-1}. Then:

$$dn/dt = (\textit{atomic vibration frequency}) \times (\textit{probability of a successful jump})$$

$$\text{or } \frac{dn}{dt} = v \exp\left(\frac{\Delta G_i}{k_B T}\right),$$

where ΔG_A is the free energy of activation for the transfer of atoms from liquid to crystal. The exponential in this equation can also be understood in terms of the fraction of atoms in the liquid, which are sufficiently activated to surmount the interface addition activation energy. Thus, the nucleation rate is proportional with the product between the probability of formation and the probability of growth of the critical nuclei:

$$I = n_L v \exp\left(-\frac{\Delta G_{cr}}{k_B T}\right) \exp\left(-\frac{\Delta G_A}{k_B T}\right) = n_L v \exp\left(-\frac{\Delta G_{cr} + \Delta G_A}{k_B T}\right). \tag{3.7}$$

For most metals the pre-exponential constant has the value $I_o = 10^{42}$ m^{-3} s^{-1}. Substituting the value of ΔG_{cr} from Eq. 3.5, one can further write:

$$I = I_o \exp\left(-\frac{K_N^{hom}}{T \Delta T^2}\right) \exp\left(-\frac{\Delta G_A}{k_B T}\right),$$

and, since at low ΔT the second exponential is very small, $\exp(-\Delta G_A/k_B T) \approx 0.01$, Eq. 3.7 simplifies to:

$$I = I_o' \exp\left(-\frac{K_N^{hom}}{T \Delta T^2}\right). \tag{3.8}$$

with $I_0' = 10^{40}$ m^{-3} s^{-1}. An alternative derivation of this equation is presented in the inset.

Alternative Derivation of Eq. 3.8 (after Flemings 1974)
The nucleation intensity can be calculated as $I = (n_i^{cr}) \cdot (v_{LS} n_s)$, where v_{LS} is the jump frequency of atoms from the liquid to the solid embryo, and n_S is the number of atoms of liquid in contact with the surface of the embryo. The jump of atoms in the bulk liquid is:

Fig. 3.2 Variation of nucleation velocity with undercooling

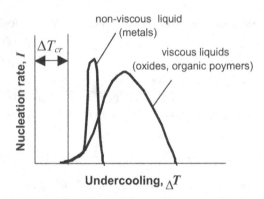

$$v_L = 6D_L/\lambda^2, \tag{3.9}$$

where D_L is the liquid diffusivity, and λ is the jump distance. Assuming a cubic lattice, the jump from liquid to solid is $v_L/6$, that is atoms can jump only at one of the 6 faces. Further assuming that the jump distance is equal to the atomic distance, a, we obtain:

$$v_{LS} = D_L/a^2. \tag{3.10}$$

The number of atoms of liquid in contact with the surface of the embryo can be evaluated as $4\pi r_{cr}^2/a^2$. The expression for the steady-state nucleation velocity is then:

$$I = \left(\frac{D_L}{a^2}\right)\left(\frac{4\pi r_{cr}^2}{a^2}\right) n_L \exp\left(-\frac{\Delta G_{cr}}{k_B T}\right). \tag{3.11}$$

For typical metals, $n_L \sim 10^{28}$ m^{-3}, $D_L \sim 10^{-9}$ m²/s, and $a \sim 3\ 10^{-8}$ m (Perepezko 1988). Substituting these values in Eq. 3.11 we obtain the sought off Eq. 3.8:

$$I \approx 10^{40} \exp\left(-\frac{\Delta G_{cr}}{k_B T}\right) = 10^{40} \exp\left(-\frac{K_N^{hom}}{T\Delta T^2}\right).$$

Equation 3.8 describes the nucleation rate (intensity of nucleation). I is expressed in m^{-3} s^{-1}. The constant K_N^{hom} can be calculated using Eq. 3.5. The nucleation rate equation shows a steep dependency of the nucleation velocity on temperature (see Fig. 3.2). ΔT_{cr} is the critical undercooling, at which nucleation occurs almost instantaneously.

Experimental data on the critical undercooling at which homogeneous nucleation occurs and Eq. 3.8 have been used by Turnbull (1950) to evaluate the solid/liquid

Table 3.1 Relationship between maximum undercooling and liquid/solid interface energy for selected metals

Metal	T_f, K	Turnbull (1950)			Kelton (1991)		
		ΔT_N, K	$\Delta T_N/T_f$	γ_{SL}, J/m^2	ΔT_N, K	$\Delta T_N/T_f$	γ_{SL}, J/m^2
Al	934	195	0.21	0.121	175	0.19	0.108
Bi	544	90	0.17	0.054	227	0.42	0.088
Co	1767	330	0.19	0.234	330	0.19	0.238
Cu	1357	236	0.17	0.177	236	0.17	0.178
Fe	1811	295	0.16	0.204	420	0.23	0.277
Ga	303	76	0.25	0.056	174	0.57	0.077
Hg	234	–	–	–	88	0.38	0.031
Mn	1519	308	0.20	0.206	308	0.20	0.216
Ni	1728	319	0.18	0.255	480	0.28	0.300
Sn	505	118	0.23	0.054	191	0.38	0.075
Pb	600	80	0.13	0.033	240	0.40	0.06
Ti	1940	–	–	–	350	0.18	0.202

(S/L) surface energy of various metals (see Table 3.1 and Application 3.2. Turnbull used the substrate technique). Note that the typical undercooling required for homogeneous nucleation derived by this method is $\Delta T_N = (0.13$ to $0.25) T_f$.

More recently, new experimental techniques such as emulsion of droplets, fluxing, and containerless solidification, produced undercoolings that are significantly higher than predicted by Turnbull (see data summarized by Kelton in Table 3.1). Thus a broader range for the $\Delta T_N/T_f$ seems to exist.

As seen from Table 3.1, Turnbull has used the undercooling data to generate values for the interface free energy of metals and proposed the equation:

$$\gamma_{SL} = \alpha \Delta H_f \rho^{2/3}, \tag{3.12}$$

where α is Turnbull's constant ($= 0.45$ for metals), ΔH_f is the latent heat of melting per atom, and ρ is the solid atomic density. Theoretical calculations, molecular dynamics (MD) simulations and experiments demonstrated that Turnbull's constant varies significantly with crystal structure. An example of results of MD calculations is presented in Fig. 3.3. Note that for FCC metals the constant is 0.55, which is in line with measurements of γ_{SL} through several techniques.

Experiments with alloys have demonstrated that they behave like pure metals, in the sense that homogeneous nucleation starts at similar undercooling. However, experimental data on the maximum undercooling are less reliable than for pure metals because the entrained droplet ("mush quenching") technique used is susceptible of allowing some weak heterogeneous nucleation on the solid solution. The undercooling must be calculated from the liquidus temperature. Some data on alloys undercooling are provided in Table 3.2.

In commercial casting alloys, homogeneous nucleation is virtually inexistent. It is, however, relevant to the new processes involving rapid solidification.

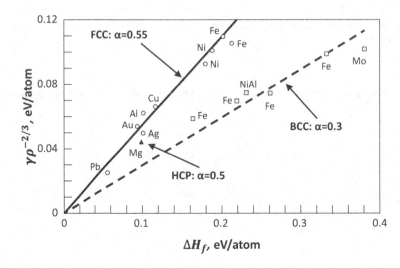

Fig. 3.3 Molecular dynamics calculated interface free energies for metals as a function of latent heat of melting. Note: for Fe and Ni more than one potential was used, resulting in different values for γ (after Hoyt et al. 2006)

Table 3.2 Maximum undercooling in alloys obtained by the entrained droplet technique

Alloy	Primary phase	ΔT_{max}, K	Reference
Cu–Pb	Cu	3	Wang and Smith (1950)
Al–10 wt%Sn	Al	99	Wang and Smith (1950)
Ni–Graphite	Ni	775	Hunter and Chadwick (1972)
Ni–Graphite (+ Mg)	Ni	800	Hunter and Chadwick (1972)
Fe–Graphite	Fe	745	Hunter and Chadwick (1972)
Fe–Graphite (+ Mg)	Fe	280	Hunter and Chadwick (1972)
Al–Si	Al	176	Southin and Chadwick (1978)

The preceding analysis was performed for pure metals. For alloys, constitutional undercooling should also be considered (see Eqs. 2.23 and 2.24).

3.1.2 Steady-State Nucleation—Heterogeneous Nucleation

From Table 3.1 it is seen that the undercooling required for homogeneous nucleation is rather high. For example, for pure iron it can be as high as 420 K. This undercooling is required because of the relatively large activation barrier required for nucleus formation (ΔG_{cr}). In fact, homogeneous nucleation is the most difficult kinetic path for crystal formation. Undercoolings of this magnitude are never observed in commercial casting alloys. The order of magnitude of the undercooling is only a few tens

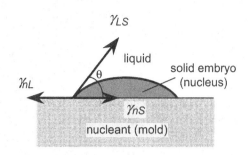

Fig. 3.4 Contact angle between embryo and substrate. Note interfacial energy relationships between substrate (n), liquid (L), and solid (S)

of degrees, at maximum. Solidification can be triggered by preexisting fragments of crystals or solid films of oxides, by the mold walls, or by additions of chemicals made on purpose (called grain refiners or inoculants). This nucleation, occurring on solid substrates foreign to the metal, is called *heterogeneous nucleation*.

Theories for homogeneous nucleation have been extended to cover heterogeneous nucleation, but the discrepancy between experimental data and theory is rather significant. This is because the mechanisms of homogeneous and heterogeneous nucleation are different (Mondolfo 1983). Homogeneous nucleation results from the stabilization of a transient grouping of atoms, so that a nucleus consisting of many atoms is formed all at once. In heterogeneous nucleation, the atoms of the metal to be nucleated attach themselves to the best locations on the nucleant, and the nucleus grows atom by atom.

The classic theory of nucleation shows that the nucleation velocity in the case of heterogeneous nucleation can be calculated with an equation similar to Eq. 3.8 using the contact angle between the growing embryo in the shape of a spherical cap and the substrate (Fig. 3.4). The angle θ is determined by the equilibrium condition between the interface energies of the three phases, liquid, growing solid embryo (nucleus), and substrate nucleant (mold). With this definition, the free energy of formation of a heterogeneous nucleus of critical radius is calculated as:

$$\Delta G^{het} = -v_S \Delta G_v + A_{LS}\gamma_{LS} + A_{nS}(\gamma_{nS} - \gamma_{nL}).$$

The surface area of spherical cap is $A_{LS} = 2\pi r^2(1 - \cos\theta)$. The area between the spherical cap and the substrate is $A_{nS} = \pi r^2 \sin^2\theta$. The volume of the spherical cap is $v_S = (\pi r^3/3)(2 - 3\cos\theta + \cos^3\theta)$. Young's equation gives $\gamma_{nL} = \gamma_{nS} + \gamma_{LS}\cos\theta$. Substituting these in the free energy of formation equation yields:

$$\Delta G^{het} = (-v_{sphere}\Delta G_v + A_{sphere}\gamma_{LS})(2 - 3\cos\theta + \cos^3\theta)/4, \text{ or simply:}$$
$$\Delta G^{het} = \Delta G^{hom} \cdot f(\theta), \quad \text{where} \quad f(\theta) = (2 + \cos\theta)(1 - \cos\theta)^2/4.$$
$$(3.13)$$

Then, the same procedure as for the derivation of the nucleation velocity for homogeneous nucleation can be used. In the calculation of the number of embryos of critical size, n_i^{cr}, the number of surface atoms of nucleation sites per unit volume,

n_a, must be substituted for the number of atoms per unit volume of liquid, n_L. An equation similar to Eq. 3.7 is derived:

$$I_{het} = n_a v \, exp\left(-\frac{\Delta G_{cr}^{het}}{k_B T}\right) \, exp\left(-\frac{\Delta G_A}{k_B T}\right) = n_a \frac{k_B T}{h} exp\left(-\frac{\Delta G_{cr}^{het} + \Delta G_A}{k_B T}\right),$$

where h is Planck's constant. Calculating the free enthalpy of formation of nuclei of critical radius from Eq. 3.13 yields:

$$I_{het} = \frac{n_a}{n_L} I_o' \, exp\left[-\frac{K_N^{hom}}{T \Delta T^2} f(\theta)\right] \tag{3.14}$$

An alternative derivation was proposed by Perepezko (1988) (see inset).

Alternative Derivation of the Nucleation Rate for Heterogeneous Nucleation

The number of atoms in contact with the surface of the embryo can be calculated as:

$$n_s = 2\pi r_{cr}^2 (1 - \cos\theta)/a^2.$$

The numerator is the surface area of the spherical cap. With these changes, Eq. 3.11 becomes:

$$I_{het} = \left(\frac{D_L}{a^2}\right)\left[\frac{2\pi r_{cr}^2 (1 - cos\theta)}{a^2}\right] n_a exp\left[\left(-\frac{\Delta G_{cr}^{hom}}{k_B T}\right) f(\theta)\right] \tag{3.15}$$

Since $n_a \sim 10^{20}$ m^{-3}, and using other numerical values as before, I_{het} can also be written as:

$$I_{het} \approx 10^{30} \, exp\left[\left(-\frac{\Delta G_{cr}^{hom}}{k_B T}\right) f(\theta)\right] = 10^{30} \, exp\left[-\frac{K_N^{hom} f(\theta)}{T \Delta T^2}\right]. \tag{3.16}$$

Note that if the angle θ is small, the nucleation velocity can be significant, even for a rather small undercooling. For the case $\theta = 0°$, $f(\theta) = 0$ and $\Delta G^{het} = 0$. This means that when complete wetting occurs, there is no nucleation barrier. If $\theta = 180°$, that is there is no wetting, $f(\theta) = 1$ and $\Delta G^{het} = \Delta G^{hom}$. If $\theta = 30°$, $f(\theta) = 0.02$, and $\Delta G^{het}/\Delta G^{hom} = 0.02$ which shows a strong influence of the substrate. Even for $\theta = 90°$, $f(\theta) = 0.5$. A schematic comparison between ΔG^{het} and ΔG^{hom} is shown in Fig. 3.5. The heterogeneous nucleation rate also increases with a higher number of available nucleation sites (or higher area of the sites), since I_{het} is proportional to n_a.

Another approach was proposed by Hunt (1984). Equation 3.7 may be rewritten for the case of heterogeneous nucleation as follows:

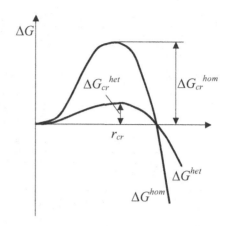

Fig. 3.5 Variation of excess free energy required for homogeneous and heterogeneous nucleation with embryo radius. Note that r_{cr} is independent of the type of nucleation

$$I_{het} = (N_s - N)K_1 \exp\left(-\frac{\Delta G_{cr}^{het}}{k_B T}\right), \tag{3.17}$$

where N_s is the number of heterogeneous substrates originally available per unit volume, N is the number of particles that have already nucleated, and K_1 is a constant. Approximating liquid diffusivity as $D_L \approx a^2 v \exp(-\Delta G_A/k_B T)$, Eq. 3.7 becomes:

$$I_{het} = \frac{n_a D_L}{a^2} \exp\left(-\frac{\Delta G_{cr}^{het}}{k_B T}\right). \tag{3.18}$$

Now, the constant K_1 can be evaluated by comparing Eqs. 3.17 and 3.18 to get $K_1 = n_a D_L a^{-2}/(N_s - N)$. Then, from Eq. 3.5 we calculate: $\frac{\Delta G_{cr}^{het}}{k_B T} = \frac{16}{3} \frac{\pi \gamma_{LS}^3 T_f^2}{\Delta H_f^2 \Delta T^2 k_B T} f(\theta)$.

For metallic systems, T is very large compared with ΔT and can be considered constant. For a given nucleant, $f(\theta)$ is also a constant; thus, one can write $\Delta G_{cr}^{het}/(k_B T) = K_2/\Delta T^2$. Equation 3.17 becomes:

$$I_{het} = (N_s - N)K_1 \exp\left(-\frac{K_2}{\Delta T^2}\right). \tag{3.19}$$

Again, following Hunt's suggestion, K_2 can be calculated if ΔT is defined as the nucleation temperature, ΔT_N, where the initial nucleation rate is $I_{het} = 1 \, \mathrm{m}^{-3} \, \mathrm{s}^{-1}$. Since for this case $N = 0$, $I_{het} = (N_s - N) K_1 \exp(-K_2/\Delta T^2)$ and the preceding equation becomes $K_2 = \Delta T_N^2 \ln(N_S K_1)$.

An example of the calculation of K_1 is given in Application 3.3. With I_{het} known, the final number of nuclei is calculated with:

$$N = \int_0^t I_{het} dt, \tag{3.20}$$

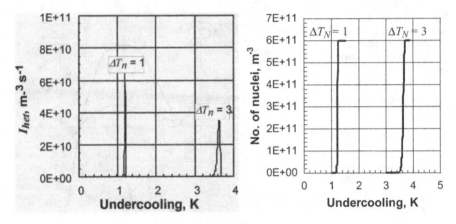

Fig. 3.6 Variation of nucleation rate, I_{het}, and N_s as a function of undercooling

where t is the time required for the number of nuclei to increase from 0 to N_s. Examples of calculation of I_{het} and N as a function of undercooling are given in Fig. 3.6 (see Application 3.4 for details).

The variation of the number of nuclei (nuclei volumetric density) as a function of undercooling is so steep for both nucleation undercoolings considered that, for all practical purposes, the complex nucleation equation, Eq. 3.20 can be substituted with a Dirac delta function (Stefanescu et al. 1990), $\delta(T - T_N)$, having the following definition:

$$\text{for } T \neq T_N, \quad \delta(T - T_N) = 0 \; and$$

$$\text{for } T = T_N, \quad \delta(T - T_N) = 1.$$

Thus, Eq. 3.19 can be written as the *instantaneous nucleation* equation (all nuclei are generated at the nucleation temperature):

$$\frac{\partial N}{\partial t} = N_s \cdot \delta(T - T_N). \tag{3.21}$$

From the preceding analysis it is clear that heterogeneous nucleation is strongly influenced by the solid metal/solid embryo (nucleus) interface energy. The value of this energy depends on the crystal structure of the two phases.

The interface between two crystals can be coherent, semicoherent, or incoherent.

Coherent interfaces may have slight deviations in the interatomic spacing, which causes lattice deformation and induces a strain in the lattice (Fig. 3.7). If the deviation in spacing is too large to be accommodated by strain, dislocations may form in distorted areas. The interface is said to be semicoherent. If there is no crystallographic matching between the two lattices, the structure changes abruptly from one crystal to the other and the interface is incoherent.

Based on the theory summarized by Eq. 3.14, an efficient heterogeneous nucleant (inoculant) should satisfy the following requirements:

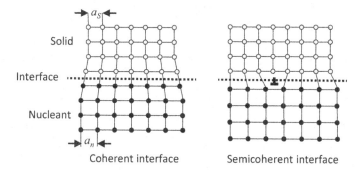

Fig. 3.7 Coherent and semicoherent interfaces

- the substrate must be solid in the melt; its melting point must be higher than the melt temperature, and it must not dissolve in the melt;
- low contact angle between metal and nucleant particles or high surface energy between the liquid and the nucleant, γ_{nL}; indeed if $\gamma_{nL} > \gamma_{nS} + \gamma_{LS} \cos\theta$, the nucleus can spread on the substrate and grow; on the other hand, if $\gamma_{nL} < \gamma_{nS} + \gamma_{LS} \cos\theta$, the nucleus must shrink and disappear;
- the nucleant must expose a large area to the liquid; this will increase n_a, and thus I; this can be achieved by producing a fine dispersion of nucleant, or by using a nucleant with a rough surface geometry;
- because the atoms are attaching to the solid lattice of the substrate, the closer the substrate lattice resembles that of the solid phase, the easier nucleation will be. This means that, ideally, the crystal structure of the substrate and the solid phase should be the same, and that their lattice parameters should be similar (*isomorphism*). They should have at least analogous crystalline planes (*epitaxy*). As the crystal structures of the solidifying alloy and the substrate may be different, the substrate must have one or more planes with atomic spacing and distribution close to that of one of the planes of the solid to be nucleated (coherent or semicoherent interface), i.e., have a low linear disregistry (Turnbull and Vonnegut 1952), δ:

$$\delta = (a_n - a_S)/a_S, \tag{3.22}$$

where a_S and a_n are the interatomic spacing along shared low-index crystal directions in the solid nucleus and the nucleant, respectively. If this is the case, γ_{nS} is very small, and some understanding of the process may be obtained by comparing γ_{LS} and γ_{nL}. Unfortunately, these data are mostly unavailable.

- low symmetry lattice (complex lattice): while it is impossible to assign numbers to lattice symmetry, to some extent the entropy of fusion can be used as a measure of lattice symmetry. In general, less symmetrical lattices have higher entropies of fusion.
- ability to nucleate at very low undercooling.

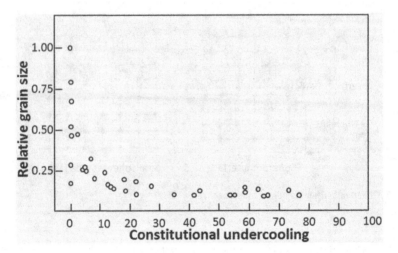

Fig. 3.8 Relative grain size as a function of constitutional undercooling for 34 Al-based binary alloys (after Spittle and Sadli 1995)

The classical theory of nucleation is adequate for small supersaturation but predicts a finite critical nucleus size and a finite work of formation at average concentrations approaching the limit of stability for the solution, i.e., the spinodal. Addressing this error, Cahn and Hilliard (1958) developed the *continuum theory of nucleation*. They derived an expression for the bulk Helmholtz free energy of a binary inhomogeneous solution which includes the contribution of composition gradients. A gradient energy coefficient plays a role in the continuum treatment analogous to the surface energy in the classical theory. Their formulation predicts that the critical nucleus corresponds to that concentration profile which yields an extreme in the free energy. Cahn and Hilliard also show that the work of formation of the critical nucleus approaches zero at the spinodal composition, and that the classical and continuum theories are equivalent in the limit of zero supersaturation.

Hoyt (1990) extended the Cahn–Hilliard continuum theory of nucleation to multicomponent solutions and derived a system of nonlinear differential equations whose solutions yielded the concentration profiles. Computations of the work of formation of a critical nucleus indicated that even small additions of a third element to a binary solution can have a significant effect on the nucleation reaction.

For Al and Ni binary alloys, Tarshis et al. (1971) found that grain size decreases with higher constitutional undercooling $\Delta T_c = -mC_o(1 - k)/k$. Experimental work by Spittle and Sadli (1995) confirmed this dependency for binary Al alloys grain refined with Al–5Ti–1B (constant 0.01 % Ti addition). As one can further observe from Fig. 3.8, it appears that with increasing values of ΔT_c, a minimum grain size is reached at $\Delta T_c \approx 15 - 20$, after which grain size remains constant. This indicates that there is a limit to the extent that grain size can be manipulated by constitutional undercooling.

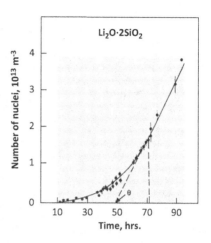

Fig. 3.9 The number of nuclei produced as a function of time at 703 K for lithium disilicate glass (after Fokin et al. 1981)

3.1.3 Time-Dependent (Transient) Nucleation in Pure Metals

To understand the concept of time-dependent nucleation, let us consider a silicate glass during annealing (Kelton 1991). The number of nuclei can be experimentally obtained by recording the number of small crystallites that appear as a function of the annealing time, as shown by the black dots on Fig. 3.9. Assuming a steady-state nucleation rate I^{st}, the number of nuclei per unit volume, N_v, will be:

$$N_v = \int_0^t I(t)dt = I^{st}t. \qquad (3.23)$$

Thus, N_v is a linear function of time with slope I^{st}. Yet, Fig. 3.9 shows a nonlinear experimental correlation between the time for the onset of nucleation, t_o, and the time to steady state, t_{st}. The time-dependent nucleation rate, $I(t)$, given by the local slope of N_v versus t, approaches I^{st} at long annealing times.

Several analytical models for transient nucleation have been proposed. Most of them can be described by an equation of the form:

$$I_{n_{cr},t} = I^{st}(1 - \exp(-t/\tau)), \qquad (3.24)$$

where τ is a function of the rate of monomer addition to a cluster of size n_{cr}, $k^+_{n_{cr}}$, and of the Zeldovich factor, $Z = (|\Delta G|/(6\pi k_B T n_{cr}))^{1/2}$. Here ΔG is the Gibbs free energy per molecule of the new phase, less than that of the initial phase. Numerical calculations (Kelton 1991) seem to indicate that $\tau = (1/2)\pi k^+_{n_{cr}} Z^2$ gives the closest approximation.

3.1.4 Inoculation and Grain Refining

The nucleation concepts introduced in the preceding paragraphs, while rather helpful from the qualitative point of view, fail to accurately predict phenomena occurring

Table 3.3 Examples of typical inoculants used for casting alloys

Metal or alloy	Inoculant
Steel	TiN, TiC
Cast iron	FeSi, SiCa, graphite
Mg alloys	Zr, carbon
Cu-base alloys	Fe, Co, Zr
Al–Si alloys	P, Ti, B
Pb alloys	As, Te
Zn-base alloys	Ti

in real alloys, frequently because of the lack of adequate data. Nevertheless, they are helpful in the understanding of the widely used *inoculation* or *grain refinement* processes. The terms inoculation and grain refinement refer to the same process. Inoculation is often used in metal casting processing in order to control the grain size, and, to a lesser extent, grain morphology. Typical inoculants (grain refiners) for different casting alloys are listed in Table 3.3. Inoculation must not be confused with modification. Modification is a process related mostly to growth and phase morphology. The main purpose of inoculation is to promote grain refinement, while modification is used to change the morphology of the eutectic aggregates.

While the inoculation processes used for specific commercial alloys will be discussed in a larger extent in later sections, some additional information on nucleating agents effective in pure iron will be introduced here, as it has relevancy to the concept of disregistry in Eq. 3.22. First, Bramfitt (1970) concluded that the Turnbull–Vonnegut equation for linear disregistry (Eq. 3.22) cannot be applied to crystallographic combinations of two phases with planes of differing atomic arrangements (e.g., cubic Fe and hexagonal WC). He modified the equation in terms of angular difference between the crystallographic directions within the plane to produce the planar disregistry equation:

$$\delta_{(hkl)_n}^{(hkl)_S} = \sum_{i=1}^{3} \frac{\left| (d_{[uvw]_S^i} \cos\theta) - d_{[uvw]_n^i} \right|}{d_{[uvw]_n^i}} 100,$$
(3.25)

where $(hkl)_S$ is a low-index plane of the substrate, $[uvw]_S$ is a low-index direction in $(hkl)_S$, $(hkl)_n$ is a low-index plane in the nucleated solid, $[uvw]_n$ is a low-index direction in $(hkl)_n$, $d_{[uvw]_n}$ is the interatomic spacing along $[uvw]_n$, $d_{[uvw]_S}$ is the interatomic spacing along $[uvw]_S$, and θ is the angle between $[uvw]_S$ and $[uvw]_n$. The effect of additions of selected carbides and nitrides to pure iron (99.95 %) was then evaluated. Their effectiveness as nucleants was estimated based on the effect of the solidification undercooling. A good nucleant produced a lower undercooling. The main results are listed in Table 3.4 together with the planar disregistry between the nucleant and iron (see Application 3.5 for example of calculation). It is observed that the highly effective inoculants have low disregistry (< 12). It must be noted that

Table 3.4 Nucleating agents for pure iron

Nucleant	Crystal structure	Undercooling, °C	Disregistry relative to ferrite, %	Effectiveness
None	Cubic	30–55	–	–
TiN	Cubic	1.7	3.9	High
TiC	Cubic	1.8	5.9	High
ZrN	Cubic	7.0	11.2	Moderate
ZrC	Cubic	13.6	14.4	Low
WC	Hexagonal	16.1	12.7	Low

for cubic metals there was no difference between the linear and planar disregistry. For the hexagonal WC the planar disregistry was much lower than the linear one (12.7 compared to 29.4). There is a parabolic relationship between the undercooling and the disregistry: $\Delta T_{cr} = 0.25\delta^2$.

The size distribution of the nucleant is critical for heterogeneous nucleation. According to the free growth theory developed by Greer et al. (2000) the undercooling required for free growth can be calculated with:

$$\Delta T_{fg} = 4\gamma_{SL}/(\Delta S_f d), \tag{3.26}$$

where γ_{SL} is the liquid/solid particle interface energy, d is diameter of the inoculant particles, and ΔS_f is the volumetric entropy of fusion. The equation states that the free growth undercooling is inversely proportional to particle size. The larger particles are more potent for heterogeneous nucleation and become active first. Only a small fraction of the particles become active before the onset of recalescence. The model implies that a narrow size distribution of particles is more efficient for heterogeneous nucleation.

Another critical parameter for heterogeneous nucleation is the wettability of the nucleant. The wetting problem can be solved practically by the formation of an intermediate phase which wets the nucleant. Multistep nucleation mechanisms have been proposed for lamellar and spheroidal graphite cast irons (e.g., Jacobs et al. 1974; Skaland et al. 1993; Riposan et al. 2009), and for Mg alloys (Fan et al. 2009).

3.1.5 Dynamic Nucleation

In many cases deliberate additions are not made to the melt, and yet equiaxed grain formation takes place in bulk liquid at small undercooling. Experiments show that dynamic conditions in the liquid may influence nucleation. At least two mechanisms have been proposed for *dynamic nucleation*, the big bang mechanism and the crystal fragmentation mechanism.

Fig. 3.10 Schematic representation of thermal convection and displacement of dendrites from the wall to the center of the mold

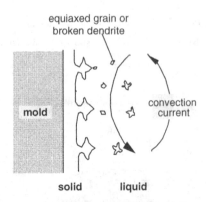

Fig. 3.11 Equiaxed grain accumulation on a steel sieve inserted in a solidifying Al–2 % Cu alloy (Ohno 1987). (With permission from Springer Science and Business Media)

Before discussing the details of these models let us analyse the behavior of the liquid metal in the proximity of the S/L interface. When liquid metal is poured into the mold, liquid motion is induced from the pouring momentum as well as from thermosolutal convection. Thermosolutal convection is generated by the difference in density within the molten metal. The metal close to the interface is colder, and thus in most cases denser than the metal in the middle of the casting. Consequently, flow in the direction of the gravity vector will develop close to the interface. Ascending flow will develop toward the middle of the casting (Fig. 3.10). This is thermal convection. Similarly, the differences in composition next to the S/L interface and in the bulk liquid will induce differences in density, which in turn will produce a flow parallel to the interface. The direction of the flow will depend on the density difference between the solute and the solvent. Due to these combined effects, the liquid at the S/L interface is in motion.

The *big bang mechanism* (Chalmers 1962) assumes that grains can grow from the predendritic nuclei formed during pouring by the initial chilling action of the mold. These grains are then carried into the bulk by fluid flow and survive until superheat has been removed. This model relies on the action of convective currents within the melt.

Compelling experimental evidence for this mechanism has been provided by Davies (1973) and by Ohno (1987). Figure 3.11 shows the results of experiments

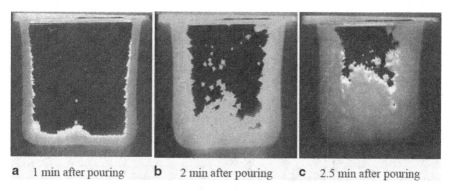

a 1 min after pouring **b** 2 min after pouring **c** 2.5 min after pouring

Fig. 3.12 Broken dendrite branches transported in the center of the ingot by liquid convection in an NH_4Cl–H_2O system ($T_L = 50\,°C$) poured at $75\,°C$ (Jackson et al. 1966)

with a steel net inserted in the middle of a crucible where an Al–2 % Cu alloy was allowed to solidify. It is seen that in the lower part only columnar grains are formed. In the upper part, a large number of small equiaxed grains are seen in the vicinity of the sieve. It is argued that these grains have originated at the mold wall, then have been carried by convection currents in the middle of the ingot, and have been prevented from sinking by the sieve.

Another line of thinking invokes the argument of fragmentation of existing crystals through ripening and local remelting of columnar dendrites (e.g., Jackson et al. 1966). Indeed, for single-phase alloys, a dendrite detachment mechanism has been shown to operate on transparent organic alloys (Fig. 3.12). Nuclei for the equiaxed zone in the middle of an ingot originate from the detached dendrite arms that are carried to the center of the mold by convection currents. If the center of the mold is still above the liquidus, the crystals swept into the center of the mold remelt (Fig. 3.12a). If the center of the mold is undercooled, these crystals act as nuclei for equiaxed grains (Fig. 3.12b). In the case of eutectics, low-gravity experiments have also shown that for regions solidified under low gravity, where convection currents are dramatically reduced, the number of eutectic grains is smaller than for the regions solidified under high gravity (Fig. 3.13). A more in-depth discussion on dendrite fragmentation is provided in Sect. 9.5.

Fragmentation can also be induced through increased convection, ultrasonic vibrations, or a pressure pulse. In the last two cases, nucleation follows because of the change in equilibrium temperature caused by the pressure changes during the collapse of cavitation bubbles. An example of the effect of increased convection through shearing under mechanical mixing for the Mg alloy is presented in Fig. 3.14.

Based on the preceding discussion, it is reasonable to assume that, even in the presence of deliberate grain refining additions, there do exist, at all times, other inherent identifiable "nuclei." It is therefore not surprising that estimation of the volumetric density of nucleation sites before and during solidification of casting

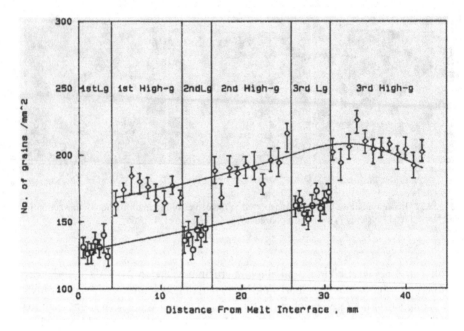

Fig. 3.13 Variation of eutectic grain density with distance from the melt interface in a directional solidification experiment conducted during parabolic flight (Tian et al. 1990). (With permission from Springer Science and Business Media)

alloys is not a trivial problem. Evaluation of nucleation laws required to calculate the volumetric density of growing grains is the weak link in the computer simulation of microstructure evolution.

More detailed analyses of nucleation can be found in the treatment by Flood and Hunt (1988) and Cantor (1997).

3.2 Growth Kinetics

3.2.1 Types of Interfaces

Once the embryo grows to reach critical radius it will spontaneously grow. The interface between the nucleus and the liquid will move (grow) toward the liquid through a mechanism that will largely depend on the nature of the S/L interface. As shown in Fig. 3.15a, two types of interfaces may be considered:

- Diffuse interface (atomically rough): the contour of the liquid/solid interface is not smooth. Each step in Fig. 3.15a corresponds to an atomic distance. This interface will advance in the liquid through *continuous growth*. Random incorporation

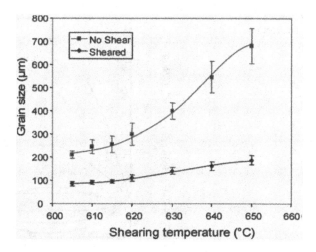

Fig. 3.14 Variation of the grain size as a function of casting temperature for Mg–8.8Al–0.67Zn–0.22Mn alloy cast with melt shearing (45 seconds shearing using a pair of screws with a fixed rotation speed of 800 rpm) and without melt shearing (Fan et al. 2009). (With permission from Maney)

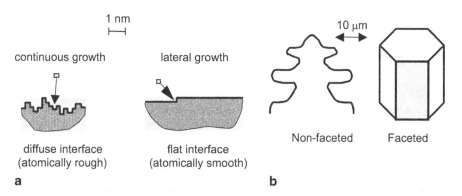

Fig. 3.15 Types of liquid/solid interfaces. **a** Atomic scale. **b** Microscale

of atoms onto the surface will occur, because incoming atoms have many nearest neighbors. This type of growth is typical for metals, and results in a *nonfaceted* liquid/solid interface

- Flat interface (atomically smooth): liquid/solid transition occurs across a single atomic layer. The interface will advance through *lateral growth*. This growth is typical for nonmetals, such as graphite growing in cast iron, or silicon in aluminum alloys, and results in a *faceted* interface

An example of nonfaceted and faceted interface at the microscale level is given in Fig. 3.15b. Whether an interface will be faceted or nonfaceted depends on the nondimensional ratio $\alpha = \Delta S_f / R$, where ΔS_f is the entropy of fusion and R is the gas constant. According to Jackson (1958), the change in relative surface free

Fig. 3.16 Relative surface free energy versus the fraction of occupied surface sites (after Jackson 1958)

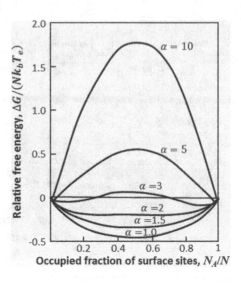

Table 3.5 Correlation between interface morphology and the entropy of fusion

Material	$\Delta S_f/R$	Morphology
Regular metals and some organics	<2	Non faceted
Semimetals and semiconductors, Bi, Sb, Ga, Si	2.2–3.2	Faceted
Most inorganics (e.g., carbides, nitrides)	>3.5	Faceted

energy when atoms are added to a smooth surface can be calculated with:

$$\frac{\Delta G}{Nk_B T_e} = \alpha N_A \frac{N - N_A}{N^2} - \ln \frac{N}{N - N_A} - \frac{N_A}{N} \ln \frac{N - N_A}{N}, \qquad (3.27)$$

where N is the number of atoms in a complete monolayer of surface and N_A is the number of atoms on the surface (see derivation in inset). The relative surface free energy is plotted against the fraction of occupied sites in Fig. 3.16. It can be seen that for $\alpha \leq 2$, the lowest free energy occurs when half of the available surface sites are filled; this means that the surface is rough. For $\alpha > 2$, the lowest free energy occurs either at a low or high fraction of occupied sites, which is when the interface is smooth. The correlation between the entropy of fusion and interface morphology is given in Table 3.5.

Derivation of Eq. 3.27

The basic assumptions are: (1) local equilibrium between solid and liquid at the interface and (2) the final structure is determined by minimization of the free energy. The approach used by Jackson (1958) was based on a nearest-neighbor bond model. The change in free energy of the interface is given by:

$$\Delta G_s = -\Delta E_o - \Delta E_1 + T\Delta S_{LS} - T\Delta S_1 - P\Delta v, \qquad (3.28)$$

where ΔE_o is the energy gained by transferring N_A atoms to the surface, ΔE_1 is the average energy gained by the N_A atoms due to the presence of other atoms on the surface, ΔS_{LS} is the entropy difference between liquid and solid, and ΔS_1 is the entropy difference due to possible randomness of atoms on the surface. Since the change in volume is very small, the last RHT is zero.

If a single atom positioned on the interface has n_S nearest neighbors in the solid before attachment and a maximum number of nearest neighbors n_1, the number of nearest neighbors to an atom in the bulk solid is: $n = 2n_S + n_1$. The energy of an atom in the solid is $2E_{LS}(n_S/n)$, where E_{LS} is the change of internal energy for the atom to transfer from liquid to solid. Then: $\Delta E_o = 2E_{LS}(n_S/n)N_A$. Each of the N_A atoms has a fraction N_A/N of its nearest neighbors sites filled, where N is the number of atoms in a complete monolayer interface. If there are n_1 nearest neighbors, $\Delta E_1 = \Delta E_{LS}(n_1/n)(N_A/N)N_A$. Also, we have $\Delta S_{LS} = (\Delta H_f/T_e)N_A$. The probability of arranging N_A atoms on then surface is $W = N!/[N_A!(N - N_A)!]$. Using Boltzman's relationship ($S = k_B \ln W$) and Stirling's approximation we write:

$$\Delta S_1 = k_B N \ln \frac{N}{N - N_A} + k_B N_A \ln \frac{N - N_A}{N_A}.$$

Introducing all these equations in Eq. 3.28 and dividing each side by $N k_B T_e$, we obtain:

$$\frac{\Delta G_s}{N k_B T_e} = -\frac{E_{LS}}{k_B T_e}\frac{N_A}{N}\left(\frac{N_A}{N}\frac{n_1}{n} + \frac{2n_S}{n}\right) + \frac{T\Delta H_f N_A}{N k_B T_e^2} - \frac{T}{T_e}\ln\left(\frac{N}{N - N_A}\right)$$
$$- \frac{T}{T_e}\frac{N_A}{N}\ln\left(\frac{N - N_A}{N_A}\right).$$

Considering the equilibrium situation, $T = T_e$. Further, since E_{LS} is the latent heat of fusion, and using the notation:

$$\frac{E_{LS}}{k_B T_e}\frac{n_1}{n} = \Delta S_f\frac{n_1}{n} = \alpha,$$

Jackson obtained Eq. 3.27. The ratio n_1/n depends on the crystallography of the interface.

As will be demonstrated in the following paragraphs, continuous growth is considerably faster than lateral growth. Indeed, it was calculated that the kinetic undercooling required for continuous growth is typically of the order of 0.01–0.05 K, while the undercooling required for lateral growth is of the order of 1–2 K.

3.2.2 Continuous Growth

Under the assumption that the rate at which atoms can move across the S/L interface to join the solid is similar to the rate at which atoms can diffuse in the melt (the *diffusion limited* growth model), the kinetics of continuous growth, that is the growth velocity, V, can be described by the equation (Turnbull 1949):

$$V = \mu_o \Delta T_k \text{ with } \mu_o = \frac{\beta D_L \Delta H_f}{a k_B T_f^2}, \tag{3.29}$$

where μ_o is the growth constant, ΔT_k is the kinetic undercooling required for atom attachment, ΔH_f is the latent heat of fusion, a is the amount the interface advances when an atom is added, and β is a correction factor that can be written as $\beta = (a/\lambda)^2 6 v_{LS}/v_L$, where λ is the jump distance for an atom in the liquid, v_{LS} is the frequency with which atoms jump across the liquid/solid interface, and v_L is the frequency with which atoms jump in the bulk liquid. The derivation of this equation is given in the inset.

Derivation of Eq. 3.29

Turnbull (1949) described kinetics of continuous growth using classic rate theory. The continuous growth velocity is $V = a \, v_{net}$, where v_{net} is the net jump frequency across the interface. In turn, the net jump frequency with which atoms can jump from liquid to solid at the melting temperature is:

$$v_{net} = v_{LS} - v_{SL} = v_{LS}[1 - \exp(-\Delta G_v/(k_B T_e))],$$

where ΔG_v is the activation energy for transport of atoms from liquid to solid (the difference in free energy between the liquid and the solid). Since $1 - \exp(-x) \approx x$, and using Eq. 2.9:

$$v_{net} = v_{LS}\frac{\Delta G_v}{k_B T_e} = v_{LS}\frac{\Delta H_f \Delta T_k}{k_B T_f^2}.$$

Substituting v_{LS} from Eq. 3.10, the continuous solidification velocity becomes:

$$V = \frac{D_L \Delta H_f}{a k_B T_f^2}\Delta T_k = \mu_o \Delta T_k.$$

However, since it is believed that λ may be as much as an order of magnitude smaller than a, a correction factor, β must be attached to μ_o.

Equation 3.29 shows a linear dependency of V on ΔT. μ_o has values between 10^{-2} and $1 \text{ m s}^{-1} \, ^\circ\text{C}^{-1}$. An example of calculation of μ_o and of the kinetic undercooling is given in Application 3.6.

In pure metals solidification can occur at very high undercoolings, where D_L is very small, and thus growth cannot be diffusion limited. It was suggested that the frequency of liquid–atom collisions with the crystal face is the rate-limiting factor (*collision-limited* growth model). Using MD simulations Broughton et al. (1982) proposed a growth law that implies a kinetic coefficient of the form:

$$\mu_o = 1.2 V_T \Delta H_f / \left(k_B T_f^2 \right) , \tag{3.30}$$

where $V_T = (k_B T / M)^{1/2}$ is the thermal velocity and M is the atomic mass. The numerical constant was derived from a fit to the MD data. An upper bound for the growth coefficient was suggested by Coriell and Turnbull (1982) as $\mu_o < V_S H_f / \left(k_B T_f^2 \right)$, where V_s is the sound velocity. Further research resulted in the development of the density-functional theory for μ (Chernov 2004), reducing the value of the numerical constant in Eq. 3.30 to 0.72 for {100} interfaces, and further suggesting that the constant is a function of crystalline anisotropy.

3.2.3 Lateral Growth

The kinetics of lateral growth can be described by different equations, depending on the growth mechanism. Two growth mechanisms will be discussed: growth by screw dislocation and growth by two-dimensional (2D) nucleation.

Growth by Screw Dislocations When growth occurs through a screw dislocation mechanism, atoms are continuously added at the step of a screw dislocation, which rotates about the point where the dislocation emerges. For one dislocation, it was demonstrated that growth kinetics could be described by the parabolic equation:

$$V = \mu_1 (\Delta T_k)^2 \text{ with } \mu_1 = \frac{(1 + 2g^{1/2})}{g} \frac{\beta D_L \Delta H_f^2}{4\pi \gamma T_f^3 k v_m}, \tag{3.31}$$

where g is the diffusness parameter and v_m is the molar volume. Difuseness is defined by the number of atomic layers, n, comprising the transition from solid to liquid at T_f for a diffuse interface. For flat interfaces $g = 1$, while for diffuse interfaces $g << 1$. The diffusness parameter can be calculated as $g = (\pi^4 / 8)n^3 \exp(-\pi^2 n/2)$.

In metals, the typical number of dislocations is 10^8 cm^{-2}. The growth equation presented above is however valid for this case also. Experimental evidence demonstrates that when dislocations were introduced into a crystal the growth rate increased rapidly.

Growth by Two-dimensional Nucleation A 2D nucleus grows by addition of atoms on the lateral sides of the nucleus (Fig. 3.17). The excess free energy required for the formation of a 2D nucleus is $\Delta G = -\pi r^2 a \Delta G_v + 2\pi r a \gamma$. Then:

Fig. 3.17 Growth of a two-dimensional nucleus on a solid substrate

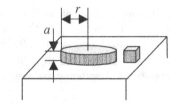

Fig. 3.18 Comparison between growth velocities by different mechanisms

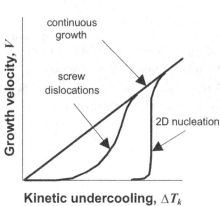

$$\frac{\partial \Delta G}{\partial r} = -2\pi r_{cr} a \Delta G_v + 2\pi a \gamma = 0 \quad \text{and} \quad r_{cr} = \frac{\gamma}{\Delta G_v} = \frac{\gamma T_f}{\Delta H_f \Delta T} \quad \text{with}$$

$$\Delta G_{cr} = \frac{\pi a \gamma^2 T_f}{\Delta H_f \Delta T}$$

Following steps as in 3D nucleation it is calculated that:

$$V = \mu_4 \exp\left(\frac{\mu_3}{3 \Delta T_k}\right) \tag{3.32}$$

with $\mu_3 = \mu_o \dfrac{\pi g B^2 a T_f^2}{\beta D_L}$ and $\mu_4 = \mu_o \left(\dfrac{\Delta H_f}{k_B T_f^2}\right)^{1/6} (\Delta T_k)^{7/6} (2 + g^{-1/2})$, where B is Turnbull's empirical relationship between γ and ΔH_f written as $B = v_m \gamma / (a \Delta H_f)$. B is equal to 0.5 for metals and 0.35 for nonmetals. Note that in this case the dependency of V on ΔT is exponential.

A comparison between the various growth kinetics is shown schematically in Fig. 3.18. It can be seen that for a given kinetic undercooling continuous growth is the fastest.

Fig. 3.19 Correlation
between nucleation velocity
and the LS interface energy

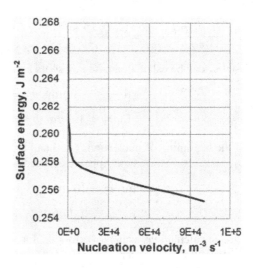

3.3 Applications

Application 3.1

Calculate the number of copper atoms included into a homogeneous nucleus of
critical size.

Answer Using Eq. 3.2 and data from Appendix B and Table 3.1 it is calculated that
the critical radius of the copper nucleus is $1.035 \cdot 10^{-9}$ m. The volume of a copper
atom is $v_a = v_m/N_{Av} = 1.18 \cdot 10^{-29}$ m^3, where v_m is the molar volume and N_{Av} is
Avogadro's number. Then, from Eq. 3.3 it is obtained that $n_{cr} = 394$.

Application 3.2

The maximum undercooling observed in liquid nickel is $319\,^{\circ}$C. Assuming that
homogeneous nucleation occurs at this temperature, calculate the liquid/solid
interface energy for nickel.

Answer From Eqs. 3.8 and 3.5 we have: $I = 10^{40} \exp\left(-\dfrac{16\pi\gamma^3 T_f^2}{3\Delta H_f^2 k_B T \Delta T^2}\right)$.

Using the data in Appendix B the correlation presented in Fig. 3.19 is obtained.
Note that for units consistency the volumetric latent heat expressed in J/m^3 must be
used. It is seen that γ_{LS} decreases rapidly at first and then much slower. Thus, it is
considered that a good number for nucleation velocity is in the range of 10^4–10^5,
which gives an interface energy of 0.255–0.256 J/m^3.

Application 3.3

Evaluate the constant K_1 for a Fe–C alloy nucleating on SiO$_2$ particles.

Answer Assuming that the substrates are spherical in shape with a radius r_s, the
number of surface atoms of substrate per unit volume of liquid can be calculated as
$n_a = 4\pi r_s^2 N_s n_s^{2/3}$. In turn, the number of atoms per unit volume of silica is:

$$n_s = \frac{\rho_{SiO_2}N_{Av}}{molar\,weight} = \frac{(2.65\cdot10^3\text{kg}\cdot\text{m}^{-3})\cdot(6\cdot10^{23}\text{atmole}^{-1})}{(28.1+32)\cdot10^3\text{kg}\cdot\text{mole}^{-1}} = 2.645\cdot10^{28}\text{atm}^{-3}.$$

Now n_a can be calculated based on some assumptions for r_s as follows:

$$\text{for } r_s = 10\ \mu\text{m}, \ n_a = 1.11\cdot10^{10}N_s$$

$$\text{for } r_s = 100\mu\text{m}, \ n_a = 1.11\cdot10^{12}N_s$$

At the beginning of nucleation $N=0$ and thus $K_1 = n_a D_L/(a^2 N_s)$. Substituting $D_L = 2\cdot10^{-8}$ m^2/s (see Appendix B) and $a=0.25$ nm, gives:

$$\text{for } r_s = 10\ \mu\text{m}, \ K_1 = 3.55\cdot10^{21}\text{ s}^{-1}$$

$$\text{for } r_s = 100\ \mu\text{m}, \ K_1 = 3.55\cdot10^{23}\text{ s}^{-1}$$

Similar calculations can be performed for Al–Si nucleating on Si crystals with results of the same order of magnitude (Stefanescu et al. 1990).

Application 3.4
Calculate the nucleation rate and the evolution of the number of nuclei for an Fe–C alloy assuming two different nucleation undercoolings of 1 and 3 K, and a final number of nuclei of $N_s = 6\cdot10^{11}$ m^{-3}. This is equivalent to the assumption of two alloys solidifying on the same number of substrates having two different chemistry. Therefore, the two types of nuclei become active at two different nucleation temperatures.

Answer The calculation is straightforward. An average value of 10^{22}s^{-1} is taken for K_1. Equation 3.19 is used to calculate I_{het} and the number of nuclei as a time summation. The results are shown in Fig. 3.6.

Application 3.5
Calculate the disregistry of WC nucleating on δ iron using the modified (planar) Turnbull–Vonnegut equation.

Answer The (0001) plane of WC is selected as the nucleating plane. Three low-index planes of δ iron, (110), (100), and (111) are selected as mating planes (Bramfitt 1970). The crystallographic relationship in the case (0001)$_{\text{WC}}\|(110)_{\text{Fe}}$ is illustrated in Fig. 3.20. Using Eq. 3.25, the planar disregistry equation for this case is:

$$\delta^{(0001)_{WC}}_{(110)_{Fe}} = \left(\frac{\left|\left(d_{[1\bar{2}\bar{1}0]_{WC}} \cos\theta\right) - d_{[001]_{Fe}}\right|}{d_{[001]_{Fe}}} + \frac{\left|\left(d_{[\bar{2}110]_{WC}} \cos\theta\right) - d_{[1\bar{1}1]_{Fe}}\right|}{d_{[1\bar{1}1]_{Fe}}} \right.$$
$$\left. + \frac{\left|\left(d_{[\bar{1}010]_{WC}} \cos\theta\right) - d_{[1\bar{1}0]_{Fe}}\right|}{d_{[1\bar{1}0]_{Fe}}} \right) \cdot 100/3. \qquad (3.33)$$

Using the values given in Table 3.6 we obtain $\delta^{(0001)_{WC}}_{(110)_{Fe}} = 12.7\%$. Similarly, it is calculated that $\delta^{(0001)_{WC}}_{(100)_{Fe}} = 35\%$ and $\delta^{(0001)_{WC}}_{(111)_{Fe}} = 29.4\%$.

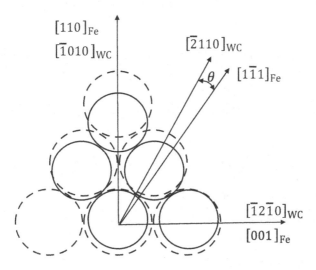

Fig. 3.20 Example of crystallographic relationships for δ iron nucleating on WC for $(0001)_{WC} \parallel (111)_{Fe}$ (after Bramfitt)

Table 3.6 Parameters for calculation of disregistry of δ iron nucleating on WC. (Bramfitt 1970)

Case	$[hkl]_S$	$[hkl]_n$	$d_{[hkl]_S}$	$d_{[hkl]_n}$	θ, deg
$(0001)_{WC} \parallel (110)_{Fe}$	$[\bar{1}2\bar{1}0]_{WC}$	$[001]_{Fe}$	2.929	2.9315	0
	$[\bar{2}110]_{WC}$	$[1\bar{1}1]_{Fe}$	2.929	2.5396	5.26
	$[\bar{1}010]_{WC}$	$[1\bar{1}0]_{Fe}$	5.073	4.1457	0
$(0001)_{WC} \parallel (100)_{Fe}$	$[\bar{1}2\bar{1}0]_{WC}$	$[010]_{Fe}$	2.929	2.9315	0
	$[\bar{2}110]_{WC}$	$[011]_{Fe}$	2.929	4.1457	15
	$[\bar{1}010]_{WC}$	$[001]_{Fe}$	5.073	2.9315	0
$(0001)_{WC} \parallel (111)_{Fe}$	$[\bar{1}2\bar{1}0]_{WC}$	$[\bar{1}11]_{Fe}$	2.929	4.1457	0
	$[\bar{1}110]_{WC}$	$[\bar{1}21]_{Fe}$	5.073	7.1804	0
	$[\bar{2}110]_{WC}$	$[001]_{Fe}$	2.929	4.1457	0

Application 3.6

Consider a pure nickel single crystal growing with a planar interface at a velocity of 10^{-5} m/s. Assuming growth by the continuous growth mechanism, calculate the growth constant and the kinetic undercooling.

Answer To conserve units consistency, the growth constant in Eq. 3.29 is written as:

$$\mu_o = \beta D_L \Delta H_f / \left(a k_B N_A T_f^2 \right),$$

where N_A is Avogadro's number. Assuming $\beta = 1$ (which is true if $a = \lambda$ and $v_{LS} = v_L/6$) and taking $D_L = 3 \ 10^{-9} m^2/s$, it is calculated that $\mu_o = 7.84 \ 10^{-3}$ m

$s^{-1}\,K^{-1}$. Note that, again for units' consistency, the latent heat must be expressed in J/mole. Then: $\Delta T_k = V/\mu_o = 1.28\ 10^{-3}$ K.

It is thus concluded that the kinetic undercooling is very small as compared with the other undercoolings discussed in this text, and therefore the solidification velocity of metals is not limited by interface kinetics.

References

Bramfitt B (1970) Metall. Trans. 1:1987–1995
Broughton JQ, Gilmer GH, Jackson KA (1982) Phys. Rev. Lett. 49:1496
Cahn JW, Hilliard JE (1958) J. Chem. Phys. 31:688
Cantor B, O'Reilly KAQ (1997) Curr. Opin. Solid State Mater. Sci. 2(3):318
Chakravery BK (1966) Surf. Sci. 4:205
Chalmers B (1962) J. Aust. Inst. Met. 8:225
Chernov AA (2004) J. Crystal Growth 264:499
Coriell SR Turnbull DH (1982) Acta Metall. 30:2135
Davies GJ (1973) in: Solidification and Casting, Chapter 6, Applied Science Publishing Co., London
Fan Z, Wang Y, Zhang ZF, Xia M, Li HT, Xu J, Granasy L, Scamans GM (2009), Int. J. Cast Metals Res. 22(1–4):318
Flemings MC (1974) Solidification Processing. Butterworths, London
Flood SC, Hunt JD (1988) Columnar to Equiaxed Transition. In: Stefanescu DM (ed) Metals Handbook Ninth Edition, vol. 15. ASM International, Metals Park, Ohio p 130–135
Fokin VM, Kalinina AM, Filipovich VN (1981) J. Crystal Growth 52:115
Greer AL, Bunn AM, Tronche A, Evans PV, Bristow DJ (2000) Acta Mater. 48:2823
Hoyt JJ (1990) Acta metal. mater. 38(8):1405–1412
Hoyt JJ, Asta M, Sun DY (2006) Philos. mag. 86:3651
Hunt JD (1984) Mater. Sci. Eng. 65:75
Hunter MJ, Chadwick GA (1972) J. Iron Steel Inst. 210:707
Jackson KA (1958) in: Liquid Metals and Solidification, ASM, Metals Park, Ohio p 174
Jackson KA, Hunt JD, Uhlmann D, Seward III TP (1966) Trans AIME 236:149
Jacobs MM, Law TJ, Melford DA, Stowell MJ (1974) Metals Technology 1(Part II Nov.):490
Kelton KF (1991) in: Ehrenreich H, Turnbull D (eds) Solid State Physics. Academic Press, Boston 45:75–178
Mondolfo LF (1983) in: Abbaschian GJ and David SA (eds) Grain Refinement in Casting and Welds. The Metallurgical Soc. of AIME, Warrendale PA p 3
Ohno A (1987) Solidification. The Separation Theory and its Practical Applications. Springer-Verlag, Berlin
Perepezko JH, Rasmussen DH, Anderson IE, Loper CR (1979) in: Solidification of Castings and Alloys. The Metals Society, London p 169
Perepezko JH (1988) Nucleation Kinetics. In: Stefanescu DM (ed) Metals Handbook Ninth Edition, vol. 15. ASM International, Metals Park, Ohio p 101–108
Riposan I, Chisamera M, Stan S, Hartung C, White D (2009), in The Carl Loper Cast Iron Symposium. CD Proceedings. Madison, WI USA p 191–200
Skaland T, Grong F, Grong T (1993) Metall. Trans. 24A: 2321 and 2347
Southin RT, Chadwick GA (1978) Acta Metall. 26:223
Spittle JA, Sadli S (1995) Mater. Sci. Techn. 11:533
Stefanescu DM, Upadhya G, Bandyopadhyay D (1990) Metall. Trans. 21A:997
Tian H, Stefanescu DM, Curreri P (1990) Metall. Trans. 21A:241
Tarshis LA, Walker JL, Rutter JW (1971) Metall. Trans. 2:2589–2597

Turnbull D (1949) Thermodynamics in Metallurgy. ASM, Metals Park, Ohio

Turnbull D, Vonnegut R (1952) Ind. Eng. Chem. 44:1292

Turnbull D (1950) J. Appl. Phys. 21:1022

Turnbull D (1956) Solid State Physics. Academic Press, New York 3:225

Turnbull D (1981) in: Christian JW, Haasen P, Massalski TB (eds) Progress in Materials Sciences. Chalmers Anniversary Volume. Pergamon Press, Oxford p 269

Wang CC, Smith CS (1950) Trans. AIME 188:136

Chapter 4
Fundamentals of Transport Phenomena as Applied to Solidification Processing

The current mathematical understanding of solidification phenomena is based largely on the differential formulation of the mechanics of the continuum and the associated conservation of mass, energy, species, and momentum equations. Thus, a short introduction to the fundamentals of transport phenomena as applied to solidification is deemed necessary to facilitate engaging in the intricacies of the subject.

4.1 General Conservation Transport Equations

At the macroscale level, the mathematical solidification problem is to solve the mass, energy, and momentum transport equations for the particular geometry and material of the casting. The general transport equation written in its standard form for advection-diffusion is:

$$\frac{\partial}{\partial t}(\rho \cdot \phi) + \nabla \cdot (\rho \cdot \mathbf{V} \cdot \phi) = \nabla \cdot (\rho \cdot \Gamma \cdot \nabla \phi) + S, \qquad (4.1)$$

where t is time, ρ is the density, ϕ is the phase quantity, $\mathbf{V}$ is the velocity vector, Γ is the general diffusion coefficient, and S is the source term. The *del* operator[1] was used to reduce the number of equations to be written.

The first left hand (LH1) term is the temporal term, the LH2 is the convective term, and the first right hand (RH1) term is the diffusive term. The specific phase quantities and diffusivity for the four basic equations required in solidification modeling are given in Table 4.1. The following notations were used: H is the sensible

---

[1] The *del* operator (or the *gradient*) of a function u is: $\nabla u \equiv \frac{\partial u}{\partial x}\mathbf{i} + \frac{\partial u}{\partial y}\mathbf{j} + \frac{\partial u}{\partial z}\mathbf{k}$. Note that ∇u is ∇ operating on u, while $\nabla \cdot \mathbf{A}$ is the vector dot product of *del* with $\mathbf{A}$ (or the *divergence*): $\nabla \cdot \mathbf{A} = \partial \mathbf{A}_x/\partial x + \partial \mathbf{A}_y/\partial y + \partial \mathbf{A}_z/\partial z$. Furthermore, $\nabla^2 u$ is the product of the *del* operator with itself, or $\nabla \cdot \nabla = \frac{\partial}{\partial x}\left(\frac{\partial}{\partial x}\right) + \frac{\partial}{\partial y}\left(\frac{\partial}{\partial y}\right) + \frac{\partial}{\partial z}\left(\frac{\partial}{\partial z}\right)$ operating on u.

© Springer International Publishing Switzerland 2015
D. M. Stefanescu, *Science and Engineering of Casting Solidification*,
DOI 10.1007/978-3-319-15693-4_4

Table 4.1 Phase quantities, diffusivities, and origin of the source term

Quantity	Mass	Energy	Species	Momentum
ϕ	1	H	C	$\mathbf{V}$
Γ	0	$\alpha = k/(\rho\,c)$	D	$v = \mu/\rho$
S	- phase motion	- phase transformation - phase motion	- phase transformation - phase motion	- phase motion - S/L interaction - natural convection - shrinkage

enthalpy, $\alpha = k/(\rho\,c)$ is the thermal diffusivity, D is the species diffusivity, $v = \mu/\rho$ is the kinematic viscosity, and μ is the dynamic viscosity.

The relevant transport equations can now be obtained from Eq. 4.1 as follows:

$$\text{Conservation of mass (continuity):} \qquad \frac{\partial \rho}{\partial t} + \nabla \cdot (\rho \mathbf{V}) = S_m. \qquad (4.2)$$

The term $\partial \rho / \delta t$, which expresses the change in density over time, describes the shrinkage-induced flow. The flow is driven toward the volume element when the averaged density of the solid and liquid phases increases with time, as is the case for most alloys.

$$\text{Conservation of energy:} \qquad \frac{\partial}{\partial t}(\rho H) + \nabla \cdot (\rho \mathbf{V} H) = \nabla \cdot \left(\frac{k}{c} \nabla H \right) + S_H, \qquad (4.3)$$

where k is the thermal conductivity and c is the specific heat.

$$\text{Conservation of species:} \qquad \frac{\partial}{\partial t}(\rho C) + \nabla \cdot (\rho \mathbf{V} C) = \nabla \cdot (\rho D \nabla C) + S_C. \qquad (4.4)$$

$$\text{Conservation of momentum:} \qquad \frac{\partial}{\partial t}(\rho V) + \nabla \cdot (\rho \mathbf{V} \cdot \mathbf{V}) = \nabla \cdot (\rho v \cdot \nabla \mathbf{V}) + S_m. \qquad (4.5)$$

Alternatively, for fluids assumed to have constant density and viscosity, conservation of momentum can be written in the form of the Navier-Stokes equation:

$$\rho \frac{D\mathbf{V}}{Dt} = \mu \nabla^2 \mathbf{V} - \nabla P + \rho g, \qquad (4.6)$$

where DV/Dt is the substantial derivative[2], P is the pressure and g is the gravitational acceleration. It simply states that the total force (left hand term) is equal to the sum of the viscous forces ($\mu \nabla^2 \mathbf{V}$), the pressure forces (ΔP), and the gravitational force ($\rho\,g$).

---

[2] *The substantial derivative of a function u is given by:* $\frac{Du}{Dt} = \frac{\partial u}{\partial t} + u_x \frac{\partial u_x}{\partial x} + u_y \frac{\partial u}{\partial y} + u_z \frac{\partial u}{\partial z}$. *It can be applied to any property of a fluid, the magnitude of which varies with time and position.*

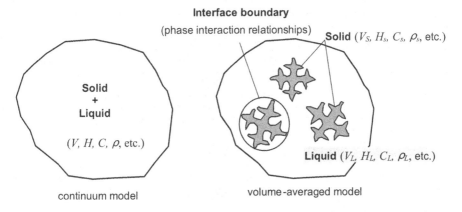

Fig. 4.1 Schematic representation of the computational domain and physical quantities for two-phase models

The source terms of these equations can be complicated. An example of their formulation will be provided later as an application of these equations to macrosegregation modeling.

There are two main difficulties in solving these equations for the problem of interest. Firstly, the application of these equations to a two-phase or multi-phase system, where all the quantities must describe not one but two phases; secondly, the formulation of the source terms for the various types of transport. Two main approaches, based on concepts from continuum mechanics, have been developed to solve the complicated problem of a two-phase system (Fig. 4.1).

In the *mixture-theory model* (continuum model) proposed by Bennon and Incropera (1987) each phase is regarded as a continuum that occupies the entire domain, and described by a set of variables that are continuous and differentiable functions of space and of time. Any location within the domain can be simultaneously occupied by all phases. The macroscopic transport equations are formulated using the classical mixture theory. Summation over the computational domain is used.

In the *volume-averaged model* proposed by Beckermann and Viskanta (1988), Ganesan and Poirier (1990), and Ni and Beckermann (1991) all phases are considered separated. Phase quantities are continuous in one phase but discontinuous over the entire domain. Discontinuities are replaced by phase interaction relationships at interface boundary. Integration of microscopic equations over a finite volume is used.

The volume averaging technique has been utilized extensively to produce solidification models that attempt to describe microstructure evolution in castings by incorporating three phases: liquid, equiaxed solid, and columnar solid (e.g. Wang and Beckermann 1996; Wu and Ludwig 2006). However, the lack of information regarding the microscopic configuration at interface boundaries is a serious complication. These models will be discussed in more details in Chaps. 17 and 18; because of space restrictions, in this section we will introduce only the mixture-theory model.

The typical mixture theory relationships used in the continuum mixture-theory model are derived based on the spatial average $\bar{\phi} = v^{-1} \int_v \phi dv$, where $\bar{\phi}$ is the averaged value of ϕ over both the solid and liquid phases, and v is the volume. In particular, the following relationships are postulated:

$$
\begin{aligned}
& g_L + g_S = 1 && f_L + f_S = 1 && f_L = g_L \rho_L / \rho && f_L = g_S \rho_S / \rho \\
& \rho = g_L \rho_L + g_S \rho_S && k = g_L k_L + g_S k_S \\
& \mathbf{V} = f_L \mathbf{V}_L + f_S \mathbf{V}_S && \mathbf{V} = V^x \cdot i + V^y \cdot j + V^z \cdot k \\
& D = f_L D_L + f_S D_S && H = g_L H_L + g_S H_S
\end{aligned}
$$

$$(4.7)$$

where the subscripts S and L denote solid and liquid, respectively, g is the volume fraction, f is the mass fraction, and k is the thermal conductivity. Additional assumptions must be made to describe the relative movement of the solid and liquid phases during solidification (e.g. Chang and Stefanescu 1996).

The simplified macroscopic transport equations as derived by Bennon and Incropera (1987) based on the continuum model and the mixture theory relationships are in principle similar to those introduced above. The source term in the continuity equation was considered zero. The conservation of momentum equation has been modified to describe flow in the mushy zone, assuming equiaxed grains floating in the liquid. For the x direction:

$$
\frac{\partial}{\partial t} (\rho V^x) + \nabla \cdot (\rho \mathbf{V}^x) = \nabla \cdot \nabla \left(\mu^* \frac{\rho}{\rho_L} \mathbf{V}^x \right) + S_m,
$$

$$(4.8)$$

where μ^* is the relative viscosity (the viscosity of the solid-liquid mixture, which is a function of the solid fraction) and V^x is the velocity component in the x direction.

4.2 Flux Laws

Diffusive transport of energy, mass, and momentum can be described through flux laws whose fundamental form is:

$$
flux = \frac{flow\ rate}{area} = transport\ property \cdot potential\ gradient.
$$

The three laws describing diffusive transport are:

Energy:	$q = -k \nabla T$	Fourier's law.	(4.9a)
Mass (species):	$J_A = -D_{AB} \nabla C_A$	Fick's law.	(4.9b)
Momentum:	$\tau_{xy} = -\mu \cdot (\partial V^y / \partial x)$	Newton's law of viscosity.	(4.9c)

where q, J_A and τ_{xy} are the heat, mass, and momentum flux, respectively, and V_y is the fluid velocity along the y axis. Analogous to mass diffusivity we can define the thermal diffusivity, $\alpha = k/(\rho\,c)$ and momentum diffusivity $v = \mu/\rho$. Hence, the flux laws can be written in their diffusion form as follows:

$$flux = diffusivity \cdot concentration\ gradient.$$

Now the three flux laws for energy, mass, and momentum transport can be written as:

$$\text{Energy:} \qquad q = -\alpha\nabla(\rho c T). \qquad (4.10a)$$

$$\text{Mass (species):} \qquad J_{Ax} = -D_{AB}\nabla C_A. \qquad (4.10b)$$

$$\text{Momentum:} \qquad \tau_{xy} = -v\partial(\rho V^y)/\partial x. \qquad (4.10c)$$

Note that the quantities at the numerators on the right hand side are energy concentration (energy/volume), mass concentration (mass/volume), and momentum concentration (momentum/volume), respectively.

The general transport equation can also describe diffusive processes by recognizing that when there is no relative movement of phases, i.e., $\mathbf{V} = 0$, the advective term in Eq. 4.1 disappears:

$$\frac{\partial}{\partial t}(\rho\varphi) = \nabla \cdot (\Gamma\rho\nabla\varphi) + S. \qquad (4.11)$$

While further discussion could be conducted for the general transport equation, it is believed that the treatment of the heat diffusion equation with appropriate examples, as provided in the following chapter, will be easier to follow. However, the discussion is equally applicable to the other forms of diffusive transport.

References

Beckermann C, Viskanta R (1988) PhysicoChem. Hydrodyn. 10:195
Bennon WD, Incropera FP (1987) Int. J. Heat Mass Transfer 30:2161, 2171
Chang S, Stefanescu DM (1996) Metall. Mater. Trans. 27A:2708
Ganesan S, Poirier DR (1990) Metall. Trans. 21B:173
Ni J, Beckermann C (1991) Metall. Trans. 22B:349
Wang CY, Beckermann C (1996) Metall. Mater. Trans. 27A:2754
Wu MG, Ludwig A (2006) Metall. Mater. Trans. 37A:1613–1631

Chapter 5
Diffusive Mass Transport at the Macroscale

Mass transport during solidification is responsible for macrosegregation. The mechanism of mass transport through species diffusion produces diffusion-controlled macrosegregation. Momentum transfer (fluid convection) generates fluid flow/solute diffusion-controlled segregation. The effects of these mechanisms on casting solidification will be discussed in the following chapters.

5.1 Solute Diffusion-Controlled Segregation

When alloys having a partition coefficient $k < 1$ solidify, solute atoms are rejected from the first region to solidify into the liquid. These atoms build up in the liquid just ahead of the solid/liquid (S/L) interface, forming a boundary layer, which has a content of solute higher than that of the bulk liquid. When $k > 1$, a boundary layer depleted in solute is formed. Thus, three zones for *mass transfer* can be defined as:

- the solid: mass transfer occurs only by chemical diffusion
- the interface: there is a boundary layer of thickness δ_c in the liquid at the S/L interface, where the solute is transported through diffusion (*diffusion boundary layer*)
- the bulk liquid: mass transport is done by diffusion and convection

The schematic evolution of composition across the interface (compositional profile) is shown in Fig. 5.1 for $k < 1$. The amount of solute rejected at the S/L interface because of partitioning is $C_L^* = C_S^*/k$. It will diffuse down the concentration gradient until the composition will be that of the bulk liquid, C_o. A diffusion boundary layer, in which the concentration is above C_o will exist. Assuming no convection in the liquid, the diffusion boundary layer can be defined as the distance from the interface, at which the diffusion rate becomes equal to the solidification rate, i.e., $D_L/\delta_c = V$. Thus, the thickness of the boundary layer is:

$$\delta_c = D_L/V. \tag{5.1}$$

© Springer International Publishing Switzerland 2015
D. M. Stefanescu, *Science and Engineering of Casting Solidification*,
DOI 10.1007/978-3-319-15693-4_5

Fig. 5.1 Composition profile
resulting from mass transfer

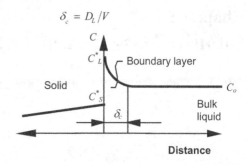

Let us derive an equation for the shape of the diffusion (solutal) boundary layer. We will assume $k < 1$, no source term, constant density, and constant diffusivity. The equation governing the diffusion process, Eq. 4.4, becomes:

$$D\nabla^2 C = \frac{\partial C}{\partial t} + \nabla \cdot (\mathbf{V}C). \tag{5.2}$$

Further assuming directional solidification, this equation can be used in one-dimensional (1D) form. Further assuming constant solidification velocity, the equation becomes:

$$\frac{\partial^2 C}{\partial x^2} - \frac{V}{D}\frac{\partial C}{\partial x} = \frac{1}{D}\frac{\partial C}{\partial t}. \tag{5.3}$$

Setting the reference point at the interface, the velocity in the advection term is the liquid velocity that compensates for shrinkage. It is equal to the solidification velocity but has the opposite sign: $V = V_L - V_{ref} \rightarrow -V$. Consequently, the advection term is subtracted from the diffusion flux. For steady state, $\partial C/\partial t = 0$ and thus:

$$\frac{\partial^2 C}{\partial x^2} + \frac{V}{D}\frac{\partial C}{\partial x} = 0 \tag{5.4}$$

This equation is known as the time independent form of the directional growth equation, or as the characteristic partial differential equation of the quasi-steady state. Its solution is (see inset for derivation):

$$C_L = C_o + \Delta C_o \exp\left(-\frac{V}{D_L}x\right) = C_o\left[1 + \frac{1-k}{k}\exp\left(-\frac{V}{D_L}x\right)\right]. \tag{5.5}$$

Derivation of the equation for the composition in the boundary layer, Eq. 5.5.

The solution of Eq. 5.4 is $C_L = A + B \exp(bx)$, where b must be a solution of the equation $b^2 + Vb/D = 0$. The solutions of this "auxiliary" equation are $b = 0$ and $b = -V/D$. Therefore, the general solution of Eq. 5.4 is:

$$C_L = A + B \exp\left(-\frac{V}{D}x\right). \tag{5.6}$$

The following boundary conditions are used:

BC1   far-field condition: for $x = \infty$ $C = C_o$

BC2   flux at the interface: for $x = 0$ $V(C_L^* - C_S^*) = -D(\partial C/\partial x)$

The second boundary condition simply states that:

rate of solute rejection at the interface = diffusional flux in the liquid

Introducing BC1 in Eq. 5.6 gives $A = C_o$. Then, since from Eq. 5.6, for $x = 0$ and $A = C_o$, we have $C_L^* = C_o + B$, the left hand side term in BC2 is:

$$V(C_L^* - C_S^*) = V(1 - k)C_L^* = V(1 - k)(C_o + B). \tag{5.7}$$

The right hand side term in BC2 is:

$$-D\left(\frac{\partial C}{\partial x}\right)_{x=0} = -D \cdot B \frac{\partial}{\partial x}\left(\exp\left(-\frac{Vx}{D}\right)\right)_{x=0} = B \cdot V. \tag{5.8}$$

Equating these last two equations we get $B = C_o(1 - k)/k = \Delta C_o$ (see Eq. 2.20). After substituting the constants A and B, Eq. 5.6 becomes Eq. 5.5.

The boundary layer given by this equation is of infinite extent, since $C_L = C_o$ at $x = \infty$. However, typically, the thickness of the boundary layer is taken to be as calculated by Eq. 5.1. To obtain a convenient practical estimate of its thickness, Kurz and Fisher (1989) have defined an equivalent boundary layer, δ_e. The equivalent boundary layer contains the same total solute amount as the infinite layer, and has constant concentration gradient G_c^e across its thickness. It is easily demonstrated that $\delta_e = 2D_L/V$.

Since the composition of the liquid and solid during solidification are different, chemical diffusion will be active during and after solidification. Thus, it is important to explore the solute redistribution resulting from this diffusion. Assuming only diffusive transport, the final composition in a solidifying casting depends on the liquid and solid diffusivity and on the partition coefficient. In our analysis, we will consider the simple case of directional solidification. This means that energy transport

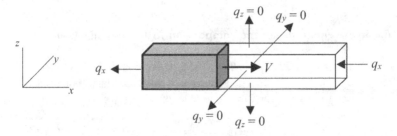

Fig. 5.2 Directional solidification

is only in the x-direction (Fig. 5.2). There is no heat flux in the y- and z-directions. By controlling the end temperatures, the solidification velocity, and the temperature gradient at the S/L interface can be maintained constant.

A rigorous complete solution of this problem can be obtained by solving the diffusion equation in three-dimensions (3D). However, this can only be done numerically, unless some simplifying assumptions are used.

Analytical solutions to the solute redistribution problem during directional solidification can be obtained on the basis of the following assumptions:

a. Equilibrium solidification: $D_S = \infty, D_L = \infty$
b. No diffusion in solid, complete diffusion in liquid: $D_S = 0, D_L = \infty$
c. No diffusion in solid, limited diffusion in liquid: $D_S = 0, 0 < D_L < \infty$
d. Partial (back) diffusion in solid, complete diffusion in liquid: $D_S > 0, D_L = \infty$
e. Limited diffusion in solid and liquid: $0 < D_S \neq D_L < \infty$
f. No diffusion in solid, partial mixing in liquid: $D_S = 0$, convection in liquid

5.2 Equilibrium Solidification

If there is enough time for solute diffusion to proceed to completion in both solid and liquid, the composition becomes uniform throughout the sample. This is equilibrium solidification, where either enough time is available for the solid and liquid solutions to become completely homogeneous from the chemical standpoint ($V_S \rightarrow 0$), or diffusion is very rapid in both the solid and the liquid ($D_S = \infty, D_L = \infty$). Such conditions require solidification times that are much higher than encountered in practical metallurgy (see Application 2.1). As solidification proceeds under equilibrium conditions, the solute composition in the solid, C_S, and in the liquid, C_L, vary along the solidus and the liquidus line of the phase diagram, respectively (Fig. 5.3).

Figure 5.4 shows the composition profile of the solute in the solid and in the liquid, in a directionally solidified sample, at three different stages: immediately after the beginning of solidification, at an intermediate time, and at the end of solidification. The initial composition of the alloy is C_o, i.e., the alloy contains C_o % solute. Thus the liquid composition at the beginning of solidification is $C_L = C_o$. As imposed by the partition coefficient, k, the first amount of solid to form will have

Fig. 5.3 Schematic phase diagram

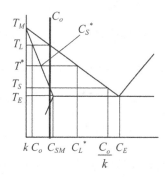

the composition $C_S = kC_o$. This means that the solid will have less solute than the liquid, and thus the solidifying solid will reject some solute in the liquid. At an intermediate time during solidification, when the interface temperature is T^*, the composition of the liquid has risen to $C_L^* > C_o$, and that of the solid to $C_S^* > kC_o$.

Writing a material balance (conservation of solute atoms) at T^* gives:

$$C_S f_S + C_L f_L = C_o \text{ with } f_S + f_L = 1,$$

where f_S and f_L are the mass fraction of solid and the fraction of liquid, respectively. Knowing that $C_S = kC_L$, the composition of the solid is:

$$C_S = \frac{kC_o}{1 - (1 - k)f_S}. \tag{5.9}$$

This equation is called *the equilibrium lever rule*. It is valid assuming $\rho_S = \rho_L$.

At the end of solidification, because of rapid solid diffusion, the composition of the solid is uniform across the volume element, and equal to the initial composition of the liquid. Note that in spite of the equilibrium nature of solidification, substantial solute redistribution occurs during solidification. The material is homogeneous only before and after solidification.

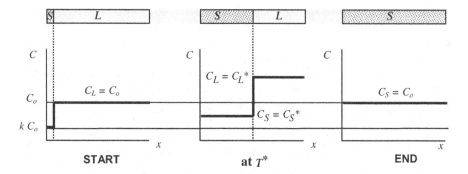

Fig. 5.4 Equilibrium solute redistribution in a directionally solidified casting

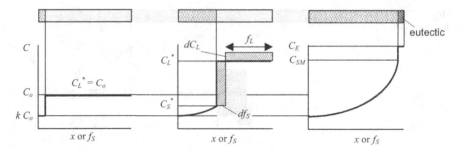

Fig. 5.5 Solute redistribution for nonequilibrium solidification for $D_S = 0$ and $D_L = \infty$

For the more general case, when $\rho_S \neq \rho_L$, a similar equation can be obtained by using volume fractions, g_S and g_L, rather than mass fractions. The relationships between volume and mass fractions are defined in Eq. 4.7. Then, when substituting f_S with $g_S \rho_S / \rho$ in Eq. 5.9, the equilibrium lever rule becomes:

$$C_S = \frac{1 - (1 - \rho_S/\rho_L)g_S}{1 - (1 - k\rho_S/\rho_L)g_S}kC_o. \tag{5.10}$$

Note that, if $\rho_S = \rho_L$, this equation reduces to the previous one.

5.3 No Diffusion in Solid, Complete Diffusion in Liquid (the Gulliver–Scheil Model)

The basic assumption is that diffusion is very rapid ($D_L = \infty$), or there is complete mixing (convection) in the liquid, but there is no diffusion in the solid ($D_S = 0$). The graphical representation of this case is given in Fig. 5.5. Note that since complete mixing in the liquid is assumed there is no diffusion boundary layer ahead of the solidifying interface.

At the beginning of solidification, the situation is identical with that for equilibrium solidification. Then, since there is no diffusion in the solid, as the solidifying liquid has increased solute, a concentration gradient will be established between the initial solid composition kC_o and the solid composition at the intermediate time (or temperature, T^*), which is C_S^*. In the liquid, the composition is homogeneous and equal to $C_L^* > C_o$, since diffusion is very rapid. The composition of the solid will continue to grow to the end of solidification, and will finally reach the maximum solubility in the solid solution on the phase diagram, C_{SM}.

To find an equation for the solid composition as a function of the solid fraction, material balance equations must be used. In the original derivation by Gulliver (1913) and Scheil (1942) a material balance at the interface was written:

solute rejected when df_S is formed = solute increase in liquid,

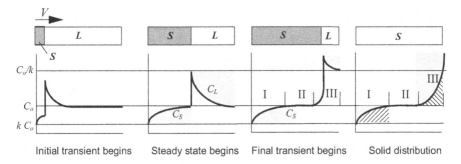

Fig. 5.6 Solute redistribution for nonequilibrium solidification for $D_S = 0$ and $0 < D_L < \infty$

that is:

$$(C_L^* - C_S^*)df_S = (1 - f_S)dC_L \tag{5.11}$$

Since $C_L = C_S/k$, and $dC_L = dC_S/k$, on integration this equation becomes:

$$\int_0^{f_S} \frac{df_S}{1 - f_S} = \frac{1}{1 - k} \int_{kC_o}^{C_S} \frac{dC_S}{C_S} \text{ or } C_S = kC_o(1 - f_S)^{k-1}. \tag{5.12}$$

This is known as the *Scheil* (more recently *Gulliver–Scheil*) equation, or the nonequilibrium lever rule. Note that, for $f_S = 1$ this equation calculates $C_S = \infty$. This is of course absurd for alloy solidification. The composition of the solid can only increase to the maximum solid solubility, C_{SM}, and that of the liquid to the eutectic composition, C_E. As solidification proceeds, the solid composition follows the solidus line from kC_o to C_{SM} and then to C_E (Fig. 5.3). The last region to solidify is of eutectic composition (see Application 5.1) The Gulliver–Scheil equation can also be derived from overall mass balance (Rappaz and Voller 1990).

5.4 No Diffusion in Solid, Limited Diffusion in Liquid

The basic assumptions are: $D_S = 0$ and $0 < D_L < \infty$. The solute redistribution for this case is shown in Fig. 5.6. A diffusion layer will exist ahead of the interface, and equations that are more complicated are used to calculate the liquid and solid composition. Three distinctive zones are seen:

1. *The initial transient, between T_L and T_S*: because of the boundary layer
2. The steady state, at T_S
3. The final transient, between T_S and T_E: buildup of solute occurs because the boundary layer reaches the end of the crucible

The initial and final transients represent chemical segregation. The shaded areas in "solid distribution" on Fig. 5.6 must be equal to conserve mass balance, so that the average composition remains C_o.

Fig. 5.7 Solute accumulation
during the final transient

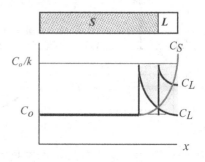

Calculation of solute redistribution during the *initial transient* can be done by using the time dependent form of the diffusion Eq. 5.3. The solution of this equation is:

$$C_S = C_o[1 - (1 - k)\exp(-k(V/D)x)]. \tag{5.13}$$

Alternatively, solute flux balance can be used to derive the equation for the initial transient, as demonstrated in the inset.

Derivation of the initial transient equation from solute flux balance (Kurz and Fisher 1989).

During nonsteady state solute flux balance at the interface gives:
solute accumulation in boundary layer = solute rejected by solid − solute diffusing in liquid

$$\text{or:} \frac{dC_L^*}{dt}\delta_c = V(C_L^* - C_S^*) - \left[-D\left(\frac{\partial C}{\partial x}\right)_{x=0}\right].$$

During steady state there is no accumulation of solute in the boundary layer. The compositional gradient can be calculated from Eq. 5.5, or it can be approximated as: $(\partial C/\partial x)_{x=0} = (C_o - C_L^*)/\delta_c = (C_o - C_L^*) \cdot V/D$. Substituting in the flux balance equation $(dC_L^*/dt)\delta_c = V(C_o - kC_L^*)$, and, since $\delta_c = D/V$, it follows that $dC_L^*/(C_o - kC_L^*) = (V^2/D)dt$. Since $V = dx/dt$:
$\int_{C_o}^{C_L^*} \frac{dC_L^*}{C_o - kC_L^*} = \frac{V}{D}\int_0^x dx$. Then, after integration: $\ln\left[\frac{C_o(1-k)}{C_o - kC_L^*}\right]^{1/k} = \frac{V}{D}x$.
Rearranging, we obtain Eq. 5.13.

During *steady state* solidification, the planar S/L interface grows at T_S. The composition of the liquid at the interface is $C_L^* = C_o/k$ and then decreases according to Eq. 5.5, and reaches C_o after a distance of approximately $2D_L/V$.

Steady state exists as long as there is enough liquid ahead of the interface for the forward diffusion of the solute to occur, and as long as solidification velocity remains constant. As the boundary of the sample is approached the first condition is not fulfilled anymore and the solute content increases above C_o (Fig. 5.7). This

Fig. 5.8 Formation of positive and negative segregation when solidification velocity is different from steady state velocity

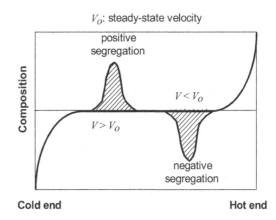

V_o: steady-state velocity

positive segregation

$V < V_o$

$V > V_o$

negative segregation

Composition

Cold end   **Hot end**

is the *final transient*. The length of the final transient is that of the solute boundary layer, D_L/V.

As shown by Smith et al. (1955), the solid composition in the final transient can be calculated with:

$$\frac{C_S}{C_o} = 1 + 3\frac{1-k}{1+k}\exp\left(-\frac{2Vx}{D}\right) + 5\frac{(1-k)(2-k)}{(1+k)(2+k)}\exp\left(-\frac{6Vx}{D}\right) +$$
$$+ \dots + (2n+1)\frac{(1-k)(2-k)\dots(n-k)}{(1+k)(2+k)\dots(n+k)}\exp\left[-\frac{n(n+1)Vx}{D}\right], \quad (5.14)$$

where $x = 0$ at the end of the specimen.

So far, it was assumed that the solidification velocity is constant during solidification. However, this condition does not hold in most solidification processes. Notable exceptions are controlled directional solidification and crystal growth. If the solidification velocity V is suddenly increased ($V > V_o$), the diffusion layer decreases, which means that the amount of solute transported forward decreases. Conservation of solute atoms requires then an increase in the composition of the solid, and a band rich in solute (positive segregation) is formed, as shown in Fig. 5.8.

On the contrary, when the solidification velocity is suddenly decreased ($V < V_o$), a band poor in solute (negative segregation) is formed. If V or δ_c varies periodically, then periodical composition changes are produced. They are called *banding*.

5.5 Limited Diffusion in Solid, Complete Diffusion in Liquid

The first model that attempted to describe this problem was proposed by Brody and Flemings (1966). The basic assumptions of the model included $D_S > 0$, $D_L \rightarrow \infty$, and some back-diffusion of solute in the solid occurs at the interface. To solve the mass balance equation an additional assumption was necessary, i.e., $V = ct$. Two cases can be considered:

- Linear growth: $f_S = t/t_f = x/l$
- Parabolic growth: $f_S = \sqrt{t/t_f}$

where t_f is the final solidification time. Unidirectional solidification typically imposes linear growth in the specimen. Solidification of dendrites is commonly assumed to follow parabolic growth.

For linear growth, it was shown (see inset for derivation) that:

$$C_S = kC_o\left(1 - \frac{f_S}{1 + k\alpha}\right)^{k-1}.$$ (5.15)

For parabolic growth the equation is:

$$C_S = kC_o[1 - (1 - 2\alpha k)f_S]^{(k-1)/(1-2\alpha k)}.$$ (5.16)

In these equations α is the dimensionless back-diffusion coefficient:

$$\alpha = 2D_St_f/l.$$ (5.17)

Simplified derivation of the *Brody–Flemings* equation (after Kurz and Fisher 1989).

When examining the shaded areas in Fig. 5.9, it is seen that mass balance at the interface requires:

$A_1 = A_2 + A_3$ $A_1 = (C_L - C_S)\,dx$ $A_2 = (l - x)\,dC_L$ $A_3 = (\delta_S/2)\,dC_S$

The approximation that allows writing this last equation is shown in Fig. 5.9. Since $x/l = f_S$ and $dx/l = df_S$, after substituting and dividing by l:

$$(C_L - C_S)df_S = (1 - f_S)dC_L + \frac{\delta_S}{2l}dC_S,$$ (5.18)

where $\delta_S = 2D_S/V = 2D_S/(dx/dt)$.

If linear growth is considered, since $dx/dt = l/t_f$, we have $\delta_S = 2D_St_f/l$. Then, substituting in Eq. 5.18 we have $C_S(1 - k)df_S = (1 - f_S)dC_S + kdC_S(D_St_f)/l^2$. A dimensionless solid-state back-diffusion coefficient is defined as $\alpha = D_St_f/l^2$. Rearranging and integrating: $\frac{1}{1-k}\int_{kC_o}^{C_S}\frac{dC_S}{C_S} = \int_0^{f_S}\frac{df_S}{1+k\alpha-f_S}$. Further manipulations produce Eq. 5.15.

Equations. 5.15 and 5.16 have been obtained without solving the "Fickian" diffusion. Because of that, when significant solid-state diffusion occurs, mass balance is violated. This can be understood by examining Fig. 5.9. Mass balance for the boundary layer δ_S is correctly described by the equation given for A_3 only as long as the boundary layer is smaller than the solidified region. Consequently, the application of these equations is limited to slow diffusion when the boundary layer is small.

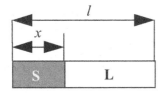

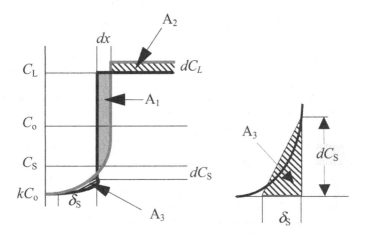

Fig. 5.9 Mass balance at the interface when complete diffusion in liquid and partial diffusion in solid

Indeed, in Eq. 5.15 for $\alpha = 0$ (i.e., $l^2 >> D_S t_f$,) this equation reduces to the Scheil equation. However, for the equilibrium condition which is $\alpha = \infty$, $l^2 << D_S t_f$. This gives $C_S = kC_o$, which is the interface equilibrium condition but not the equation for equilibrium solidification. Similarly, Eq. 5.16 reduces to the Scheil equation for $\alpha = 0$, and the lever rule is obtained for $\alpha = 0.5$. Unfortunately, $\alpha = 0.5$ does not describes the physics of equilibrium. Thus, it is not surprising to find that, while the Brody–Flemings equation predicts compositions closer to the experimental ones, it is still quite inaccurate (Fig. 5.10).

Another problem is the solutal profile shown in Fig. 5.9. If the diffusion in solid is finite, the solutal profile should be intermediate between that predicted by Scheil and equilibrium, as shown in Fig. 5.10. C_S^* should decrease which in turn will determine a lower C_L^*.

Clyne and Kurz (1981) have used the Brody–Flemings model and added a spline fit to match predictions by Scheil equation and the equilibrium equation for infinitesimal and infinite diffusion coefficient, respectively. This relation has no physical basis.

Fig. 5.10 Comparison between experiment (Gungor 1989), Scheil, and Brody–Flemings (Sigworth 2014) profiles for an Al-4.5 %Cu alloy

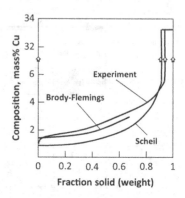

Kobayashi (1988) obtained an exact solution (Laguerre polynomial). Solidification rate and physical properties were considered constant. Parabolic solidification was assumed for the planar geometry. A large number of terms (20,000 for a $Fo = 0.05$) is required for convergence. However, calculations with the second order approximate solution were very close to the exact solution. This approximation is:

$$C_S = kC_o\xi^{\frac{k-1}{1-\beta k}}\left\{1 + \Gamma\left[\frac{1}{2}\left(\xi^{-2} - 1\right) - 2(\xi^{-1} - 1) - \ln\xi\right]\right\}, \qquad (5.19)$$

where $\gamma = 2\alpha$ for planar geometry and:

$$\xi = 1 - (1 - \beta k)f_S \qquad \beta = \frac{2\gamma}{1 + 2\gamma}$$

$$\Gamma = \beta^3 k(k - 1)[(1 + \beta)k - 2](4\gamma)^{-1}(1 - \beta k)^{-3}.$$

Note that, for $D_S = 0$ we have $\gamma = 0$, $\beta = 0$, $\Gamma = 0$, and $\xi = 1 - f_S$ and this equation reduces to the Scheil equation. In addition, for $D_S \to \infty$ it gives $C_S = kC_o$, which upon integration gives the equilibrium composition. Kobayashi has also demonstrated that the Brody–Flemings and Clyne–Kurz solutions underestimate segregation by overestimating the effect of D_S, and are particularly inaccurate for low values of k and α.

Himemiya and Umeda (1998) developed an integral profile method that can consider all significant diffusion cases. For finite diffusion in solid and complete diffusion in liquid, a second order differential equation was obtained.

For linear growth the equation is:

$$f_S^2(1 - f_S)\frac{d^2C_L}{df_S^2} + \left[(k - 4)\,f_S^2 + (3k\alpha - 3\alpha + 2)\,f_S + 3\alpha\right]\frac{dC_L}{df_S}$$
$$= (1 - k)(2f_S + 3\alpha)C_LC_o.$$

For parabolic growth the equation is:

$$f_S^2(1 - f_S)\frac{d^2C_L}{df_S^2} + [(k - 4 - 6\alpha)f_S + (3k\alpha + 6\alpha + 2)]\frac{dC_L}{df_S}$$
$$= (1 - k)(2 + 6\alpha)C_LC_o.$$

The Runge–Kutta method was used to solve these equations.

Fig. 5.11 Boundary layer
when convection in the liquid
is assumed

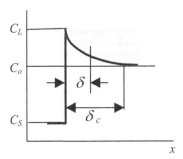

5.6 Limited Diffusion in Solid and Liquid

The Himemiya–Umeda model is applicable to this problem. However, complicated equations describing an initial value problem must be solved. A simpler analytical model proposed by Nastac and Stefanescu (1992) is only valid at the microscale because of some of the assumptions made during derivation. This model is described in detail in Sect. 8.1.

5.7 Partial Mixing in Liquid, No Diffusion in Solid

The segregation measured in solids is, in most cases, intermediate between that for complete mixing and no mixing. When a temperature gradient exists in the liquid, thermal convection will occur, because of the difference in density between the cold and hot metal. Therefore, mass transport is not only by diffusion but also by fluid flow. A more complicated situation must be considered, and an additional assumption is necessary.

As discussed earlier, within the diffusion layer of thickness δ_c (Fig. 5.11), mass transport is by diffusion only, while outside this layer convection insures homogeneity within the liquid. In terms of hydrodynamics, the diffusion layer is stagnant. The diffusion layer for partial mixing in the liquid is treated by using an *effective distribution coefficient, k_{ef}*. It can be shown from the boundary layer theory (Burton et al. 1953) that k_{ef} is related to k by the equation:

$$k_{ef} = k/[k + (1 - k)\exp(-V\delta/D_L)] \quad \text{with} \quad 1 \geq k_{ef} \geq k \qquad (5.20)$$

An equation similar to the Scheil equation is derived for the calculation of the solid composition as a function of fraction of solid:

$$C_S = k_{ef}C_o(1 - f_S)^{k_{ef}-1}. \qquad (5.21)$$

Note that for $D_L \rightarrow \infty$, $k_{ef} = k$, and the equation for complete mixing (Scheil) is obtained. For $D_L \rightarrow 0$, $k_{ef} = 1$, $C_S = C_o$, which means that no mass transport occurs.

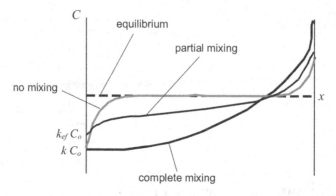

Fig. 5.12 Summary of solute redistribution

For limited diffusion in liquid ($0 < D_L < \infty$) and no diffusion in solid, Nastac (2004) derived the following equation for the liquid concentration profile:

$$C_L = C_o + \frac{C_o(1-k)}{2k}\exp\left(-\frac{Vx}{D_L}\right)erfc\left(\frac{x}{2\sqrt{D_Lt}} - \frac{V}{2}\sqrt{\frac{t}{D_L}}\right)$$
$$-\frac{C_o}{2}erfc\left(\frac{x}{2\sqrt{D_Lt}} + \frac{V}{2}\sqrt{\frac{t}{D_L}}\right) - \frac{C_o(1-2k)}{2k}$$
$$\times \exp\left[(k-1)\left(k\frac{V^2t}{D_L} + \frac{Vx}{D_L}\right)\right]erfc\left(\frac{x}{2\sqrt{D_Lt}} - V\left(\frac{1}{2}-k\right)\sqrt{\frac{t}{D_L}}\right)$$

5.8 Summary of Diffusion-Controlled Macrosegregation

Finally, a graphic summary of the various solute redistribution analytical models discussed in this chapter is presented in Fig. 5.12. Note that complete mixing (Scheil model) occurs when considerable convection exists in the liquid. This is the case for most directional solidification experiments performed in the earth's gravitational field. For experiments conducted in a microgravity environment the no-mixing model gives a more realistic description of reality.

It must be noted that, when using analytical models to evaluate segregation, it must be assumed that all physical properties are constant. The solid-state concentration can only be calculated at the interface, and cannot be modified by subsequent solid diffusion. In other words, only the trace of the solid-state concentration can be plotted. Thus, the equilibrium and Scheil model predict the solid concentration across the whole length as well as at the interface, while the Brody–Flemings and the Kobayashi models can only calculate the interface solid composition. The applicability of various models presented in this section can be understood by studying Application 5.2.

Many other analytical and numerical models have been proposed. Some of them will be reviewed as part of the discussion on microsegregation.

The study of Fig. 5.12 reveals that the last part of the casting to solidify is richer in solute than the initial one. This difference in composition at the macroscopic level is called *macrosegregation*. It can alter mechanical properties dramatically. It can be easily understood to occur in single crystal castings that are solidified directionally. When considering three-dimensional solidification of castings with columnar structure, the middle of the casting will be richer in solute than the skin that has solidified in contact with the mold walls.

Macrosegregation is significant in large castings, but can become a factor also in small or medium size castings, when the partition coefficient is relatively high. Typical alloys that will exhibit such a behavior are some aluminum and copper alloys.

When the occurrence of macrosegregation is governed by a law such as the Scheil equation, it is termed *normal segregation*. Note that normal segregation is in fact a positive segregation, because the last solid to form is richer in solute than the average composition. The degree of normal segregation increases as the solidification velocity, V, or the solute boundary layer, δ_c, decrease.

The occurrence of macrosegregation is more complicated than the solute diffusion models previously discussed. Other effects must be considered in order to obtain an accurate description of the process, as follows (Ohnaka 1992):

- Gravity effect on density differences caused by phase, compositional, or thermal variations (natural convection)
- Solidification contraction
- Capillary forces
- External centrifugal or electromagnetic forces
- Deformation of solid phases due to thermal stress and static pressure

Most of these effects are related to the fluid flow and will be discussed later.

5.9 Zone Melting

The understanding of the segregation phenomena has led to the development of solidification techniques for metal purification. Metal purification through solidification processing can be performed in two ways:

- Successive directional solidification of the alloy, and rejecting the last part after each cycle; it is not practical, but it is possible; a typical set-up is the Bridgman-type furnace in Fig. 5.13a
- Moving a short molten zone along a solid bar; this is zone melting; a typical set-up is illustrated in Fig. 5.13b; the liquid is held in place by a crucible or by surface tension (floating zone).

Let us consider a molten zone of length ℓ that is moved along the crucible (Fig. 5.14). Assuming no convection in the liquid, flux balance at the interface gives $(C_o - C_S)\, dx = \ell\, dC_L$, where ℓ is the length of the molten zone, assumed to be

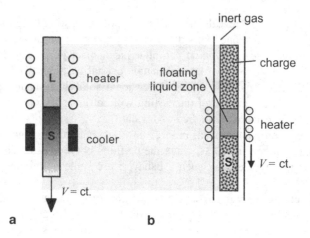

Fig. 5.13 Single crystal growth processes. **a** the Bridgman method. **b** zone melting (floating zone)

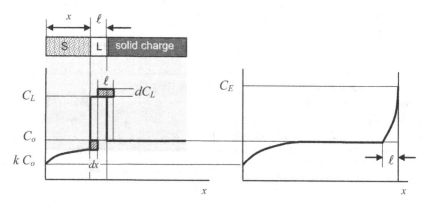

Fig. 5.14 Solute redistribution during zone melting after a single path

constant and small. Integrating between kC_o and C_S, and 0 and x, we obtain the equation for the solid concentration after a single path:

$$C_S = C_o \left[1 - (1 - k) \exp\left(-\frac{k}{\ell} x \right) \right]. \tag{5.22}$$

This equation was originally derived by Pfann (1952). If convection is present, k_{ef} rather than k are used in the above equation. This distribution is true with the exception of the terminal transient region where a rapid increase in concentration occurs. The net result of the process is transport of solute from one end to the other of the crucible (Fig. 5.14). The composition in the final transient can be calculated with Scheil's equation.

If several passes (directional solidification cycles) are executed, further purification becomes possible, as shown in Fig. 5.15. If the partition coefficient is low,

Fig. 5.15 Metal purification
as a result of successive passes

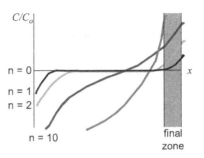

concentration is reduced fast in a few passes. On the contrary, if the partition coefficient is close to unity, a large number of passes are required for purification.

Commercial applications of zone melting include refining of metals, crystal growth, and fabrication of superconductors. Depending on the goal, zone melting can be used in two different ways:

- To achieve maximum uniformity; this is called zone leveling, and one pass is sufficient
- To achieve maximum purity; this is called zone refining; since maximum transport of solute is needed, a large number of passes is used (see Application 5.5)

5.10 Applications

Application 5.1 Calculate the amount of eutectic at the end of solidification in an Al-51 wt%Zn alloy. Hint: Use Scheil equation and the maximum solubility in solid, 82.2 wt%.

Answer From the Al–Zn phase diagram, at the eutectic temperature, we obtain the partition coefficient : $k = 0.822/0.95 = 0.865$. We further know that $C_o = 0.51$, and that $C_{SM} = 0.822$. The governing equation is the Scheil equation. At the eutectic temperature $C_s = C_{SM}$. Substituting into the Scheil equation gives: $C_{SM} = kC_o(1 - f_s)^{k-1} = kC_o(f_L)^{k-1}$. Noting that $f_L = f_E$ we now solve for the fraction of eutectic:

$$f_E = (C_{SM}/kC_o)^{\frac{1}{k-1}} = (0.822/(0.865 \cdot 0.51))^{\frac{1}{0.865-1}} = 0.0099.$$

Application 5.2 Cylindrical steel rods, 60 mm in length, were solidified in a Bridgman directional solidification furnace at velocity 2.8 μm/s. An induction coil was used for heating. Consequently, complete mixing in the liquid can be assumed. The composition of the steel was as follows: 0.25 %C, 0.57 %Cr. Solidification was interrupted by quenching the samples in water and the interface solid composition was measured using a scanning electron microprobe. The following results were obtained (Pershing 1997):

Solidified length, m	Fraction of solid	%C	%Cr
0.01	0.167	0.12	0.52
0.03	0.5	0.135	0.56
0.06	1	0.23	0.72

Element	Materials constants	
	D_S in austenite	k
C	$1 \cdot 10^{-9}$	0.196 or 0.495
Cr	$1.2 \cdot 10^{-12}$	0.915

The solid diffusivity in austenite rather than in δ-ferrite was chosen because it is the rate controlling value for diffusion near the interface in the peritectic transformation in steel. There are two partition coefficients for C because the alloy is of hyper-peritectic composition. The partition coefficient will change as the temperature decreases under the peritectic temperature. Calculate the solid composition using the equilibrium, Scheil, Brody–Flemings, and Kobayashi models. Compare with the experimental results.

Answer Calculations for the equilibrium, Scheil and Brody–Flemings models are straightforward. For the Kobayashi model the local solidification time is calculated as $t_f = l/V = 2.14 \times 10^6$ s. Then, some terms of the equation are calculated on the Excel spreadsheet as follows:

	D_S	K	C_o	α	β	Γ
	Data	Data	Data	Eq. 5.17	$4\alpha/(1+4\alpha)$	From Eq. 5.19
for C	1.00E-09	0.43	0.25	5.95E-03	2.3E-02	6.9E-05
for Cr	1.20E-12	0.915	0.57	7.14E-06	2.9E-05	3.4E-11

These terms are then used to calculate the change in composition as a function of fraction solid. Different values for the fraction of solid are used as input on the Excel spreadsheet.

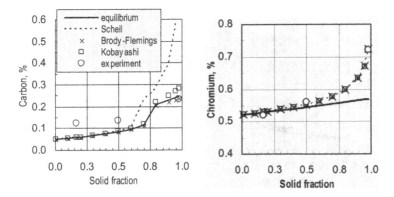

Fig. 5.16 Comparison of segregation calculations with various analytical models for $V_S = 2.8 \times 10^{-6}$ m/s.

A	B	C	D	H	J	N
f_S	0.01	0.100	0.167	0.500	0.700	0.980
ξ	Eq. 5.19	Eq. 5.19	Eq. 5.19	Eq. 5.19	Eq. 5.19	Eq. 5.19
ξ_C	0.990	0.900	0.834	0.502	0.303	0.024
ξ_{Cr}	0.990	0.900	0.833	0.500	0.300	0.020
C_S	Eq. 5.19	Eq. 5.19	Eq. 5.19	Eq. 5.19	Eq. 5.19	Eq. 5.19
C_S for C	0.049	0.053	0.057	0.085	0.226	0.751
C_S for Cr	0.522	0.526	0.530	0.553	0.578	0.727

Note that as the C content increases above 0.1 % (max. solubility in the δ phase) the peritectic temperature the partition coefficient of C changes (see the Fe–C diagram in Appendix C). Thus, and IF statement must be used when calculating ξ or C_S. For example, for column C in the calculation of C_S: IF(C_S in column B < 0.1, Eq. 5.19 with $k = 0.196$, Eq. 5.19 with $k = 0.495$). Similar calculations are performed for equilibrium, Eq. 5.9, Scheil, Eq. 5.12, and Brody–Flemings, Eq. 5.16 models.

The calculation and experimental results are presented in Fig. 5.16 for C and Cr, respectively. It is seen that for C, that has a relatively high solid diffusivity (interstitial solution), the equilibrium assumption gives the closest results. Yet, the experimental value is only predicted at 0.98 fraction of solid. This is because the equilibrium equation calculates the C concentration at the interface only, while the experimental data include some significant solid diffusion. Also, note the change in slope on the curves, as the C content increases above 0.1 %. The Brody–Flemings and the Kobayashi equations predict close results to the equilibrium equation.

For Cr, that has a much lower solid diffusivity (substitutional solution), the Scheil equation works well. There is no difference between the prediction of Scheil, Brody–Flemings, and Kobayashi equations. This means that the solid diffusivity of Cr is negligible.

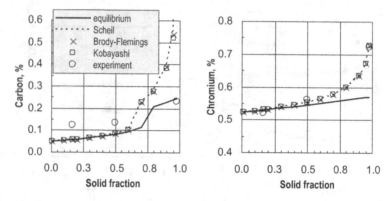

Fig. 5.17 Comparison of segregation calculations with various analytical models for $V_S = 2.8 \times 10^{-8}$ m/s.

Application 5.3 Consider the problem in Application 5.2. Compare calculation results for equilibrium, Scheil, Brody–Flemings, and Kobayashi assuming a solidification velocity of 0.028 μm/s. Discuss the differences.

Answer Using the same calculation procedure as in Application 5.2 we obtain the results plotted in Fig. 5.17. It is seen that for C the Brody–Flemings and Kobayashi models are very close to the Scheil calculation. This is because the solidification velocity is now very close to the diffusion velocity ($D_S/l = 1.67 \times 10^{-8}$) and C diffusivity cannot be considered infinite. However, for Cr there are no differences between the three nonequilibrium models. This is because for Cr the diffusion velocity (2×10^{-11}) is still much smaller than the solidification velocity.

Application 5.4 A vanadium–8 wt%C alloy is directionally solidified at a velocity of 10 μm/s. The heating is by induction, so that considerable convection occurs in the liquid. Microprobe evaluation of the boundary layer produced a value of $\delta = 2 \times 10^{-4}$ m. Calculate the solid composition at 0.5 solid fraction for the case of transport in liquid by diffusion only (complete diffusion) and for partial mixing in liquid. The phase diagram is provided in Appendix C.

Answer From the phase diagram $k_{max} = 1.88$ is calculated at the eutectic temperature of 1653 °C. A $k_{min} = 1.16$ is calculated at the temperature of 2180 °C. The average value, $k = 1.52$, will be used in subsequent calculation.

For the case of complete diffusion in liquid, no diffusion in solid Scheil Eq. 5.12 is used. For $f_S = 0.5$ it is obtained $C_S = 8.48\%$. For the case of partial mixing first, the effective partition coefficient must be calculated with Eq. 5.20. Taking $D_L = 10^{-8}$ m²/s we obtain $k_{ef} = 1.39$. Then, equation Eq. 5.21 is used and we obtain $C_S = 8.49\%$. This is a slight increase over the Scheil value.

Application 5.5 Consider zone melting. Calculate the composition profile (C-x graph) for a steel rod of length 60 mm having the composition 0.21 % C and 0.9 % Si, for one and two passes. Assume the length of the molten zone to be 3 mm.

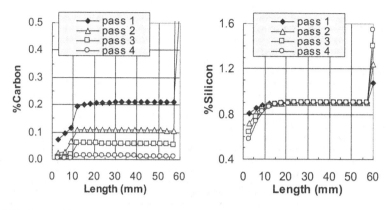

Fig. 5.18 Composition evolution during zone refining

Note: for the last cell composition is calculated from mass balance.

Answer The steel is of hypereutectic composition. This means that as the C content of the solid increases above 0.1 % the partition coefficient will change from 0.196 to 0.495. This will require the implementation of an IF statement.

Select the computational grid so that the total sample length is a multiple of the length of the molten zone. The Excel spreadsheet could be organized as shown in the following table. Equation 5.22 is used in cell B3 with $\ell = 3$ and x = value in cell A3. For all other cells the following IF statement must be implemented: IF (%C in preceding cell < 0.1, Eq. 5.22 with $k = 0.196$, Eq. 5.22 with $k = 0.495$). In all cells in row 24 a mass balance equation is implemented.

	A	B	C	D	E
1		First pass		Second pass	
2	Length (mm)	%C	%Si	%C	%Si
3	3	0.0712	0.8042	0.0242	0.7185
4	6	0.0959	0.8556	0.0260	0.8134
5	9	0.1163	0.8795	0.0665	0.8594
6	12	0.1954	0.8905	0.1083	0.8811
23	57	0.2100	0.9000	0.1067	0.9000
24	60	0.5940	1.0785	2.3523	1.2440

Results of calculations with this scheme are presented in Fig. 5.18. This calculation can be run until the eutectic composition is reached. Then the composition in the section next to the last one should be allowed to increase. Note that for Si that has a relatively high partition coefficient a large zone leveling region occurs.

References

Brody HD, Flemings MC (1966) Trans. TMS-AIME 236:615

Burton JA, Prim RC, Slichter WP (1953) J. Chem. Phys. 21:1987

Clyne TW, Kurz W (1981), Metall. Trans. 12A:965

Gulliver GH (1913) J. Inst. Met. 9:120

Gungor MN (1989) Metall. Trans. 20A:2529–2533

Himemiya T, Umeda T (1998) ISIJ Intern. 38:730

Kobayashi S (1988) Trans. Iron Steel Inst. Jpn. 28:728

Kurz W, Fisher DJ (1989) Fundamentals of Solidification. Trans Tech Publications, Switzerland

Nastac L (2004) Modeling and Simulation of Microstructure Evolution in Solidifying Alloys. Kluwer Academic Publishers, Boston

Ohnaka I (1992) Microsegregation and Macrosegregation. In: Stefanescu DM (ed) ASM Handbook vol.15 Casting. ASM International, Metals Park, OH

Pershing MA (1997) An assessment of some models for micro and macrosegregation as applied to cast steel. MS Thesis, The University of Alabama, Tuscaloosa

Pfann WG (1952) Trans. AIME 194:747

Rappaz M, Voller V (1990) Metall. Trans. 21A:749

Scheil E (1942) Zeitschrift Metallkde. 34:70

Sigworth GK (2014) Int. J. Metalcasting 8:7–20

Smith VG, Tiller WA, Rutter JW (1955) Canadian J. of Physics 33:723

Chapter 6
Diffusive Energy Transport at the Macroscale

Energy transport from the casting to the environment is ultimately responsible for solidification. There are three forms of energy transport: conduction (diffusive transport), convection (heat transmitted by the mechanical motion of the fluid), and/or radiation (through space). All three are active during solidification of a casting. Energy diffusion and convection occurs within the casting, at the metal/mold interface, and within the mold. Energy is transported by radiation from the mold to its environment, which is typically the air.

The corresponding equations for the heat flux are:

conduction: $q = -\alpha \frac{\partial(\rho c_P T)}{\partial x}$. Fourier's law

convection: $q = h\left(T(t) - T(0)\right)$. Newton's law

radiation: $q = \varepsilon\sigma\left(T_1^4 - T_2^4\right)$. Stefan–Boltzman's law

where α is the thermal diffusivity, h is the heat transfer coefficient, σ is the Stefan–Boltzman constant, and ε is the emissivity factor.

Steady state equations can be derived and solved based on the above equations, for the case of no phase change. However, solidification of castings is a process that can be either steady state or nonsteady state, and involves phase transformation. Thus, it is a transient problem involving partial differential equations (PDEs). In turn, PDEs can be solved either analytically or through numerical approximation methods. Specific boundary conditions (BC) must be used to describe various casting processes. In the following sections, we will discuss the governing equation with the source term and the most common boundary conditions.

6.1 Governing Equation for Diffusive Energy Transport

The governing equation for solidification of a casting is the conservation of energy equation written in its advection–diffusion form, Eq. 4.3, repeated here for convenience:

$$\frac{\partial}{\partial t}(\rho H) + \nabla \cdot (\rho \mathbf{V} H) = \nabla \cdot \left(\frac{k}{c}\nabla H\right) + S_H \tag{6.1}$$

© Springer International Publishing Switzerland 2015
D. M. Stefanescu, *Science and Engineering of Casting Solidification*,
DOI 10.1007/978-3-319-15693-4_6

The macroscopic heat flow equation in terms of temperature rather than enthalpy can be obtained from the conservation of energy. Indeed, for $V = 0$, assuming constant ρ and c, and since the enthalpy is $H = c \cdot T$:

$$\partial T / \partial t = \alpha \nabla^2 T + S_H / \rho c \tag{6.2}$$

Further assuming that the source term is the heat flow rate resulting from the latent heat of solidification, $S_H = \dot{Q}_{gen}$, the heat flow equation becomes:

$$\frac{\partial T}{\partial t} = \alpha \nabla^2 T + \frac{\dot{Q}_{gen}}{\rho c} \tag{6.3}$$

To solve this partial differential equation, we need an initial condition (because it has one time derivative) and two BC (because it has two spatial derivatives). Typically, the initial condition at $t = 0$ is the initial temperature distribution:

$$T(x, 0) = f(x) \tag{6.4}$$

A number of simplifications of Eq. 6.3 are possible. If there is no phase transformation or heat generation (no source term):

$$\partial T / \partial t = \alpha \nabla^2 T \tag{6.5}$$

The steady-state solution, when BC and the source term are independent of time, is:

$$\nabla^2 T + \dot{Q}_{gen} / k = 0 \quad \text{Poisson's equation} \tag{6.6}$$

This equation gives the equilibrium temperature distribution. If there is no heat generation this equation is reduced to:

$$\nabla^2 T = 0 \quad \text{Laplace's equation} \tag{6.7}$$

The problem becomes considerably more complicated for a two-phase system. For this case, the source term as derived by Benon and Incropera (1987) is:

$$S = \nabla \cdot \left[\frac{k}{c_S} \nabla (H_S - H) \right] - \nabla \cdot \left[\rho f_S (V - V_S)(H_L - H_S) \right] \tag{6.8}$$

The first right hand term (RH1) in the source term equation is the energy flux associated with phase transformation, and the second right hand term (RH2) is the energy flux associated with phase motion.

The energy equation for a two-phase system can be expressed in terms of temperature instead of enthalpy on the basis of the following relationships:

$$H = g_L H_L + g_s H_S \qquad H_S = c_S T \qquad H_L = c_L T + (c_S - c_L)T_e + \Delta H_f$$

Assuming that $c_S = c_L = c_p$, the energy equation becomes:

$$\frac{\partial}{\partial t} \left(\rho c_p T \right) + \nabla \cdot \left(\rho c_p V T \right) = \nabla \cdot (k \nabla T) + S \tag{6.9}$$

where $S = -\frac{\partial}{\partial t}(\Delta H_f \rho_L g_L) - \nabla \cdot [\Delta H_f \rho g_L (\mathbf{V} - \mathbf{V}_S)] - \nabla \cdot (\Delta H_f \rho_L g_L \mathbf{V})$

In the last equation, ΔH_f is the latent heat of fusion of the alloy. In the source term equation, RH1 is the latent heat from generation of solid fraction and RH2 and RH3 are the energy flux associated with phase motion. Note that when the advective terms are ignored, this equation reduces to:

$$\frac{\partial}{\partial t}\left(\rho c_p T\right) = \nabla \cdot (k\nabla T) - \frac{\partial}{\partial t}\left(\Delta H_f \rho_L g_L\right) \quad \text{or} \quad \frac{\partial T}{\partial t} = \nabla \cdot (\alpha \nabla T)$$
$$+ \frac{\Delta H_f}{\rho c_p}\frac{\partial}{\partial t}\left(\rho_S g_S - \rho\right)$$

If it is assumed that $\rho_S = \rho_L = \rho$, then $g_S = f_S$, $\partial \rho / \partial t = 0$ and the source term in Eq. 6.1 becomes:

$$\dot{Q}_{gen} = \rho_S \Delta H_f \left(\partial f_S / \partial t\right) \tag{6.10}$$

For the first stage of equiaxed solidification, when the grains move freely with the liquid, $\mathbf{V} = \mathbf{V}_S = \mathbf{V}_L$, and Eq. 6.9 can be simplified to:

$$\frac{\partial}{\partial t}\left(\rho c_p T\right) + \nabla \cdot \left(\rho c_p \mathbf{V}T\right) = \nabla \cdot (k\nabla T) - \frac{\partial}{\partial t}\left(\Delta H_f \rho_L g_L\right) - \nabla \cdot \left(\Delta H_f \rho_L g_L \mathbf{V}\right) \tag{6.11}$$

where, RH3 is the convected latent heat.

In the second stage of equiaxed solidification, when after dendrite coherency occurs the solid is fixed, $\mathbf{V}_S = 0$. The energy equation can then be written as:

$$\frac{\partial}{\partial t}\left(\rho c_p T\right) + \nabla \cdot \left(\rho c_p \mathbf{V}T\right) = \nabla \cdot (k\nabla T) + \frac{\partial}{\partial t}\left(\Delta H_f \rho_S g_S\right) \tag{6.12}$$

Since g_S and g_L are function of time and temperature, appropriate expressions must be provided for these quantities in order to solve the energy equation. The energy equations derived above are second order PDEs. The complete equations cannot be solved analytically without further simplifying assumptions. Thus, numerical solutions have been proposed. Regardless of the solution, BC and initial conditions are necessary. They will be discussed in the following section.

6.2 Boundary Conditions

Typical BC used for computational modeling of casting solidification are shown in Fig. 6.1. These BCs can be expressed as follows:

known heat flux (Newtons law of cooling, convective BC):

$$-k\frac{\partial T}{\partial x}(0, t) = h(T(t) - T(0)) \tag{6.13a}$$

Fig. 6.1 Typical BC for castings

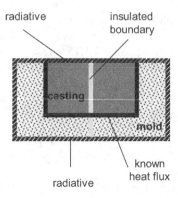

insulated boundary (Neumann problem): $-k\dfrac{\partial T}{\partial x}(0, t) = 0$ (6.13b)

prescribed temperature (Dirichlet problem): $T(0, t) = T_1$ (6.13c)

where, h is the heat transfer coefficient.

The radiation loss requires additional discussion. As summarized by Upadhya and Paul (1994), heat loss through a surface in the general case is by radiation and convection, and the heat transfer coefficient becomes:

$$h = h_r + h_c = \sigma \varepsilon F_{m-a} \left(T_S^2 + T_o^2\right)(T_S + T_o) + c(T_S + T_o)^{1/3},$$

where h_r and h_c are the radiative and convective heat transfer coefficients, respectively, $F_{m\text{-}a}$ is the view factor of the mold with respect to air, and c is a constant dependent on the surface geometry. The view factor is defined as the fraction of the radiation that leaves surface i in all directions and is intercepted by surface j. When two surfaces, dA_1 and dA_2, undergo radiation exchange, the view factor can be mathematically expressed as (Siegell and Howell 1981):

$$F_{1-2} = \iint\limits_{A_1 A_2} \frac{\cos\theta \cos\phi}{R_{1-2}^2} dA_1 dA_2,$$

where, R_{1-2} is the distance between the two surfaces, and θ and ϕ are the angles of the two surface normals with the line joining the two surfaces.

The calculation of the view factor may become rather complicated when multiple surfaces are involved in the radiation process. A code for view factor calculation has been developed by Lawrence Livermore National Laboratory (LLNL) (Shapiro 1983).

For sand and die casting the first two BC are generally used. However, for investment casting the third one is necessary, so that the extension of solidification models for sand castings to investment castings is not trivial.

In the following sections, we will discuss some analytical solutions of the energy transport equation applied to particular cases for castings. It is convenient to discuss these solutions first for steady-state and then for nonsteady-state energy transport.

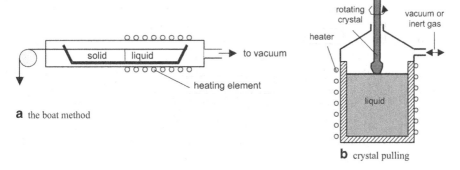

Fig. 6.2 Single crystal growth processes

6.3 Analytical Solutions for Steady-State Solidification of Castings

Steady-state solidification occurs during controlled directional solidification. Directional solidification is widely used for industrial applications. Steady state solidification is used mainly for the production of single crystals.

Theprocesses used for growth of single crystals include:

- Normal freezing: the boat method (Fig. 6.2a) or the Bridgman method (Fig. 5.13a)
- Crystal pulling (Czochralski) (Fig. 6.2b)
- Zone melting and zone freezing: with crucible or crucibless (surface held by surface tension and/or magnetic forces) (Fig. 5.13b)

Two solutions of the steady-state heat conduction equation (Laplace equation) are given in Appendix A. Unfortunately, they are not useful for the study of directional solidification because they are not applicable to the moving boundary problem that is directional solidification. Thus, other solutions must be developed.

The basic heat flow objectives are to obtain a constant thermal gradient across the solid/liquid (S/L) interface and to move the temperature gradient at a controlled rate. Let us consider an S/L interface. Heat flux balance at the interface is:

$$q_S = q_L \quad (\text{in J m}^{-2}\text{s}^{-1} = \text{W m}^{-2}), \qquad (6.14)$$

with $q_S = -k_S G_S$ and $q_L = -k_L G_L - \Delta H_f V$, where k is the thermal conductivity (J m^{-1} s^{-1} K^{-1}) and G is the gradient (K m^{-1}). Note the heat generation term in the second equation, where the latent heat is expressed in J m^{-3}. Solving for V we obtain:

$$V = (k_S G_S - k_L G_L)/\Delta H_f \qquad (6.15)$$

The maximum solidification velocity can then be calculated for $G_L \to 0$, as $V = k_S G_S/\Delta H_f$. G_S can be evaluated from experiments or from heat flow calculations.

Let us try to calculate G_S assuming that in the directionally solidified sample energy transport is carried out by conduction alone. This would be a good approximation for Bridgman-type solidification. The problem is one of steady state heat flow with moving boundary. The governing equation is Eq. 6.9 simplified for the case of steady state and no source terms. In one-dimensional form it becomes:

$$\frac{\partial^2 T}{\partial x^2} - \frac{V}{\alpha}\frac{\partial T}{\partial x} = 0. \qquad (6.16)$$

Setting the reference system at the S/L interface, the advective term is added to the energy flow (V is positive). Note that for the case of energy transport, the diffusion problem is solved for the solid, whereas for solute transport it was solved for the liquid. This is a linear, homogeneous second-order ordinary differential equation. Its solution is:

$$T = A + B \exp\left(\frac{V}{\alpha_S}x\right), \qquad (6.17)$$

where, A and B are constants. The BC are as follows:

BC1   at $x = 0$ $T = T_f$
BC2   at $x \to -\infty$ $T = T_o$

From BC1, $T_f = A + B$, and from BC2, $T_o = A$. Substituting in Eq. 6.17:

$$T = T_o + (T_f - T_o)\exp(Vx/\alpha_S). \qquad (6.18)$$

The interface thermal gradient in the solid is:

$$G_S = (dT/dx)_{x=0} = (T_f - T_o)\; V/\alpha_S. \qquad (6.19)$$

Typical experimental gradients are 1 K/cm for high melting point crystals, and 5 K/cm for low melting point crystals (Flemings 1974).

6.4 Analytical Solutions for Non-Steady-State Solidification of Castings

Nonsteady state heat transport is typical for some directional solidification (DS) processes, such as chill-casting or controlled directional solidification of superalloys, as well as for the vast majority of casting processes, including sand casting, die casting, investment casting, continuous casting, ingot casting, etc. Typical nonsteady-state directional solidification processes are presented schematically in Fig. 6.3. The main difference between the two processes is that in chill casting the solidification velocity decreases continuously during solidification, while it is typically maintained constant during DS. In all nonsteady-state processes, the solidification

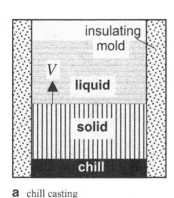

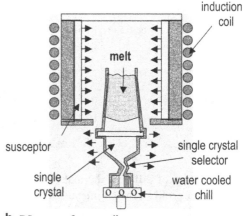

a chill casting   **b** DS process for superalloys

Fig. 6.3 Nonsteady state directional solidification processes. **a** chill casting. **b** DS process for superalloys

Fig. 6.4 Temperature profile in a casting-mold assembly

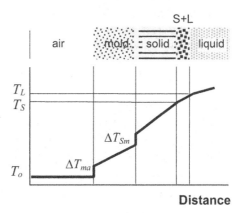

velocity and/or the temperature gradient at the interface change continuously during solidification. Accordingly, steady-state energy transport can no longer be assumed.

The PDE that must be solved is Eq. 6.9. This equation cannot be solved analytically without further simplifying assumptions. The general simplifying assumptions include:

- The metal is poured without superheat, that is at T_f
- Pure metal (has a melting temperature not a melting range)
- Mold is semi-infinite (outside temperature is that of the ambient, T_o)
- No heat generation

The temperature profile in the mold is shown in Fig. 6.4 for the general case. The temperature decrease within a domain, or at the interface between two domains, can be considered an interface resistance. Additional assumptions are made on the

Table 6.1 Assumptions on thermal resistance in a casting-mold system

Casting process	Fig. 6.5	Resistance			
		In solid	At solid/mold interface	In mold	At mold/air interface
Insulating molds	a	0	0	High	0
Permanent molds	b	0	High	0	0
Ingot molds	c	High	0	0	0
Ingot molds	d	High	0	High	0

relative values of different thermal resistances, as summarized in Table 6.1. The temperature profile across a metal—mold section following these assumptions is shown in Fig. 6.5.

The governing equation derived from Eq. 6.9 assuming $\mathbf{V}=0$ and $S=0$ is:

$$\partial T/\partial t = \alpha\left(\partial^2 T/\partial x^2\right). \tag{6.20}$$

Several methods may be used to solve this equation. Three Fourier series solutions for different BC are given in Appendix A. When solving with the method of combination of variables the solution is:

$$T = C_1 + C_2 \mathrm{erf}(u) \quad \text{where} \quad u = x/\left(2\sqrt{\alpha t}\right), \tag{6.21}$$

erf(u) is the error function discussed in detail in the inset. The constants in the preceding equation can be found if appropriate BC are available. Unfortunately, this is not possible for the general case. The simplifying BC listed in Table 6.1 must be used.

The error function.
The error function is derived from the "bell" curve shown in Fig. 6.6. The curve is generated by the equation $\int_0^\infty \exp\left(-x^2\right)dx = \sqrt{\pi}/2$. The dashed area under the curve has the property $\int_0^\infty \exp\left(-x^2\right)dx = \sqrt{\pi}/2$.

The error function of x, erf(x), is defined as the ratio between the area under the curve $y|_{x=0}^{x=x}$ and $y|_{x=0}^{x=\infty}$. That is $\mathrm{erf}(x) = \frac{2}{\sqrt{\pi}}\int_0^x \exp\left(-x^2\right)dx$

Hence, some properties of the error function are:

$$\mathrm{erf}(0) = 0 \quad \mathrm{erf}(1) = 0.842 \quad \mathrm{erf}(2) = 1.$$
$$\mathrm{erf}(\infty) = 1 \quad \mathrm{erf}(-\infty) = -\mathrm{erf}(\infty).$$

The series representation of the error function is
$$erf(x) = \frac{2}{\sqrt{\pi}}\left(x - \frac{x^3}{3\cdot 1!} + \frac{x^5}{5\cdot 2!} - \frac{x^7}{7\cdot 3!}\cdots\right).$$

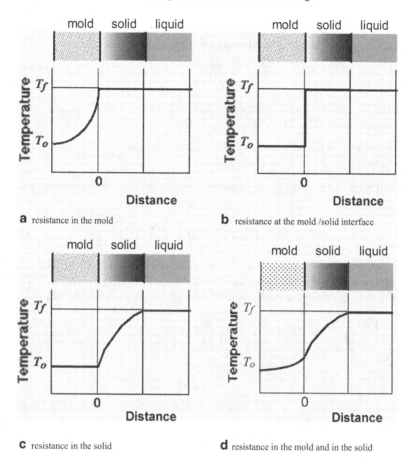

a resistance in the mold

b resistance at the mold/solid interface

c resistance in the solid

d resistance in the mold and in the solid

Fig. 6.5 Temperature profile in the mold and in the casting for different assumptions

Fig. 6.6 Bell curve used for definition of error function

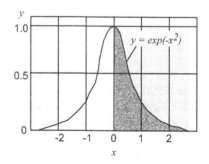

6.4.1 Resistance in the Mold

To establish some BC, let us assume that the metal is poured in an insulating mold (*e.g.*, sand, shell, or investment mold) for which the main resistance is in the mold.

Let us derive an equation for the temperature distribution in the mold. If the zero of the x-axis chosen at the solid/mold interface (Fig. 6.5a), the BC are:

BC1   at $x = -\infty$ $T = T_o$,
BC2   at $x = 0$ $T = T_f$,

where, T_f is the fusion temperature of the metal, and T_o is the ambient temperature. Using these BCs in Eq. 6.21:

from BC1: $T_o = C_1 + C_2 erf(-\infty) = C_1 - C_2 erf(\infty) = C_1 - C_2$,
from BC2: $T_f = C_1 + C_2 erf(0) = C_1$.

Substituting in Eq. 6.21, the temperature distribution in the mold can be calculated with:

$$T = T_f + (T_f - T_o)\, erf\left(x/\left(2\sqrt{\alpha t}\right)\right). \qquad (6.22)$$

Let us now evaluate the thickness of casting, L, solidified over some time t. Heat flux balance at the solid/mold interface gives $-k_m(\partial T/\partial x)_{x=0} = -\rho_S \Delta H_f dx/dt$. From Eq. 6.22 the diffusive heat transport through the mold is calculated as:

$$-k_m \left(\frac{\partial T}{\partial x}\right)_{x=0} = -k_m (T_f - T_o)\left(\frac{\partial}{\partial x} erf \frac{x}{2\sqrt{\alpha t}}\right)_{x=0} = -\frac{k_m(T_f - T_o)}{\sqrt{\pi \alpha t}}$$

$$= -\sqrt{\frac{k_m \rho_m c_m}{\pi t}}\,(T_f - T_o).$$

By analogy with Ohm's law, the resistance in the mold can be defined from $q = k \cdot \Delta T/\Delta x = \Delta T/(\Delta x/k)$ as:

$$R_m = A^{-1}\sqrt{\pi t/(k_m \rho_m c_m)}.$$

Substituting the diffusive heat transport through the mold in the heat flux balance equation, rearranging and integrating between $t = 0$ and $t = t$, and $x = 0$ and $x = L$ results in:

$$L = \frac{2}{\pi}\left(\frac{T_f - T_o}{\rho_S \Delta H_f}\right)\sqrt{k_m \rho_m c_m}\; t^{1/2}. \qquad (6.23)$$

Note that the terms in parenthesis are constants referring to the metal, while those under the square root are constants referring to the mold. This equation does not take into account the superheating required for pouring. However, the heat generated during solidification has been included in the heat flux balance. This equation can be used to calculate the thickness solidified for non-divergent heat flow (Fig. 6.7b).

In the preceding equations, L is the conductive path length, and is the characteristic linear dimension. In general, a characteristic linear dimension may be obtained by dividing the volume of the solid by its surface area, $L = v/A$. This is exact for plate-type castings, and approximate within 5 % error for cylinders and spheres. Taking the characteristic length of the casting to be $L = v/A$, where v is the volume

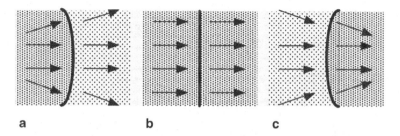

Fig. 6.7 Types of heat flow. Divergent heat flow results in heat dissipation, while convergent heat flow results in heat accumulation. **a** divergent. **b** non-divergent **c** convergent

of the casting and A is its cooling surface area, Eq. 6.23 can be rewritten to give the final solidification time for a casting poured into an insulating mold:

$$t_f = \frac{\pi}{4} \left(\frac{1}{k_m \rho_m c_m} \right) \left(\frac{\rho_S \Delta H_f}{T_f - T_o} \right)^2 \left(\frac{v}{A} \right)^2 = ct. \cdot \left(\frac{v}{A} \right)^2 = ct. \cdot M^2, \qquad (6.24)$$

where M is the *casting modulus*. This equation is known as the *Chvorinov* equation (Chvorinov 1940).

The effect of the superheating temperature, ΔT_{super}, on the final solidification time can also be included by writing the heat to be removed from the casting as:

$$Q = \rho_S \Delta H_f v + \rho_L c_L \Delta T_{super} v.$$

Assuming that $\rho_S = \rho_L = \rho$ the heat to be removed is written as:

$$Q = \rho v \left(\Delta H_f + c_L \Delta T_{super} \right) = \rho v \Delta H_{eff},$$

where ΔH_{eff} is the effective heat of fusion. Then, Eq. 6.24 becomes:

$$t_f = \frac{\pi}{4} \left(\frac{1}{k_m \rho_m c_m} \right) \left(\frac{\rho \Delta H_{eff}}{T_f - T_o} \right)^2 \left(\frac{v}{A} \right)^2. \qquad (6.25)$$

An example of the use of the *Chvorinov* Eqs. 6.24 and 6.25 to calculate the final solidification time of a casting is given in Application 6.1. The two Chvorinov equations do not fit exactly the experimental data, in particular at the beginning of solidification. Experimental data are better described by an equation having the shape:

$$L = C_1 \sqrt{t} - C_2, \qquad (6.26)$$

where C_1 is a constant made of the first three terms of Eq. 6.24, and C_2 is a constant resulting from convection in the liquid that removes superheat and delays solidification, which introduces a finite mold—metal resistance to heat transfer.

A graphical representation of this equation is given in Fig. 6.8 for experiments in which aluminum was poured in steel and dry sand molds. It can be seen that the

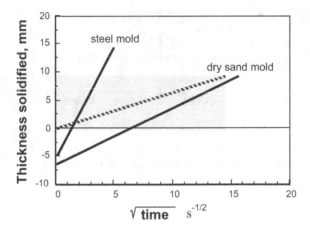

Fig. 6.8 Influence of time and cooling rate (mold material) on the solidified thickness

slope of the curve increases from the sand mold to the steel mold. This is a result of the increase in the cooling rate (CR), which is a consequence of higher thermal conductivity of the steel mold as compared with the dry sand mold. The dotted line represents schematically the correlation calculated with Eq. 6.23.

More accurate calculation with the Chvorinov equation can be performed using a time-stepping procedure. An example is given in Application 6.2 where the following equation was derived assuming resistance in the mold:

$$T^{new} = T^{old} - \frac{A}{v\rho c}\sqrt{\frac{k_m \rho_m c_m}{\pi t^{new}}}\left(T^{old} - T_m\right)\Delta t + \frac{\Delta H_f}{c}\Delta f_S. \tag{6.27}$$

This equation allows calculation of the temperature at the *new* time step, if the temperature is known at the previous time step, *old*. This approach is called a time-stepping procedure.

6.4.2 Resistance at the Mold/Solid Interface

A rather simple case to analyze is that when the resistance is at the mold/solid (*mS*) interface (Fig. 6.5b). The basic assumption is that heat transfer from the solid to the mold is by convection. This is particularly valid when an air gap forms at the interface in such processes as permanent molding or die-casting. Other processes, such as splat cooling and atomization can also be described under such an assumption.

Heat flux balance at the mold/solid (*mS*) interface can be written as $h\left(T_f - T_o\right) = -\rho_S \Delta H_f dx/dt$. The interface resistance is: $R_{mS} = 1/(A \cdot h)$. After the integration of the flux balance equation, between $t = 0$ and $t = t$, and between $x =$

Fig. 6.9 Temperature profile
assuming solid resistance and
interface temperature lower
than the melting point

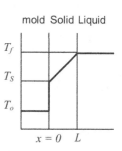

0 and $x = l$, it is obtained that $l = (T_f - T_o)\, ht/(\rho_S \Delta H_f)$. Then, if non-divergent heat transfer is assumed, and since $l = v/A$:

$$t_f = \frac{\rho_S \Delta H_f}{h\,(T_f - T_o)} \frac{v}{A} \tag{6.28}$$

This equation is valid when ΔT_{mS} is large compared with ΔT_S or ΔT_m. That is, $\Delta T_{mS} \gg \Delta T_S$ when $1/h \gg L/k_S$, or $\Delta T_{mS} \gg \Delta T_m$ when $1/h \gg [\pi t/(k_m \rho_m c_m)]^{1/2}$.

A more complicated case in which $T_S \neq T_f$ ($\Delta T_S > 0$) can be considered (Fig. 6.9). The flux into the mold/solid interface is $q|_{x=0} = -k_S\,(T_f - T_S)/L$. The flux out of the mold/solid interface is $q|_{x=0} = -h\,(T_S - T_o)$. Eliminating T_S between these two equations and recognizing that the flux at $x = 0$ must be equal to the flux at $x = L$ to satisfy energy conservation, as well as that the flux into the S/L interface from the liquid is the result of the latent heat evolved (remember that there is no superheat) we have:

$$q|_{x=L} = \frac{T_f - T_o}{1/h + L/k_S} = \rho_S \Delta H_f \frac{dL}{dt}.$$

Integrating between 0 and L, and between 0 and t it is found that:

$$L + \frac{h}{2k_S} L^2 = \frac{h\,(T_f - T_o)}{\rho_S \Delta H_f} t. \tag{6.29}$$

The problem has been solved by Adams in a more rigorous manner, by assuming that the temperature profile is not necessarily linear. The following equation was obtained:

$$L + \frac{h}{2k_S} L^2 = \frac{h\,(T_f - T_o)}{\rho_S \Delta H_f a} t \quad \text{where} \quad a = \frac{1}{2} + \sqrt{\frac{1}{4} + \frac{c_S\,(T_f - T_o)}{3\Delta H_f}}. \tag{6.30}$$

Heat flux balance at the interface can be used to evaluate the temperature gradient during casting solidification, as shown in Application 6.3.

Another approach is the so-called *lump analysis*. If the Biot number (Bi $= \bar{h}L/k$, where $\bar{h}$ is the average heat transfer coefficient for convection for the entire surface, and L is the conduction length) is small, *e.g.*, Bi < 0.1, the internal temperature gradients are also small. The object of analysis is considered to have a single mass averaged temperature and the transient problem can be treated as follows. Consider a macro-volume element of volume v and surface area A. Heat flow rate balance for this element requires:

$$\bar{h}A\,[T(t) - T_o] = v\dot{Q}_{gen} - v\,\rho c\frac{dT}{dt}. \tag{6.31}$$

As shown in Eq. 6.10, $\dot{Q}_{gen} = \rho\,\Delta H_f\,(df_S/dt)$, and thus:

$$\bar{h}A\,[T(t) - T_o] = v\rho\Delta H_f\frac{df_S}{dt} - v\,\rho c\frac{dT}{dt}. \tag{6.32}$$

This equation can be integrated only if f_S is a function of temperature, *e.g.*, $f_S(T) = a + b \cdot T$. Rearranging:

$$\frac{dT - \left(\Delta H_f/c\right)df_S}{T(t) - T_o} = -\frac{\bar{h}A}{v\,\rho c}dt.$$

Integrating and applying the initial condition gives:

$$T(t) = T_o + (T_i - T_o)\exp\left(-\frac{\bar{h}At}{v\,\rho c\left(1 - (b/c)\Delta H_f\right)}\right). \tag{6.33}$$

From this equation the solidification time of a casting of volume v is:

$$t_f = -\frac{\rho c}{\bar{h}}\frac{v}{A}\left(1 - \frac{b}{c}\Delta H_f\right)\ln\left(\frac{T_f - T_o}{T_i - T_o}\right), \tag{6.34}$$

where T_i is the initial (superheat) temperature. This equation is valid in the solidification interval. Above T_L and under T_S, $\Delta H_f = 0$.

Eq. 6.32 can be rewritten to describe the CR in the casting, which is the derivative of the cooling curve:

$$\frac{dT}{dt} = -\frac{\bar{h}A\,[T(t) - T_o]}{\rho c v} + \frac{\Delta H_f}{c}\frac{df_S}{dt}. \tag{6.35}$$

If in this equation (RH2) > (RH1), heating occurs that shows up on the cooling curve as recalescence. An example of the use of lump analysis is provided in Application 6.4.

When solidification of an alloy must be described, since the fraction of solid is a function of temperature, it can be expressed as $df_S/dt = (df_S/dT)(dT/dt)$, and the CR is:

$$\frac{dT}{dt} = -\frac{\bar{h}A}{\rho c v}\,[T(t) - T_o]\left(1 - \frac{\Delta H_f}{c}\frac{df_S}{dt}\right)^{-1}.$$

Fig. 6.10 Possible metal/mold contacts at the metal/mold interface

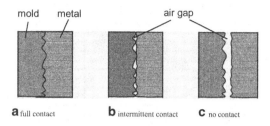

a full contact   **b** intermittent contact   **c** no contact

From this equation, it is seen that solidification decreases the CR since $df_S/dT < 0$. If lump analysis is not acceptable, or f_S as a function of temperature is not available, numerical methods must be used.

Alternatively, an equation for time-stepping calculation can be derived straight-forward from Eq. 6.32 (see also Application 6.4):

$$T^{new} = T^{old} + \frac{\rho \Delta H_f v \Delta f_S - hA \left(T^{old} - T_o\right) \Delta t}{\rho c v}. \tag{6.36}$$

6.4.3 The Heat Transfer Coefficient

The treatment of the solidification problem in the previous section is based on the simple assumption that the heat transfer is either constant or a continuous inverse function of time. However, the heat transfer coefficient at the mold/metal interface can vary over a wide range, depending whether an air gap is formed at the interface or not.

The issue of the numbers to use for the heat transfer coefficient is an open one, since it depends on a number of factors including temperature, casting geometry, and gap formation. In principle, the value of the heat transfer coefficient depends on the mechanism of heat transport at the solid/mold interface, which in turn is a function of the metal-mold contact. Three stages can be rationalized (Ho and Pehlke 1984; Trovant and Argyropoulos 2000) as summarized in Fig. 6.10.

In stage I, at the beginning of solidification, the contact between the liquid metal and the mold can be assumed good. Heat transport is through conduction from liquid metal to the mold wall.

However, as a solid layer forms, the metal will shrink away from the mold and a discontinuous air gap will result (stage II). The mold and solid metal will have par-tial contact at the asperities of the surfaces. Heat transport is now through mold/solid metal conduction at the points of contact and through gas conduction and radiation through the metal/mold gap.

In stage III, the metal will pull away completely from the mold and heat transport is only through the gap. Ho and Pehlke (1984) have demonstrated that it is possible to calculate the interfacial heat transfer for this case by a simple superposition of gas conduction and radiation via the quasi steady-state approximation.

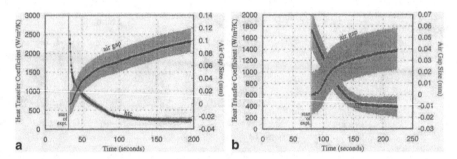

Fig. 6.11 Heat transfer coefficient and air gap width measurement for pure aluminum and A356 poured in graphite molds (Trovant and Argyropoulos 2000). With permission from Springer Science and Business Media

Thus, the heat transfer coefficient across the interface may be written as the sum of three components:

$$h = h_{mc} + h_{gc} + h_{gr}, \tag{6.37}$$

where h_{mc} accounts for metal-to-mold conduction, h_{gc} for conduction through the air gap, and h_{gr} for radiation through the air gap. The first term can be formulated straightforward for normal casting conditions, assuming good solid contact (see for example Application 6.4). However, it is a function of pressure. Thus, for squeeze casting, large heat transfer coefficients of the order of 50,000 W m^{-2} K^{-1} were reported (Nishida and Matsubara 1976). Note that the calculation of h_{mc} in stage II must take into account the reduced surface area of contact, which is not trivial. The second term can be calculated as $h_{gc} = k_{air}/l$, where k_{air} is the thermal conductivity of air and l is the gap thickness.

Finally, the third term of Eq. 6.37 can be calculated with the analytical equation:

$$h_{gr} = \frac{\sigma \left(T_{Si}^2 + T_{mi}^2\right)(T_{Si} + T_{mi})}{\varepsilon_{Si}^{-1} + \varepsilon_{mi}^{-1} - 1},$$

where σ is Stefan-Boltzmann constant, ε is emissivity, and the subscripts Si and mi, mean solid metal interface and mold interface, respectively.

Some examples of the dependency of the heat transfer coefficient on the gap width for pure aluminum and for the aluminum alloy A356 poured in graphite molds are given in Fig. 6.11. Typical average values of heat transfer coefficients for particular solidification processes are given in Appendix B, Table B6.

6.4.4 Resistance in the Solid

The typical example for this assumption (Fig. 6.5c) is an ingot mold where the mold is water-cooled. The governing equation is again the PDE Eq. 6.20, whose solution is Eq. 6.21. The following BC apply:

BC1 at $x = 0$ $T = T_o$,

BC2 at $x = L$ $T = T_f$.

From BC1 we derive $C_1 = T_o$. From BC2 we have $C_2 = (T_f - T_o)/erf\left(L/2\sqrt{\alpha_S t}\right)$. Using the notation $\gamma = L/2\sqrt{\alpha_S t}$ it is seen that, since C_2, T_f, T_o, are constant, γ is also constant. Thus, the distance solidified is:

$$L = 2\gamma\sqrt{\alpha_S t}, \tag{6.38}$$

with γ constant and unknown. When substituting the constants C_1, C_2 in the solution, the temperature distribution in the solid is obtained as:

$$T = T_o + \frac{T_f - T_o}{erf\,\gamma}erf\frac{x}{2\sqrt{\alpha_S t}}. \tag{6.39}$$

The heat flux at the liquid/solid interface $k_S(\partial T/\partial t)_{x=L} = \rho_S \Delta H_f \partial L/\partial t$ is used to find γ. The temperature gradient can be obtained from the preceding equation:

$$\left(\frac{\partial T}{\partial x}\right)_{x=L} = \frac{\partial}{\partial x}\left[\frac{T_f - T_o}{erf\,\gamma}erf\frac{x}{2\sqrt{\alpha_S t}}\right]_{x=L} = \frac{T_f - T_o}{erf\,\gamma}\frac{\exp\left(-\gamma^2\right)}{\sqrt{\pi\alpha_S t}}.$$

The solidification velocity can be calculated from Eq. 6.38 as: $\partial L/\partial t = \gamma\sqrt{\alpha_S/t}$. Substituting in the heat flux balance equation it is found that:

$$\gamma\,erf\,\gamma\,\exp\gamma^2 = (T_f - T_o)\frac{c_S}{\Delta H_f\sqrt{\pi}}. \tag{6.40}$$

To obtain γ this equation must be solved by iterations. Then, the solidified thickness is calculated with Eq. 6.38. Note that the solidified thickness is a parabolic function of time $(L \sim t^{-1/2})$. This is consistent with the shape of the solidified thickness curve in the solid region of Fig. 6.5c.

Analytical solutions to more complicated scenarios including resistance in the solid, the mold and at the interface can be found in other references such as Flemings (1974); Stefanescu (2002) or Poirier and Geiger (1994). However, analytical solutions are rarely used today, as numerical solutions can deal with more complicated problems, including complex geometry, and give more accurate results.

6.5 Thermal Analysis

Thermal analysis (TA) entails recording and interpreting the temperature variation in time of a cooling or heating material. In its simplest form, as applied to metal casting, the cooling curve of a metal solidifying in a mold is recorded and analyzed. Its interpretation is based on the understanding that all the events occurring during solidification leave their mark on the shape of the cooling curve. Le Chatelier was the first scientist that recorded temperature as a function of time for heating curves in 1887.

Differential thermal analysis (DTA), performed first by Roberts–Austen in 1899, consists in time and temperature difference measurements. It can be conducted with, or without, a reference body and using one or two thermocouples in the same test mold. The reference body is a material, real or virtual (computer-generated) that has no phase transformation over the temperature interval of interest.

TA is a powerful tool for casting process control. The application of TA in metal casting appears to have started in 1931 by Esser and Lautenbusch, who showed that increasing the superheating of gray iron depresses the eutectic arrest. Today, TA can be used to predict alloy composition, grain refining in steel, aluminum, magnesium and other alloys, eutectic morphology (*e.g.*, graphite morphology in cast irons or degree of modification in Al alloys), and shrinkage propensity. Computer analysis of the cooling curve can provide quantitative information on solidification, such as latent heat of solidification, evolution of fraction solid, amounts of phases, dendrite coherency, and dendrite arm spacing.

6.5.1 Direct Thermal Analysis

In its simplest form TA uses one thermocouple inserted in the test mold. The shape of the cooling curve is determined by the balance between the latent heat liberated during solidification and the heat lost to surroundings (the test cup and the atmosphere). The standard terminology used in TA showed in Fig. 6.12 is from an example of cast iron.

Because of the direct correlation between transformation temperatures and composition, TA has been used extensively to build phase diagrams. Its first application in process control, which was to evaluate the carbon equivalent of cast iron and the silicon content in Al–Si alloys, is based on the correlation between the liquidus temperature of an alloy (T_L) and its chemical composition (%C), as illustrated in Fig. 6.13. However, because solidification is a non-equilibrium process, the real cooling curve will depart from the theoretical one and exhibit undercooling with respect to the equilibrium temperatures. Thus, the thermodynamic equilibrium liquidus temperature will be in most cases higher than the actual non-equilibrium temperature measured by the thermocouple. Applications of direct TA for specific casting alloys will be discussed in Chap. 19.

6.5.2 Differential Thermal Analysis

ClassicDTA is performed with a reference body (Fig. 6.14). The sample and the reference body are cooled in a furnace at a controlled speed. At the beginning of cooling the temperatures of the sample (T_{sam}) and of the reference (T_{ref}), follow parallel paths. When an exothermic reaction, such as a phase change, occurs,

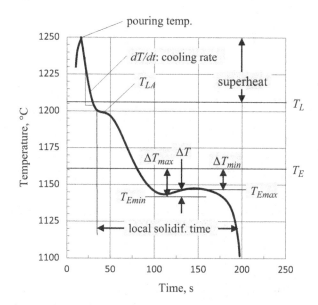

Fig. 6.12 Typical terminology used in TA for casting process control: T_L—equilibrium liquidus temperature, T_E—equilibrium eutectic temperature, T_{LA}—temperature of liquidus arrest, T_{Emin}—temperature of eutectic undercooling, T_{Emax}—temperature of eutectic recalescence, ΔT—recalescence, ΔT_{max}—maximum undercooling, ΔT_{min}—minimum undercooling (Stefanescu 2015). With permission from the American Foundry Soc

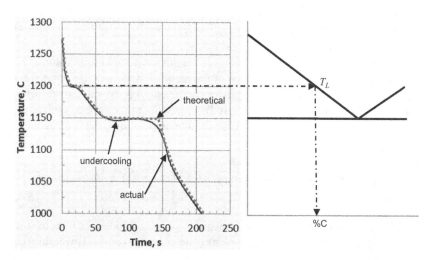

Fig. 6.13 Theoretical (equilibrium) and experimental (non-equilibrium) cooling curves for a hypoeutectic alloy (Stefanescu 2015). With permission of the American Foundry Soc

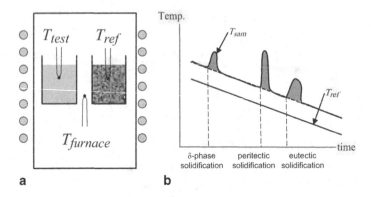

Fig. 6.14 Schematic representations of the principles of DTA (Stefanescu 2015). With permission from the American Foundry Soc. **a** equipment for DTA. **b** interpretation of DTA for high speed steel

the temperature will increase and a peak will be formed on the sample temperature graph because of the solidification of the δ-phase, the peritectic reaction, and the eutectic solidification. The shaded areas are proportional with the latent heat released during the phase transformation. An endothermic peak may occur during cooling when a crystalline phase change occurs.

The latent heat produced during the transformation can be calculated from the heat flow rate balance of the test-casting/crucible system:

$$\frac{dQ_f}{dt} - v\rho c_p (dT/dt)_{sam} = hA\,(T_{sam} - T_o) + \varepsilon\sigma\left(T_{sam}^4 - T_o^4\right), \qquad (6.41)$$

where Q_f is the solidification latent heat, t is the time, v is the sample volume, ρ is the metal density, c_p is the specific heat of the metal, T is the average temperature in the sample, h is the effective heat transfer coefficient, A is the surface area of the sample, T_o is the furnace temperature, ε is the emissivity of the surface sample, and σ is the Stefan-Boltzmann constant. After some manipulations, the time evolution of the latent heat can be calculated as:

$$\frac{dQ_f}{dt} = v\rho c_p\left[(dT/dt)_{sam} - (dT/dt)_{ref}\right] + hA\,(T_{sam} - T_o) + \varepsilon\sigma\left(T_{sam}^4 - T_o^4\right).$$
$$(6.42)$$

The total latent heat evolved during solidification, Q_f, can be obtained from the time integration of this equation. The reaction rate (rate of fraction of solid evolution) can then be calculated through the numerical integration of:

$$f_S^i = f_S^{i-1} + Q_f^i/Q_f. \qquad (6.43)$$

where, f_S^i is the fraction solid generated from time *zero* to time t_i, and Q_f^i is the total heat generated till time i.

This method is widely applied in research laboratories, but it is not directly used in metal casting processing because it cannot provide timely data useful in operation.

Integral calculus applied to the temperature-time data allows evaluation of areas that are directly related to the energy evolution. DTA lends itself to straightforward integral calculus analysis and thus it can be conducted without a reference sample. A test cup with one thermocouple is used. The problem is then to calculate the cooling of the virtual reference sample. There are in principle two major approaches: (i) Newtonian analysis and (ii) Fourier analysis. Newtonian analysis requires only one thermocouple and is the most widely used. The mathematics is straight forward (see for example Ekpoom and Heine 1981; Chen and Stefanescu 1984; Bäckerud and Sigworth 1989; Upadhya et al. 1989; Kierkus and Sokolowski 1999). Fourier analysis is a more accurate treatment of the heat transfer problem, but it requires two thermocouples and the mathematics is more cumbersome (see for example Fras et al. 1993; Barlow and Stefanescu 1997; Dioszegi and Hattel 2004).

Newtonian Analysis In the Newtonian analysis, it is assumed that the thermal gradient across the sample is zero (reasonable for small Biot numbers) and that heat transfer between the casting and the mold occurs by convection. The mathematical background (Ekpoom and Heine 1981, Chen and Stefanescu 1984) is similar with that for the DTA with a reference sample, with some simplifications. Assuming that the heat loss by radiation is negligible during solidification, Eq. 6.41 becomes:

$$\frac{dQ_f}{dt} - v\rho c_p \left(\frac{dT}{dt}\right)_{cc} = h_{cc} A \left(T_{cc} - T_o\right), \tag{6.44}$$

where, the subscript cc designates the cooling curve and T_0 is the ambient temperature. The thermophysical quantities are assumed constant. The equation can be rearranged to describe the CR:

$$\left(\frac{dT}{dt}\right)_{cc} = \frac{1}{v\rho c_p} \left[\frac{dQ_f}{dt} - h_{cc} A \left(T_{cc} - T_o\right)\right]. \tag{6.45}$$

Assuming no phase transformation during cooling, $Q_f = 0$ and the equation becomes:

$$\left(\frac{dT}{dt}\right)_{zc} = -\frac{h_{zc} A \left(T_{cc} - T_o\right)}{v\rho c_p}. \tag{6.46}$$

The subscript zc denotes the zero curve (the time evolution of the CR of the alloy assuming no phase transformation). This is the CR of the virtual reference (neutral) body. Assuming that $h_{cc} = h_{zc}$ the time evolution of the latent heat can be calculated as:

$$\frac{dQ_f}{dt} = v\rho c_p \left[\left(\frac{dT}{dt}\right)_{cc} - \left(\frac{dT}{dt}\right)_{zc}\right],$$

Fig. 6.15 Calculation of the amount of phases from the areas under the CR and the zero curves for cast iron (Stefanescu 2015). With permission from the American Foundry Soc

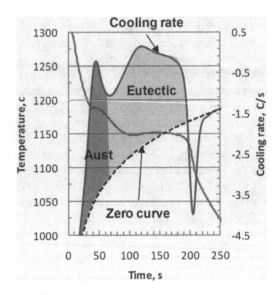

and the total heat evolution during solidification is:

$$Q_f = v\rho c_p \int_{t_{start}}^{t_{end}} \left[(dT/dt)_{cc} - (dT/dt)_{zc}\right] dt.$$

The fraction solid at time, i, is then calculated with Eq. 6.43, or as:

$$f_S^t = \frac{Q_f^t}{Q_f} = \frac{v\rho c_p \int_{t_{start}}^{t} \left[(dT/dt)_{cc} - (dT/dt)_{zc}\right] dt}{Q_f}. \tag{6.47}$$

The method consists in generating the first derivative of the cooling curve with respect to time (the CR), generating a ZC, and then subtracting the area under the ZC from the area under the CR. Once the zero-curve in known, the time evolution of the phase transformation is obtained by dividing the area corresponding to the phase formation to the total area (for example area corresponding to the solidification of primary austenite in Figure 6.15 to the total area). The evolution in time of the amount of the phase of interest is thus obtained. If the beginning of eutectic solidification is known, the amount of primary austenite and the amount of eutectic can be calculated. Then, the evolution of fraction solid over the solidification time can be plotted.

There are two problems in the Newtonian analysis: (i) establishing the beginning and the end of the austenite and eutectic solidification, and (ii) calculating the ZC.

Whereas many papers suggest the use of the first derivative to find the beginning and end of transformation, this is inaccurate. These critical points are best obtained from the second derivative of the cooling curve (Alonso et al. 2012). Fig. 6.16 shows a cooling curve in the region of the eutectic transformation of a lamellar graphite

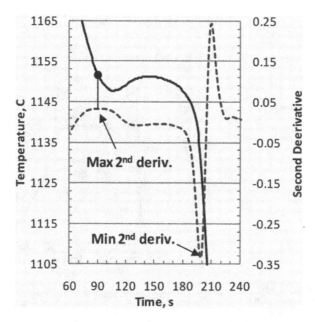

Fig. 6.16 The use of the second derivative to establish the beginning and the end of eutectic solidification (Stefanescu 2015). With permission from the American Foundry Soc

iron. The maximum of the second derivative corresponds to the beginning of the eutectic solidification, while the minimum on the second derivative corresponds to the end of the eutectic solidification.

To generate the ZC, the heat transfer coefficient for the cooling without transformation is needed. Ekpoom and Heine (1991) calculated a heat transfer coefficient for the liquid stage, h_L, and one for the solid stage, h_S. In the mushy zone linear interpolation between h_L and h_S was used. Alternatively, the heat transfer coefficient can be generated from the cooling curve or its first derivative.

In principle, three approaches can be used to compute the zero line (Barlow and Stefanescu 1997) as shown in Fig. 6.17:

1. ZC1h: Logarithmic trend line for points chosen at the beginning and the end of the CR curve; this method uses only one heat transfer coefficient for the entire cooling curve (Upadhya et al 1989).
2. ZC2h: Logarithmic trend line for one point on the CR corresponding to the beginning of the austenite solidification (the maximum of the second derivative) and points at the end of the CR curve; while only one h is needed for the calculation of the transformation region of the ZC, two are needed for the whole ZC.
3. ZC3h: Logarithmic trend line for points on the CR corresponding to the beginning and end of the austenite solidification.

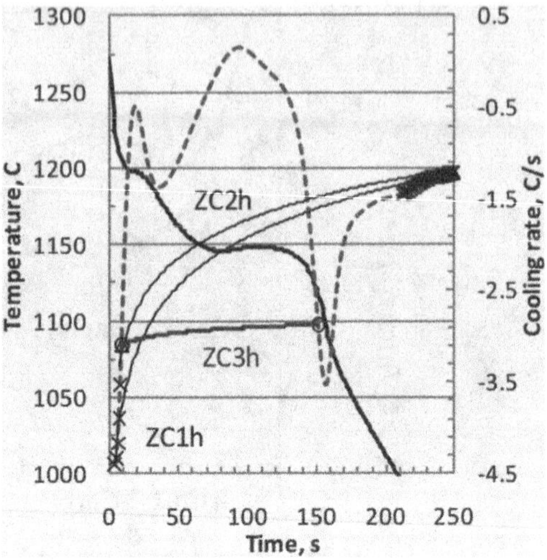

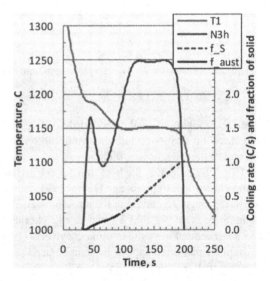

Fig. 6.17 Three methods of generation of the zero curve (Alonso et al. 2012). With permission from the American Foundry Soc

Fig. 6.18 Normalized CR (N3h) and the evolution of the solid and primary austenite fractions (Alonso et al. 2012). With permission from the American Foundry Soc

Once the ZC is generated, the normalized CR (the difference between the area under the CR curve and that under the ZC) is obtained (Fig. 6.18). The normalized CR covers only the solidification part of the cooling curve. The area under the normalized cooling curve is proportional to the latent heat of solidification. The evolution of the fraction austenite and total fraction solid can then be calculated and plotted.

Calculated fraction solid obtained from a single-thermocouple Newtonian TA could have significant departure from data obtained through computational modeling of the specimen when the specimen cooled at high Biot numbers.

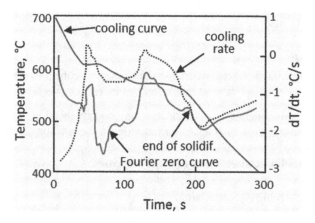

Fig. 6.19 Cooling curve, CR, and Fourier ZC for an aluminum alloy (Barlow and Stefanescu 1997). With permission from the American Foundry Soc

Fourier Analysis Fourier analysis assumes that heat transfer takes place by conduction only. The following analysis (Barlow and Stefanescu 1997) closely follows the method suggested by Fras et al. (1993). The Fourier equation with a heat source term is:

$$\frac{\partial t}{\partial t} = \alpha \nabla^2 T + \frac{1}{c_p}\frac{\partial Q_f}{\partial t} \qquad \text{or} \qquad \frac{\partial Q_f}{\partial t} = C_p\left(\frac{\partial T}{\partial t} - Z_F\right), \qquad (6.48)$$

with the ZC given by $Z_F = \alpha \nabla^2 T$. To calculate this curve we must know the temperature filed, which for a cylindrical mold can be calculated as $\nabla^2 T = 4(T_2 - T_1)/(r_2^2 - r_1^2)$, where T_2 and T_1 are the temperatures at radii r_2 and r_1, respectively. This introduces the need for two thermocouples. The thermophysical quantities are time and temperature dependent. The latent heat and fraction solid evolution are calculated as described for the Newtonian analysis. Typical results are shown in Fig. 6.19 for an aluminum alloy. The time close to the end of solidification, when the cooling curve and the Fourier ZC start coinciding, is considered the end of solidification. Note that this occurs earlier than the minimum at the end of the cooling curve.

Mathematically, the Fourier analysis is more accurate than the Newtonian analysis, but its experimental application is more onerous because the two thermocouples must be positioned accurately in the measuring cup, an almost impossible task for sand cups. A more detailed analysis of the Fourier method through inverse heat conduction analysis was offered by Diószegi and Hattel (2004).

6.6 Applications

Application 6.1

Consider a simple 0.6 % C steel casting in the shape of a cube, having a volume of 0.001 m^3, poured into a silica sand mold. Neglecting corner effects, calculate:

a. The solidification time of this casting assuming no superheating
b. The solidification time of the casting assuming a superheating temperature of 1550 °C
c. The average solidification velocity for the two cases

Answer The governing equations are:

    for case (a): Chvorinov's equation, Eq. 6.24
    for case (b): Chvorinov's equation, Eq. 6.25
    for case (c): $V = (v/A)^2/t_f$

The required data are found in Appendix B. It is calculated that the solidification time without superheat is of 444 s, and with superheat of 619 s. The average solidification velocity is 3.8×10^{-5} m/s for no-superheat, and 2.7×10^{-5} m/s with superheat.

Application 6.2

Calculate the solidification time of a 0.6 % C steel cube of volume 0.001 m^3 poured into a silica sand mold from a superheating temperature of 1550 °C. Use a time-stepping analysis assuming resistance in the mold and ignoring the corner effects.

Answer Equating the heat flow rate into the mold on the mold side at the mold/metal interface (derived after Eq. 6.22) with the flux coming from the metal (see Eq. 6.10 for heat generation) we have:

$$\sqrt{\frac{k_m \rho_m c_m}{\pi t}} (T - T_m) A = -v\rho c \frac{dT}{dt} + v\rho \Delta H_f \frac{df_S}{dt}. \tag{6.49}$$

Rearranging and writing the equation in time-stepping format, that is $dT = \Delta T = T^{n+1} - T^n$, $dt = \Delta t$, etc.:

$$T^{n+1} = T^n - \frac{A}{v\rho c}\sqrt{\frac{k_m \rho_m c_m}{\pi t^{n+1}}} (T^n - T_m) \Delta t + \frac{\Delta H_f}{c} \Delta f_S, \tag{6.50}$$

where T^n and T^{n+1} are the temperatures at times n and $n+1$ respectively, and T_m is the mold temperature. To find an expression for Δf_S, linear evolution of the fraction of solid over the solidification interval ΔT_o is assumed, that is $f_S = a + b \cdot T$.

    Since   at: $T = T_L - \Delta T_o$ $f_S = 1$
    and   at: $T = T_L$ $f_S = 0$
    we have $f_S = (T_L - T)/\Delta T_o$. Then, $f_S^n = (T_L - T^n)/\Delta T_o$ and $f_S^{n+1} = (T_L - T^{n+1})/\Delta T_o$.

Table 6.2 Organization of spreadsheet

	A	B	C	D
	Data		Calculations	
1	T_{init}	1550	Time	Temp.
2	v	0.001	0	1550
3	A	0.06	10	1522
4	ρ_S	7210	20	1502
5	c_S	794	30	1487
6	ΔH_f	2.72E + 05	40	1484
7	ΔT_o	72	50	1482
8	k_m	0.52	60	1480
9	ρ_m	1600	70	1479
10	c_m	1170	80	1477
11	Δt	10	90	1476
12	T_L	1490	100	1474

Thus, $\Delta f_S = (T^n - T^{n+1})/\Delta T_o$. Substituting in the temperature equation, the final equation is:

$$T^{n+1} = T^n - \frac{A}{v}\frac{T^n - T_m}{\rho\,(\Delta H_f/\Delta T_o + c)}\sqrt{\frac{k_m\rho_m c_m}{\pi t^{n+1}}}\Delta t. \tag{6.51}$$

The Excel spreadsheet is organized as shown in the Table 6.2, where the data in Appendix B have been used. The temperature in cell D2 is the initial temperature. Eq. 6.51 was implemented in cell D3. Note that an IF statement must be used since for $f_S = 0$ and $f_S = 1$, the term containing the latent heat evolution is zero. The form of the IF statement is:

IF(OR($T^n > T_L$, $T^n < (T_L - \Delta T_o)$), Eq. 6.51 with $\Delta H_f = 0$, Eq. 6.51)

The cooling curve presented in Fig. 6.20 is obtained. From the graph, the solidification time is approximately 880 s. This is significantly different from Chvorinov's equation results (619s with superheat).

Application 6.3
Calculate the average temperature gradient at the end of solidification in the steel casting described in Application 6.1.

Answer A flux balance at the L/S interface is written as: $\rho_S \Delta H_f \frac{dL}{dt} = k_S \frac{T_f - T_S}{L}$.

The solidification velocity was calculated in Application 6.1 to be $V = dL/dt = 3.8 \times 10^{-5}$ m/s. Then, calculating the gradient at the time when the center of the casting has solidified, and using the data in Appendix B:

$$G_T = (T_f - T_S)/L = (\rho_S \Delta H_f V)/k_S = (7210)\left(2.72 \cdot 10^5\right)\left(3.8 \cdot 10^{-5}\right)/40$$

$$= 1863\,\text{K/m}.$$

Fig. 6.20 Cooling curve of
0.6 % C steel casting

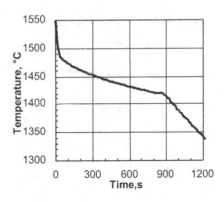

Application 6.4

Consider the casting in Application 6.2. Using the assumption of thermal resistance
at the mold/metal interface (small Bi number, no gradient in the casting) calculate
the heat transfer coefficient at the interface.

Answer We will use a time-stepping procedure in conjunction with lump analysis.
The governing equation is flux balance at the interface, Eq. 6.32. In time-stepping
format this equation can be written as:

$$hA\left(T^n - T_o\right)\Delta t = -\rho c v\left(T^{n+1} - T^n\right) + \rho\Delta H_f v \Delta f_S. \qquad (6.52)$$

Assuming a linear dependence of the solid fraction on temperature as in Application
6.2, $\Delta f_S = \left(T^n - T^{n+1}\right)/\Delta T_o$, and this equation becomes:

$$T^{n+1} = T^n - \frac{hA}{\rho v}\frac{T^n - T_o}{\Delta H_f/T_o + c}\Delta t.$$

Note that this equation is similar to that developed in Application 6.2, except for
h. By implementing this equation in an Excel spreadsheet and using the data in
Appendix B, the cooling curve given in Fig. 6.21a is obtained. A heat transfer coefficient of 315 J/m$^2 \cdot$ K $\cdot$ s was used (the value for cast iron since no value for steel
was available). Note that the solidification time is about 70 s, which is much smaller
than that which is calculated in Application 6.2. This is because of the faster heat
transfer for resistance at the interface as compared with resistance in the mold.

Comparing the time-stepping equation in this application with that in Application
6.2, it is found that for the case of a sand mold the heat transfer coefficient can be
calculated as:

$$h = \sqrt{(k_m \rho_m c_m)/(\pi t^{n+1})}.$$

The time dependency of h is shown in the Fig. 6.21b, where the cooling curve from
Application 6.2 is also included. The average value of the coefficient from pouring
to the end of solidification is $\bar{h} = 34.7$ J/(m$^2 \cdot$ K $\cdot$ s). When this value is used, the
same solidification time of 880 s as in Application 6.2 is obtained.

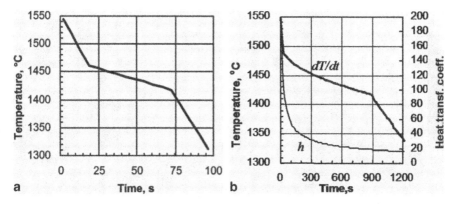

Fig. 6.21 Calculated cooling curves for 0.6 % C steel under different cooling conditions. **a** cooling curve of 0.6 % C steel assuming resistance at the mold /metal interface. **b** cooling curve and heat transfer coefficient of 0.6 % C steel assuming resistance in the sand mold

Application 6.5
Examine the influence of the time step, Δt, on the results of Applications 6.2 and 6.4.

References

Alonso G, Larrañaga P, Sertucha J., Suárez R., Stefanescu DM (2012) Trans. AFS 120:329
Barlow JO, Stefanescu DM (1997) AFS Trans., 105:349–354
Bäckerud SL, Sigworth GK (1989) Trans. AFS 97:459
Bennon W D, Incropera FP (1987) Int. J. Heat Mass Transfer 30:2161, 2171
Chen IG, Stefanescu DM (1984) Trans. AFS 92:947
Chvorinov N (1940) Giesserei 27:177
Ekpoom L, Heine RW (1981) Trans. AFS 89:27
Flemings MC (1974) Solidification Processing. McGraw-Hill, New York
Ho K, Pehlke RD (1984) AFS Trans. 92:587
Kierkus WT, Sokolowski JH (1999) Trans. AFS 107:161
Nishida Y, Matsubara H (1976) British Foundryman 69:274
Poirier DR, Geiger GH (1994) Transport Phenomena in Materials Processing. TMS, Warrendale PA
Siegel R, Howell JR (1981) Thermal radiation heat transfer. Hemisphere Pub. Corp.
Shapiro A (1983) FACET-A radiation View Factor Computer Code for Axy-symmetric 2D and 3D Geometries with Shadowing. Lawrence Livermore National Lab., California
Stefanescu DM (2002) Science and Engineering of Casting Solidification. Kluwer Academic/Plenum Publishers, New York
Stefanescu DM (2015) Int. J. Metalcasting 9(1):7–22
Trovant M, Argyropoulos S (2000) Metall. and Mater. Trans. 31B:75
Upadhya KG, Stefanescu DM, Lieu K, Yeger DP (1989) Trans. AFS 97:61
Upadhya G, Paul AJ (1994) Trans. AFS. 102:69

Chapter 7
Momentum Mass Transport at the Macroscale

During casting solidification, significant flow of the molten metal will affect the local composition. The driving forces for fluid flow can be internal or external. The internal sources of fluid flow include shrinkage (solidification contraction) flow, natural convection, capillary forces, formation of gas bubbles, and deformation of solid phases because of thermal stress and static pressure. The external driving forces may include centrifugal and/or electromagnetic forces.

7.1 Shrinkage Flow

During solidification, the vast majority of metals and alloys shrink. Solidification contraction can be calculated as:

$$\beta = \frac{v_L - v_S}{v_L} = \frac{\rho_L - \rho_S}{\rho_L}, \tag{7.1}$$

where v_L and v_S are the specific volumes of the liquid and solid, respectively.

The macroscopic transport equation that describes the shrinkage flow is the mass conservation (continuity; Eq. 4.2). Note that, if in this equation the change of density over time is ignored, the equation simply states that the velocity gradient must be constant. Thus, any model that attempts to describe shrinkage flow should use the continuity equation and assume different liquid and solid densities.

7.2 Natural Convection

Natural convection is the flow that results from the effect of gravity on density differences caused by phase or solute variations in the liquid. As the liquid is colder next to the interface (thus denser) than in the bulk, a downward flow driven by the temperature gradient will occur next to the interface (Fig. 7.1). As a consequence,

© Springer International Publishing Switzerland 2015
D. M. Stefanescu, *Science and Engineering of Casting Solidification*,
DOI 10.1007/978-3-319-15693-4_7

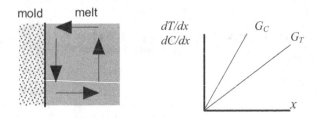

Fig. 7.1 Natural convection induced by the flow of denser metal at the mold wall

an upward flow will occur in the bulk. When the solute rejected at the interface is denser than the solvent, a downward flow, driven by the composition gradient (more solute at the interface than in the bulk liquid), will also occur next to the interface (Fig. 7.1). In the opposite case, more complex flow patterns will result.

In a first simplified analysis, it can be stated that a necessary condition for stability is that the liquid has throughout a negative gradient of liquid density upward (Flemings 1974). Thermal expansion is produced by changes in temperature and/or density. Thus, a thermal (β_T) and solutal (compositional) (β_C) expansion coefficient can be defined as:

$$\beta_T = -\frac{1}{\rho}\frac{\partial \rho}{\partial T}\bigg|_C \quad \text{and} \quad \beta_C = -\frac{1}{\rho}\frac{\partial \rho}{\partial C}\bigg|_T. \tag{7.2}$$

Then, since $\beta_T = -\frac{1}{\rho}\frac{\partial \rho}{\partial x}\frac{\partial x}{\partial T}$ and $\beta_C = -\frac{1}{\rho}\frac{\partial \rho}{\partial x}\frac{\partial x}{\partial C}$, the gradient of liquid density is:

$$\frac{\partial \rho_L}{\partial x} = -\rho_L\left(\beta_T \frac{\partial T}{\partial x} + \beta_C \frac{\partial C_L}{\partial x}\right).$$

A more complete analysis requires expansion of the momentum equation, Eq. 4.5. The source term S_m will include the hydrostatic pressure gradient, P, an additional body force term ρg induced by gravitational acceleration, and an additional viscous term S_{visc}:

$$S = -\nabla P - \rho g - S_{visc}. \tag{7.3}$$

For the time being, we will ignore the viscous term, S_{visc}. Then, since $\rho = \rho_L$, and assuming that the body force is oriented in the y direction, the momentum equation, Eq. 4.8, in the y direction becomes:

$$\frac{\partial}{\partial t}(\rho V^y) + \nabla \cdot (\rho V V^y) = \nabla \cdot \nabla(\mu V^y) - \nabla P - \rho g. \tag{7.4}$$

For a static fluid, this equation simplifies to $P_o + \rho_o g = 0$, where P_o is the hydrostatic pressure corresponding to density ρ_o and temperature T_o (in this case, the liquidus temperature is T_L). Combining these last two equations:

$$\frac{\partial}{\partial t}(\rho V^y) + \nabla \cdot (\rho V V^y) = \nabla \cdot \nabla(\mu V^y) - \nabla(P - P_o) - (\rho - \rho_o)g. \tag{7.5}$$

To solve this equation, it is now necessary to formulate the pressure gradient, P, and the density as a function of composition and temperature.

Several methods have been proposed for the calculation of the unknown pressure field. We will only introduce the principles of the *semi-implicit method for pressure-linked equations* (SIMPLE; for details, see Patankar 1980). In this method, the correct pressure, P, is assumed to be described by $P = P^* + P'$, where P^* is the guessed pressure and P' is the pressure correction. Then, a sequence of calculation is performed to evaluate in order guessed values for V^x, V^y, V^z, then P', then P, then the corrected values for V^x, V^y, V^z. The corrected pressure P is then treated as the new guessed pressure P^*, and the whole procedure is repeated until a converged solution is obtained.

For certain alloys, data for $\rho = f(C, T)$ are obtainable. In this case, all the information required to solve the momentum equation is now available. If these data are not available, we can make use of the coefficients of thermal expansion. Ignoring the variation of all fluid properties other than density (the *Boussinesq approximation*), the last term of Eq. 7.5 can be formulated as follows:

$$d\rho = \left.\frac{\partial \rho}{\partial C}\right|_T dC + \left.\frac{\partial \rho}{\partial T}\right|_C dT = \rho\left[-\beta_C\left(C - C_o\right) - \beta_T\left(T - T_o\right)\right], \qquad (7.6)$$

where C_o is the initial composition. In general, β_C and β_T vary with both density and temperature. Assuming the coefficients of thermal expansion to be constant over certain ranges of temperature and composition, the previous equation can be integrated to $\rho = \rho_o \exp\left[-\beta_C^o\left(C - C_o\right) - \beta_T^o(T - T_o)\right]$, where T_o is the liquidus temperature, and C_o is the liquid concentration at the liquidus temperature. Using only the first two terms of the series expansion, we have $\rho = \rho_o\left[1 - \beta_C^o\left(C - C_o\right) - \beta_T^o(T - T_o)\right]$, where β_C^o and β_T^o have the same expressions as above with ρ substituted by ρo. Both are assumed constant throughout solidification. Then, the buoyancy source term in Eq. 7.5 becomes (the *Boussinesq approximation for natural convection*):

$$S_b = \rho_o g\left[\beta_C^o\left(C - C_o\right) + \beta_T^o(T - T_o)\right]. \qquad (7.7)$$

A characteristic temperature difference, ΔT, and a characteristic length scale, l_b, can be defined to characterize the horizontal thermal gradient, and consequently the horizontal density gradient. In addition, a characteristic concentration difference, ΔC, can be defined. The resulting nondimensional parameters are:

Thermal Grashof number: $\mathrm{Gr}_T = g\beta_T\Delta T l_b^3 v^{-2}$
Solutal Grashof number: $\mathrm{Gr}_C = g\beta_C\Delta C l_b^3 v^{-2}$
Prandtl number: $\mathrm{Pr} = v\alpha^{-1}$
Schmidt number: $\mathrm{Sc} = vD^{-1}$

where α is the thermal diffusivity, μ is the dynamic viscosity, and v is the kinematic viscosity. The Rayleigh number can be defined as the sum of the thermal and the solutal Rayleigh numbers as follows:

$$\mathrm{Ra} = \mathrm{Ra}_T + \mathrm{Ra}_C = \mathrm{Gr}_T \cdot \mathrm{Pr} + \mathrm{Gr}_C \cdot \mathrm{Sc}. \qquad (7.8)$$

The Rayleigh number compares the buoyancy forces to the viscous forces. For small Rayleigh numbers, the viscous forces dominate. The conduction regime is maintained with negligible thermosolutal convection effects on macrosegregation. Above a critical Rayleigh number ($\sim$ 2000) buoyancy forces become important, a convection regime is established, and the effect of thermosolutal convection on macrosegregation becomes significant (Nastac 2004).

From this analysis, it is clear that a model that attempts to include the effect of fluid on segregation must include at least the density difference in the continuity equation and the body force term in the momentum equation.

7.3 Surface-Tension-Driven (Marangoni) Convection

When free or deformable interfaces exist in the system (e.g., liquid–liquid or liquid–gas interfaces), the temperature and compositional gradient will impose a surface tension gradient along the interface, which exerts a shear stress on the fluid. This induces a flow toward regions with higher values of γ-termed *Marangoni convection*. The force balance can be written as (Shy et al. 1996):

$$\nabla(\mathbf{V} \cdot \mathbf{t}) \cdot \mathbf{n} = \frac{\partial \gamma}{\partial T} \nabla T \cdot \mathbf{t}, \tag{7.9}$$

where $\mathbf{n}$ is the unit normal vector to the free surface, and $\mathbf{t}$ is the unit tangent vector to the free surface. This equation states that the surface tension gradient is proportional to the temperature gradient and gives rise to the normal derivative of the tangential velocity at the free surface. To describe surface convection strength, the nondimensional Marangoni number can be used:

$$\mathrm{Ma} = \left| \frac{\mathrm{d}\gamma}{\mathrm{d}T} \right| \frac{\Delta T l_b}{\mu \alpha}. \tag{7.10}$$

The governing equation describing both the thermal and solutal Marangoni convection is:

$$\tau = \mu \left(\frac{\partial V_y}{\partial y} \right) = \left(\frac{\partial \gamma}{\partial T} \right) \left(\frac{\partial T}{\partial x} \right) + \sum_i \left(\frac{\partial \gamma}{\partial a_i} \right) \left(\frac{\partial a_i}{\partial x} \right),$$

where τ is the shear stress caused by the surface tension gradients, V_y is the velocity component parallel to the surface, x and y are the coordinates parallel and perpendicular to the surface, and a_i is the thermodynamic activity of alloying element i.

7.4 Flow Through the Mushy Zone

The mushy zone is the region where solid and liquid coexist as a mixture. As long as the equiaxed dendrites are free to flow with the liquid, it may be assumed that the flow velocity is affected only by the change in viscosity. However, when dendrite coherency is reached and a fixed solid network is formed, or if the dendrites are columnar, the flow will be considerably influenced by the morphology of the mushy zone. In most solidification models, the flow through the mushy zone is treated as flow through porous media, and *Darcy*'s law is used for its mathematical description. The standard form of Darcy's law as applied to flow through a fixed dendritic network is (e.g., Poirier 1987):

$$\mathbf{V}_L = -(K/\mu g_L) \cdot (\nabla P - \rho g), \tag{7.11}$$

where $\mathbf{V}_L$ is the velocity of the interdendritic liquid, K is the specific permeability of the mushy zone, and g_L is the volume fraction of the interdendritic liquid. Darcy's law is valid under the following assumptions (Ganesan and Poirier 1990):

- Slow flow ($\mathbf{V}_L \rightarrow 0$); this allows ignoring inertial effects
- Steady flow
- Uniform and constant volume fraction of liquid
- Negligible liquid–liquid interaction forces.

Now, the permeability must be defined. Two models may be used for this purpose: the *Hagen–Poiseuille* model or the *Blake–Kozeny* model.

7.4.1 The Hagen–Poiseuille Model

Following the analysis by Poirier (1987), the Hagen–Poiseuille law gives the velocity of an incompressible fluid under laminar flow conditions through a tube as:

$$V = -\frac{r^2}{8\mu} \left(\frac{\partial P}{\partial y} - \rho g \right), \tag{7.12}$$

where the flow is in the direction of gravity (the y direction), and r is the tube radius. Applying this law to the flow through the interdendritic network, we derive:

$$V = - \left(g_L \lambda_I^2 / 8\pi \mu \right) (\mathrm{d}P/\mathrm{d}y - \rho g),$$

where λ_I is the primary dendrite arm spacing (DAS). Comparing this equation with Eq. 7.11, the permeability is derived to be:

$$K = C_1 \lambda_I^2 g_L^2 \tag{7.13}$$

where C_1 is a parameter that depends on the geometry (tortuosity) of the flow channels. For Pb–Sn alloys, experimental data give 3.75×10^{-4} for this parameter.

7.4.2 The Blake–Kozeny Model

For flow through porous media in the vertical direction, the Blake–Kozeny equation gives:

$$V = -C_2 \frac{d^2 g_L^3}{\mu(1-g_L)^2} \left(\frac{dP}{dy} - \rho g\right). \tag{7.14}$$

where d is the characteristic dimension of the solid phase. For spheres, d is the diameter of the sphere.

By comparing Eqs. 7.11 and 7.14, the permeability is (Poirier 1987) $K = C_2 d^2 g_L^3/(1-g_L)^2$. Assuming further that the characteristic dimension is related to the volume fraction of liquid by $(1-g_L) \propto (D/\lambda_I)^2$, the final equation for permeability is:

$$K = C_2 \lambda_I^2 g_L^3/(1-g_L) . \tag{7.15}$$

For the Pb–Sn system, C_2 is of the order of 1.43×10^{-3}.

7.5 Segregation Controlled by Fluid Flow

Gravity plays an important role in the formation of segregation. Settling or flotation of liquid or solid phases having a different composition, and therefore a different density than the bulk liquid, will produce *gravity segregation*. Typical examples are dendrites settling at the bottom of ingots, and coarse primary lamellar graphite (kish) or spheroidal graphite floating on top of large cast iron castings. Centrifugal forces enhance the gravitational forces applied on the casting, and they have significant effects on segregation.

Figure 7.2 shows the types of macrosegregation formed in a killed steel ingot (an ingot procured from a melt which has been deoxidized). The "+" and "−" signs denote positive and negative segregation, respectively. The type of segregation has also been traditionally defined in terms of the shape or the location of segregation. The streaks arranged in a V-pattern at the center of the casting are called *channel, centerline,* or *V-type* segregation. The *A-type* segregation, also called *freckles*, refers to the streaks oriented almost vertically in an A-pattern at the upper and outer regions of the ingot. These are all positive segregations. Negative segregation is distributed in a cone at the base of the ingot.

The main driving force of the fluid flow during solidification is *solidification contraction*. Interdendritic liquid flow can cause solute concentration to be higher than the average concentration in the earlier solidified regions. This is opposite to the distribution shown in Fig. 5.12 for the beginning of solidification. This is termed *inverse segregation*. Such complex segregation patterns cannot be explained through solute diffusion alone.

Fig. 7.2 Macrosegregation
in a killed steel ingot. (After
Derge 1964)

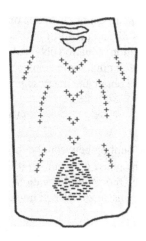

The physics of macrosegregation formation can be summarized as follows. Seg-regation starts at the microscopic level as solidification proceeds. During solidifica-tion, solute is rejected ($k < 1$) or depleted ($k > 1$) continuously from the precipitated solid, and the composition of the surrounding liquid is consequently affected. If significant concentration gradients are developed at the interface, the interdendritic liquid can be driven simultaneously by thermal and solutal buoyancy, as well as by solidification contraction. The induced flow will wash away the liquid next to the interface, resulting in segregation at the macroscopic level (macrosegregation). Let us first discuss the case where segregation arises mainly from fluid flow rather than from solute diffusion at the macroscopic scale. This particular problem is of inter-est, because the solute diffusion layer is much smaller than the typical dimension of the volume element. In other words, the volume element is open to fluid flow, but closed to diffusion. However, diffusion is active within the volume element. The governing equation is that of conservation of species, i.e., Eq. 4.4. Assuming no source term, fast solute diffusion ($\nabla C = 0$) and constant density, the conservation equation reduces to :

$$\frac{\partial C}{\partial t} + \nabla \cdot (VC) = 0 \qquad (7.16)$$

The second term represents the solute transport that is associated with fluid flow. Assuming that the solid does not move (e.g., fixed dendrite skeleton), the average velocity within the volume element can be obtained from the real velocity of the liquid phase, V_L: $V = (1 - f_S)V_L$. Note that for $f_S = 0$, $V = V_L$, and for $f_S = 1$, $V = 0$.

The average composition over the volume element is:

$$C = \int_v C_L df_L + \int_v C_S df_S. \qquad (7.17)$$

The assumptions made on the type of solute redistribution will determine the form of C, and the form of the governing fluid-flow-controlled segregation (macrosegregation) equation obtained when substituting C in Eq. 7.16 (Rappaz and Voller 1990). Assuming, for example, complete mixing in both liquid and solid, i.e., the lever rule, we obtain the macrosegregation equation:

$$\nabla(VC_L) + \frac{\partial C_L}{\partial t} - \frac{\partial[f_S(1-k)C_L]}{\partial t} = 0. \tag{7.18}$$

Coupling between the thermal and solutal field is done through the liquidus line in the phase diagram, $C_L(t)$. Since V is unknown, this equation must be coupled with the momentum equation. If complete mixing in the liquid and no diffusion in solid are assumed (Scheil model), Eq. 7.17 takes the form:

$$C = C_L f_L + \int_v C_S \mathrm{d} f_S = C_L(1 - f_S) + \int_o^{f_S} C_S(f_S) \mathrm{d} f_S.$$

Substituting in Eq. 7.16 and after derivation with respect to time, we obtain:

$$\nabla(VC_L) + \frac{\partial[(1-f_S)C_L]}{\partial t} + C_S^* \frac{\partial f_S}{\partial t} = 0. \tag{7.19}$$

Eliminating C_S^* by using the equilibrium condition is no longer trivial for an open system.

7.6 Segregation Controlled by Fluid Flow and Solute Diffusion

From the preceding discussion, it follows that macroscale segregation is controlled simultaneously by fluid flow and solute diffusion. Accordingly, macrosegregation models must describe both phenomena. This is a rather complicated problem, since fluid flow through the liquid/solid mixture must be described.

Analytical models cannot tackle the intricacies of flow through the mushy zone. Nevertheless, they contribute to the understanding of the physics of macrosegregation. A first analytical model for macrosegregation was proposed by Flemings and Nereo (1967). First, Eq. 7.16 was written as:

$$\nabla(\rho_L g_L V C_L) + \partial(\bar{\rho}\bar{C})/\partial t = 0,$$

where $\bar{\rho}$ and $\bar{C}$ are the average density and composition, respectively, in the volume element. Manipulation of this equation yields:

$$\frac{\partial C_L}{\partial t} = -\left(\frac{1-k}{1-\beta}\right) \frac{C_L}{g_L} \frac{\partial g_L}{\partial t} - V \cdot \nabla C_L, \tag{7.20}$$

where β is given by Eq. 7.1. This is the local solute redistribution equation used to calculate macrosegregation (Flemings 1974). It describes the influence of shrinkage

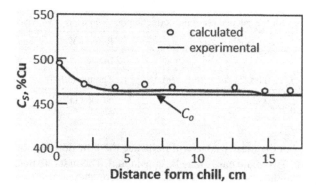

Fig. 7.3 Macrosegregation in a directionally solidified Al-4.6 % Cu alloy. (After Flemings 1974)

flow on the composition of the solid at each fraction solid. However, it does not include the effect of natural convection.

A slightly modified form of this equation was suggested by Ohnaka (1992):

$$\frac{\partial C_L}{\partial t} = \left(\frac{1-k}{1-\beta}\right)\left(1 + A - \frac{V_n}{dx/dt}\right)^{-1}\frac{C_L}{1 - g_S}\frac{\partial g_S}{\partial t}, \qquad (7.21)$$

where V_n is the flow velocity normal to the isotherms (equivalent to V_L), and dx/dt is the velocity of the isotherms. Also, $A = 0$ for no diffusion in solid and $A = kg_S$ $(1 - \beta)^{-1}(1 - g_S)^{-1}$ for complete diffusion in solid. Since

$$V \cdot \nabla C_L = V_L\frac{\partial C_L}{\partial t}\frac{\partial t}{\partial x} = -\frac{V_L}{V_S}\frac{\partial C_L}{\partial t} \quad (for\ ID),$$

where V_S is the solidification velocity, Eq. 7.20 can be integrated to give:

$$C_S = kC_o(1 - g_S)^{\overline{(1-\beta)[1-V_L/V_S]}} = kC_o(1 - g_S)^{\frac{k-1}{\xi}}. \qquad (7.22)$$

Analysis of this equation suggests three possible scenarios:

a. $\xi = 1$, e.g., no solidification shrinkage ($\beta = 0$) and no flow ($V_L = 0$); the equation becomes Scheil equation, and negative macrosegregation ($C_S < C_o$) is calculated for the first part of solidification, while positive segregation ($C_S > C_o$) is predicted for the second part of solidification.

b. $0 < \xi < 1$, e.g., no flow ($V_L = 0$), C_S is larger than predicted by the Scheil equation, which means that a strong tendency toward positive segregation exists. For example, since at the mold wall $V_L = 0$, positive segregation, i.e., inverse segregation, can occur next to the mold wall. This is illustrated in Fig. 7.3 for the case of directional solidification against a chill.

c. $\xi > 1$ or $V_L/V_S < -\beta/(1-\beta)$, C_S is smaller than predicted by the Scheil equation, which means that a strong tendency toward negative segregation exists.

Table 7.1 Equations for the calculation of the correction factor in the relative viscosity equation

Correction factor φ	Reference
$2.5\,f_S$	Einstein
$2.5 \cdot f_S + 10.05 \cdot f_S^2 + 0.00273 \cdot \exp(16.6 \cdot f_S)$	Thomas (1965)
$\left(1 - f_S/f_S^{cr}\right)^{-2.5 f_S^{cr}} - 1$	Krieger (1972)

The morphology of the mushy zone will influence the flow during solidification in a complex manner. Two main cases may be considered. The first one involves the flow of the liquid through a fixed solid network ($V_S = 0$). This will be a good approximation for columnar solidification, or for equiaxed solidification after dendrite coherency is reached. This flow can be treated as flow through a porous medium (*Darcy flow*).

The second case involves the flow through a solid/liquid mixture when the solid can move with the liquid ($V_S = V_L$). This will be true for equiaxed solidification *before coherency*. Here, the concept of *relative viscosity* is applied to describe the viscosity of the liquid/solid mixture. The relative viscosity can be calculated with:

$$\mu^* = \mu(1 + \varphi), \tag{7.23}$$

where φ is a correction factor. Numerous expressions were proposed for the correction factor (see review by Kaptay 2000). Selected expressions for the correction factor are given in Table 7.1. In this table, f_S is the fraction solid, and f_S^{cr} is the critical fraction solid at which coherency is reached.

The complex macrosegregation problem can only be solved numerically. This will be discussed in some detail in Chap. 17.

7.7 Macroshrinkage

During cooling and solidification in the mold, most metals and alloys shrink. The combined effect of metal shrinkage and mold behavior during casting solidification dictates casting soundness. Improper management of heat flow may result in casting defects such as cold shuts or shrinkage defects. These defects are responsible for considerable financial loss in the metal casting industry.

7.7.1 Metal Shrinkage and Feeding

It is convenient to distinguish three types of shrinkage: liquid, solidification, and solid shrinkage (Fig. 7.4). *Liquid shrinkage*, occurring from the pouring temperature to the liquidus temperature, is usually compensated by flow of liquid from the gating systems and the risers. *Solidification shrinkage* can also be compensated through

Fig. 7.4 Shrinkage regimes

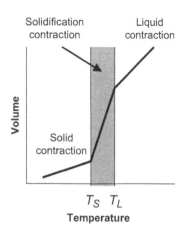

Fig. 7.5 Mechanisms of feeding

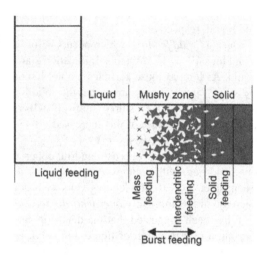

liquid feeding from the risers. However, since feeding channels may be interrupted during solidification before all parts of the casting are fully solid, local shrinkage cavities may occur. *Solid shrinkage* (also called *patternmaker shrinkage*) is accommodated by allowing suitable corrections to the dimensions of the pattern used for making the mold. It is apparent that because of liquid and solidification shrinkage, a mass deficit may result in certain regions of the casting. This mass deficit translates into shrinkage cavities that may cause rejection of the castings. Therefore, it is important to understand and control the feeding of regions of mass deficit. The feeding mechanisms are summarized in Fig. 7.5 as suggested by Campbell (1969), for alloys with freezing range.

Let us try to evaluate the feeding velocity, V_f, which is the average velocity of the mass moving to fill the mass deficit. As explained earlier (see Eq. 4.7), it can be

Table 7.2 Solidification contraction of various metals and alloys

Material	Volumetric solidification contraction (%)	Material	Volumetric solidification contraction (%)
Carbon steel	2.5–3	Cu-30 % Zn	4.5
1 % carbon steel	4	Cu-10 % Al	4
White iron	4–5.5	Aluminum	6.6
Gray iron	− 2.5 (expansion) to 1.6	Al-4.5 % Cu	6.3
Ductile iron	− 4.5 (expansion) to 2.7	Al-12 % Si	3.8
Copper	4.9	Magnesium	4.2
		Zinc	6.5

expressed as $V_f = f_L V_L + f_S V_S$, where V_S and V_L are the velocities of the solid and liquid, respectively.

During *liquid feeding*, which occurs before the beginning of solidification, $f_S = 0$ and thus $V_f = V_L$. When solidification starts, solid particles (grains) form in the liquid. As long as these particles are not in contact with one another, that is, when $f_S < f_S^{cr}$, it may be assumed that the solid moves with the liquid ($V_S = V_L$), and the metal behaves like a slurry (semisolid). Its relative viscosity is increased (fluidity is decreased). Because of this increased viscosity, during *mass* (semisolid) *feeding*, the flow velocity decreases to $V_f < V_L$.

As solidification proceeds, dendrite coherency (i.e., a rigid network of contiguous dendrites) will occur when $f_S < f_S^{cr}$, and a fixed solid network will form. Then, since $V_S = 0$, the feeding velocity becomes $V_f = f_L V_L$, meaning a further decrease in feeding. Only *interdendritic feeding* is possible at this point.

It has been suggested that the dendritic network collapses during solidification, causing a redistribution of liquid and solid, which has been termed *burst feeding* (Campbell 1969). More recent research by Dahle et al. (1997, 1999) seems to confirm this hypothesis. Indeed, their measurements suggest that interdendritic fluid flow can develop stresses in the mushy region that are of similar magnitude to the shear strength of the interdendritic network.

Once solidification is complete and $f_S = 1$, only limited solid feeding through elastic and plastic deformation of the metal is possible.

For effective feeding to occur during solidification, four main requirements must be satisfied:

- A feeding source (riser) that solidifies after the region to be fed
- Sufficient liquid must be available to feed the shrinkage
- Unrestricted feeding channels (path of flow from the feeder to the shrinkage)
- Sufficient pressure on the liquid to make it flow toward the shrinkage region.

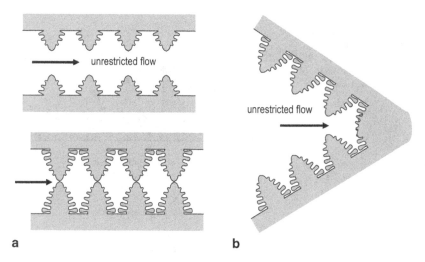

Fig. 7.6 Plate- and wedge-type solidification (**a**) parallel solidification fronts (**b**) converging solidification fronts

Satisfaction of the first requirement is dictated by the overall heat transport during solidification and was discussed in detail Section 6.4 as the solidification time criterion. This criterion allows finding the last region to solidify in the casting and then attaching a riser that can feed this region and solidify after it.

The amount of liquid required to feed the shrinkage depends on the type of alloy, as various metals and alloys have significantly different shrinkage coefficients. Liquid shrinkage for carbon steel is, for example, $1.6 - 1.8\,\%/100\,^{\circ}\mathrm{C}$ of superheat, while for graphitic cast irons it is $0.68 - 1.8\,\%/100\,^{\circ}\mathrm{C}$ (Plutshack and Suschil 1988). Some typical values for solidification shrinkage are given in Table 7.2 (Flinn 1963). It is seen that for graphitic cast iron, expansion may occur during solidification. This is because the graphite formed during solidification has a lower density than the liquid from which it is formed.

Gray and ductile iron expand during solidification because of graphite precipitation. However, poured in nonrigid green sand molds an additional 15 % feed metal requirement above that needed to satisfy the calculated liquid and solidification shrinkage may be required (Plutshack and Suschil 1988). In copper-base alloys, an additional 1 % volumetric shrinkage may be expected under similar conditions.

The third requirement amounts to efficient feeding channels. The efficiency of the feeding channels is affected by the type of alloy as well as by the geometry of the casting. The type of alloy influences the width of the mushy zone (the solidification interval). Wide mushy zone alloys ($T_L - T_S > 110\,^{\circ}\mathrm{C}$) that solidify typically with equiaxed grains rely heavily on semisolid and interdendritic feeding. Thus, their feeding velocity is small, and significant difficulties are experienced in feeding the numerous tortuous channels. The resistance to flow is relatively high. Alloys with narrow mushy zone ($T_L - T_S < 50\,^{\circ}\mathrm{C}$) that exhibit columnar structure rely mostly

Fig. 7.7 Directional (progressive) solidification in an L-shaped casting resulting from increasing temperature gradient from the extremity of the casting to the riser

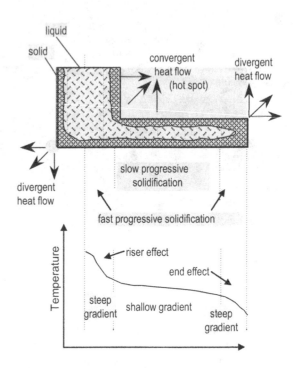

on liquid feeding, and therefore their feeding velocity is high. They are called *skin-forming alloys*.

The local geometry of the solidifying volume can also affect significantly the feeding efficiency. Consider the solidifying plate in Fig. 7.6a. There is no temperature gradient along the plate, as heat is only conducted perpendicular to the plate sides. Parallel solidification fronts will move from the mold wall to the center of the casting. The flow of liquid metal is gradually restricted, because liquid feeding is gradually replaced by interdendritic feeding. Eventually, the feeding channel will be closed, and porosity, known as dispersed centerline shrinkage, will occur between the dendrites. In the case of a solidifying wedge (Fig. 7.6b), a steep temperature gradient from the center to the edge of the casting exists. Liquid feeding is possible until the end of solidification.

Consider now the case of the L-shaped casting presented in Fig. 7.7. At the corners of the casting, either convergent or divergent heat flow may occur. When the heat flow is divergent, solidification will occur at a faster rate, since heat is lost faster. The contrary is true for convergent heat flow which results in the formation of hot spots. From the study of the drawing, it can be seen that at the extremity of the plate casting, as well as at the bottom of the riser, wedge-type solidification occurs. Therefore, dispersed shrinkage is unlikely to appear in these regions. The casting exhibits an *end effect* and a *riser effect*, respectively. In the long, horizontal part of the casting, parallel solidification fronts converge toward the center of the plate, and centerline dispersed shrinkage is to be expected.

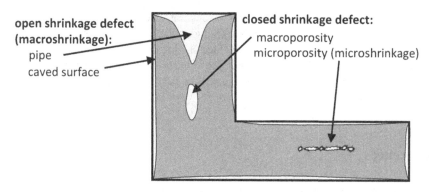

Fig. 7.8 Definition and classification of shrinkage defects. (Stefanescu 2005)

The temperature profile at some time during solidification is also shown in Fig. 7.7. From this plot, it can be seen that steep gradients are associated with the riser and end effects, while shallow gradients occur in the plate region of the casting where centerline shrinkage is expected. This is why criteria attempting to predict the position of centerline shrinkage include the thermal gradient in their formulation.

Finally, sufficient pressure should be applied on the liquid to move it from the riser to the last region to solidify. This is discussed comprehensively under Sect. 17.5, where a number of criteria functions for shrinkage defect prediction are introduced.

7.7.2 Shrinkage Defects

The soundness of the casting depends on uninterrupted flow of liquid metal to the region that solidifies, to feed the mass deficit resulting from solidification contraction. Failure to feed the mass deficit will produce shrinkage defects. Since the terminology is rather ambiguous, in this book, we use the classification and definitions presented in Fig. 7.8. Shrinkage defects that are open to the atmosphere (also called shrinkage cavities) are a consequence of metal contraction, while cooling in liquid state and during solidification. This defect is a macroscale defect that can also be termed *macroshrinkage*. The mass deficit produced by shrinkage is compensated by atmospheric gasses, a process that is independent of the gas content of the metal and which does not require gas pores nucleation and growth. On the contrary, closed shrinkage defects correlate well with pores nucleation and growth in the mushy zone or the amount of bifilms, and thus seem to depend on the impurity level and the amount of gas dissolved in the metal. They can be either *macroporosity* or *microshrinkage* (microporosity) defects. In summary, shrinkage cavities are driven only by metal contraction (shrinkage flow), while shrinkage porosity is driven by both metal contraction and pore nucleation and growth or bifilms growth.

From this analysis, it is apparent that the accurate prediction of shrinkage cavity formation must be based on three-phase (liquid, solid, and gas) mass conservation, coupled with energy conservation. To include casting distortion, stress must also be modeled. This is a problem that has not yet found a satisfactory answer. The state of the art in modeling of macroshrinkage will be presented in Chap. 17, while that of shrinkage porosity in Chap. 18.

References

Campbell J (1969) AFS Cast Metals Research J. March:1.
Dahle AK, Arnberg L, Apelian D (1997) AFS Trans. 105:963
Dahle AK, Thevik HJ, Arnberg L, StJohn DH (1999) Met. Mater. Trans. B 30B:287
Derge G (ed) (1964) Basic Open Hearth Steel Making, 3d ed. AIME, New York
Flemings MC (1974) Solidification Processing. McGraw-Hill, New York
Flemings MC, Nereo GE (1967) Trans. AIME 239:1449
Flinn RA (1963) Fundamentals of Metal Casting. Addison-Wesley, London
Ganesan S, Poirier DR (1990) Metall. Trans. 21B:173
Kaptay G, Kelemen KK (2000) in: Rohatgi PK (ed) State of the Art in Cast Metal Matrix Composites in the Next Millenium. TMS, Warrendale, Pa.
Krieger IM (1972) Adv. in Coll. Interf. Sci. 3:111
Nastac L (2004) Modeling and Simulation of Microstructure Evolution in Solidifying Alloys. Kluwer Academic Publishers, Boston
Ohnaka I (1992) Microsegregation and Macrosegregation. In: Stefanescu DM (ed) ASM Handbook vol.15 Casting. ASM International, Metals Park, OH p 136–141
Patankar SV (1980) Numerical Heat Transfer and Fluid Flow. Hemisphere Publ. Corp., New York
Plutshack LA, Suschil AL (1988) Riser Design. In: Stefanescu DM (ed) Metals Handbook Ninth Edition, vol. 15 Casting. ASM International, Ohio p 576–588
Poirier DR (1987) Metall. Trans. 18B:245
Rappaz M, Voller V (1990) Metall. Trans. 21A:749
Shy W, Udaykumar HS, Rao MM, Smith RW (1996) Computational Fluid Dynamics with Moving Boundaries. Taylor & Francis, Washington DC
Stefanescu DM (2005) Int. J. Cast Metals Res. 18(3):129–143
Thomas DG (1965) J. of Colloid Science 20:267

Chapter 8
Diffusive Mass Transport at the Microscale; Microsolute Redistribution and Microsegregation

As solutes partition differently between the solid and the liquid phases, compositional nonhomogeneities occur at the macro- and microscale level. The microscale nonhomogeneities are termed as microsegregations. Accurate tracking of the solutal field during solidification is required for a comprehensive theoretical treatment of interface morphology. Assessment of microsegregation occurring in solidifying alloys is also important because it influences mechanical properties.

The concept of solute redistribution, discussed previously at the macroscale level, can be extended straightforward to the microscale. Consider, for example, a micro-volume element extending from the middle of a dendrite to the middle of an interdendritic liquid region (Fig. 8.1). Similarly, the case of equiaxed grains can be considered. Further, assume that the curvature of the solid/liquid (S/L) interface is infinite, i.e., the interface is planar. Then, the concepts used to develop equations for macroscopic solute redistribution can also be used for the evaluation of microscopic solute redistribution, i.e., microsegregation.

8.1 Summary of Microsegregation Models

Since the dendrite geometry is rather complicated, all models start by assuming some simplified volume element over which calculations are performed (Fig. 8.2). A summary of major assumptions used in some analytical microsegregation models is given in Table 8.1. Quantitative evaluation of the extent of microsegregation can be done using some of the equations previously derived for the macroscale. The basic equations for microsegregation models are given in Table 8.2. Many of these equations have been previously discussed and are repeated here for convenience.

The Scheil equation (Eq. 5.12) can be used to calculate microsegregation when solid diffusivity is very small. However, the diffusion of solute into the solid phase can affect microsegregation significantly, especially toward the end of solidification. For example, calculations by Brooks et al. (1991) showed that little solid-state diffusion occurs during the solidification and cooling of primary austenite solidified

© Springer International Publishing Switzerland 2015   135
D. M. Stefanescu, *Science and Engineering of Casting Solidification*,
DOI 10.1007/978-3-319-15693-4_8

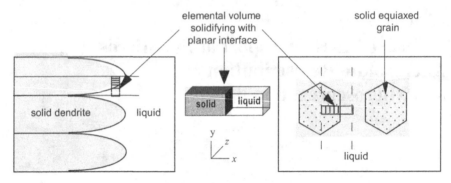

Fig. 8.1 Schematic drawing for the calculation of microsolute redistribution during solidification

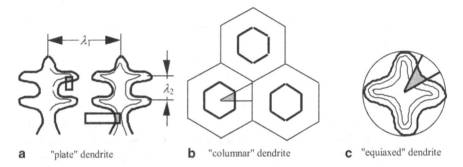

Fig. 8.2 Schematic representation of models for microsegregation. A "plate dendrite" is essentially a 1D dendrite, a "columnar dendrite" is a 2D dendrite, and "equiaxed dendrite" is a 3D dendrite

welds of Fe–Ni–Cr ternary alloys, whereas structures that solidify as ferrite may become almost completely homogenized because of diffusion.

Note that for $f_S = 1$, the Scheil equation predicts $C_S = \infty$, which is impossible. While examining, for example Fig. 5.3, it is seen that when the composition of the solid reaches the maximum solubility in the solid, C_{SM}, the composition of the liquid reaches the eutectic composition, C_E. Consequently, the remaining liquid will solidify as eutectic, and thus, the Scheil equation is not valid anymore. However, the *Scheil* equation can be used to calculate the amount of eutectic that will form at the end of single-phase alloys solidification, as $f_E = 1 - f_{SM}$, where f_E is the fraction of eutectic and f_{SM} is the fraction of primary phase formed when the composition of the solid C_{SM} (see Application 5.9).

Clyne and Kurz (CK) used the Brody–Flemings (BF) model and added a spline fit (the term Ω in Eq. (8.4)) forcing the predictions of the BF model to match the predictions by Scheil equation and the equilibrium equation for infinitesimal and infinite diffusion coefficient, respectively. This relation has no physical basis. The Scheil, CK, and BF models are 1D Cartesian and describe "plate" dendrite solidification.

Table 8.1 Major assumptions used in analytical microsegregation models

Model	Geometry	Solid diffusion	Liquid diffusion	Partition coefficient	Growth	Coarsening
Lever rule	No restriction	Complete	Complete	Variable	No restriction	No
Scheil (1942)	No restriction	No	Complete	Constant	No restriction	No
Brody-Flemings (1966)	No restriction	Incomplete	Complete	Constant	No restriction	No
Clyne-Kurz (1981)	No restriction	Spline fit	Complete	Constant	No restriction	No
Ohnaka (1986)	Linear, columnar	Quadratic equation	Complete	Constant	Linear parabolic	No
Sarreal-Abbaschian (1986)	No restriction	Limited	Complete	Constant	No restriction	No
Kobayashi (1988)	Columnar	Limited	Complete	Constant	Linear	No
Nastac-Stefanescu (1993)	Plate, columnar, equiaxed	Limited	Limited	Variable	No restriction	Yes

Table 8.2 Equations for models in Table 8.1

Model	Equation	Equation number
Lever rule	$C_S = kC_o/[(1 - f_S) + k f_S]$	(8.1) and (5.1)
Scheil (1942)	$C_S = kC_o(1 - f_S)^{k-1}$	(8.2) and (5.12)
Brody-Flemings (1966)	$C_S = kC_o[1 - (1 - 2\alpha k)f_S]^{(k-1)/(1-2\alpha k)}$ with $\alpha = 4D_S t_f/\lambda^2$	(8.3) and (5.16)
Clyne-Kurz (1981)	$C_S = kC_o[1 - (1 - 2\Omega k)f_S]^{(k-1)/(1-2\Omega k)}$ with $\Omega = \alpha[1 - \exp(-1/\alpha)] - 0.5\exp(-1/2\alpha)$	(8.4)
Ohnaka (1986)	$C_S = kC_o[1 - (1 - 2\beta k)f_S]^{(k-1)/(1-2\beta k)}$ with $\beta = 2\gamma/(1 + 2\gamma)$ $\gamma = 8D_S t_f/\lambda_f^2$	(8.5)
Kobayashi (1988)	$C_S = kC_o\xi^{(k-1)/(1-\beta k)}\{1 + \Gamma[0.5(\xi^{-2} - 1) - 2(\xi^{-1} - 1) - \ln\xi]\}$ with $\xi = 1 - (1 - \beta k)f_S$ $\Gamma = \beta^3 k(k-1)[(1 + \beta)k - 2](1 + \beta k)^{-1}(1 - \beta k)^{-3}$	(8.6)
Nastac-Stefanescu (1993)	$C_S^* = kC_o\left[1 - \dfrac{(1-k)f_S}{1-(m+1)\left(kf_S^{(m+1)}+I_L^{(m+1)}\right)}\right]^{-1}$ $f_S = (r^*/r_f)^{m+1}$ see text for I_S and I_L	(8.7)

The BF and CK analyses were used to explain microsegregation in Al–Cu and Al–Si alloys at cooling rate up to 200 K/s (Sarreal and Abbaschian 1986). For higher cooling rate, a new equation based on the BF model that includes the effects of dendrite tip, undercooling, and eutectic temperature depression was developed (Eq. (8.4)).

Ohnaka (1986) proposed a model for a "columnar" dendrite (Fig. 8.2b, Eq. (8.5)). Complete mixing in the liquid and parabolic growth was assumed. Based on an assumed profile $(C_S = A + Bx + Cx^2)$, an equation for solute redistribution in the solid that includes the CK equation, was derived. Note that for $D_S = 0$ this equation reduces to the Scheil equation, and for $D_S \rightarrow \infty$ it becomes the equilibrium equation. However, the diffusion equation was not directly solved. Prior knowledge of the final solidification time is required.

Nastac and Stefanescu (1993) (NS) have proposed a complete analytical and a numerical model for "Fickian" diffusion with time-independent diffusion coefficients and zero-flux boundary condition in systems solidifying with equiaxed morphology (Fig. 8.2c, Eq. (8.7)). The model can be used for "plate" or "equiaxed" dendrites. It takes into account solute transport in the solid and liquid phases and includes overall solute balance. The model allows a comprehensive treatment of dendritic solidification through calculation of the fraction of solid with an MT–TK model (MT—macro-transport, TK—transformation kinetics). The main features of the model are as follows:

- Solute transport in the solid and liquid phases is by diffusion with diffusion coefficients independent of concentration; diffusion depends only on the radial coordinate, r; Fick's second law must be satisfied in each phase:

$$\frac{\partial C_S}{\partial t} = \frac{1}{r^m} \frac{\partial}{\partial r} \left(r^m D_S \frac{\partial C_S}{\partial r} \right) \quad \text{and} \quad \frac{\partial C_L}{\partial t} = \frac{1}{r^m} \frac{\partial}{\partial r} \left(r^m D_L \frac{\partial C_L}{\partial r} \right),$$

where $m = 1, 2$, or 3, for plate, cylindrical, and spherical geometry, respectively;

- Closed system is assumed (no solute flow into or out of the volume element); the overall solute balance can then be written in integral form as:

$$\frac{1}{\rho v} \int_v \rho C(r, t) dv = C_o.$$

- The boundary conditions are as follows:
  at the interface $C_S^* = k C_L^*$
  at $r = 0$ $D_S \cdot \partial C_S / \partial r = 0$
  at $r = R_f$ $D_L \cdot \partial C_L / \partial r = 0$, where R_f is the final radius of the domain.

The final exact analytical solution obtained through the method of separation of variables is Eq. (8.7) in Table 8.2. The values of the coefficients I_S and I_L are as follows:

• For spherical geometry:

$$I_S^{(3)} = \frac{2 f_S}{\pi^2} \sum_{n=1}^{\infty} \frac{1}{n^2} \exp\left[-\left(\frac{n\pi}{f_S^{1/3}}\right)^2 \frac{D_S t}{r_f^2}\right]$$

$$\text{and }\ I_L^{(3)} = 2 f_S^{2/3} \left(1 - f_S^{1/3}\right) \sum_{n=1}^{\infty} \frac{1}{\alpha_n^2} \exp\left[-\left(\frac{\alpha_n}{1 - f_S^{1/3}}\right)^2 \frac{D_L t}{r_f^2}\right],$$

where α_n is the n-th root of the equation $\alpha_n/\tan(\alpha_n) = 1 - f_S^{1/3}$;

• For plate geometry:

$$I_S^{(1)} = \frac{2 f_S}{\pi^2} \sum_{n-1}^{\infty} \frac{1}{(n - 0.5)^2} \exp\left[-\left(\frac{(n - 0.5)\pi}{f_S}\right)^2 \frac{D_S t}{r_f^2}\right]$$

$$\text{and }\ I_L^{(1)} = \frac{2}{\pi^2} (1 - f_S) \sum_{n-1}^{\infty} \frac{1}{(n - 0.5)^2} \exp\left[-\left(\frac{(n - 0.5)\pi}{1 - f_S}\right)^2 \frac{D_S t}{r_f^2}\right]$$

Note that for D_S, $D_L \rightarrow \infty$, Eq. (8.7) becomes the equation for equilibrium solidification. Also, for $D_S = 0$ and $D_L \rightarrow \infty$ at $f_S = 0$ it yields $C_S^* = k\,C_o$, and at $f_S = 1$ it predicts $C_S^* \rightarrow \infty$. However, for this last set of conditions the Scheil equation is not obtained. The reason is that the Fickian diffusion equation exhibits a singularity at $D_S = 0$. Thus, Eq. (8.7) should not be used for the particular case of $D_S = 0$.

Another limitation of the model comes from the use of the method of separation of variables. It can be demonstrated that the method holds only when the solidification velocity is much smaller than the diffusion velocity (or the diffusion time is much smaller than the solidification time). This amounts to assuming a solid-state back-diffusion coefficient $\left(\alpha = D_S t_f/l^2\right)$ larger than one. Conversely, the condition can be expressed as a Péclet number smaller than one. While this is typically the case at the microscale level, the condition is not necessarily met for the macroscale.

A comparison of predictions of niobium redistribution in Inconel 718 by various models is presented in Fig. 8.3. Less than 2 % by volume, Laves phases were measured by Thompson et al. (1991). Further discussion of the applicability of various models is offered through Application 8.1.

As pointed out by Battle (1992) in his excellent review of the modeling of solute segregation, when using analytical models to evaluate microsegregation, it must be assumed that all physical properties are constant. The solid-state concentration can only be calculated at the interface, and cannot be modified by subsequent solid diffusion. In other words, only the trace of the solid-state concentration can be plotted. To obtain the average composition in the liquid and solid, respectively, the overall mass balance equations for liquid and solid must be used. Thus, a numerical scheme is required.

Many numerical microsegregation models have also been proposed. However, the use of numerical segregation models in macro–micro solidification codes is

Fig. 8.3 Comparison of various models for Nb redistribution in Inconel 718 solidified with equiaxed morphology. Initial Nb content was 5.25 wt.% (Nastac and Stefanescu 1993). With permission of Springer Science and Business Media

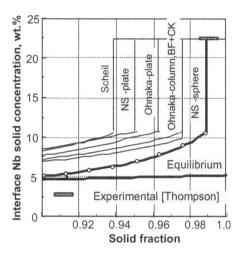

cumbersome because the computational time is significantly increased. Ogilvy and Kirkwood (1987) further developed the BF model to allow for dendrite arm coarsening in binary and multicomponent alloys. For binary systems the basic equation is:

$$C_L^* (1 - k) \frac{dX}{dt} = D \frac{\partial C}{\partial x} + \frac{dC_L}{dt} \left(\frac{\lambda}{2} - X \right) + \frac{C_L - C_o}{2} \frac{d\lambda}{dt} \qquad (8.1)$$

Here, X is the distance solidified. Thus, $f_S = 2X/\lambda$. The end term represents the increase in the size of the element due to arm coarsening, which brings in liquid of average composition that requires to be raised to the composition of the existing liquid. This equation was solved numerically under the additional assumptions of constant cooling rate and liquidus slope. Also, a correction factor for fast diffusing species was added.

Matsumiya et al. (1984) developed a 1D multicomponent numerical model in which both diffusions in liquid and solid were considered. Toward the end of solidification, especially for small partition coefficients, a lower liquid concentration than predicted by the analytical models was obtained.

Yeum et al. (1989) proposed a finite difference method to describe microsegregation in a "plate" dendrite that allowed the use of variable k, D, and growth velocity. However, complete mixing in liquid was assumed.

Battle and Pehlke (1990) developed a 1D numerical model for "plate" dendrites that can be used either for the primary or for the secondary arm spacing. Diffusion was calculated in both liquid and solid, and dendrite arm coarsening was considered.

An integral profile method was used by Himemyia and Umeda (1998) to develop a numerical model that can calculate microsegregation assuming finite diffusion in both the liquid and the solid. The ordinary differential equations can be easily solved, for example, through the Runge–Kutta method.

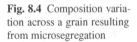

Fig. 8.4 Composition variation across a grain resulting
from microsegregation

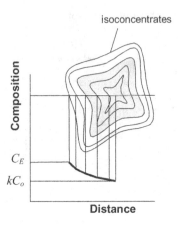

Further, complications arise when multicomponent systems are considered. Chen
and Chang (1992) have proposed a numerical model for the geometrical description of the solid phases formed along the liquidus valley of a ternary system for
plate dendrites. Constant growth velocity, variable partition coefficients, and the BF
model for diffusion were used.

In many solidification models, the microsegregation problem is simplified by
assuming infinite diffusivity in the liquid for all elements, and no diffusivity in the
solid for substitutionally dissolved elements. Nevertheless, experiments conducted
by Hillert et al. (1999) on Fe–Cr–C alloys demonstrate that while indeed C and Cr
have infinite diffusivity in the liquid, and C can be assumed as having infinite diffusivity in the face-centered cubic (fcc) solid, the back-diffusion of chromium cannot
be ignored. In addition, back-diffusion during cooling after the end of solidification
should not be neglected either.

To decrease computational time in complex solidification simulation packages,
it is preferably to use analytical models for microsegregation in conjunction with
numerical schemes for energy and mass transport.

Let us now analyze the implications of microsegregation on microstructure. Consider, for example, the solidification of a spherical grain. Solidification starts with
composition $k\,C_o$ and ends with composition C_E when $k < 1$, $C_o < C_E$. In other
words, if solid diffusivity is relatively small, the center (core) of the grain that solidifies first is poorer in solute than the outside shell, (Fig. 8.4). If conversely $k > 1$ (right
hand corner of a phase diagram), it is also possible to have the core of the grain richer
in solute than its shell. This phenomenon is called *coring*. Rapid solid diffusion or
extended exposure to high temperature will decrease the extent of microsegregation,
since the solid will become increasingly chemically homogeneous.

In industrial applications, microsegregation is evaluated by the *microsegregation
ratio* (C_{Smax}/C_{Smin}), and by the amount of nonequilibrium second phase in the case
of alloys that form eutectic compounds.

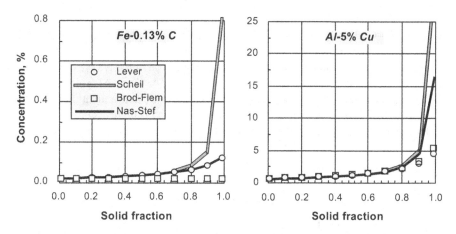

Fig. 8.5 Comparison of the interface solid concentration calculated with different models

The intensity of microsegregation depends on the value of the partition coefficient. The farther away the value is from unity, the larger is the segregation.

8.2 Applications

Application 8.1

Consider a micro-volume element in a casting. Calculate and compare the compositional profiles in the solid after solidification assuming two different materials for the casting, Fe–0.13 % C and Al–5 % Cu, for the following segregation models, lever rule, Scheil, BF, and NS. The data required for calculation are given in the following table (SI units).

Alloy	k	D_S	D_L	t_f	R_f
Fe–0.13 % C	0.17	$1 \cdot 10^{-9}$	$2 \cdot 10^{-8}$	180	$5 \cdot 10^{-5}$
Al–5 % Cu	0.145	$5.5 \cdot 10^{-13}$	$3 \cdot 10^{-9}$	93	$2.3 \cdot 10^{-5}$

Answer:
Linear solidification and plate geometry will be assumed. Then, $t = f_S \cdot t_f$. Alternatively, $t = f_S^2 \cdot t_f$ could be used for parabolic solidification. The Excel spreadsheet was used for calculation. It will be seen that five terms are sufficient for I_S to converge. I_L seems to have little influence for the diffusivity values used here. Calculation results are presented in Fig. 8.5. For the Fe–0.13 % C alloy, the lever rule and the NS model predict the same amount of segregation. This is not surprising, since the solid diffusivity of carbon is very high (interstitial diffusion), for

practical purposes—infinite. Thus, the equilibrium lever rule is satisfactory when modeling Fe–C alloys. The Scheil model predicts much higher segregation, while the BF model much lower. They both have shortcomings.

On the contrary, for the Al–5 % Cu alloy, the prediction of the NS is closer to the Scheil model because of the very low solid diffusivity of Cu in Al (solid solution diffusion). This infers that the Scheil model is a reasonable approximation for slow diffusing substitutional elements. However, the NS model is more accurate. The BF model predicts slightly higher segregation than the equilibrium equation.

References

Battle TP (1992) Int. Materials Reviews 37(6):249
Battle TP, Pehlke RD (1990), Metall. Trans. 21B:357
Brody HD, Flemings MC (1966) Trans. Met. Soc. AIME 236:615
Brooks JA, Baskes MI, Greulich FA (1991) Metall. Trans. 22A:915
Chen SW, Chang YA (1992) Metall. Trans. 23 A:965
Clyne TW, Kurz W (1981), Metall. Trans. 12A:965
Himemyia T, Umeda T (1998) ISIJ International 38:730
Hillert M, Höglund L, Schlan M (1999) Metall. and Mater. Trans. 30A:1653
Kobayashi S (1988) Trans. Iron Steel Inst. Jpn. 28:728
Matsumiya T, Kajioka H, Mizoguchi S, Ueshima Y, Esaka H (1984) Trans. Iron and Steel Inst. of Japan 24:873
Nastac L, Stefanescu DM (1993) Metall. Trans. 24A:2107
Ohnaka I (1986) Trans. Iron Steel Inst. Jpn. 26:1045
Ogilvy AJW, Kirkwood DH (1987) Applied Scientific Research 44:43
Sarreal JA, Abbaschian GJ (1986) Metall. Trans. 17A:2863
Scheil E (1942) Z. Metallk. 34:70
Thompson RG, Mayo DE, Radhakrishnan B (1991) Metall. Trans. 22A:557
Yeum KS, Laxmanan V, Poirier DR (1989) Metall Trans. 20A:2847

Chapter 9
Solidification of Single-Phase Alloys; Cells and Dendrites

9.1 Interface Stability

In the preceding discussion we have not concerned ourselves with the morphology of the solid/liquid (S/L) interface. For calculation purposes it has been considered to be reasonably smooth. However, this is seldom the case in solidification of castings. It will be demonstrated that the thermal and compositional field ahead of the solidifying interface determines its morphology. If such influences are not considered, there is no reason for the interface to lose its planar morphology and become unstable. To evaluate the evolution of interface morphology, interface stability arguments are used. A perturbation is assumed to form at the interface. Then, if the perturbation is damped out in time, the interface is considered to be stable. If the perturbation is amplified in time, the interface is unstable.

9.1.1 Thermal Instability

Since in pure substances there is no constitutional undercooling, only the instabilities resulting from the thermal field must be considered. In the case shown in Fig. 9.1a, the temperature decreases continuously from the liquid to the solid. It is said that a positive thermal gradient exists.

Solidification will start at the mold/liquid interface, on some nuclei on the mold wall, and proceed toward the liquid. If the temperature at the S/L interface is equal to the solidification temperature, the interface is at equilibrium and cannot move. A small kinetic undercooling is required to drive the process. If a thermal instability (a local perturbation) should form and grow at the interface, it will find itself in an environment where the temperature is higher than its melting point. Consequently, this perturbation cannot grow, will disappear, and the interface will remain stable. Its morphology will be flat (planar). Note that on Fig. 9.1a several grains are shown in the solid region. The number of grains that will form in the solid is a function of the *nucleation potential* at the mold/liquid interface.

© Springer International Publishing Switzerland 2015   145
D. M. Stefanescu, *Science and Engineering of Casting Solidification*,
DOI 10.1007/978-3-319-15693-4_9

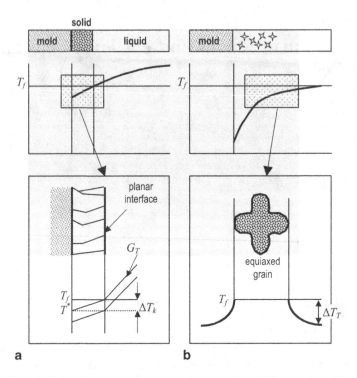

Fig. 9.1 Formation of various types of structures because of instabilities generated by the thermal field. **a** Low thermal undercooling, positive gradient, planar interface. **b** High thermal undercooling, negative gradient, equiaxed solidification

If the melt becomes highly undercooled because of lack of nucleation sites at the mold wall, solidification can start on nuclei forming in the bulk liquid when the nucleation undercooling is reached, away from the interface (Fig. 9.1b). Since growth conditions in the liquid are isotropic, the new crystals will have a spherical shape at the beginning of solidification. The crystals growing in the liquid from these spherical nuclei are called *equiaxed crystals*. Because the latent heat of fusion is evolved at the grain surface, the temperature will be higher at the L/S interface than in the bulk liquid. It is said that a negative thermal gradient has occurred. The crystal surface will find itself in an undercooled environment and will continue to grow. Local surface instabilities will also grow at the interface, and the final shape of the equiaxed crystal will not be spherical but dendritic.

Thus, in a pure metal where only thermal instabilities can occur, there are two types of possible structures:

- Planar (the interface is stable) when the thermal gradient is positive
- Dendritic equiaxed (the interface is unstable) when the thermal gradient is negative

Fig. 9.2 Correlation between
the thermal gradient at the
interface and the interface
morphology

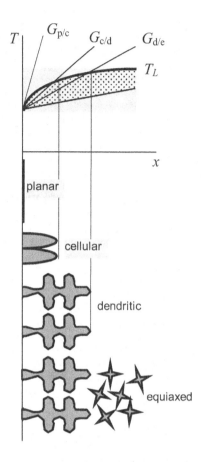

9.1.2 Solutal Instability

Interface instability can also be promoted by the evolution of the compositional field ahead of the growing interface. It has been demonstrated (see for example Figs. 2.9 and 2.10) that when the thermal gradient in the liquid at the S/L interface is smaller than the liquidus temperature gradient, i.e., $G_T < G_L$, the liquid at the interface is at a lower temperature than its liquidus. This liquid is constitutionally undercooled. Instabilities growing in this region will become stable because they will find themselves at a temperature lower than their equilibrium temperature. They will continue to grow. On the contrary, if $G_T > G_L$, the interface will remain planar.

For small constitutional undercooling, the instabilities will only grow in the solidification direction, and a cellular interface will result. This is shown schematically in Fig. 9.2 and supported with pictures resulting from experimental work in Fig. 9.3b and c. The planar-to-cellular transition occurs at a gradient $G_{p/c}$.

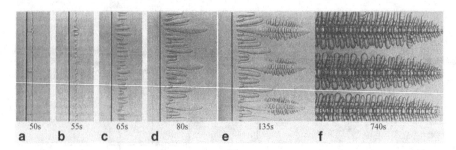

Fig. 9.3 Change in interface morphology of a succinonitrile—4 % acetone solution when increasing the solidification velocity from 0 to 3.4 μm/s at a temperature gradient of 6.7 K/mm. Magnification × 30. (Trivedi and Somboonsuk 1984; with permission from Elsevier)

Fig. 9.4 Definition of solutal supersaturation and of constitutional undercooling

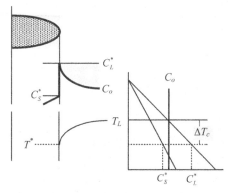

However, as the constitutional undercooling increases because of lower thermal gradient, the spacing between the cells increases (this will be demonstrated later), and constitutional undercooling may also occur in the y-direction, perpendicular to the growth direction. Instabilities will develop on the sides of the cells, resulting in the formation of dendrites. This is the cellular-to-dendrite transition. It takes place at a temperature gradient $G_{c/d}$ (Figs. 9.2 and 9.3d). Both cellular and dendritic growth occurring from the wall in the direction opposite to the heat transport can be described as *columnar growth*. If constitutional undercooling is even higher, equiaxed grains can be nucleated in the liquid away from the interface. The dendritic-to-equiaxed transition occurs at $G_{d/e}$. If the thermal gradient is almost flat, i.e., $G_T = 0$, the driving force for the columnar front will be extremely small. A complete equiaxed structure is expected.

Figure 9.3 also illustrates grain growth competition and the survival of the dendrites oriented preferentially in the direction of heat extraction.

Let us now have a closer look at the diffusion field at the tip of the perturbation. The driving force for the growth of the solutal perturbation is the composition gradient at the tip, $C_L^* - C_o$. Using the notations in Fig. 9.4, this driving force can be expressed in a nondimensional form as:

$$\Omega_c = \frac{C_L^* - C_o}{C_L^* - C_S^*} = \frac{C_L^* - C_o}{C_L^*(1 - k)}, \tag{9.1}$$

where Ω_c is the solutal supersaturation. Note that the supersaturation can vary between 0 (no transformation) and 1 (maximum transformation rate). ΔT_c is the constitutional undercooling. It can vary between 0 and ΔT_o.

Let us now try to quantify the conditions under which the planar-to-columnar transition occurs. The criterion for instability formation is $G_T < G_L$. The temperature gradient along the liquidus line at the interface is:

$$(G_L)_{x=0} = \left(\frac{dT_L}{dx}\right)_{x=0} = m_L \left(\frac{dC_L}{dx}\right)_{x=0}. \tag{9.2}$$

The concentration gradient at the interface can be evaluated from the composition of the boundary layer, Eq. 5.5, to yield:

- For steady state $\quad \left(\frac{dC_L}{dx}\right)_{x=0} = -C_o \frac{1-k}{k} \frac{V}{D_L}$
- For non-steady state $\left(\frac{dC_L}{dx}\right)_{x=0} = -C_L^*(1-k)\frac{V}{D_L}$

Substituting these equations in the criterion for instability formation, Eq. 9.2, one obtains the *criterion for constitutional undercooling:*

$$\text{for steady state:} \frac{G_T}{V} < -\frac{m_L C_o (1-k)}{k D_L} = \frac{\Delta T_o}{D_L} \tag{9.3a}$$

$$\text{for non-steady state:} \frac{G_T}{V} < -\frac{m_L C_L^*(1-k)}{D_L}. \tag{9.3b}$$

The criterion for the cellular-to-dendritic transition was derived by Laxmanan (1987) as:

$$G_T/V < \Delta T_o/(2D_L). \tag{9.4}$$

Trivedi et al. (2003) derived a different criterion for the cellular-to-dendritic transition based on experiments with the succinonitrile-salol system. They argued that there is a critical cell spacing, λ_{cd}, above which cells transform to dendrites:

$$\lambda_{cd} = \left(\frac{10.8}{C_o}\right)\left(\frac{\Delta T_o}{G_T} \frac{D}{V} \frac{\Gamma}{\Delta T_o}\right). \tag{9.5}$$

Note that the ratios in the second parenthesis are the characteristic thermal, solutal, and capillary length. The numerical constant contains system constants and is nondimensional, as is the composition expressed in weight %.

This simple analysis of interface stability explains the four possible S/L interfaces found in the experimental solidification of metal and alloys, under low and moderate growth velocities, as summarized in Fig. 9.5. Note that by changing either the thermal gradient or the solidification velocity, the interface morphology can change from planar to equiaxed, or vice versa, for a fixed composition C_o.

In commercial alloys, cast in sand or even some metal molds, interface instability will occur because the thermal gradient is typically very small (see Application 9.1).

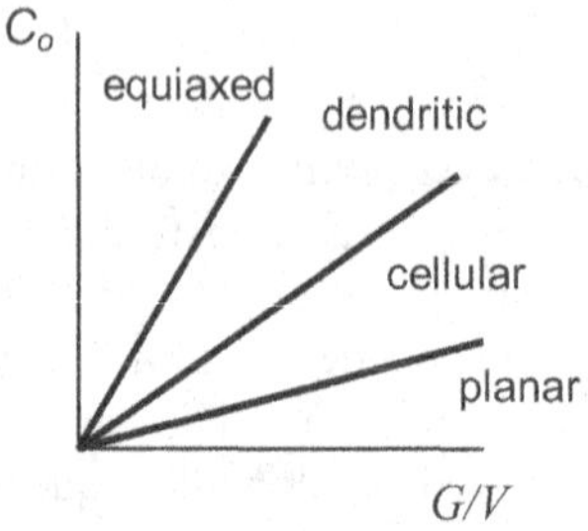

Fig. 9.5 Influence of composition, thermal gradient, and growth velocity on interface morphology

Fig. 9.6 The combined effects of the temperature and solutal field on interface stability. **a** Constrained growth. **b** Unconstrained growth

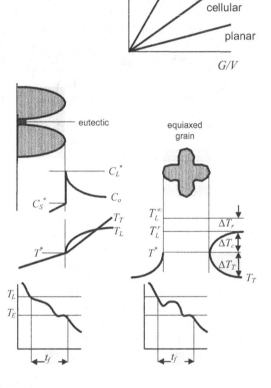

9.1.3 Thermal, Solutal, and Surface Energy Driven Morphological Instability

When the combined effects of the thermal and solutal field are considered (Fig. 9.6), two types of growth can be defined for the perturbation:

- Constrained : the growth rate is controlled by the temperature gradient ahead of the interface (there is no thermal undercooling) and by constitutional undercooling; heat flows from the melt, to the perturbation, to the mold, that is, the melt is the hottest; if the perturbation is stable, columnar structure results; no recalescence will be seen on the cooling curve
- Unconstrained : the growth rate is controlled by the thermal and constitutional undercooling; heat flows from the perturbation (grain), to the melt, to the mold, that is, the grain is the hottest; if the liquid is undercooled, the grain grows freely in the liquid; equiaxed structure results; recalescence will be seen on the cooling curve

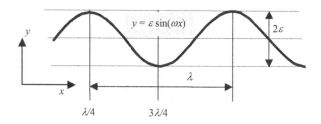

Fig. 9.7 Elements of sinusoidal perturbations

Thus, in constrained growth the driving force is the constitutional undercooling, ΔT_c, which is determined by the $G_L - G_T$ difference. Unconstrained growth is controlled by both solutal and thermal undercooling, $\Delta T_c + \Delta T_T$. In addition, in both cases the curvature of the perturbation will have to be considered through the change in local surface energy. An additional undercooling, ΔT_r, will be added. The stability of a perturbation must be evaluated as a function of the thermal and solutal field and of the local surface energy (curvature).

Following the approach introduced by Mullins and Sekerka (1964) and modified by Kurz and Fisher (1989), first a planar, unperturbed interface is considered. The general form of the governing equation to solve is:

$$\frac{\partial^2 \varphi}{\partial y^2} + \frac{V}{\Gamma} \frac{\partial \varphi}{\partial y} = 0, \tag{9.6}$$

where ϕ is the phase quantity and Γ is the general diffusion coefficient. This equation must be solved for both solute and heat diffusion. Ignoring solute diffusion in the solid because it is very small, the variables in the above equation are $\phi = C_L, T_L, T_S$, and the diffusivities are $\Gamma = D_L, \alpha_L, \alpha_S$. Note that this equation written for solute diffusion (Eq. 5.4) has been solved before to obtain the shape of the diffusion boundary layer (Eq. 5.5). The solutions for the planar interface are:

$$C_L = C_o - \frac{D_L G_L}{mV} \exp\left(-\frac{Vy}{D_L}\right) \tag{9.7a}$$

$$T_L = T_o + \frac{\alpha_L G_T^L}{V}\left[1 - \exp\left(-\frac{Vy}{\alpha_L}\right)\right] \tag{9.7b}$$

$$T_S = T_o + \frac{\alpha_S G_T^S}{V}\left[1 - \exp\left(-\frac{Vy}{\alpha_S}\right)\right], \tag{9.7c}$$

where C_o is the concentration at infinity, while T_o is the interface temperature.

If the interface loses planarity, it can be assumed that small sinusoidal perturbations are formed (Fig. 9.7). The interface can then be described as a sinusoidal function, for example, $y = \varepsilon \sin(\omega x)$, where ε is the amplitude and $\omega = 2\pi/\lambda$ is the wave number. The problem is now to evaluate whether the perturbation is stable or not.

Using the perturbation technique, a term having the same form as the perturbation in Fig. 9.7 is added to the three equations that represent the solution for the stable interface:

$$C_L = C_o - \frac{D_L G_L}{mV} \exp\left(-\frac{Vy}{D_L}\right) + C_1 \varepsilon \sin(\omega x) \exp(-b_c y) \qquad (9.8a)$$

$$T_L = T_o + \frac{\alpha_L G_T^L}{V}\left[1 - \exp\left(-\frac{Vy}{\alpha_L}\right)\right] + C_2 \varepsilon \sin(\omega x) \exp(-b_L y) \qquad (9.8b)$$

$$T_S = T_o + \frac{\alpha_S G_T^S}{V}\left[1 - \exp\left(-\frac{Vy}{\alpha_S}\right)\right] + C_3 \varepsilon \sin(\omega x) \exp(-b_S y), \qquad (9.8c)$$

where we must have $b = V/2\alpha + [(V/2\alpha)^2 + \omega^2]^{1/2}$ to satisfy Eq. 9.6 and C_1, C_2, C_3, b_c, b_L, and b_S must be determined. Assuming that the velocity of the perturbed interface is $V = \dot{\varepsilon} \sin(\omega x)$, after further manipulations and further simplifications, we obtain:

$$\frac{\dot{\varepsilon}}{\varepsilon} = -\Gamma\omega^2 - G_T \xi_T + G_L \xi_c. \qquad (9.9)$$

Here, G_T is the thermal conductivity weighted temperature gradient, given by:

$$G_T = \frac{k_S G_T^S + k_L G_T^L}{k_S + k_L}. \qquad (9.10)$$

The quantities ξ_T and ξ_c are:

$$\xi_T = 1 - (1 + 4\pi^2 P_T^{-2})^{-1/2} \quad \xi_c = 1 - 2k_V[(1 + 4\pi^2 P_c^{-2})^{1/2} - 1 + 2k_V]^{-1}. \qquad (9.11)$$

In this last equation, the quantities $P_T = V\lambda/2\alpha$ and $P_c = V\lambda/2D$ are called the thermal and solutal *Péclet numbers*, respectively. k_V is given by Eq. 2.34.

Equation 9.9 is plotted in Fig. 9.8. For $\dot{\varepsilon}/\varepsilon < 0$, since $d\varepsilon/dt$ must be negative, i.e., the perturbation amplitude decreases, the perturbation is unstable. For the smaller gradient, G_1, it is seen that there is a range where $\dot{\varepsilon}/\varepsilon > 0$, and thus the perturbation is stable. At $\lambda \geq \lambda_i$ the perturbation becomes stable and continues to grow. Let us try to evaluate λ_i

Under the assumption of small Péclet numbers, $\omega \gg V/D$ or $\gg V/\alpha$, the parameters ξ_T and ξ_c are taken as unity. This is the case for typical casting processes. Then, for $\dot{\varepsilon}/\varepsilon = 0$, Eq. 9.9 reduces to $-\Gamma\omega^2 - G_T + G_L = 0$, or:

$$\lambda_i = 2\pi \sqrt{\Gamma/(G_L - G_T)}. \qquad (9.12)$$

When $G_L = G_T$, the perturbation becomes $\lambda_i = \infty$, which means that the interface is planar and thus stable. This is the previously discussed criterion for constitutional undercooling. If $G_T \ll G_L = V\Delta T_o/D_L$, then:

Fig. 9.8 Relative growth velocity of instability as a function of interspace

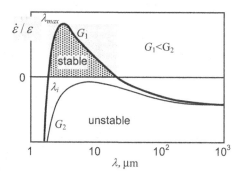

$$\lambda_i = 2\pi \sqrt{\frac{D}{V}\frac{\Gamma}{\Delta T_o}}, \tag{9.13}$$

where $\delta_c = D/V$ is the diffusion length and $d_o = \Gamma/\Delta T_o$ is the capillary length.

At low solidification velocity, the solute gradient destabilizes the interface (the diffusion length is high). When, because of increased solidification velocity, the diffusion length becomes of the size of the capillary length, i.e., $\delta_c = d_0$, the interface is again stable. This is because the boundary layer has approached atomic dimensions, and the surface energy has a stabilizing effect. The *absolute limit of stability* has been reached. Thus, the condition for absolute stability is $D/V = \Gamma/\Delta T_o$. In the formal derivation of the absolute stability criterion (Huntley and Davis 1993), the partition coefficient appears in the equation because of nonequilibrium effects. The solidification velocity required to attain absolute stability, V_{as}, can then be calculated as:

$$V_{as} = D\,\Delta T_o(V)/(\Gamma\,k_V). \tag{9.14}$$

$\Delta T_o(V)$ is a function of velocity. However, note that for rapid solidification $k_V \to 1$. When $V > V_{as}$, diffusionless solidification occurs.

The correlation between the various parameters influencing interface stability is shown graphically in Fig. 9.9. It is seen that increasing C_o increases the region of instability. A higher G increases the chances for planar interface solidification but does not change the absolute stability line. Finally, a higher solidification velocity is conducive first to interface destabilization. However, further increase in velocity beyond the limit of absolute stability stabilizes the interface.

9.1.4 Influence of Convection on Interface Stability

In the analysis of interface stability presented so far, the role of thermosolutal convection has been ignored. The results of numerical analysis of coupled thermosolutal convection and morphological stability for a lead–tin alloy, solidified in the vertical-stabilized Bridgman configuration, are summarized in Fig. 9.10.

While the constitutional undercooling criterion predicts stability in the entire field under the constitutional line, when thermosolutal convection is taken into

Fig. 9.9 Influence of composition, thermal gradient, and solidification velocity on interface stability

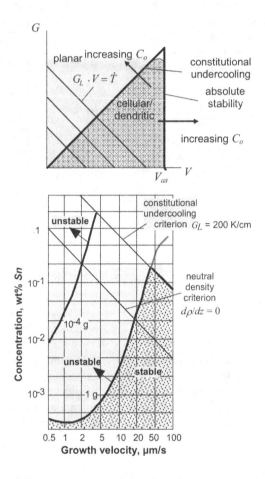

Fig. 9.10 Concentration–growth velocity stability diagram for a Pb–Sn alloy. (After Coriell et al. 1980)

account, the stability field is decreased to the shaded region. In low gravity (10^{-4}g), where convection is lower, calculation predicts higher stability for constant boundary conditions (the instability curve moves to the left).

9.2 Morphology of Primary Phases

The interface morphology of primary phases can be classified in faceted and non-faceted. Whether a phase grows faceted or non-faceted depends mostly on its entropy of fusion. A discussion of the criterion for faceting during growth was provided in Sect. 3.2.1. In general, if $\Delta S_f/R < 2$, where R is the gas constant, non-faceted growth is expected. This is mostly the case for metals. If $\Delta S_f/R > 2$, faceted growth will occur. This is common for nonmetals. Some typical examples of faceted growth in metal/nonmetal systems are given in Fig. 9.11. The faceting behavior is also common in some transparent organic materials such as salol (Fig. 9.12).

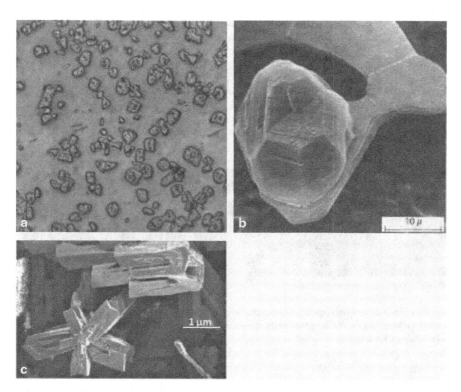

Fig. 9.11 Faceted growth in metal/nonmetal systems. **a** Vanadium carbides in a Fe–C–V alloy. **b** Graphite crystal in Ni–C alloy (Lux et al. 1975). **c** Star-like primary Si crystal in Al–25Si alloy. (Ullah and Carlberg 2011; with permission of Elsevier)

Fig. 9.12 Faceted cells in salol. (Hunt and Jackson 1966)

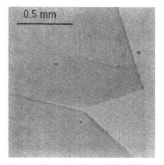

Regardless of morphology, the driving force for growth is the undercooling. Uneven undercooling on the growing surfaces of the crystal will determine dramatic changes in its morphology. A schematic sequence of the shape change of a faceted primary phase growing in the liquid from a cubic crystal to a dendrite is presented in Fig. 9.13. At the corners of the cube divergent transport occurs, and the thermal

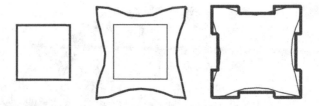

Fig. 9.13 Schematic representation of the growth of a faceted dendrite

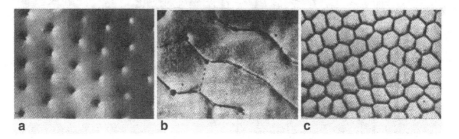

Fig. 9.14 Evolution of segregation substructure as a function of constitutional undercooling; cross-section view. **a** Nodes. **b** Elongated cells. **c** Hexagonal cells. (Biloni and Boettinger 1996; with permission of Elsevier)

as well as the solutal undercooling are larger than on the facets. Consequently, the corners will grow faster, resulting in the degeneration of the cube into a dendrite. Thus, for a faceted phase the {111} planes are the preferred growth direction.

The more interesting problem is that of the morphology of primary non-faceted phases as encountered in commercial alloys such as steel, cast iron, aluminum alloys, and superalloys. At the onset of constitutional undercooling, instabilities appear on the interface as segregations associated with depressions (nodes; Fig. 9.14a). As the undercooling increases, these nodes become interconnected by interface depressions, forming first elongated cells (Fig. 9.14b) and eventually a hexagonal cellular substructure (Fig. 9.14c).

According to the theory of constitutional undercooling, as the undercooling increases, the cells should gradually change into dendrites. The question is how does this transition occur? Regular cells grow in the direction of heat extraction, which is typically perpendicular to the S/L interface. When the solidification velocity is increased, because of higher requirements for atomic transport, the main growth direction becomes the preferred crystallographic growth direction of the crystal (Fig. 9.15a). The preferred crystallographic growth directions for some typical crystals are given in Table 9.1. At the same time with the change in growth direction, the cross section of the cell deviates from a circle to a Maltese cross, and eventually secondary arms are formed (Fig. 9.15d).

The orientation of the growing dendrite with respect to the direction of the heat flow can affect significantly the dendrite morphology, as exemplified in Fig. 9.16.

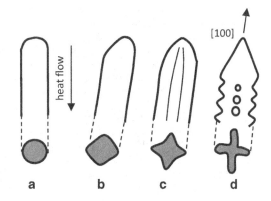

Fig. 9.15 Sequential change of interface morphology as the solidification velocity increases: cell growing in the direction of heat extraction (**a**), cell growing in the [100] direction (**b**), cell/dendrite (**c**), dendrite (**d**). (After Morris and Winegard 1969)

9.3 Analytical Tip Velocity Models for Cells and Dendrites

Cells are a relatively simple periodic pattern of the S/L interface. Dendrites that evolve during solidification of metals are complex patterns characterized by side branches (primary, secondary, tertiary, etc.). Describing mathematically the temporal evolution of such patterns is a challenging endeavor. Both analytical and numerical models have been proposed to describe dendritic growth. Only analytical models will be discussed in this chapter.

The analytical models are limited in scope, attempting to describe solely dendrite tip kinetics, as determined by the thermal and solutal field and by capillarity. Dendritic array models include also an analysis of transport from the root to the tip.

9.3.1 Solute Diffusion-Controlled Growth (Isothermal Growth) of Needle-Like Crystals and Dendrites Tip

Consider a needle-like crystal growing in the liquid. Assume diffusion-controlled growth, which means that the only driving force for growth is the concentration gradient (curvature and thermal undercooling ignored). In a first approximation,

Table 9.1 Preferred crystallographic growth directions

Crystal structure	Growth direction	Example
fcc	[100]	Al
bcc	[100]	δFe
bc tetragonal	[110]	Sn
hcp	[10$\bar{1}$0]	Ice flakes, graphite
	[0001]	$Co_{17}Sm_2$

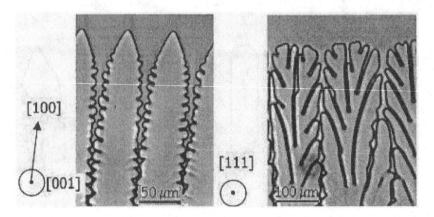

Fig. 9.16 Effect of crystalline anisotropy on the morphology of directionally solidified dendrites; growth velocity 35 μm/s, heat extraction upward; thin films of a CBr₄–8 mol% C₂Cl₆ alloy (Akamatsu et al. 1995; with permission from Phys. Rev.; copyright 1995 by the American Physical Soc.)

Fig. 9.17 Diffusion field ahead of a hemispherical needle

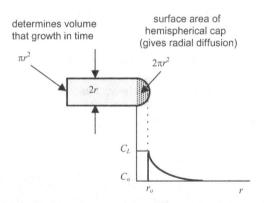

let us assume that the tip of the crystal is a hemispherical cap (hemi-spherical approximation), as shown in Fig. 9.17. Flux balance at the interface gives:

$$\pi r^2 V (C_L - C_S) = -2\pi r^2 D \left(\frac{dC}{dr}\right)_{tip} \quad \text{or} \quad V C_L (1 - k) = -2D \left(\frac{dC}{dr}\right)_{r=r_o}.$$
$$(9.15)$$

The solution of this equation is (see inset for derivation):

$$P_c = \Omega_c,\qquad\qquad(9.16)$$

where the solutal Péclet number as defined earlier is:

$$P_c = Vr/2D,\qquad\qquad(9.17)$$

and the solutal supersaturation, Ω_c, is given by Eq. 9.1. Substituting in Eq. 9.17 we obtain:

$$V = 2D_L \Omega_c / r. \qquad (9.18)$$

Derivation of the growth velocity of the hemispherical needle, Eq. 9.16.
To find the composition gradient at the tip of the crystal, it is necessary to solve the steady-state diffusion equation in radial coordinates with no tangential diffusion:

$$\frac{d^2C}{dr^2} + \frac{2}{r}\frac{dC}{dr} = 0 \quad \text{or} \quad r^2\frac{d^2C}{dr^2} + 2r\frac{dC}{dr} = 0 \quad \text{or} \quad \frac{d}{dr}\left(r^2\frac{dC}{dr}\right) = 0.$$

The general solution of this equation is $C = C_1 + C_2/r$, where C_1 and C_2 are constants. The following boundary conditions are used:
at $r \to \infty$ $C = C_o$ thus $C_1 = C_o$
at $r = r_o$ $C = C_L$ thus $C_2 = r_o (C_L - C_o)$.
Then:

$$\left(\frac{dc}{dr}\right)_{r-r_o} = \left[-\frac{r_o}{r^2}(C_L - C_o)\right] = -\frac{C_L - C_o}{r_o}$$

Substituting in Eq. 9.15 we obtain: $\frac{Vr_o}{2D} = \frac{C_L - C_o}{C_L(1-k)}$, and since the solutal supersaturation is $\Omega_c = (C_L - C_o)/[C_L(1 - k)]$ and the solutal Péclet number is $P_c = Vr_o/(2D)$, we obtain the final solution Eq. 9.16.

Equation 9.18 gives the growth velocity of the hemispherical needle. It indicates that velocity depends on the tip radius, r, and on supersaturation, Ω_c, which is the driving force. However, velocity is not uniquely defined since this equation does not have a unique solution for V, but rather pairs of solutions for V and r. In other words, the solution of the diffusion equation does not specify whether a dendrite will grow fast or slow but only relates the tip curvature to the dendrite rate of propagation.

The other problem with this solution is that the shape defined by this velocity is not self-preserving. In other words, the hemispherical cap does not grow only in the x-direction but also in all r-directions, meaning that the needle thickens as it grows. Experimental work on dendrite growth has demonstrated that the dendrite tip preserves its shape. Consequently, another solution must be found for the diffusion problem.

If it is assumed that the dendrite tip has the shape of a paraboloid of revolution (Fig. 9.18), which is self-preserving, the solution to the steady-state diffusion equation given by Ivantsov (1947) is:

$$I(P_c) = \Omega_c \quad \text{where} \quad I(P_c) = P_c \exp{(Pc)}E_1(P_c) = P_c \exp{(Pc)}\int_P^\infty \frac{\exp{(-x)}}{x}dx. \qquad (9.19)$$

Fig. 9.18 Dendrite tip having the shape of a paraboloid of revolution

Here, $E_1(P_c)$ is the exponential integral function. This solution is valid for both the solutal diffusion (P_c and Ω_c) and the thermal diffusion (P_T and Ω_T).

There are several approximations of the Ivantsov number, $I(P)$, that can be used in numerical or analytical calculations (see inset). Since $I(P)$ is a function of both V and r, the problem of evaluating an unique velocity is still to be solved.

Approximation of the Ivantsov number.
The continued fraction approximation is:

$$I(P) = \cfrac{P}{P + \cfrac{1}{1 + \cfrac{1}{P + \cfrac{2}{1 + \frac{2}{P + \dots}}}}}$$

Note that the zeroth approximation of the continued fraction approximation of the Ivantsov function is $I_o(P) = P$, that is, the hemispherical approximation. Typically, for casting solidification $P < 1$. For this case, the following approximation can be used (Kurz and Fisher 1989):

$$I(P) = P \exp(P)[a_0 + a_1 P + a_2 P^2 + a_3 P^3 + a_4 P^4 + a_5 P^5 - \ln(P)],$$

where $a_0 = -0.57721566$, $a_1 = 0.99999193$, $a_2 = -0.24991055$, $a_3 = 0.05519968$, $a_4 = -0.00976004$, $a_5 = 0.00107857$.

For limiting values of the Péclet number, the Ivantsov function for a paraboloid of revolution can be approximated as (Trivedi and Kurz 1994):

$$\text{for} \quad P \ll 1: \quad I(P) \approx -P \ln P - 0.5772P$$

$$\text{for} \quad P \gg 1: \quad I(P) \approx 1 - 1/P + 2/P^2.$$

9.3.2 Thermal Diffusion-Controlled Growth

During solidification, a thermal gradient is imposed over the system. Thermal diffusion will drive the process. In pure metals, this will be the only driving force for growth. If it is assumed that the driving force for perturbation growth is only the thermal gradient (thermal dendrite), similar equations to those obtained for the diffusion-controlled growth can be derived:

$$P_T = \Omega_T \quad \text{with} \quad P_T = V r/(2\alpha) \quad \text{and} \quad \Omega_T = \Delta T_T/(\Delta H_f/c) \qquad (9.20a)$$

and alternatively:

$$I(P_T) = \Omega_T, \qquad (9.20b)$$

where P_T is the thermal Péclet number and Ω_T is the thermal supersaturation. For the hemispherical approximation a derivation of the particular case of Eq. 9.20b is given in the inset.

Derivation of the correlation between the thermal Péclet number and thermal supersaturation.

Temperature flux balance at the interface gives $\pi r^2 V \Delta H_f/c = -2\pi r^2 \alpha_L (dT/dr)_{r=r_o}$. To calculate the thermal gradient at the interface, we need the temperature of the tip. The solution of the steady-state diffusion equation in radial coordinates is $T = C_1 + C_2/r$. Applying the boundary conditions: i) at $r \to \infty; T = T_{bulk}$ and ii) at $r = r_o; T = T_f$, the solution becomes $T = T_{bulk} + \Delta T \cdot r_o/r$.

Thus, the temperature gradient at the tip is $(dT/dt)_{r=r_o} = -\Delta T_T/r_o$. Substituting in the flux balance equation and rearranging:

$$\frac{Vr}{2\alpha} = \frac{\Delta T_T}{\Delta H_f/c} \quad \text{or} \quad P_T = \Omega_T.$$

All the diffusion models discussed above conclude that at steady state the tip of the dendrite will advance in the liquid following the simple law $Vr = $ const. This means that there is no unique solution since multiple pairs of V and r satisfy this relationship. However, experimental work has demonstrated that for each undercooling a unique value of tip velocity and radius is obtained. The problem is then to find the additional constrains that impose a unique dendrite tip radius from the multiple solutions offered by the diffusion models.

Fig. 9.19 Growth velocity–
tip radius correlation; the *full
line* is the *V–r* correlation

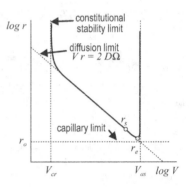

9.3.3 Solutal, Thermal, and Capillary-Controlled Growth

To obtain a unique solution, it is necessary to find additional criteria that define the
tip radius. Several models have been proposed:

- The *extremum* criterion
- The marginal stability criterion
- The microsolvability theory

The extremum criterion

As discussed earlier, at high solidification velocities, when the diffusion length
reaches the size of the solute capillary length ($D/V = \Gamma/\Delta T_o$), the interface
becomes planar. Thus, as shown in Fig. 9.19, the maximum velocity of a dendrite
tip is limited by the absolute stability. The *extremum* criterion implies that the perturbation will grow at the maximum possible velocity and the minimum possible
undercooling. These conditions are satisfied by the velocity corresponding to the
radius tip of the dendrite r_e.

An expression for V can be obtained, for example, for a perturbation driven only
by the solutal and curvature undercooling (solutal perturbation), starting from $\Delta T =
\Delta T_c + \Delta T_r$ where:

$$\Delta T_c = -m(C_L - C_o) = -m(C_L - C_S)\Omega_c = -m(1-k)C_L\Omega_c \quad \text{and} \quad \Delta T_r = \frac{2\Gamma}{r}$$

$$= m(k-1)C_L P_c = -m(1-k)C_L\frac{Vr}{2D}. \tag{9.21}$$

Substituting these last two equations in the total undercooling equation we have:

$$\Delta T = -m(1-k)C_L^*\frac{Vr}{2D} + \frac{2\Gamma}{r}. \tag{9.22}$$

According to the *extremum* criterion, it will be assumed that growth proceeds at the
minimum undercooling (the maximum of the curve), which can be obtained from
$\partial\Delta T/\partial r = 0$, that is:

Fig. 9.20 Tip of a growing succinonitrile dendrite. (Huang and Glicksman 1981; with permission from Elsevier)

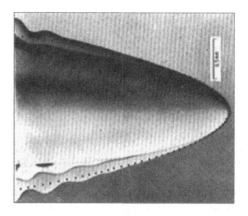

$$r_e = \left[\frac{4D\Gamma}{m\,(k-1)\,C_L^*} \right]^{1/2} V^{-1/2} = \sqrt{\frac{4\Gamma}{G_L}}. \qquad (9.23)$$

Substituting in Eq. 9.22, an equation for tip velocity is obtained:

$$V = \mu \Delta T^2 \quad \text{with} \quad \mu = \frac{D_L}{4\Gamma m\,(k-1)\,C_L^*}. \qquad (9.24)$$

Note that for steady-state solidification, $C_L^* = C_o/k$.

Many other models have been proposed based on the extremum criterion (see, for example, Burden and Hunt 1974). However, experimental work by Nash and Glicksman (1974), who measured concomitantly r, V, and ΔT during the growth of a succinonitrile dendrite, demonstrates that such velocities are considerable higher than the measured ones. This is shown schematically in Fig. 9.19, where r_s is the position of the experimental point and r_e is the value calculated from the *extremum* criterion. They also proved that the tip of the dendrite fits a parabolic curve (dotted line in Fig. 9.20).

The marginal stability criterion Langer and Müller-Krumbhaar (1978) performed a linear stability analysis for an Ivantsov parabola dendrite tip region in a pure undercooled melt. A small departure from the parabolic shape, caused by interface energy, was introduced in the system. It was concluded that dendrite tip radii are not stable at values smaller than predicted by the extremum criterion, or larger than a certain critical value. At such large radii tip splitting will occur to decrease the radius. They proposed that this largest radius is selected by the dendrite during its growth (*marginal stability* criterion). A number of models discussed in the following paragraphs use this criterion to obtain a unique dendrite tip radius.

Kurz and Fisher (1981) were the first to apply the marginal stability criterion, originally proposed by Langer and Muller-Krumbhaar for dendritic growth in a supercooled pure melt, to the problem of dendritic growth in an alloy for dendrites with hemispherical tip. Later, the model was improved (Esaka and Kurz 1985) to describe the growth of dendrites with parabolic tip. The governing equation of this model is given by the following equation which is difficult to implement in a numerical code:

$$V^2 - \frac{D_L P^2 m_L (1-k) C_o}{P^2 G[I_v(P)(1-k)-1]} V + \frac{D_L^2 P^2 G}{\pi^2 \Gamma} = 0. \tag{9.25}$$

To formulate the growth velocity of the dendrite tip for the most general case, it is assumed that the tip is a growth instability driven by the kinetic, solutal, thermal, and capillary undercooling:

$$\Delta T = \Delta T_k + \Delta T_c + \Delta T_T + \Delta T_r. \tag{9.26}$$

A comprehensive treatment of this problem applicable to a wide range of Péclet numbers was given by Boettinger et al. (1988). ΔT_k was formulated through Eq. 2.29, ΔT_T through Eq. 9.20a, and ΔT_r was taken as $2\Gamma/r$. The constitutional undercooling was written as:

$$\Delta T_c = -m_L(V)(C_L - C_o) = m_L C_o \left[1 - \frac{m_L(V)/m_L}{1-(1-k_e)\,\Omega_c} \right]. \tag{9.27}$$

Here, $m_L(V)$ is the velocity dependent liquidus slope given in Eq. 2.35. When all these equations are introduced in Eq. 9.26, the values of dendrite tip velocity and tip radius for a given ΔT can be calculated. A dependency as shown in Fig. 9.19 is obtained. To obtain a unique value for dendrite tip velocity, it was assumed that growth occurs at the limit of stability (marginal stability criterion). In other words, the perturbation will grow with the shortest stable wavelength, i.e., $r_s = \lambda_i$. This implies that if the tip radius of the perturbation is smaller than λ_i, the radius will tend to increase, while if it is larger than λ_i, additional instabilities will form and the radius will decrease. Then, for $\dot{\varepsilon}/\varepsilon = 0$, Eq. 9.9 gives:

$$r^2 = \frac{\Gamma/\sigma^*}{G_L}\xi_c - G_T\xi_T \quad \text{with} \quad \sigma^* = (4\pi^2)^{-1}. \tag{9.28}$$

Note that this equation can be used to derive dendrite growth velocity equations for both slow solidification rates typical for castings ($P < 1$) and rapid solidification rates ($P < 1$). The modified Eqs. 9.26 and 9.28 can now be concomitantly solved numerically to give a unique solution for the dendrite tip velocity.

In their model, Lipton et al. (1984) also started from Eq. 9.26, ignored the kinetic undercooling, and used the formulations for ΔT_c, ΔT_T, and ΔT_r as before:

$$\Delta T = \Delta T_T + \Delta T_c + \Delta T_r$$
$$= \frac{\Delta H_f}{c_p} I(P_T) + m C_o \left[1 - \frac{1}{1-(1-k)I(P_c)} \right] + \frac{2\Gamma}{r}. \tag{9.29}$$

The dendrite tip radius is derived from the marginal stability theory as:

$$r = \frac{\Gamma}{\sigma^*}\left[\frac{\Delta H_f}{c_p} P_T - P_c \frac{mC_o(1-k)}{1-(1-k)I(P_c)} \right]^{-1}. \tag{9.30}$$

From the solution of the diffusion field the tip velocity is formulated as:

$$V = 2\alpha P_T / r. \tag{9.31}$$

Finally, the solutal and thermal Péclet numbers are correlated by:

$$P_c = P_T \alpha / D. \tag{9.32}$$

Here, σ^* is the dendrite tip selection parameter $\approx 1/(4\pi^2)$. This parameter will be discussed later in more detail. Since velocity is introduced through the thermal and solutal supersaturations, either the hemispherical ($P = \Omega$) or the paraboloid ($I(P) = \Omega$) approximations can be used.

The dendrite tip velocity for equiaxed dendrites growing at small undercooling can be calculated from the preceding four equations which are solved by numerical iterations.

A more complete solution was derived by Trivedi and Kurz (1994). They started with Eq. 9.26 and obtained an equation similar to Eq. 9.29, but with a different formulation for the solutal undercooling (second RH term) as follows:

$$\Delta T_c = \frac{k\Delta T_o I(P_c)}{1 - (1 - k)I(P_c)}. \tag{9.33}$$

The capillary term in Eq. 9.29 is generally negligible for metals at low undercooling (the case of shaped castings) but is significant under rapid solidification conditions. Substituting the values for the thermal and constitutional gradients in Eq. 9.28, the general dendrite tip radius selection criterion in undercooled alloys is obtained:

$$Vr^2 \left(\frac{k_V \Delta T_o(V)}{\Gamma D} \right) \left(\frac{1}{1 - (1 - k_V)I(P_c)} \right) \xi_c + Vr^2 \left(\frac{\Delta H_f/c_L}{2\Gamma\alpha_L\beta} \right) \xi_L = \frac{1}{\sigma^*}, \tag{9.34}$$

where $\Delta T_o(V)$ is the velocity dependent solidification interval and $\beta = 0.5[1 + (k_S/k_L)]$. This equation is valid for slow as well as rapid cooling unconstrained growth. The modified Eqs. 9.29 and 9.34 completely describe the dendrite growth problem and can be solved numerically to obtain the growth velocity.

A simple analytical solution can be obtained for a solutal dendrite under the assumption of small Péclet number ($P_c = 0$, $\xi_c = 1$) and using Eq. 9.28 for σ^*:

$$Vr^2 = \frac{4\pi^2\Gamma D_L}{k\,\Delta T_o} = \frac{4\pi^2\Gamma D_L}{m(k-1)C_o}. \tag{9.35}$$

For constrained growth (directional solidification), Trivedi and Kurz proposed the equation:

$$V \left(\frac{k\Delta T_o}{D} \right) \left(\frac{C_L^*}{C_o} \right) \xi_c - G_T = \frac{\Gamma}{\sigma^2 r^2}. \tag{9.36}$$

Purely analytical solutions can be obtained with further simplifying assumptions. Following the derivation proposed by Nastac and Stefanescu (NS) (1993) for unconstrained growth (equiaxed dendrites), ignoring kinetic undercooling and assuming that the effect of surface energy (curvature) is introduced through the limit of stability criterion, only the solutal and thermal undercooling must be considered. The dendrite tip velocity equation is (see inset for derivation):

$$V = \mu_{eq} \Delta T^2 \quad \text{with} \quad \mu_{eq} = \left[2\pi^2 \Gamma \left(\frac{m(k-1)C_L^*}{D_L} + \frac{\Delta H_f}{c\alpha_L} \right) \right]^{-1}. \tag{9.37}$$

The growth coefficient can also be written as:

$$\mu_{eq} = \left(\mu_c^{-1} + \mu_T^{-1} \right)^{-1} \quad \text{with} \quad \mu_c = \frac{D_L}{2\pi^2 \Gamma m (k-1) C_L^*} \quad \text{and} \quad \mu_T = \frac{c\alpha_L}{2\pi^2 \Gamma \Delta H_f}.$$

For steady state, C_L^* will be substituted with C_o/k. The growth coefficient is a constant only for steady state since C_L^* is constant only for steady state. An equation similar to Eq. 9.37 can be obtained if a paraboloid of revolution-shaped tip is assumed. Then, the $I(P) = \Omega$ relationships must be used in the derivation.

This equation describes unconstrained growth (equiaxed dendrites) since it includes thermal undercooling ahead of the interface. For some metallic alloys, it can be calculated that the contribution of thermal undercooling is negligible as compared to that of solutal undercooling (see Application 9.2). Then, the velocity equation can be simplified. Since $\Delta T_T = 0$ and thus $G_T = 0$, the tip velocity is given by Eq. 9.37 with the growth coefficient given by the equation for μ_c.

Derivation of the Nastac-Stefanescu equation.
When using the assumptions used by the model, the total undercooling is:

$$\Delta T = \Delta T_c + \Delta T_T, \tag{9.38}$$

where ΔT_c is given by Eq. 9.21 and ΔT_T is calculated from Eq. 9.20a. If it is further assumed that the tip of the instability is of hemispherical shape and substituting the value of P_c and $P_T = \Omega_T$, it is obtained that:

$$\Delta T = \frac{V r}{2} \left(\frac{m (k-1) C_L^*}{D_L} + \frac{\Delta H_f}{c \alpha_L} \right). \tag{9.39}$$

The liquidus and thermal gradients are:

$$G_L = -m \frac{\partial C_L}{\partial r} = \frac{V}{2D_L} m C_L^*(k-1) = \frac{P_c \Delta T_c}{r \Omega_c} = \frac{\Delta T_c}{r}$$

$$G_T = -\frac{\partial T_T}{\partial r} = \frac{V}{2\alpha_L} \frac{\Delta H_f}{c} = -\frac{P_T \Delta T_T}{r \Omega_T} = -\frac{\Delta T_T}{r}.$$

The first of these two equations was obtained through the use of Eq. 5.5. A factor of 1/2 was introduced to describe the flux at the hemispherical tip. The last equation is valid for a negative gradient, which occurs during equiaxed solidification. Then, assuming that growth occurs at the limit of stability $(r = \lambda_i)$, Eq. 9.13 gives $Vr^2 = 2\Gamma D/(\sigma_c^* \Delta T_o)$. Substituting the expressions for G_L and G_T and using Eq. 9.38, the tip radius is:

$$r = 4\pi^2 \Gamma / \Delta T. \tag{9.40}$$

Substituting this expression for r in Eq. 9.39, the equation for the hemispherical tip velocity, Eq. 9.37, is obtained.

For the case of columnar dendrites (constrained growth), there is no thermal undercooling. Thus, ignoring the kinetic undercooling, the basic undercooling equation simplifies to $\Delta T = \Delta T_c + \Delta T_r$.

Using similar formulations for the undercooling as above, and the hemispherical approximation, Kurz, Giovanola, and Trivedi (KGT) (1986) derived the following equation for columnar dendrites:

$$V = \frac{m(k-1)C_o D_L \Omega_c^2}{\pi^2 \Gamma} = \frac{D_L}{\pi^2 \Gamma m(k-1)C_o} \Delta T_c^2. \tag{9.41}$$

Both the KGT and NS models are only valid for small Péclet numbers since this assumption was used to derive Eq. 9.13, which is adopted for the limit of stability criterion.

Let us now evaluate to what degree the simplifications introduced in the analytical NS model produce deviations from the more accurate semi-analytical Trivedi-Kurz (TK) model. To this effect, a comparison of calculated V–r correlations with the two models and experimental data is shown in Fig. 9.21 (for details of the calculation see Application 9.4). Both models are very close to the experimental data on the linear part of the log-log graph. As the growth velocity decreases and a dendritic-to-cellular transition occurs, the radius increases very fast (it should tend to infinity when planar solidification occurs). Only the TK model for columnar growth (simplified Eq. 9.35) follows well the experimental data in the velocity range smaller than 1 μm/s. However, as typical velocities in casting solidification are in the range of 0.01–0.5 m/s, it is apparent that both models can be used for conventional castings. It should also be noted that in the original paper by Tian and Stefanescu (1992) the Esaka-Kurz model described by Eq. 9.25 produced the closest data to the experimental ones.

A columnar dendrite operates in a constrained environment since the temperature gradient ahead of the interface is always positive. Its growth is driven by the temperature gradient. The interface temperature of the columnar dendrite is between T_L and T_S, and the undercooling is mostly constitutional, ΔT_c. The assumption of negligible thermal undercooling is reasonable for most metallic alloys, so that its growth velocity can be calculated with $V_c = \mu_c \Delta T_c^2$. At steady state the columnar

Fig. 9.21 Measured and calculated tip radii of cells and dendrites in a Fe–3.08 % C–2.01 % Si alloys solidified under a thermal gradient of 5000 K/m at various velocities. (Experimental data are from Tian and Stefanescu 1992)

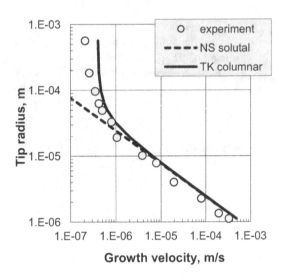

dendrite can grow at a maximum velocity corresponding to the maximum undercooling $\Delta T_c = \Delta T_o$. However, steady state can be reached also at an undercooling smaller than ΔT_o, when the solutal velocity, V_c, is equal to the thermal velocity, V_T, calculated from macrotransport considerations. As long as $V_c < V_T$, the tip velocity is simply V_c. If, on the contrary, $V_c > V_T$, dendritic growth is constrained at V_T. An equiaxed dendrite operates in an unconstrained environment since the temperature gradient ahead of the interface is negative. Its velocity is again mostly dictated by the constitutional undercooling and can be calculated with Eq. 9.37, as long as $\Delta T_{bulk} < \Delta T_o$ (see example in Application 9.5). If $\Delta T_{bulk} > \Delta T_o$ ($T^* < T_S$), then interface equilibrium does not apply anymore, and the partition coefficient is velocity dependent.

Koseki and Flemings (1995) developed a model that includes the combined effects of the undercooled melt and heat extraction through the solid, applicable to chill-casting. It is a hybrid of the models for constrained and unconstrained growth.

The microsolvability theory While, as discussed, the marginal stability criterion gives an excellent agreement with most experimental results, there is no physical reason to accept the marginally stable state over the other stable states.

Kessler and Levine (1986) and Bensimon et al. (1987) have found a unique, self-consistent solution to the steady-state dendrite problem (the interface shape obtained from the thermal and solutal field equations with the boundary conditions that includes the effect of surface energy satisfies the shape preserving condition) taking into account the anisotropy of the interface energy around the dendrite tip. This unique solution, known as the microsolvability condition, gives a unique value for the dendrite tip radius.

The concept that the capillary effect is a singular perturbation which destroys Ivantsov's continuous family of solutions has been confirmed by analytical and numerical studies. While the solvability theory has achieved notable theoretical successes, its quantitative relevance to the interpretation of experimental data has not been established (Barbieri and Langer 1987).

9.3.4 Interface Anisotropy and the Dendrite Tip Selection Parameter σ*

In their analysis, Langer and Müller-Krumbhaar (1978) introduced the dendrite tip selection parameter, σ^*, through the relationship:

$$r = (\delta_c d_o / \sigma_c^*)^{1/2}, \tag{9.42}$$

where $\delta_c = 2D_L/V$ is the diffusion length, $d_o = \Gamma/\Delta T_o$ is the capillary length, and σ_c^* is the parameter for the solutal case (with $D_S \ll D_L$). Substituting these relationships, this equation can be rewritten as:

$$Vr^2 = 2\Gamma D/(\sigma_c^* \Delta T_o). \tag{9.43}$$

Note that this equation is identical with Eq. 9.13 when assuming $\lambda_i = r$ and using the notation $\sigma^* = 1/4\pi^2$. In the marginal stability theory, this parameter is considered to be a constant equal to 0.02533.

Similarly, a parameter for the purely thermal case (for $\alpha_S = \alpha_L$) can be derived, using for example Eqs. 9.20a and 9.40, to obtain:

$$\frac{1}{\sigma_T^*} = Vr^2 \left(\frac{\Delta H_f / c_L}{2\Gamma \alpha_L} \right). \tag{9.44}$$

In order to analyze the dendrite tip selection parameter for undercooled alloys, Trivedi and Kurz (1994) substituted Eqs. 9.43 and 9.44 in Eq. 9.34 to obtain:

$$\left(\frac{2}{\sigma_c^*} \right) \left(\frac{C_L^*}{C_o} \right) \xi_c + \left(\frac{1}{\sigma_T^*} \right) \left(\frac{\xi_L}{\beta} \right) = \frac{1}{\sigma^*}. \tag{9.45}$$

They stated that for an alloy system, σ^* is constant, but σ_c^* and σ_T^* are not.

The microsolvability theory produces equations similar in form as those derived through the marginal stability criterion. The main difference is that σ^* is not a numerical constant any more, but rather a function of the interface energy anisotropy parameter, ε, as follows:

$$\sigma^* = \sigma_0 \varepsilon^{1.75}, \tag{9.46}$$

where σ_o is of the order of unity and the definition of ε is discussed in the following paragraphs.

Crystalline materials are characterized by anisotropic S/L interface energy. For cubic crystals, the variation of the interfacial energy with orientation θ can be expanded about the dendrite tip orientation ($\theta = 0$), which has fourfold symmetry. This expansion up to the first order term is (Trivedi and Kurz 1994) $\gamma(\theta) = \gamma_0(1 + \delta \cos 4\theta)$, where δ is the interface anisotropy parameter.

Using the Gibbs-Thomson equation for anisotropic materials, the capillary undercooling is:

$$\Delta T_r = (\gamma + d^2\gamma/d\theta^2)K/\Delta S_f = \gamma_0(1 - \varepsilon \cos 4\theta)K/\Delta S_f, \qquad (9.47)$$

where $\varepsilon = 15\delta$ and is known as the anisotropy coefficient. The anisotropy in the S/L interface energy strongly affects the tip radius. Higher anisotropy reduces the tip radius (Lu and Liu 2007).

In a recent paper, Glicksman (2012) challenged the current theories of dendritic crystal growth. According to Glicksman, all theories of dendrite formation have relegated the Gibbs-Thomson temperature distribution as an interface boundary condition on the external field. He suggested that the equilibrium temperature distribution is an active interface energy field, albeit an extremely weak one. As such, capillarity acting on a crystal interface, which is well away from its equilibrium configuration, produces energy gradients and fluxes and, consequently, divergences of those weak vector fields.

9.3.5 Effect of Fluid Flow on Dendrite Tip Velocity

The rather intricate picture of dendritic growth presented so far is made even more complicated by buoyancy-driven convection, which is unavoidable during solidification under terrestrial conditions. It was calculated (Miyata 1995) that the growth velocity of the dendrite increases with the forced melt flow at a given undercooling. The effect of melt flow becomes particularly significant in the low velocity regime of both dendrite growth and forced flow.

For the case of uniform fluid velocity, u_∞, directed opposite to the crystal growth direction, Ananth and Gill (1991) have proposed a simplified approximation for the effect of fluid flow on the velocity–undercooling relation. Their solution of the three-dimensional Navier-Stokes equation under the approximation of low Reynolds number, Re, resulted in the following expression for the thermal melt undercooling:

$$\Delta T_t = \frac{u_\infty r \Delta H_f}{2\alpha c_L} \int_1^\infty \exp\left[-\int_1^z f(\eta)d\eta\right] dz, \text{ with the function } f(\eta) \text{ given by:}$$

$$f(\eta) = \frac{r}{2\alpha}(u_\infty + V) + \frac{1}{\eta} - \frac{Vr(\exp[-Re^*/2] - \exp[-Re^*\eta/2])}{Re^*\eta\alpha E_1(Re^{*/2})}$$

$$- \frac{VrE_1(Re^*\eta/2)}{2\alpha E_1(Re^*/2)}.$$

Fig. 9.22 Dendrite tip velocity as a function of undercooling for terrestrial and microgravity conditions. *IDGE* isothermal dendritic growth experiment. (After Glicksman et al. 1995)

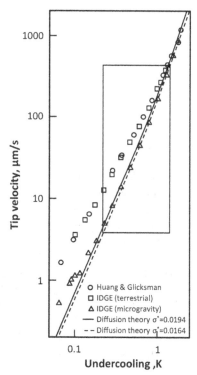

Re* is defined by the relation $Re^* = V r/\nu + Re$, with the Reynolds number $Re = u_\infty r/\nu$. E_1 denotes the exponential integral function and ν is the kinematic viscosity.

Glicksman et al. (1995) measured the dendritic growth velocities and tip radii of succinonitrile in microgravity environment experiments (space shuttle Columbia) during isothermal solidification (Fig. 9.22). It was observed that convective effects under terrestrial conditions increase the growth velocity by a factor of two at lower undercooling (< 0.5 K). In the undercooling range of 0.47–1.7 K, the data remained virtually free of convective effects. A diffusion solution to the dendrite problem was not consistent with the experimental data.

9.3.6 Multicomponent Alloys

Extending these binary phase models to multicomponent alloys is not a trivial endeavor. A more in-depth discussion is presented in Chap. 11. A simple approach is discussed in the following paragraphs.

Equation 9.37 can be extended to dilute multicomponent systems, by writing the growth constant as (Nastac and Stefanescu 1995):

$$\mu = \frac{1}{2\pi^2} \left[\sum_{i=1}^{n} \left[\Gamma_i \left(\frac{m_i \, (k_i - 1) \, C_{Li}^{*2} \rho_i}{D_{Li} \sum_{j=1}^{n} C_{Lj}^{*} \rho_j} + \frac{\Delta H_f}{c \alpha_L} \right) \right] \right]^{-1}, \tag{9.48}$$

where i is the component, and n is the number of components in the alloy that have a significant contribution upon the tip growth velocity.

The solidification parameters of interest for calculation of dendrite tip velocity of multicomponent alloys, i.e., the equilibrium liquidus temperature, T_L, the liquidus slope, m_L, and the partition coefficient, k, can be obtained by using thermodynamic calculation, as described by Boettinger et al. (1988). However, when these parameters are used one needs to solve the mass transport equation for each species.

To avoid excessive computational time, a combined multicomponent/pseudo-binary approach can be used (Nastac et al. 1999). First, thermodynamic calculations are used to obtain the slope of the liquidus line and the partition coefficient for each element at successive temperatures. The algorithm for such calculations is now incorporated in some solidification simulation software (e.g., Pandat, Thermo-Calc). Then, an equivalent slope, $\overline{m}_L$, and partition coefficient, $\overline{k}$, are calculated for each temperature using the equations:

$$\overline{m}_L = \sum_{i=1}^{n} \left(m_L^i C_L^i \right) / \overline{C}_L \quad \text{and} \quad \overline{k} = \sum_{i=1}^{n} \left(m_L^i C_L^i k^i \right) / \sum_{i=1}^{n} \left(m_L^i C_L^i \right), \tag{9.49}$$

where $\overline{C}_L$ is the sum of all the elements in the liquid, and m_L^i, C_L^i, and k_L^i are the slope, liquid composition, and partition coefficient of individual elements, respectively. Then regression equations were fitted through the m_L–T and k_L–T curves to obtain the temperature dependence of m_L and of k_L:

$$\overline{m}_L = a + b \cdot T + c \cdot T^2 \quad \text{and} \quad \overline{k} = a' + b' \cdot T + c' \cdot T^2,$$

where a, b, etc. are known coefficients.

The liquidus temperature can be calculated using the equivalent slope:

$$T_L = T_f + \overline{m}_L \cdot \overline{C}_L \tag{9.50}$$

or from the weighted average of the slope and liquid composition of each element:

$$T_L = T_f + \sum_i m_L^i C_L^i, \tag{9.51}$$

where T_f is the melting temperature of the pure solvent.

Fig. 9.23 Array of instabilities

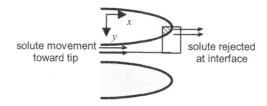

solute movement
toward tip

solute rejected
at interface

9.4 Dendritic Arm Spacing and Coarsening

The dendritic arm spacing (DAS) is a morphological parameter directly related to the mechanical properties of the alloy. In general, the finer is the arm spacing, the higher the mechanical properties. In columnar solidification, both primary and secondary arm spacing (SDAS) can be measured through metallographic analysis. In equiaxed solidification, only the SDAS is an issue. In the characterization of the fineness of the microstructure, the primary arm spacing is replaced by the number of grains.

9.4.1 Primary Arm Spacing

The relationship that allows calculation of the primary spacing, λ_1, is a complicated dependency of solidification velocity and temperature gradient. Two simpler relationships for the primary DAS are discussed here.

A first relationship can be obtained based on *Flemings* array (Fig. 9.23). Ignoring solute diffusion in the x-direction, material balance dictates:

$$D_L \frac{\partial^2 C_L}{\partial y^2} = \frac{\partial C_L}{\partial t} \quad \text{where} \quad \frac{dC_L}{dt} = \frac{dC}{dx}\frac{dx}{dt} = -\frac{V_y G_L}{m}.$$

Upon substitution in the governing equation, we have $(d^2 C_L/dy^2) = -(V/D_L)$ (G/m). Integrating between 0 and dC/dy and between 0 and y yields $dC/dy = -VGy/Dm$. Integrating again between C_o/k and C_L^{max} (in the interdendritic spacing) and between 0 and $\lambda/2$ gives:

$$\lambda^2 = -\frac{8mD}{VG}\left(C_L^{max} - \frac{C_o}{k}\right) \quad \text{or:} \tag{9.52}$$

$$\lambda_1 = ct.(GV)^{-1/2} \quad \text{or, more general,} \quad \lambda_1 = ct.(\dot{T})^{-n}. \tag{9.53}$$

Examples of cooling rates and dendrite arm spacing are given in Table 9.2.

Kurz and Fisher (1989) have derived a more complex relationship. Assume that the dendrites are half of ellipsoids of revolution (Fig. 9.24). Then, the dendritic tip radius is $r = b^2/a$. For a hexagonal arrangement of dendrites, $b = 0.58\lambda_1$. From

Table 9.2 Range of cooling rates in solidification processes. (Cohen and Flemings 1985)

Cooling rate, K/s	Production processes	Dendrite arm spacing, μm
10^{-4}–10^{-2}	Large castings	5000–200
10^{-2}–10^{3}	Small castings, continuous castings, die castings, strip castings, coarse powder atomization	200–5
10^{3}–10^{9}	Fine powder atomization, melt spinning, spray deposition, electron beam or laser surface melting	5–0.05

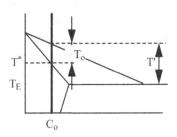 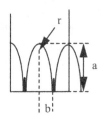

Fig. 9.24 Assumptions for calculation of primary DAS. (After Kurz and Fisher 1986)

the phase diagram it can be approximated that $\alpha = \Delta T^*/G = (T^* - T_E)/G \approx \Delta T_o/G$.

This approximation is increasingly valid, as the composition of the alloy is closer to the eutectic. Then: $(0.58\lambda_1)^2 = r\Delta T_o/G$, or $\lambda_1 = \sqrt{3\Delta T_o r/G}$. Substituting r from Eq. 9.35, we obtain:

$$\lambda_1 = \mu_{\lambda 1} \cdot V^{-1/4} \cdot G^{-1/2}, \quad \text{where} \quad \mu_{\lambda 1} = 4.3(\Delta T_o D_L \Gamma/k)^{1/4} \qquad (9.54)$$

The constant is valid for a single-phase alloy.

Earlier, Hunt (1979) has derived a similar equation for primary spacing different only through the numerical constant, which was 2.83 rather than 4.3. Note that all models introduced here demonstrate that the primary spacing is a function of G and V.

Bouchard and Kirkaldy (1997) tested these equations against experimental data for steady-state solidification of cells (28 alloys) and dendrites (21 alloys including Al–Cu, Pb–Sn, Pb–Au, Pb–Sb, Sn–Pb, Fe–Ni) in binary alloys. The experimental data summarized by the following equations agree reasonable well with theoretical predictions:

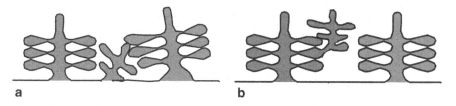

Fig. 9.25 Mechanisms for primary DAS adjustment. **a** Engulfing. **b** Branching

for cells: $\lambda_1 = ct \cdot \dot{T}^{-0.36\pm0.05}$
for dendrites: $\lambda_1 = ct \cdot V^{-0.28\pm0.04} \cdot G^{-0.42\pm0.04}$ and $\lambda_1 = ct \cdot \dot{T}^{-0.3\pm0.03}$.

However, for unsteady-state flow all equations failed to perform adequately.

Once the primary spacing is established, it will remain constant throughout steady-state solidification and during cooling in solid state. If nonsteady-state solidification occurs, the primary spacing will change. Two typical mechanisms for adjustment of primary spacing are presented in Fig. 9.25. Engulfing results in the increase of DAS while branching decreases DAS.

Changes in growth velocity will affect the primary spacing. Examining the effects of deceleration on dendrite array growth on succinonitrile-5.6 % H$_2$O, Liu et al. (2002) found that the primary spacing increases slowly, while the tip curvature increases rapidly after the onset of deceleration. Side arm detachment was very slow under steady-state conditions but accelerated with increasing rates of deceleration. The accelerated detachment of side arms was attributed to higher interdendritic solute concentration, which causes accelerated necking at the connection secondary-primary arm.

9.4.2 Secondary Arm Spacing

In the early understanding of dendrite growth, it was assumed that the secondary dendrite arm spacing is formed at the beginning of solidification. Then, arms thicken and grow as solidification proceeds. Thus, the final arm spacing, λ_f, was thought to be the same as the initial spacing, λ_o.

Later it was realized that as solidification proceeds, only the larger arms grow. The smaller arms remelt (dissolve) and eventually disappear. Consequently, throughout solidification the SDAS increases and $\lambda_f > \lambda_o$. This is the dynamic coarsening of dendrites. The effect of coarsening on the SDAS of a transparent organic material is shown in Fig. 9.3f. It is seen that the secondary DAS increases with the distance behind the tip.

Many mathematical models have been developed for dendrite coarsening based on the concept that dendrite coarsening is diffusion controlled, the diffusive species under consideration being the solvent. Assuming isothermal coarsening, the growth rate of the distance, λ, between two spherical particles must be proportional to the compositional gradient:

$$d\lambda/dt = ct \cdot (\Delta C_L/\lambda). \tag{9.55}$$

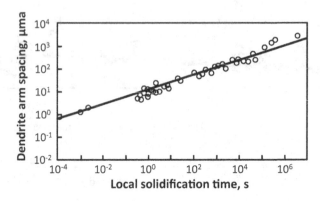

Fig. 9.26 Relation between SDAS and solidification time for Al–4.5 % Cu alloys. (After Flemings et al. 1991)

The liquid temperature and composition in equilibrium with a solid surface depend on the curvature of that surface. Indeed, the curvature undercooling at the tip of the dendrite is $\Delta T_r = 2\Gamma/r$. Since $\Delta T_r = m\,\Delta C_r$, $\Delta C_r = 2\Gamma/(mr) = ct \cdot r^{-1}$. Curvature and local curvature differences must increase approximately proportionally with the inverse of the spacing λ. Thus, $r = ct \cdot \lambda$. It follows that $\Delta C_r = ct \cdot \lambda^{-1}$, and also $\Delta C_L = C_{r1} - C_{r2} = ct \cdot \lambda^{-1}$. Substituting in Eq. 9.55 yields $d\lambda/dt = ct \cdot \lambda^{-2}$. Rearranging and integrating between an initial arm spacing, λ_o, and λ_f and between zero and the final local solidification time (the solidification time interval between the liquidus and the solidus), t_f, gives:

$$\lambda_f^3 - \lambda_o^3 = \mu_o\, t_f. \tag{9.56}$$

Assuming that $\lambda_o \ll \lambda_f = \lambda_2$ results in a final SDAS of:

$$\lambda_2 = (\mu_o\, t_f)^{1/3} \text{ or, more general} \quad \lambda_2 = (\mu_o\, t_f)^n, \tag{9.57}$$

where t_f is the local solidification time, and n is an exponent with values between 0.33 and 0.5. μ_o is a function of alloy concentration and constituents. At long holding times, dendrite arm coarsening produces grains spheroidization. Using the experimental data presented in Fig. 9.26, it can be calculated that for Al–4.5 % Cu alloys the constant in the equation is $\mu_o = 2.7 \times 10^{-15}$ m³/s.

Equation 9.57 can be rewritten as a function of cooling rate (GV) if the local solidification time is written as $t_f = (T_L - T_S)(GV)^{-1} = \Delta T_o(GV)^{-1}$:

$$\lambda_2 = (\mu_o\, \Delta T_o)^{1/3}(GV)^{-1/3}. \tag{9.58}$$

The coarsening constant μ_o in Eq. 9.57 has been derived by several researchers. Some typical formulations and their basic assumptions are given in Table 9.3 for *isothermal coarsening* (the only driving force is solute diffusion and the fraction solid is constant) as well as for *dynamic coarsening* (during solidification the temperature decreases, the fraction solid increases, and an additional driving force, thermal diffusion, must be considered).

Table 9.3 Coarsening constants

Model	Coarsening constant μ_o	Basic assumptions
(Kattamis-Flemings 1965)	$\dfrac{20\,D_L\Gamma}{m_L\,C_L\,(k-1)}\dfrac{ln(C_E/C_o)}{C_E-C_o}$	Isothermal coarsening, of spheres; see inset for derivation
(Ardell 1991)	$\dfrac{v_m^2\gamma_{SL}C_oD_L}{RT}(1-C_o)\dfrac{1-f}{f}$ C_o in fraction	Dynamic coarsening of spheres, diffusion of solute depends on a characteristic distance defined by the main free path
(Voorhees-Glicksman 1984)	$\dfrac{8}{9}\dfrac{v_m^2\gamma_{SL}C_oD_L}{RT}\dfrac{\alpha^3}{1-f^{1/3}}$ α: fct. of f_S given in tabulated form	Solution of dynamic multiparticle diffusion problem; random pattern of precipitates generated by Monte-Carlo simulation
(Kirkwood 1985)	$\dfrac{128\,D_L\Gamma}{m_L\,(k-1)}\dfrac{ln(C_E/C_o)}{C_E-C_o}$	Isothermal coarsening of cylinders with hemispherical tip
(Mortensen 1991)	$\dfrac{27D_L\Gamma}{2m_LC_L(k-1)g_S^{2/3}(1-g_S^{1/3})}$	Dynamic coarsening of spheres
	$\dfrac{27D_L\Gamma}{4m_LC_L(k-1)g_S(1-g_S^{1/2})}$	Dynamic coarsening of cylinders
(Kurz and Fisher 1989)	$\dfrac{166\,D_L\Gamma}{m_L\,(k-1)}\dfrac{ln(C_E/C_o)}{C_E-C_o}$	Isothermal coarsening

The first derivation by Kattamis et al. (1965) assumed spherical particles (see derivation in the inset). Kirkwood (1985) and then Kurz and Fisher (1989) extended the derivation to cylindrical dendrite arms with hemispherical tip. Kirkwood calculations predict the axial remelting of the thinner arm.

Derivation of coarsening constant—Kattamis/Flemings (1965) model.
The basic assumptions of the model are as follows: isotropic growth of two spherical dispersoids of constant total volume, constant concentration gradient, unidirectional diffusion, the radius of the larger dispersoid is much larger than that of the smaller dispersoid ($r_2 \gg r_1$).

Both spheres are at the same undercooling and separated by a distance λ. As discussed before, the difference in the interface concentration of the two spheres is: $\Delta C = C_L^{r_1} - C_L^{r_2} = \frac{\Gamma}{m}(1/r_1 - 1/r_2)$.

The flux of solute from r_2 to r_1 is: $J = -D \Delta C/\lambda = -(D\Gamma/\lambda m)(1/r_1 - 1/r_2)$.

The flux of solvent from r_1 to r_2 is: $J = ((1 - C_L^{r_1}) - (1 - C_S)) \dfrac{dr}{dt} = -C_L^{r_1}(1 - k)\dfrac{dr}{dt}$.

Flux balance gives: $\dfrac{dr}{dt} = \dfrac{D\Gamma}{mC_L^{r_1}(1 - k)\lambda} \left(\dfrac{1}{r_1} - \dfrac{1}{r_2} \right)$.

It is further assumed that the interface concentration of the large particle is equal to the average liquid concentration due to segregation, $C_L^{r_1} \approx C_L$, and that the liquid concentration is a linear function of time, from an initial concentration C_o to the final eutectic concentration C_E:

$$C_L = C_o + (C_E - C_o)\, t/t_f.$$

If it is also assumed that since $r_2 \gg r_1$, $r_2 = $ ct. and $r_1 = f(r)$, the last equation can be integrated between the initial time (0) when the small particle has a radius $r_1 = r_o$, and the time t_f when the particle has vanished:

$$\int_{r_1 = r_o}^{0} \frac{r_1 r_2}{r_2 - r_1} dr = \int_{0}^{t_f} \frac{D\Gamma}{mC_L(1 - k)\lambda} d.$$

The final coarsening time is then: $t_f = \dfrac{mC_L(k-1)(C_E - C_o)}{D\Gamma \ln(C_E/C_o)} \lambda r_2^2 \left(\dfrac{r_o}{r_2} + \ln\left(1 - \dfrac{r_o}{r_2}\right) \right)$.

It is not surprising that the coarsening time depends on the initial size of the particles. To obtain an order of magnitude of the coarsening constant, it can be further assumed that $r_2 = 0.5\lambda$ and $r_o = 0.5r_2$. Then:

$$\lambda_f^3 = \frac{20\, D_L \Gamma}{m_L\, C_L\, (k - 1)} \frac{ln(C_E/C_o)}{C_E - C_o} t_f.$$

The most widely used equation for the calculation of SDAS, including in some commercial software (e.g., ProCAST), is that proposed by Kurz and Fisher (1989), repeated here for convenience:

$$\lambda_2 = \left(\frac{166.4\ \ D_L\ \Gamma}{m_L\ (k - 1)} \frac{ln\,(C_E/C_o)}{C_E - C_o} t_f \right)^{\frac{1}{3}}. \tag{9.59}$$

Note that the coarsening constant λ_o in the Kurz and Fisher equation above is only slightly different than that of Kirkwood in Table 9.3 in the numerical constant (166.4 versus 128).

Other models consider the dendrite arms to be tear shaped. If a tear-shaped arm is surrounded by two cylindrical arms (Chernov 1956), material is transported from the base, where the radius of curvature r_1 is small, to the tip, where the radius r_2 is

Fig. 9.27 Material transport for one tear-shaped arm surrounded by cylindrical arms (**a**) or two tear-shaped arms surrounded by cylindrical arms (**b**). (After Mendoza et al. 2003; with permission from Springer Science and Business Media)

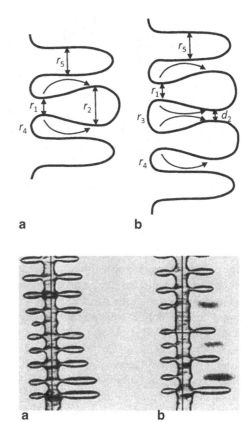

Fig. 9.28 Separation of dendrite arms in NH_4Cl (Papapetrou 1935). After separation the detached arms move out of focus. **a** After 18 min. **b** After 32 min

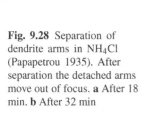

large (Fig. 9.27a). Eventually r_2 becomes zero, and the arm detaches from the stem of the dendrite as demonstrated experimentally by Papapetrou (1935; see Fig. 9.28).

If case b in Fig. 9.27 is considered (Young and Kirkwood 1975), material transported away from the base accumulates at the tip and the distance d_2 decreases until coalescence occurs. Experimental evidence of this mechanism is shown in Fig. 9.29.

Rappaz and Boettinger (1990) extended the applicability of Eq. 9.59 to multicomponent alloys where the diffusion coefficients in the liquid of the various solute elements are different:

$$\lambda_2 = \frac{164\ \Gamma}{\sum_{j=1}^{n} m_j \left(k_j - 1\right) \left(C_{E,j} - C_{o,j}\right)/D_j}\ ln\left[\frac{\sum_{j=1}^{n} m_j \left(k_j - 1\right) C_{E,j}/D_j}{\sum_{j=1}^{n} m_j \left(k_j - 1\right) C_{o,j}/D_j}\right], \tag{9.60}$$

where the sums extends over all the n elements of the alloy. If the values of all D_j are equal to D_L, the equation simplifies to:

$$\lambda_2 = \frac{164\ \Gamma D}{\sum_{j=1}^{n} m_j \left(k_j - 1\right) \left(C_{E,j} - C_{o,j}\right)}\ ln\left[\frac{\sum_{j=1}^{n} m_j \left(k_j - 1\right) C_{E,j}}{\sum_{j=1}^{n} m_j \left(k_j - 1\right) C_{o,j}}\right]. \tag{9.61}$$

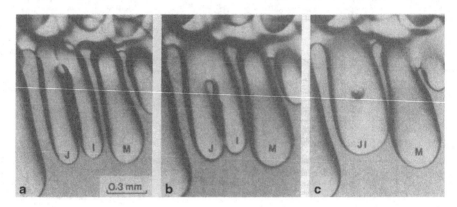

Fig. 9.29 Coalescence of arms *J* and *I* in succinonitrile. (Huang and Glicksman 1981; with permission from Elsevier)

The compositional factor that most affects SDAS is $Q = \sum_{j=1}^{n} m_j c_j (k_j - 1)$. The value Q_o for the initial alloy concentration correlates with large changes in grain refinement but has little effect on SDAS. On the contrary, the value Q_f for the liquid composition at which SDAS is established correlates strongly with SDAS (Easton et al. 2010).

Kundin et al. (2014) extended the Rappaz/Boettinger model by taking into account the cross dependencies between the components in multicomponent diffusion and the case of slow diffusion in the solid phase. They used the model to calibrate their phase field model.

In the preceding analysis a number of effects, such as convection, finite volume fraction, dendritic arrangement, direction of growth, and growth fluctuations were ignored. Theoretical models that included the effect of convection on solute redistribution predict an increased SDAS and *n* values of up to 0.5 rather than 1/3 (Ratke and Thieringer 1985; Diepers et al. 1999). The increase in *n* was confirmed with experiments where forced convection was induced by magnetic fields (Steinbach and Ratke 2005).

Analyzing the effect of the finite volume fraction on mass transport between the dendrite arms, Ratke (2009) concluded that the finite volume fraction significantly alters the transport fluxes producing a much higher coarsening rate.

Because of solute pileup followed by settling (during downward solidification) or accumulation between dendrites (during upward solidification), fluctuations in growth rate will occur. Conducting upward and downward experiments in Al–30 % Cu and Al–20 % Cu alloys, Ruvalcaba et al. (2009) demonstrated that dendrite arrangement, direction of growth, and growth fluctuation affect solute distribution and thus solute gradients as illustrated in Fig. 9.30. Growth of high-order branches increases SDAS due to an increase in constitutional undercooling because of higher solute gradients. This mechanism can affect greatly the coarsening exponent *n* by increasing it up to *n* = 0.73 for the reported experiments. In regions with small solute gradients and lower undercooling, a reduced growth of high-order branches is reflected in lower *n* values that are within those reported in literature.

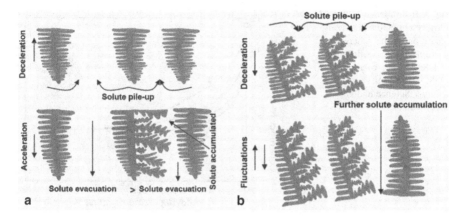

Fig. 9.30 Influence of dendrite growth direction and arrangement with the solute-rich liquid flow and its effect on morphology development. **a** Downward growth. **b** Upward growth. (Ruvalcaba et al. 2009; with permission from Maney)

Experimental data on SDAS have also been reported to fit an equation similar to Eq. 9.53 proposed for primary spacing. Analysis of 60 experimental data on two alloys (Al–Cu, Pb–Au) by Bouchard and Kirkaldy (1997) produced the following dependency on the cooling rate $\dot{T}$:

$$\lambda_2 = ct \cdot (\dot{T})^{-0.34 \pm 0.02}. \tag{9.62}$$

For alloys solidifying with equiaxed structure, it is not possible to define a primary arm spacing. To evaluate the length scale of the microstructure, the average grain size (average diameter of grains on the metallographic sample) or volumetric grain density (number of grains per unit volume) is used. These numbers are primarily functions of the nucleation potential of the melt. SDAS can also be used to evaluate the fineness of equiaxed structures. However, as noted above, it represents thermal conditions during solidification, not nucleation conditions. Thus, it is possible to have a coarse grained casting with fine secondary arm spacing.

9.4.3 Dendrite Coherency

In the early stages of the equiaxed dendritic solidification, the dendrites are not in contact with one another and can move freely through the melt, which behaves like a slurry. Its viscosity increases with the fraction solid of dendrites. Later in the solidification process, when the dendrites touch and impinge on one another, the system acquires the characteristics of a solid. The transition is associated with an increase in shear strength, and mass feeding comes to an end. This transition point is called *dendrite coherency*. At this time, a number of undesired consequences, such as macrosegregation, macro- and microshrinkage, and hot tearing, begin to occur.

Chai et al. (1995) studied the onset of dendrite coherency on four different Al alloys using a viscometer. The evolution of fraction solid was calculated from the

Fig. 9.31 Influence of the
alloy system and solute concentration on the coherency
fraction solid at cooling rates
of 1 K/s. (After Chai et al.
1995; with permission from
Springer Science and Business Media)

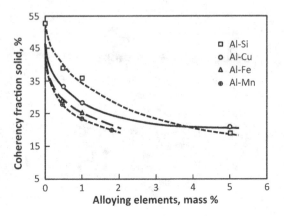

cooling curves. From their results (Fig. 9.31), it is seen that coherency fraction solid, f_S^{coh}, varied between about 18 and 56 %. However, as only columnar structure was obtained for pure aluminum, the upper limit for equiaxed structures is at about 38 %.

The same authors also derived the following equation for the coherency grain radius:

$$R^{coh} = ct.\Delta D_L [\pi^2 \Gamma \sqrt{dT/dt}\, m C_o (k-1) t^{coh}]^{-1}. \tag{9.63}$$

This equation suggests that the coherency radius (the final grain size) is inversely proportional with the cooling rate, the solute content and the coherency time.

9.5 The Columnar-to-Equiaxed Transition

In many applications, either a columnar or an equiaxed structure is desired for the casting. If, because of lack of adequate process control, a sudden columnar-to-equiaxed transition (CET) occurs in these castings, they will be rejected having an unacceptable structure. Therefore, it is important to understand the conditions under which a CET can occur in a given casting.

An example of such a transition in an ingot is shown in Fig. 9.32. This is not a schematic drawing but a modeled microstructure, which correctly describes the real microstructure. The details of the model will be discussed in Sect. 18.4.1. The results of the calculations are very realistic. Three different structural regions are shown on the middle ingot: a chill zone, made of small equiaxed grains resulting from rapid cooling against the mold wall; a columnar zone; and an equiaxed zone toward the middle of the casting. The figure also indicates that as the undercooling increases the structure in the bulk of the ingot changes from fully columnar to mixed columnar/equiaxed. Thus, a CET occurs as the undercooling is increased. A typical structure showing the CET in an Al–5 % Cu ingot is presented in Fig. 9.33. The subject of CET seems to have exercised a special fascination on researchers if

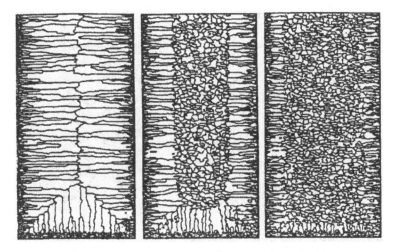

Fig. 9.32 Typical structural regions as a function of undercooling. (Zhu and Smith 1992; with permission from Elsevier)

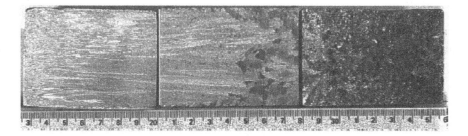

Fig. 9.33 CET occurring in an Al–5 wt.% Cu ingot solidified against a chill placed at its base, *left* on the picture (Guo and Stefanescu 1992). Solidification is from *left to right*. The CET occurred when the temperature gradient in the melt ahead of the interface decreased in the range of 113–234 K/m. In the equiaxed zone, the number of grains was 5×10^6 m^{-3}. The average solidification velocity at the CET was measured to be 3.5×10^{-4} m/s. (With permission from the American Foundry Soc.)

the vast amount of often conflicting experimental data is to be considered (see the review papers by Flood and Hunt 1990 and Spittle 2006).

Several mechanisms have been suggested as contributors to CET. A first hypothesis was based on the change in constitutional undercooling during solidification (Winegard and Chalmers 1954). At the beginning of solidification, the temperature gradient in the liquid is rather high, and constitutional undercooling is limited (Fig. 9.34a). As solidification continues, the mold is heated. The temperature gradient in the liquid decreases, and the constitutional undercooling may reach the middle of the casting (Fig. 9.34b). If nucleation of equiaxed grains occurs, they will have favorable conditions and will grow ahead of the columnar interface (see also Fig. 9.1). The CET then occurs by the impingement of the columnar dendritic

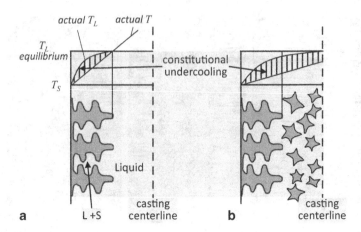

Fig. 9.34 Occurrence of CET because of increased constitutional undercooling resulting from lower temperature gradient (actual T on the figure). **a** No constitutional undercooling in the middle of the casting—only columnar growth is possible. **b** Constitutional undercooling in the middle of the casting—if nucleation occurs, equiaxed growth is possible

interface upon the dendritic skeleton of floating equiaxed crystals after the equiaxed dendrites have reached coherency.

While the role of constitutional undercooling in the CET is not under dispute, it does not seem to be the only mechanism responsible for it. Chalmers (1963) has shown that when the center of a casting is isolated with a cylinder, fewer and coarser grains grow than in the absence of isolation. This is in spite of the fact that the center is constitutionally undercooled in both cases. A similar experiment is shown in Fig. 3.11. Chalmers then proposed the *big bang mechanism*, which postulates that equiaxed grains are generated from the nuclei formed during pouring by the initial chilling action of the mold. The grains are then carried into the bulk liquid. If they survive until the superheating is removed, a CET occurs.

Jackson et al. (1965) noticed that increased convection during solidification of organic alloys produced a large number of nuclei in the liquid. They postulated the *dendrite detachment mechanism* which stipulates that dendrite arms remelt because of recalescence, detach from the dendrite stem, and then float into the center of the casting where they serve as nuclei for equiaxed grains. Alternatively, local remelting can be caused by solute pileup in the mushy zone (e.g., Yasuda et al. 2004; Mathiesen et al. 2006). Another argument for the dendrite detachment mechanism is that the mechanical strength of the dendrite is negligible close to its melting point, and thus convection currents can simply break the dendrite (O'Hara and Tiller 1967). Furthermore, melt flow during solidification is expected to affect fragmentation, as it generates temperature oscillations and redistribution of the local solute concentration.

Mahapatra and Weinberg (1987) argued that the columnar ndendrite tips become unstable when the temperature gradient ahead of the tips falls below a critical

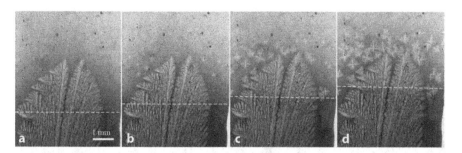

Fig. 9.35 CET induced by a sharp pulling rate jump from 1.5 to 15 μm/s during solidification of an Al–3.5 wt.% Ni; $G = 20$ K/cm: $t = t_o + 42$ s (**a**), $t = t_o + 63$ s (**b**), $t = t_o + 87$ s (**c**), and $t = t_o + 111$ s (**d**). The solid mainly constituted of Al appears in *grey* while the Ni enriched liquid is *dark*. The *dash line* underlines the eutectic front position. (Reinhart et al. 2005; with permission of Elsevier)

value. The ensuing solute accumulation at the dendrite tips constrains the growth of the columnar front, which in turn causes cooling of the liquid creating favorable conditions for equiaxed growth.

Recently, the X-ray radioscopic technique became the method of choice for real time and in situ observations of the solidification front with a spatial resolution of a few micrometers (e.g., Curreri and Kaukler 1996). As apparent on synchrotron X-ray radiography pictures in Fig. 9.35, an increase in the solidification velocity will trigger the CET, since when enough grains nucleate and grow, they may block the columnar growth. As the size of equiaxed grains increased, the velocity of the columnar front gradually decreases toward zero. The fact that columnar as well as the equiaxed dendrites close to the columnar dendrites are blocked before the grains are touching each other (Fig. 9.35c, 9.35d) suggests that the blocking mechanism is mostly solutal (Reinhart et al. 2005).

Using synchrotron radiation on in situ directionally solidified Al–Cu alloys, Yasuda et al. (2009) reported that nucleation and fragmentation of dendrite arms were often observed in the 15 and 10 % Cu alloys (Fig. 9.36), but rarely observed in the 5 % Cu alloys. This may be because of the lower degree of constitutional undercooling in the dilute alloys. In addition, as the buoyancy forces due to the inhomogeneity of melt density in the dilute alloys are relatively small, compared with that in the rich alloys, the convection in the mushy region of the dilute alloys is expected to be too low to contribute to fragmentation. As a result of fragmentation and nucleation, CET was observed in the alloy with 15 and 10 % Cu, but not in that with 5 % Cu (Fig. 9.37). Unrelated to CET, notice evolution of gas bubbles deep inside the dendrites, toward the roots of the secondary arms.

A first analytical model for CET was developed by Tiller (1962). The main assumptions included: (i) nucleation of equiaxed grains in the constitutionally undercooled region, (ii) CET occurs at a critical nucleation frequency corresponding to a critical maximum undercooling value for a particular alloy, and (iii) this critical condition is determined by G/R alone. The expression derived for the ratio

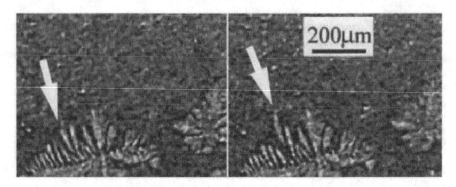

Fig. 9.36 Example of a fragmentation of dendrite arm in Al–15 % Cu alloy (pulling rate 10 mm s^{-1}). (Yasuda et al. 2009; with permission from Maney)

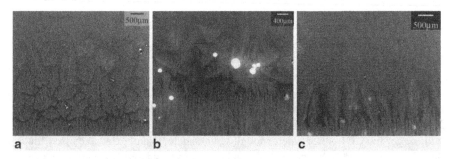

Fig. 9.37 X-ray transmission images from Al–Cu alloy (pulling rate 10 mm s^{-1}). **a** Al–15 % Cu alloy after 173 s. **b** Al–10 % Cu alloy after 220 s. **c** Al–5 % Cu alloy after 220 s. (Yasuda et al. 2009; with permission from Maney)

of equiaxed zone length to columnar zone length in a finite ingot predicts that the columnar zone will decrease with higher superheating, solidification range, the number of nuclei, and liquid stirring.

The most widely analytical model used is that by Hunt (1984). His one-dimensional (1D) analytical model is based on the assumption that equiaxed grains are formed by heterogeneous nucleation and do not move with the liquid and that steady state is possible for a fully columnar, columnar + equiaxed, or fully equiaxed growth. It was further assumed that a fully equiaxed structure results when the fraction of equiaxed grains is higher than a critical fraction of solid, $f_S^e > f_S^{cr} = 0.49$ and that a fully columnar structure is produced when $f_S^c < 10^{-2}$ $f_S^{cr} = 0.0049$. Using a model of hemispherical dendrite growth and assuming small thermal undercooling, the following criteria were derived:

- Fully equiaxed growth occurs when:

$$G_T < 0.49 \left(\frac{N}{f_S^{cr}} \right)^{1/3} \left[1 - \frac{(\Delta T_N)^3}{(\Delta T_c)^3} \right] \Delta T_c \qquad (9.64a)$$

- Fully columnar growth develops when:

$$G_T > 0.49 \left(\frac{100 \cdot N}{f_S^{cr}} \right)^{1/3} \left[1 - \frac{(\Delta T_N)^3}{(\Delta T_c)^3} \right] \Delta T_c, \qquad (9.64b)$$

where N is the volumetric nuclei density, ΔT_N is the undercooling required for heterogeneous nucleation, and ΔT_c is the undercooling at the columnar front calculated as:

$$\Delta T_c = [-8\Gamma m_L (1 - k) C_o V / D]^{1/2}. \qquad (9.65)$$

The selection of the thresholds in the derivation of the above relationships is debatable. It seems difficult to accept that columnar grains can grow beyond the point when dendrite coherency is established, f_S^{coh}. Accordingly, a more reasonable upper limit is $f_S^{cr} = f_S^{coh}$. Typically $f_S^{coh} = 0.2$–0.4. Also, mixed equiaxed-columnar structures are seldom observed in castings. In most cases an abrupt CET is seen. Indeed, microgravity work performed by Dupouy et al. (1998) on Al–4 % Cu alloys demonstrated that while a smooth (mixed structure) CET is obtained in microgravity (no thermosolutal convection), an abrupt CET is seen on the same sample solidified under terrestrial conditions. Thus, it is reasonable to assume that the CET occurs simply when coherency is reached.

Another weak assumption is that of stationary equiaxed grains. Indeed, because of the thermosolutal and shrinkage convection, the equiaxed grains will move with the liquid, unless coherency is reached. As discussed previously, one of the main reasons for the CET is the presence of thermosolutal convection.

Based on the preceding discussion another simple analytical model can be construed. Consider a volume element of length l that extends from some arbitrary point in the columnar region to a region where the temperature is equal to the nucleation temperature ΔT_N that is very close to the liquidus temperature (Fig. 9.38). Consequently, the grains moving away out of the volume element in the bulk liquid in the x-direction will not survive, and no grains are advected from the bulk liquid into the element in the x-direction. It is further assumed that the net contribution of the flow in the y-direction to the number of grains is zero. Within the volume element, the CET occurs if the equiaxed grains can reach f_S^{coh} before the columnar front traverses the element. Thus, condition for the CET is:

$$t_c \geq t_e^{coh}, \qquad (9.66)$$

where t_c is the time required for the columnar front to move across the volume element, and t_e^{coh} is the time required for the equiaxed grains to reach coherency.

Fig. 9.38 Volume element for calculation of CET

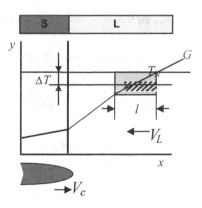

Then: $t_c = l/V_c = 2\Delta T/(GV_c) = 2\Delta T/(G\mu_c\Delta T^2)$, where G and ΔT are the average thermal gradient and undercooling in the volume element, respectively, V_c is the growth velocity of the columnar dendrites, and μ_c is the growth coefficient of the columnar dendrites.

Similarly, since when assuming spherical grains, the fraction of solid in the volume element can be calculated as $f_S = (4/3)\pi\,\bar{r}^3\bar{N}$, where $\bar{r}$ and $\bar{N}$ are the average grain radius and the average volumetric grain density, respectively:

$$t_e^{coh} = \frac{\bar{r}^{coh}}{V_e} = \left(\frac{3}{4\pi}\frac{f_S^{coh}}{\bar{N}}\right)^{1/3}(\mu_e\Delta T^2)^{-1},$$

where V_e is the growth velocity of the equiaxed dendrites and μ_e is the growth coefficient of the equiaxed dendrites.

Introducing the last two equations in the CET criterion of Eq. 9.66, the CET will occur when:

$$G_T \leq 3.22\left(\frac{\bar{N}}{f_S^{coh}}\right)^{1/3}\frac{\mu_e}{\mu_c}\Delta T. \tag{9.67}$$

Assuming that the outer limit of the element is at a temperature below the nucleation temperature, the average volumetric grain density can be written as the sum between the active heterogeneous nuclei, N, and the grains entering the volume element because of fluid flow into the volume element, $\bar{N} = N + (V_L/V)N$, where V_L is the flow velocity in the x-direction and V is the solidification velocity. Substituting in Eq. 9.67 the final CET condition becomes:

$$G_T \leq 3.22\left[\frac{\bar{N}}{f_S^{coh}}\left(1+\frac{V_L}{V}\right)\right]^{1/3}\frac{\mu_e}{\mu_c}\Delta T. \tag{9.68}$$

This equation suggests that the probability of formation of an equiaxed structure increases as the nucleation potential, liquid convection, and the undercooling increase and as the coherency solid fraction decreases. Note that for the case when convection is ignored ($V_L = 0$), this equation is very similar to those derived by Hunt.

9.6 Applications

Application 9.1
Calculate the critical thermal gradient for the planar to cellular transition for a 0.6% carbon steel cast in a cube having the volume of 0.001m^3, poured into a silica sand mold.

Answer: The database required for this calculation is taken from Appendix B as follows: $C_o = 0.6$, m $= -65$, $k = 0.35$, $D_L = 2 \cdot 10^{-8}$ m/s. If the solidification velocity calculated in Application 5.1 is used ($2.7 \cdot 10^{-5}$ m/s), from Eq. 9.3a it is calculated that $G_L = 9.78 \cdot 10^5$ K/m. This is a very large number, much higher than the thermal gradient typically existing in the casting. Compare for example with the gradient $G_T = 1.86 \cdot 10^3$ K/m calculated in Application **6.3**. Thus, clearly $G_T < G_L$.

The constitutional undercooling criterion in Eq. 9.3 ignores the effect of interfacial energy which should inhibit the formation of perturbations since an additional energy is required if the interface area is increased. In addition, the influence of the temperature gradient in the solid has also been ignored. A more complete analysis will now be introduced.

Application 9.2
Compare the solutal and thermal undercooling for an Al–4.5 % Cu alloy and for an Fe–0.09 % C alloy.

Answer Since the expressions for undercooling include both velocity and tip radius, it is not possible to calculate the undercooling without additional data. However, a comparison can be made by calculating the ratio $\Delta T / Vr$ for the two cases. From Eqs. 9.21 and 9.20a and using data in Appendix B:

$\Delta T_c / Vr = m(k - 1)C_o / 2kD = 1.03 \times 10^{10}$ K·s·m^{-2} for the Al–Cu alloy, and $= 8.9 \times 10^8$ K·s·m^{-2} for the Fe–C alloy

$\Delta T_T / Vr = \Delta H_f / 2\alpha c = 4.94 \times 10^6$ K·s·m^{-2} for the Al–Cu alloy, and $= 2.53 \times 10^7$ K·s·m^{-2} for the Fe–C alloy

It is obvious that the thermal undercooling is very small as compared with the solutal undercooling for the Al–Cu alloy, but within an order of magnitude for the Fe–C alloy. Thus, the thermal undercooling cannot always be neglected.

Application 9.3
Compare the tip radius–growth velocity correlation for solutal dendrites and solutal-thermal dendrites for a Fe–0.09 % C alloy using the NS (1974) and the TK (1994) models. Most of the data required in this calculation can be obtained from Appendix B: $T_L = 1531$ °C, $\Delta T_o = 36$ °C, m $= -81$, $k = 0.17$, $\Delta H_f = 2.72 \times 10^5$, $c_S = 880$, $D_L = 2 \times 10^{-8}$, $\alpha_L = 6.1 \times 10^{-6}$, $\Gamma = 1.9 \times 10^{-7}$. Assume a thermal gradient $G_T = 5000$. All units are in SI.

Answer Combining Eqs. 9.38 and 9.39 and assuming steady state, we obtain the V–r correlation for the NS model as follows:

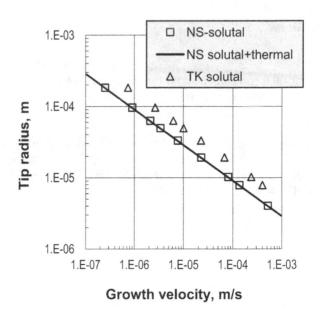

Fig. 9.39 Calculated dendrite tip radii–growth velocity correlation for solutal and solutal-thermal dendrites for a Fe-0.09 % alloy

$$V r^2 = 8\pi^2 \Gamma \left(\frac{m(k-1)C_o}{kD_L} + \frac{\Delta H_f}{c\alpha_L} \right)^{-1}. \tag{9.69}$$

For the TK model, assuming steady state and low Péclet number ($<<1$) we can obtain a similar equation from Eq. 9.34:

$$V r^2 = 4\pi^2 \Gamma \left(\frac{m(k-1)C_o}{D_L} + \frac{\Delta H_f}{2c_L\alpha_L} \right)^{-1}. \tag{9.70}$$

For the solutal dendrite, that is a dendrite whose growth is controlled solely by the solutal field, only the first term in the parenthesis is used. Using the given data base, we obtain the graph in Fig. 9.39. There is little difference between the solutal and the solutal + thermal NS in the range of velocities used in the calculation. The solutal NS and TK are different by a factor of 1.41 resulting from the difference in the numerical constants in the two equations.

Application 9.4

Compare the dendrite tip radius - growth velocity relationship calculated with the Trivedi-Kurz (1994) and Nastac-Stefanescu (1993) models, with the experimental data on the Fe-3.08% C-2.01% Si - 0.104% Mn - 0.016% S - 0.029% P alloys obtained by Tian and Stefanescu (1992). The experimental thermal gradient was $G_T = 5000$K/m. The experimentally measured dendrite tip radii for various velocities were as follows:

V	2.1e-7	2.6e-7	3.5e-7	5e-7	8e-7	1.05e-6	3.83e-6	7.83e-6	1.93e-5	7.83e-5	3.27e-4
r	5.66e-4	1.83e-4	9.64e-5	4.96e-5	3.31e-5	1.92e-5	1.02e-5	7.87e-6	4.04e-6	2.28e-6	1.12e-6

For complete data see Table 1 in Tian and Stefanescu (1992). Most of the other data can be obtained from Appendix B: $T_L = 1224\,°C$, $\Delta T_o = 70\,°C$, $T_E = 1154\,°C$, $m = -130$, $k = 0.49$, $\Delta H_f = 2.610^5$, $k_L = 27.5$, $c_S = 880$, $\rho_S = 7000$, $D_L = 2 \times 10^{-8}$, $\alpha_L = 6.1 \times 10^{-6}$, $\Gamma = 1.910^{-7}$.

Answer For the solutal equiaxed dendrites only the solutal parts of Eqs. 9.69 (the NS model) and 9.70 (the TK model) will be used. For columnar dendrite, Eq. 9.36 (the TK model) will be used. Again, assuming steady state ($C_L^* = C_o/k$) and low Péclet number ($\xi_c = 1$), the TK columnar dendrite growth velocity becomes:

$$V = \left(\frac{4\pi^2\Gamma}{r^2} + G_T \right) \frac{D_L}{m(k-1)C_o}. \tag{9.71}$$

Since the alloy is a multicomponent alloy, the average composition C_o must be expressed as a carbon equivalent to reduce the multicomponent alloy to a pseudo-binary one. The following relationship is used: $C_o = \%C + 0.31\%Si + 0.33\%P - 0.27\%Mn + 0.4\%S = 3.72$.

The predicted and experimental results are plotted in Fig. 9.40. Note that while both the NS and TK solutal models fail at growth velocities smaller than $\sim 1\ \mu m/s$, which is in the range of cellular growth, they describe growth reasonable well within the range of velocities typical for dendrites and castings. The TK columnar model catches the slope change occurring at low velocities. It is apparent that the thermal gradient is important only in the cellular solidification range.

Application 9.5
Calculate the temperature and solid fraction evolution during the solidification of an equiaxed dendrite of an Fe–0.6 % C alloy that has a volumetric grain density of 1 grain/mm^3. Assume that the alloy is cooled at constant heat extraction rate of $Q = 3 \times 10^8\ J\ m^{-3}\ s^{-1}$ and an initial temperature of $1520\,°C$.

Answer The governing heat transport equation is: $\dot{Q} = \rho \Delta H_f d f_S/dt - \rho c dT/dt$. Rearranging and discretizing for time stepping:

$$T^{new} = T^{old} - \frac{\dot{Q}}{\rho c}\Delta t + \frac{\Delta H_f}{c}\Delta f_S^{new}. \tag{a}$$

Note that this equation is independent of volume. Assuming a spherical equiaxed dendrite, $f_S = (4/3)\,\pi\,r_S^3 N$, and $df_S/dt = 4\pi r_S^2 N dr_S/dt = 4\pi r_S^2 N V_s$. The number of nuclei, N, is equal to one. The time-discretized equation is:

$$\Delta f_S^{new} = 4\pi(r_S^{new})^2 V_S^{new} \Delta t. \tag{b}$$

Further, the grain size is:

$$r_S^{new} = r_S^{old} + V_S^{new}\Delta t. \tag{c}$$

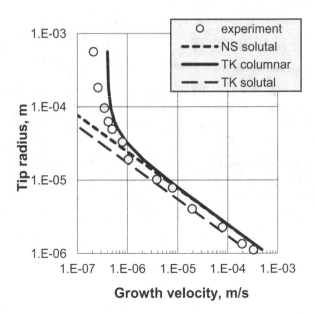

Fig. 9.40 Comparison of calculation of cells and dendrite tip radius–velocity correlation with experimental data (from Tian and Stefanescu 1992) for a cast iron of 3.72 carbon equivalent

The solidification velocity is $V_S = \mu\Delta T^2$. The growth coefficient can be calculated with Eq. 9.37. Assuming a solutal dendrite, the discretized equation for the solidification velocity is:

$$V_S^{new} = \frac{D_L}{2\pi^2\Gamma m(k-1)\langle C_L\rangle^{new}}(\Delta T^{new})^2, \quad (d)$$

where $\langle C_L\rangle$ is the average liquid composition.

To calculate the average liquid composition needed in this equation, a diffusion model must be used. We will compare the Scheil and equilibrium diffusion models. The time-discretized equations for the two models are:

Scheil: $C_L^{new} = C_o\,(1-f_S^{old})^{k-1}$ Equilibrium: $C_L^{new} = \dfrac{C_o}{1-(1-k)f_S^{old}}.$ (e)

The evolution of the fraction solid is: $f_S^{new} = f_S^{old} + \Delta f_S^{new}.$ (f)

Finally, the undercooling is calculated as:

$$\Delta T = \Delta T_c + \Delta T_T = m(C_o - C_L) + T^* - T_{bulk} = T_f + mC_o - T_{bulk},$$

where T_{bulk} is the average temperature in the volume element (the macrotemperature). C_o is the average liquid composition, $\langle C_L\rangle$. In discretized form this is:

$$\Delta T^{new} = T_f + m\langle C_L\rangle^{new} - T^{old}. \quad (g)$$

Table 9.4 Program implementation on the excel spreadsheet

Time	$\langle C_L \rangle$	ΔT	v_S	r_S	Δf_S	f_S	T
Eq.	(e)	(g)	(d)	(c)	(b)	(f)	(a)
0				1.00E-07			1520
0.01	0.600	00	0.00E+00	1.00E-07	0.00E+00	0.00	1519.5
0.59	0.6	0.394	2.62E-05	3.62E-07	4.31E-19	4.31E-19	1489.1
1.47	0.602	44.7	3.35E-01	1.07E-01	4.82E-04	0.01	1444.8
4.63	0.891	19.4	4.29E-02	4.90E-01	1.29E-03	0.50	1447.2
9.17	1.746	17.6	1.79E-02	6.19E-01	8.62E-04	1.00	1380.5

Fig. 9.41 Evolution of temperature and solid fraction during the solidification of a condensed dendrite

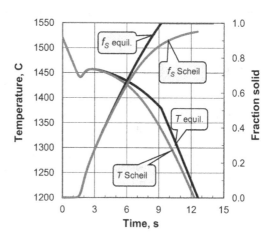

Since above the liquidus temperature there is no undercooling, an IF statement must be included which allows this equation to become effective only at $T^{new} < T_L$. These equations are then implemented in the Excel spreadsheet, for example, as shown in Table 9.4 for the case of equilibrium. An initial radius at time zero is assumed. The initial temperature is 1520 °C.

The calculated results are plotted in Fig. 9.41. It is seen that significant differences exist in both temperature and solid fraction evolution as a function of the chosen diffusion model. When the *Scheil* model is used, longer time is needed for completion of solidification, and a lower solidus temperature is reached (1187 °C as compared to 1381 °C for equilibrium). Dendritic solidification ends when the liquid composition becomes $C_L = C_E = 4.26$.

Application 9.6

Calculate the critical gradient for CET in the Al–5 % Cu ingot presented in Fig. 9.33 ($N = 5 \times 10^6 \text{m}^{-3}$ and $V = 3.5 \times 10^{-4} \text{m/s}$). Assume $f_S^{coh} = 0.3$. The other data required for calculation are given in Appendix B.

Answer Let us use the Hunt model first. From Eq. 9.65 it is calculated that $\Delta T_c = 1.93$ K. Then, assuming that $\Delta T_N \gg \Delta T_c$ (which is not necessarily true) and using Eq. 9.64a, it is calculated that the critical gradient for CET is $G_T = 205$ K/m for the $f_S^{cr} = 0.49$ postulated but Hunt. This is within the range determined experimentally, which was 113 to 234 K/m.

Let us now use the model described by Eq. 9.68. The steady-state growth coefficient for columnar growth can be calculated from Eq. 9.41 as $\mu_c = D_L(\pi^2 \, \Gamma \, m(k-1) \, C_o)^{-1}$. We obtain $\mu_c = 7.64 \times 10^{-5}$. Then, from the same equation, assuming that all undercooling is constitutional, the undercooling is $\Delta T = \sqrt{V/\mu} = 2.14$ K. The growth coefficient for the equiaxed grains can be calculated with Eq. 9.37. When the thermal undercooling is ignored, we obtain $\mu_e = 5.35 \times 10^{-5}$. Introducing these values in Eq. 9.68, it is obtained that $G_T = 123$ K/m, which is in the range of the experimental data.

References

Akamatsu S, Faivre G, Ihle Th (1995) Phys. Rev. E51:4751–4773
Ananth R, Gill WN (1988) J. Crystal Growth 91:587
Ardell AJ (1972) Acta Metall. 20:61
Barbieri A, Langer JS (1989) Physical Review A 10:5314–5325
Bensimon D, Pelce P, Shraiman B I (1987) J. Phys. A 48:2081
Biloni H, Boettinger WJ (1996) in: Cahn RW, Haasen P (eds) Physical Metallurgy, fourth edition. Elsevier Science BV p 670
Boden S, Eckert S, Willers B, Gerbeth G (2008) Metall. Mater. Trans. A 39A:613–623
Boettinger WJ, Coriell SR, Trivedi R (1988), in: Mehrabian R, Parrish PA (eds) Rapid Solidification Processing: Principles and Technologies. Claitor's Publishing, Baton Rouge, LA p 13
Bouchard D, Kirkaldy JS (1997) Metall. and Mater. Trans. 28B:651
Burden NH, Hunt JD (1974) J. Cryst. Growth 22
Chai G, Bäckerud L, Roland T, Arnberg L (1995) Metall. and Mater. Trans. 26A:965–970
Chalmers B (1963) J. Aust. Inst. Met. 8:225
Chernov AA (1956) Kristallographya 65:583
Cohen M, Flemings MC (1985) in: Das SK, Kear BH, Adam CM (eds) Rapidly Solidified Crystalline Alloys. TMS, Warrendale, PA p 3
Coriell SR, Cordes MR, Boetinger WJ, Sekerka RF (1980) J. Crystal Growth 49:22
Curreri P, Kaukler WF (1996) Metall. Mater. Trans. A 27A:801–808
Diepers HJ, Beckerman C, Steinbach I (1999) Acta Mater. 47(13):3663–78
Dupouy MD, Camel D, Botalla F, Abadie J, Favier JJ (1998) Microgravity sci. technol. XI(1):2
Easton M, Davidson C, St John D (2010) Metall. Mater. Trans. A 41A:1528
Elliott R (1983) Eutectic Solidification Processing. Butterworths, London
Esaka H, Kurz W (1985) J. Cryst. Growth 72:578–84
Flemings MC, Kattamis TZ, Bardes BP (1991) AFS Trans. 99:501
Flood SC, Hunt JD (1990) Columnar to Equiaxed Transition. In: Stefanescu DM (ed) ASM Handbook vol. 15 Casting, ASM International, Ohio 130–136
Glicksman ME, Koss MB, Bushnell LT, Lacombe JC, Winsa EA (1995) ISIJ International 35:604
Glicksman ME (2012) Metall. and Mater. Trans. 43A:391
Guo X, Stefanescu DM (1992) AFS Trans. 100:273
Huang SC, Glicksman ME (1981) Acta Metall. 29:701
Hunt JD (1979) in: Solidification and Casting of Metals, The Metals Society, London p 3
Hunt JD (1984) Mat. Sci. and Eng. 65:75

Hunt JD, Jackson KA (1966) Trans. Met. Soc. AIME 236:843

Huntley DA, Davis SH (1993) Acta metal. mater. 41:2025

Ivantsov GP (1947) Doklady Akademii Nauk SSSR 58:695

Jackson KA, Hunt JD, Uhlmann D, Seward TP (1966) Trans. AIME 236:149

Kattamis TZ, Flemings MC (1965) Trans. Met. Soc. AIME 233:992

Kessler DA, Levine H (1986) Phys. Rev. Lett. 57:3069

Kirkwood DH (1985) Mat. Sci. Eng. 73:L1-L4

Koseki T, Flemings MC (1995) ISIJ International 35:611

Kundin J, Rrezende JLL, Emmerich H (2014) Metall. Mater. Trans. 45A:1068

Kurz W, Fisher DJ (1981) Acta Metall. 29:11-20

Kurz W, Giovanola B, Trivedi R (1986) Acta Metall. 34:823

Kurz W, Fisher DJ (1989) Fundamentals of Solidification, 3rd ed. Trans Tech Publications, Switzerland

Langer J S, Müller-Krumbhaar H (1978) Acta Metall. 26:1681

Laxmanan V (1987) J. Crystal Growth 83:391

Lipton J, Glicksman ME, Kurz W (1984) Mat. Sci. Eng. 65:57

Liu S, Lu SZ, Hellawell A (2002) J. Crystal Growth 234:740–750

Lu SZ, Liu S (2007) Metall. and Mater. Trans. 38A:1378-1387

Lux B, Minkoff I, Mollard F, Thury E (1975) Branching of graphite crystals growing from a metallic solution. In: Lux B, Minkoff I, Mollard F (eds) The Metallurgy of Cast Iron. Georgi Publ. Co., St Saphorin, Switzerland p 497–508

Mathiesen RH, Arnberg L, Bleuet P, Somogyi A (2006) Metall. Mater. Trans. A 37A:2515–2524

Mahapatra RB, Weinberg F (1987) Metall. Trans. B 18B:425

Mendoza R, Alkemper J, Voorhees PW (2003) Metall. and Mater. Trans. 34A:481

Miyata Y (1995) ISIJ International 35:600

Morris LR, Winegard WC (1969) J. Crystal Growth 6:61

Mullins WW, Sekerka RF (1964) J. Appl. Phys. 35:444

Mortensen A (1991) Metall. Trans. 22A:569

Nash GE, Glicksman ME (1974) Acta Metall. 22:1283

Nastac L, Stefanescu DM (1993) Metall. Trans. 24A:2107

Nastac L, Stefanescu DM (1996) Metall. Trans. 27A:4061

Nastac L, Chou JS, Pang Y (1999) in: Symp. Liquid Metal Processing and Casting, Santa Fe, New Mexico

O'Hara S, Tiller WA (1967) Trans. AIME 239:497

Papapetrou A (1935) Z. Kristall. 92:89

Rappaz M, Boettinger WJ (1990) Acta Mater. 47(11):3205-3219

Ratke L (2009) Int. J. Cast Metals Res. 22(1–4):268

Ratke L, Thieringer W (1985) Acta Metall. 33(10):1793–1802

Reinhart G, Mangelinck-Noel N, Nguyen-Thi H, Schenk T, Gastaldi J, Billia B, Pino P, Hartwig J, Baruchel J (2005) Mater. Sci. Eng. A 413–414:384

Ruvalcaba D, Eskin DG, Mathiesen RH, Arnberg L, Katgerman L (2009) Int. J. Cast Metals Res. 22(1–4):271

Spittle JA (2006) Int. Materials Reviews 51(4):247

Steinbach S, Ratke L (2005) Mater. Sci. Eng. A 413–414:200–04

Tian H, Stefanescu DM (1992) Metall. Trans. 23A:681

Tiller WA (1962) Trans. Met. Soc. AIME 224:448

Trivedi R. Somboonsuk K (1984) Mat. Sci. and Eng. 65:65–74

Trivedi R, Kurz W (1988) Solidification of Single Phase Alloys. In: Stefanescu DM (ed) Metals Handbook Ninth Edition vol. 15 Casting. ASM International, Metals Park, Ohio p 115–119

Trivedi R, Kurz W (1994) International Materials Reviews 39,2:49

Trivedi R, Shen YX, Liu S (2003) Metall. Mater. Trans. 34A:395

Ullah MW, Carlberg T (2011) J. Crystal Growth 318:212–218

Voorhees PW, Glicksman EM (1984) Metall. Trans. 15A:1081

Winegard WC, Chalmers B (1954) Trans. ASM. 216:1214

Yasuda H, Ohnaka I, Kawasaki K, Sugiyama A, Ohmichi T, Iwane J, Umetani K (2004) J. Cryst.
    Growth 262:645–652
Yasuda H, Yamamoto Y, Nakatsuka N, Yoshiya M, Nagira T, Sugiyama A, Ohnaka I, Uesugi K,
    Umetani K (2009) Int. J. Cast Metals Res. 22(1–4):15
Young KP, Kirkwood DH (1975) Metall. Trans. 6A:197
Zhu P, Smith RW (1992) Acta metall. mater. 40:683 and 3369

Chapter 10
Solidification of Two-Phase Alloys—Micro-Scale Solidification

The two-phase alloys types discussed in this chapter include eutectics, peritectics, and monotectics. Some of the most important casting alloys in terms of tonnage as well as applications, such as cast iron and aluminum–silicon alloys, are essentially eutectic two-phase alloys.

Peritectic solidification is very common in the solidification of metallic, ceramic, and organic materials. Typical examples of systems with peritectic solidification include Fe–C (steel), Fe–Ni, Cu–Sn (bronze), Cu–Zn (brass), Al–Ti, lanthanide magnets (Nd–Fe–B; Umeda et al. 1996), and ceramic superconductors (Y–Ba–Cu–O; Izumi and Shiohara 2005). Some of these materials have exotic applications. For example, the naphthalene-capric acid system is a potential latent heat storage material (Jin and Xiao 2004).

10.1 Eutectic Solidification

The eutectic has a fixed composition in terms of species A and B, and is in fact a two-phase solid ($\alpha + \beta$). Solidification of a liquid of eutectic composition proceeds by transformation of the liquid into a two-phase solid.

10.1.1 Classification of Eutectics

Many eutectic classifications have been proposed, based on different criteria. A first classification of eutectics is based on their growth mechanism:

- *Cooperative growth*: the two phases of the eutectic grow together as a diffusion couple
- *Divorced growth*: the two phases of the eutectic grow separately; there is no direct exchange of solute between the two solid phases and no tri-junction

Another classification was proposed (Hunt and Jackson 1966) based on the interface kinetics of the component phases of the eutectic. As discussed earlier, phases

© Springer International Publishing Switzerland 2015
D. M. Stefanescu, *Science and Engineering of Casting Solidification,*
DOI 10.1007/978-3-319-15693-4_10

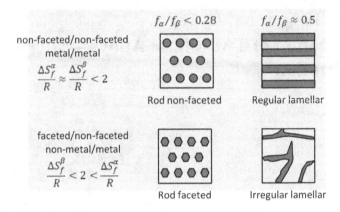

Fig. 10.1 Types of cooperative eutectics. (After Kurz and Fisher 1989)

having low entropies of fusion solidify with a non-faceted interface, while phases having high entropy of fusion solidify with faceted interface. Thus, the following classification was proposed:

- Non-faceted/ non-faceted eutectics (nf/nf)
- Non-faceted/ faceted (nf/f)
- Faceted/ faceted (f/f)

The first two categories are common and have commercial applications. The f/f eutectics are less studied. Croker et al. (1973) suggested that in addition to the entropy of fusion, the volume fraction of the two phases plays a significant role in the resulting microstructure. Depending on the ratio between the fractions of the two phases of the eutectic, f_α and f_β, and on the morphology of the liquid – solid interface, several types of cooperative eutectics may form (Fig. 10.1). The nondimensional entropy of fusion, $\Delta S_f/R$, where R is the gas constant, is used to distinguish between faceted and non-faceted morphologies. The classification in Fig. 10.1 is a simplification of the rather extensive one proposed by Croker et al. (1973), which includes a large number of irregular (anomalous) structures.

Alloys such as Pb–Sn and Al–Al$_2$Cu (Fig. 10.2a), where there are approximately equal volume fractions of non-faceted phases solidify as regular, lamellar eutectics. If one of the phases is non-faceted the morphology becomes irregular, because the faceted phase grows preferentially in a direction determined by specific atomic planes. A typical example is the Mg–Mg$_2$Sn eutectic shown in Fig. 10.2b.

When the volume fraction of one phase is significantly lower than that of the other (typically lower than 0.25), a fibrous structure will result (example the Ni–NbC eutectic shown in Fig. 10.2c). This is because of the tendency of the system to minimize its interfacial energy by selecting the morphology that is associated with the smallest interfacial area. Fibers have smaller interfacial area than lamellae. However, when the minor phase is faceted, a lamellar structure may form even at

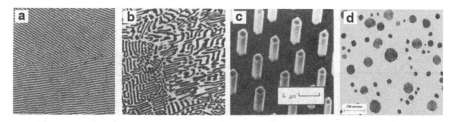

Fig. 10.2 Eutectic microstructures. **a** Regular nf/nf eutectic (Al–Al$_2$Cu; Magnin and Kurz 1988a). **b** Irregular f/nf eutectic (Mg–Mg$_2$Sn). The dark phase is the faceted Mg$_2$Sn (Magnin and Kurz 1988a). **c** Rod f/nf eutectic (Ni–NbC; Cooper and Billingham 1980). **d** Divorced eutectic (Fe–spheroidal graphite). (**a** and **b** reprinted with permission of ASM International. All rights reserved. www.asminternational.org)

a very low volume fraction, because specific planes may have the lowest interfacial energy. The minor phase will then grow such as to expose these planes even when lamellae rather than fibers are formed. The two commercially most significant eutectics, Al–Si and Fe–graphite (Gr) fall into this category. Note that in the Fe–Gr eutectic $f_{Gr} = 0.07$. The Fe–Gr eutectic can be either cooperative, irregular, as is the case for lamellar graphite cast iron, or divorced, as for spheroidal graphite cast iron (Fig. 10.2d). In this last case, at the beginning of solidification the two phases, graphite and austenite dendrites, grow independently from the liquid without establishing a diffusion couple.

10.1.2 Cooperative Eutectics

During the solidification of single-phase alloys, the solute is rejected from the growing tip of the dendrite and from the dendrite sides, and accumulates in the interdendritic regions. During the solidification of two-phase eutectic alloys, two solutes are rejected. Solute A accumulates in front of the β phase, while solute B accumulates in front of the α phase. The solute only needs to diffuse along the solid/liquid (S/L) interface from one phase to the other. Accordingly, the diffusion boundary layer for eutectics should be much smaller than for single-phase alloys (see Application 10.1). Each of these solutes is then incorporated in the growing solid solutions. Sideways diffusion is thus responsible for eutectic solidification. Thus, the morphology of the eutectic is made of alternative lamellae of phases α and β (Fig. 10.3a). The smaller is the lamellar spacing, the smaller the solute buildup. In other words, solute diffusion tends to decrease the lamellar spacing.

Since excess free energy is associated with grain boundaries, a lower equilibrium temperature will exist for the tri-junction $\alpha/\beta/L$. At the three-phase junction $\alpha/\beta/L$, the surface energies must be balanced to insure mechanical equilibrium. This imposes fixed contact angles, and in turn induces a curvature of the S/L interface of each lamella (Fig. 10.3b). Because the contact angles are material constants, this

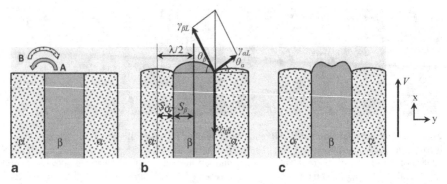

Fig. 10.3 Schematic representation of interface morphology for eutectic solidification, explaining the effects of boundary energy and solute accumulation. **a** Boundary energy and solute accumulation at interface ignored. **b** Excess energy associated with grain boundaries considered. **c** Boundary energy and solute accumulation considered

Fig. 10.4 Correlation between undercooling and composition

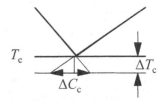

curvature is smaller when the lamellar spacing is higher. Accordingly, curvature acts to increase the lamellar spacing.

The scale of the eutectic structure is therefore determined by a compromise between two opposing factors:

• Solute diffusion which tends to decrease the spacing
• Surface energy (interface curvature) which tends to increase the spacing

The dominant variables of eutectic solidification are undercooling, ΔT, growth velocity, V, and interlamellar spacing, λ. Their quantitative evaluation is performed starting again with the equation of the interface undercooling:

$$\Delta T = \Delta T_k + \Delta T_r + \Delta T_c + \Delta T_T,$$

where the kinetic and the thermal undercooling are neglected. Since $\Delta T_r = 2\Gamma/r$ and, $r = ct \cdot \lambda$, as discussed for secondary DAS, we have $\Delta T_r = \mu_r \lambda^{-1}$, where μ_r is a material constant depending on curvature at the tip of the lamella. The constitutional undercooling is a function of composition, $\Delta T_c = m \Delta C_c$, as derived from Fig. 10.4 The solidification velocity will increase with diffusivity and difference in composition, but decrease with higher lamellar spacing: $V = ct \cdot D \Delta C_c/\lambda$. Thus, $\Delta C_c = ct \cdot \lambda V$ and the constitutional undercooling becomes $\Delta T_c = \mu_c \lambda V$, where μ_c is a material constant depending on composition. Substituting ΔT_r and ΔT_c in the equation for the total undercooling results in:

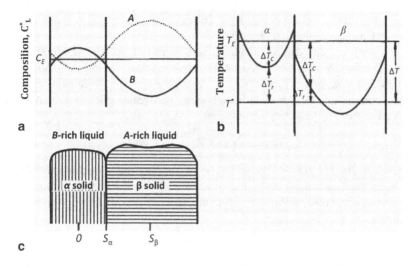

Fig. 10.5 Morphology of eutectic front. (After Hunt and Jackson 1966)

$$\Delta T = \frac{\mu_r}{\lambda} + \mu_c \lambda V. \tag{10.1}$$

To obtain the specific values of μ_r and μ_c a more detailed analysis must be conducted.

It must be further noted that the solute concentration is higher at the middle than at the edge of a lamella. Consequently, the liquidus temperature will be lower in the middle, and a lamella may have a negative curvature in the middle, as shown in Fig. 10.3c for phase β. To demonstrate this, let us consider the interface composition (Fig. 10.5a) and the undercooling (Fig. 10.5b) ahead of the eutectic. The interface undercooling, T^*, is the same for both lamellae. Thus, the interface is maintained isothermal at T^* by the lamellae adjusting their radii of curvature. If ΔT_r becomes negative, as is the case for the β lamella, curvature and the radius of curvature becomes negative (Fig. 10.5c). This means that the β lamella will have three curvatures rather than one.

10.1.3 Models for Regular Eutectic Growth

A formal theory of directionally solidified eutectics was proposed as early as 1946 by Zener and then modified by Tiller (1958). It is based on the free energy change for eutectic solidification of unit volume of liquid written as $\Delta G_v^\lambda = \Delta G_v^\infty + 2\gamma_{\alpha\beta}/\lambda$, where the free energy change for ∞ spacing is $\Delta G_v^\infty = \Delta H_f \Delta T_c / T_e$, ΔH_f is the latent heat, T_e is the equilibrium temperature, λ is the lamellar spacing, $2/\lambda$ is the α/β interface per unit volume, $\gamma_{\alpha\beta}$ is the α/β interface energy, and the significance of other quantities are given in Fig. 10.4. The minimum spacing λ_{min} is obtained for $\Delta G_v^\lambda \to 0$, as $\lambda_{min} = 2\gamma_{\alpha\beta}/(\Delta H_f \Delta T_c)$.

The growth velocity can be calculated as $V = ct \cdot D_L \Delta C_c/\lambda$, where D_L is the liquid diffusivity. Since from Fig. 10.4 it is clear that $\Delta C_c = ct \cdot \Delta T_c$, a relationship between V, λ, and ΔT_c is obtained. Assuming now that solidification occurs at the maximum rate or minimum undercooling (extremum criterion), correlations between the various quantities are obtained:

$$\lambda_{min}^2 V_{max} = ct \quad \text{and} \quad V_{max} = ct \cdot \Delta T_c^2. \tag{10.2}$$

The Jackson–Hunt (JH) (1966) Model A more complete mathematical analysis was done by Jackson and Hunt (1966). For steady-state growth, with the coordinate system moving with velocity V in the x direction (Fig. 10.3), the diffusion equation to be solved is:

$$\nabla^2 C + \frac{V}{D}\frac{\partial C}{\partial x}.$$

The following boundary conditions apply:

      B.C.1 at $x \to \infty$ $C = C_o = C_E + \Delta C_\infty$

      B.C.2 at $y = 0$ and at $y = S_\alpha + S_\beta$ $\partial C/\partial y = 0$

The general solution of this diffusion equation is (see derivation in inset):

$$C(x, y) = B_o \exp\left(-\frac{V}{D}x\right) + \sum_{n=1}^{\infty} B_n \cos\frac{n\pi y}{L}$$

$$\cdot \exp\left\{\left[-\frac{V}{2D} - \left[\left(\frac{V}{2D}\right)^2 + \left(\frac{n\pi}{L}\right)^2\right]^{1/2}\right]x\right\} \quad \text{with } L = S_\alpha + S_\beta. \tag{10.3}$$

Derivation of the JH Eutectic Model

The governing equation is:

$$\frac{\partial^2 u}{\partial x^2} + \frac{\partial^2 u}{\partial y^2} + \frac{V}{D}\frac{\partial u}{\partial x} = 0.$$

Using the method of separation of variables, $u(x, y) = P(x)\,Q(y)$, which, after substitution in the governing equation and appropriate manipulations results in:

$$\frac{1}{P}\frac{\partial^2 P}{\partial x^2} + \frac{1}{Q}\frac{\partial^2 Q}{\partial y^2} + \frac{V}{D}\frac{1}{P}\frac{\partial P}{\partial x} = 0 \quad \text{or} \quad \frac{1}{P}\frac{\partial^2 P}{\partial x^2} + \frac{V}{D}\frac{1}{P}\frac{\partial P}{\partial x} = -\frac{1}{Q}\frac{\partial^2 Q}{\partial y^2} = \lambda,$$

where λ is the separation variable.

The first equation, $\frac{\partial^2 P}{\partial x^2} + \frac{V}{D}\frac{\partial P}{\partial x} - P\lambda = 0$, has the general solution $P = C_1 exp(-C_2 x)$, where C_2 must be the root of the equation $C_2^2 + (V/D)C_2 - \lambda = 0$, i.e., $C_2 = -(V/2D) - [(V/2D)^2 + \lambda]^{1/2}$. Note that we have kept only the root with negative sign to avoid having a constant zero at $\lambda = 0$. Thus, the solution of the P function is:

$$P = C_1 \exp\left\{\left[-\frac{V}{2D} - \left[\left(\frac{V}{2D}\right)^2 + \lambda\right]^{1/2}\right]x\right\}.$$

The second equation, $\partial^2 Q/\partial y^2 + \lambda Q = 0$, has the solutions $Q = C_3 \cos(C_4 y)$ and $Q = C_5 \cos(C_6 y)$, which gives:

$$Q = C_3 \cos(\lambda y) + C_5 \sin(\lambda y) \text{ and } dQ/dy$$
$$= -C_3 \sqrt{\lambda} \sin(\sqrt{\lambda} y) + C_5 \sqrt{\lambda} \cos(\sqrt{\lambda} y).$$

The applicable boundary conditions are:

$$Q(0, t) = dQ/dy = 0 \quad \text{and} \quad Q(L, t) = dQ/dy = 0, \quad \text{where } L = S_a + S_b.$$

Applying these conditions in the expression of Q, we have $C_5 = 0$ and $(dQ/dy)_L = -C_3 \sqrt{\lambda} \sin(\sqrt{\lambda} L) = 0$.

For $C_3 \neq 0$ we obtain $\sin(\sqrt{\lambda} L) = \sin(n\pi)$ which gives the eigenvalue $\lambda = (n\pi/L)^2$ and the eigenfunction $Q = C_n \cos(n\pi y/\lambda)$. Thus, a general solution of the governing equation is:

$$u(x, y) = P(x)Q(y) = C_1 \exp\left\{\left[-\frac{V}{2D} - \left[\left(\frac{V}{2D}\right)^2 + \lambda\right]^{1/2}\right]x\right\} \cos\frac{n\pi x}{L},$$

and the complete solution is:

$$u(x, y) = \sum_{n=0}^{\infty} B_n \cos\frac{n\pi y}{L} \exp\left\{\left[-\frac{V}{2D} - \left[\left(\frac{V}{2D}\right)^2 + \left(\frac{n\pi}{L}\right)^2\right]^{1/2}\right]x\right\}$$

$$= B_o \exp\left(-\frac{V}{D}x\right) + \sum_{n=1}^{\infty} B_n \cos\frac{n\pi y}{L}$$

$$\cdot \exp\left\{\left[-\frac{V}{2D} - \left[\left(\frac{V}{2D}\right)^2 + \left(\frac{n\pi}{L}\right)^2\right]^{1/2}\right]x\right\}.$$

Fig. 10.6 Definition of $\overline{C}_o^\alpha$
and $\overline{C}_o^\beta$

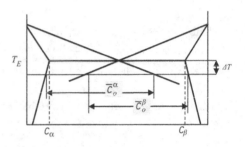

The exponential term contains two quantities: $V/2D$ which is the inverse of the primary phase boundary layer, and $n\pi/L$ which is the inverse of the eutectic boundary layer. For the case of eutectic solidification $L = S_\alpha + S_\beta = \lambda/2$. Thus, the eutectic boundary layer is $\lambda/2\pi$, and the primary phase boundary layer is $2D/V$. Now $V/2D << 2\pi/\lambda$ because $\delta = D/V$ is large at slow growth rates, while λ is small (see Application 10.1). Thus, $V/2D$ will be neglected.

The solution given by Eq. 10.3 satisfies BC2, but not BC1. Since the solution also has to satisfy BC1, it becomes:

$$C_L = C_E + \Delta C_\infty + B_0\,exp(-Vx/D) + \sum_{n=1}^{\infty} B_n \cos(2n\pi y/\lambda) \cdot \exp(-2n\pi x/\lambda),$$
$$(10.4)$$

where ΔC_∞ is the difference between the initial composition of the liquid and the eutectic composition, C_E.

To evaluate the Fourier coefficients in Eq. 10.4, conservation of mass at the interface is applied:

$$-D_L \frac{\partial C}{\partial x}\bigg|_{x=0} = \begin{cases} V(C_E - C_{\alpha M}) & \text{for} \quad 0 \le y \le S_\alpha \\ V(C_E - C_{\beta M}) & \text{for} \quad S_\alpha \le y \le S_\alpha + S_\beta \end{cases} \quad (10.5)$$

The Fourier coefficients are $B_0 = \overline{C}_o^\alpha f_\alpha - \overline{C}_o^\beta f_\beta$ and, $B_n[\lambda/(n\pi)^2](V/D)(\overline{C}_o^\alpha + \overline{C}_o^\beta) \sin(n\pi f_\alpha)$ where $f_\alpha = S_\alpha/(S_\alpha + S_\beta)$, $f_\beta = S_\beta/(S_\alpha + S_\beta)$, and $\overline{C}_o^\alpha$ and $\overline{C}_o^\beta$ are defined in Fig. 10.6. If the liquid is of exact eutectic composition, Eq. 10.4 simplifies to:

$$C_L = C_E + \sum_{n=1}^{\infty} B_n \cos(2n\pi y/\lambda) \cdot \exp(-2n\pi x/\lambda). \quad (10.6)$$

Eq. 10.4 is valid when it is assumed that the density difference between the two eutectic phases is negligible. A more complete analysis that included the density difference was done by Magnin and Trivedi (1991).

At a distance ahead of the interface $x = \lambda/2$, $\exp(-n\pi)$ is: $\exp(-\pi) = 0.04$, $\exp(-2\pi) = 0.001$, and $\exp(-3\pi) = 0.00008$. Thus, from Eq. 10.6, $C_L \approx C_E$. This means that composition deviation of liquid from eutectic is damped out fast ahead of the interface. The eutectic diffusion layer is thus $\lambda/2$. Calculations show that maximum composition deviation from eutectic is small, $\sim 0.1\,\%$.

Fig. 10.7 Comparison of calculated and observed shapes for a transparent organic eutectic. (Jackson and Hunt 1966)

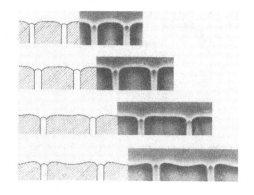

To find the particular values of the constants in Eq. 10.1 it is necessary to formulate both ΔT_c and ΔT_r. ΔT_c can be obtained from Eq. 10.6 by specifying V and λ, since $\Delta T_c = m(C_L - C_E) = m \cdot f(\lambda, V)$. To obtain ΔT_r it is necessary to use the contact angles at the tri-junction (Fig. 10.3). From the definition of curvature in Cartesian coordinates, it can be shown that $\Delta T_r = \Gamma K = 2 \sin \theta_i / f_i \lambda$, where the subscript i stands for the phase α or β. By manipulating these last two equations, and Eq. 10.6, and substituting in Eq. 10.1, the values of the constants are found to be:

$$\mu_c = \overline{m} \frac{C_\beta - C_\alpha}{D} F(f) \quad \text{and} \quad \mu_r = 2\overline{m}\delta \left(\frac{\Gamma_\alpha \sin \theta_\alpha}{f_\alpha |m_\alpha|} + \frac{\Gamma_\beta \sin \theta_\beta}{f_\beta m_\beta} \right) \quad \text{with}$$

$$\overline{m} = \frac{|m_\alpha| m_\beta}{|m_\alpha| + m_\beta}, \tag{10.7}$$

where C_α and C_β are the compositions of the α and β phases, respectively, $|m_\alpha|$ and $m\beta$ are the slopes of the liquidus lines of the α (non-faceted) and β (faceted) phases, respectively, $F(f) = 0.335(f_\alpha f_\beta)^{0.65}$, $f\alpha$ and f_β are the volume fraction of solids of the eutectic phases, respectively, Γ_α and Γ_β are the Gibbs–Thomson coefficients of the two phases, θ_α and θ_β are the contact angles, and δ is a parameter equal to 1 for lamellar eutectics. This equation is valid for lamellar eutectics. For α-fibrous eutectics the equation is:

$$F(f_\alpha) \approx 4.908 \cdot 10^{-3} + 0.3122 f_\alpha + 0.6918 f_\alpha^2 - 2.604 f_\alpha^3 + 3.238 f_\alpha^4 - 1.619 f_\alpha^5$$

A comparison between calculated and experimental results is shown in Fig. 10.7.

Equation 10.1 is represented graphically in Fig. 10.8. It is seen that the curve representing the total undercooling, ΔT, is the result of the sum of the solutal and curvature undercoolings. The total undercooling is high for small and large values of the lamellar spacing, and reaches a minimum at a certain value of the spacing, which we will call extremum, λ_{ex}. For $\lambda < \lambda_{ex}$, capillarity effects are controlling the process ($\Delta T_c < \Delta T_r$). For $\lambda > \lambda_{ex}$ diffusion is the controlling mechanism ($\Delta T_c > \Delta T_r$). It is also seen that no unique solution is available. Indeed, pairs of λ and ΔT will satisfy the equation. Nevertheless, experiments demonstrate that for a given system solidifying at constant velocity, for regular eutectics, the lamellar spacing varies

Fig. 10.8 Extremum and branching lamellar spacing in eutectics

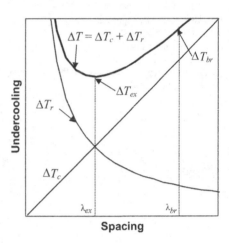

within a very narrow range. In the case of irregular eutectics, the range of variation of the spacing is considerably larger. It is generally accepted that the lamellar spacing will vary between λ_{ex} and the branching spacing, λ_{br}.

For the case of regular eutectics, the most common approach is to use the extremum criterion, proposed by Zener. It states that the eutectic will grow at the minimum undercooling possible. This criterion has been found to yield correct results for regular eutectics such as Pb–Sn.

Applying the extremum criterion to Eq. 10.1 (the first derivative with respect to λ is equated to zero) equations correlating the eutectic growth parameters are obtained:

$$\lambda_{ex}^2 V_{ex} = \mu_r/\mu_c \quad \text{and} \quad \lambda_{ex}\Delta T_{ex} = 2\mu_r \quad \text{and} \quad \Delta T_{ex}^2 V_{ex}^{-1} = 4\mu_c\mu_r. \quad (10.8)$$

These equations can also be written as:

$$\lambda_{ex} = \mu_\lambda V^{-1/2} \quad \text{and} \quad V = \mu_V \Delta T_{min}^2, \quad (10.9)$$

with $\mu_\lambda = \sqrt{\mu_r/\mu_c}$ and $\mu_V = 1/(4\mu_c\mu_r)$. In these equations μ_λ and μ_V are material constants. Some typical values for these constants are given in Appendix B, Tables B1–B4.

However, as demonstrated by Seetharaman and Trivedi (1988), even for regular eutectics, for a given velocity the spacing is not fixed, but rather occupies a range whose lower limit is indeed λ_{ex}, but whose upper limit is about 20 % higher than the extremum value. In other words, an average spacing exists, that is higher than λ_{ex}.

For irregular eutectics such as Al–Si, Fe–C, Fe–Fe$_3$C, the experiments revealed that the average spacing is much higher than predictions made by the JH model (Toloui and Hellawell 1976; Elliot and Glenister 1980; Jones and Kurz 1981; Hogan and Song 1987; Magnin et al. 1991). Typical values are shown in Fig. 10.9.

Fig. 10.9 $\lambda-V$ correlation for various diffusion couples

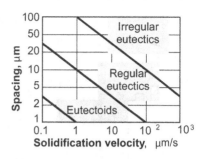

10.1.4 Models for Irregular Eutectic Growth

The irregular faceted/non-faceted structure occurs because the specific surface energy between phases is very anisotropic. Certain lowest-energy crystallographic orientations develop between phases to minimize the interfacial energy. The lamellar spacing of irregular eutectics does not obey Eqs. 10.8 or 10.9. In other words, the extremum criterion does not explain the mechanism of occurrence of lamellar spacing, or rather is not sufficient to explain it. While local spacing corresponding to the extremum condition can be found, the mean spacing is considerably larger. This results from the fact that in these eutectics the crystallographic growth directions of the two phases do not coincide.

The irregular nature of the lamellar graphite–austenite eutectic was not confronted until 1974, when Sato and Sayama introduced the concept of partially cooperative growth of the non-faceted/faceted phases. Although the calculated spacings were larger than the JH predictions, they still failed to approach the even larger experimental value.

The Fisher–Kurz (1980) and Magnin–Kurz (1987) Models

According to Kurz and Fisher (1989) the smallest spacing will be the one dictated by the extremum criterion, λ_{ex}, but a larger spacing will also exist, dictated by a branching condition, λ_{br}. The two phases can grow convergent (one toward the other) until λ_{ex} is reached, then they will grow divergently until λ_{br} is reached, as shown in Fig. 10.8 and Fig. 10.10. Because of this growth, irregular eutectics solidify with a non-isothermal interface. The lamellar graphite–austenite eutectic formed in gray iron, or the silicon plates–aluminum eutectic are typical examples.

In their irregular faceted/non-faceted eutectic models, Fisher, Magnin and Kurz used the following assumptions: (i) non-isothermal interface; (ii) the γ phase that has a diffuse interface grows faster than the graphite phase that is faceted; (iii) branching occurs when a depression forms on the faceted phase. To impose a non-isothermal coupling condition over the interface, they ascribed a cubic function. They demonstrated that the smallest spacing of the lamellar eutectic is dictated by the extremum condition, but that a larger spacing will also exist, λ_{br}, dictated by a branching condition. λ_{br} can be calculated as the product between a function of the physical constants of the faceted phase and a material constant. This constant must

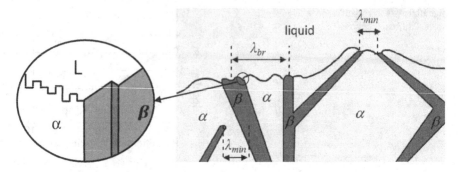

Fig. 10.10 Growth of irregular eutectics (after Kurz and Fisher 1998). Note: non-isothermal S/L interface; the diffuse α interface can grow easily, while the faceted β phase grows more difficult

be postulated to match the theoretical predictions with the experimental measurements, which limits the generality of the model. The model predicts that the average spacing in eutectics with non-isothermal S/L interface obeys relationships similar to those developed by JH for the isothermal interface:

$$\lambda_{av}^2 V = \phi^2 \mu_r / \mu_c \quad \text{or} \quad \lambda_{av} \Delta T_{av} = (\phi^2 + 1)\mu_r \quad \text{or}$$
$$V^{-1} \Delta T_{av}^2 = (\phi + 1/\phi)^2 \mu_r \mu_c, \tag{10.10}$$

where the average spacing, λ_{av}, and the operating point factor, ϕ, are defined as follows:

$$\lambda_{av} = (\lambda_{ex} + \lambda_{br})/2 = \phi \lambda_{ex} \quad \text{and} \quad \phi = (1 + \lambda_{br}/\lambda_{ex})/2.$$

Note that since $\lambda_{br} > \lambda_{ex}$, $\phi > 1$, and $\lambda_{av} > \lambda_{ex}$. This is shown in Fig. 10.9 where the λ-V line for irregular eutectics is positioned above the one for regular eutectics. For eutectoids, diffusion occurs only through solid phases and it is slower. Thus, the diffusion distance will be lowered by decreasing the spacing.

Equation 10.10 describes the growth of both regular ($\phi = 1$) and irregular ($\phi > 1$) eutectics. Assuming that β is the faceted phase, for irregular eutectics:

$$\phi = 0.5 + \left[F'(f_\beta) \left(1 + \frac{f_\beta m_\beta}{f_\alpha m_\alpha} \frac{\Gamma_\alpha \sin \theta_\alpha}{\Gamma_\beta \sin \theta_\beta} \right) \right]^{-1/2}, \tag{10.11}$$

where $F'(f_\beta) \approx 0.03917 + 0.6047 f_\beta - 1.413 f_\beta^2 + 2.171 f_\beta^3 - 1.236 f_\beta^4$. The following values for ϕ were postulated: 3.2 for the Al–Si system 5.4 for Fe–C and 1.8 for Fe–Fe$_3$C.

Attempting to understand the spacing selection mechanism, Magnin et al. (1991) performed an analysis on the Al–Si system. They found that the position of the three characteristic measured spacings (λ_{min}, λ_{av}, λ_{br}) on the theoretical V-λ curve line up as shown in Fig. 10.11. They concluded that the growth mechanism involves a continuous adjustment of λ and V while the S/L interface is maintaining its undercooling at a certain constant value, as illustrated in Fig. 10.12.

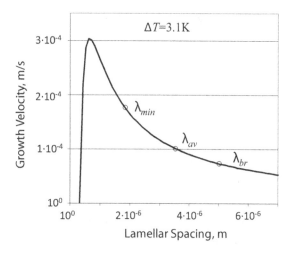

Fig. 10.11 The operating range of irregular eutectics for the Al–Si system. (After Magnin et al. 1991)

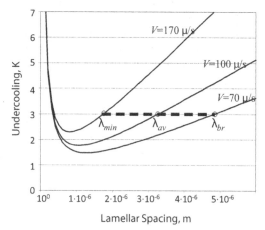

Fig. 10.12 The mechanism of spacing selection at constant growth undercooling for the Al–Si system. (After Magnin et al. 1991)

It should be pointed out that an S/L interface must be planar in order to maintain a constant undercooling during its growth. However, given the need for the continuous adjustment of λ and V required by the selected undercooling, the interface planarity cannot be preserved. Furthermore, the disruption of the interface planarity implies local changes of the growth undercooling therefore invalidating the assumptions of the proposed mechanism. Also, this mechanism is unable to establish a theoretical basis that would explain the magnitude of the spacing range λ_{min}-to-λ_{br}.

The Catalina–Stefanescu (CS) (2010) Modified JH Model

Their analysis was restricted to alloys of eutectic composition. Therefore, the term ΔC_∞ in Eq. 10.4 was set to zero. As significant differences between the densities ρ_α and ρ_β are probable in irregular eutectics, an *effective* liquid diffusion coefficient that accounts for the fluid motion induced at the S/L interface by the density difference was adopted as suggested by Magnin and Trivedi (1991):

$$D_{eff} \cong D_L/[1 - 100 \cdot (\rho_\beta - \rho_\alpha)(\rho_\beta C_\beta - \rho_\alpha C_\alpha)^{-1}]. \qquad (10.12)$$

To solve for the Fourier coefficients B_o and B_n the mass balance at the S/L interface was formulated as:

$$-\frac{D_{eff}}{V} \frac{\partial C}{\partial x}\bigg|_{x=0} = \begin{cases} C_1(y) - \eta_\alpha C_\alpha & \text{for } \alpha \text{ phase} \\ C_1(y) - \eta_\beta C_\beta & \text{for } \beta \text{ phase} \end{cases}, \qquad (10.13)$$

where $C_I(y)$ is the liquid composition at the S/L interface, and η_α and η_β account for the density effect and are defined as:

$$\eta_i = \rho_i/\rho_L \quad \text{with} \quad i = \alpha, \beta, \qquad (10.14)$$

where ρ_L is the density of the liquid phase which is not necessarily equal to that of the solid eutectic as assumed in the Magnin and Trivedi (1991) formulation. As the composition of the faceted phase (e.g., $\beta = $ graphite, Si, or Fe$_3$C) is fixed, and therefore does not depend on that of the liquid at the S/L interface, Eq. 10.13 does not make use of the solute partition coefficients. Therefore, the quantities C_α and C_β in Eq. 10.13 are the compositions of the α- and β-phase, respectively, as given by the phase diagram.

Direct integration of Eq. 10.13 gives:

$$\rho_\alpha f_\alpha C_\alpha + \rho_\beta f_\beta C_\beta = \rho_L C_E. \qquad (10.15)$$

This expression allows calculating the volume fractions by accounting for the volume change during the L-to-S transformation. Then, using the orthogonality of $\cos\left(\frac{2n\pi}{\lambda} \cdot y\right)$:

$$B_n = \frac{\lambda V}{D_{eff}}(\eta_\beta C_\beta - \eta_\alpha C_\alpha)\frac{\sin(n\pi f_\alpha)}{(n\pi)^2}. \qquad (10.16)$$

Further, by following the same procedure as in the JH treatment, the average undercooling, $\Delta\overline{T}$, of the α- and β-lamellae can be expressed as:

$$\Delta\overline{T}_\alpha = -m_\alpha(B_o + S_n/f_\alpha) + K_2^\alpha/\lambda \text{ and } \Delta\overline{T}_\beta = -m_\beta(B_o - S_n/f_\beta) + K_2^\beta/\lambda, \qquad (10.17)$$

where $S_n = \frac{\lambda V}{D_{eff}}(\eta_\beta C_\beta - \eta_\alpha C_\alpha)\sum_1^\infty \frac{\sin^2(n\pi f_\alpha)}{(n\pi)^3}$, $K_2^i = 2\Gamma \sin(\theta_i)/f_i$ and $i = \alpha, \beta$.

By imposing the condition that the adjacent α- and β-lamellae grow at the same undercooling (i.e., $\Delta\overline{T} = \Delta\overline{T}_\alpha = \Delta\overline{T}_\beta$), the coefficient B_o can be determined from Eq. 10.17:

$$B_o = (|m_\alpha| + m_\beta)^{-1}[(m_\beta/f_\beta - |m_\alpha|/f\alpha)S_n + \lambda^{-1}(K_2^\beta - K_2^\alpha)]. \qquad (10.18)$$

The isothermal assumption imposed on Eq. 10.17 yields Eq. 10.1, just as in the JH treatment. The constants are:

$$\mu_r = \overline{m}\left(\frac{K_2^\alpha}{|m_\alpha|} + \frac{K_2^\beta}{m_\beta}\right) \qquad \mu_c = \overline{m}\frac{-\eta_\beta C_\beta - \eta_\alpha C_\alpha}{f_\alpha f_\beta D_{eff}} \sum_1^\infty \frac{\sin^2(n\pi f_\alpha)}{(n\pi)^3} \quad (10.19)$$

While μ_r is identical to that in the JH model, μ_c is different, mainly because it accounts for the density effect. Moreover, μ_c is also different from that of Magnin and Trivedi (1991) because of the different formulation of the mass balance at the S/L interface. However, the CS model consisting of Eqs. 10.1 and 10.19 is still unable to match the experimental values of λ, V, and $\Delta\overline{T}$ for the irregular eutectics. Thus, a spacing selection mechanism other than the growth at extremum must be in effect for this type of eutectics.

The faceted phase of irregular eutectics does not have the ability to quickly adjust their growth direction with that of the heat flow. This precludes the selection of the average operating point at the point of extremum. Furthermore, in order to preserve a stable front growth, the selection is also precluded in regions characterized by high absolute values of $dV/d\lambda$, as shown in Fig. 10.11. In order to find a critical value of $dV/d\lambda$ one may consider taking the derivative with respect to λ of Eq. 10.1, which, at constant $\Delta\overline{T}$ gives:

$$\frac{\lambda}{V}\frac{dV}{d\lambda} + 1 = \frac{\mu_r}{\mu_c}\frac{1}{\lambda^2 V}. \qquad (10.20)$$

In the original CS formulation this equation was used to derive equations for the average operating point. A similar approach will be used here to find the minimum and maximum operating points. Referring to Fig. 10.11, let us evaluate λ_{min}. Assume now that on this curve, at constant undercooling, ΔT_{ex}, the minimum operating point, different than extremum, is characterized by a parameter ξ_{min} such as:

$$\left(\frac{\lambda}{V}\frac{dV}{d\lambda}\right)_{min} = \xi_{min} - 1 \quad \text{and} \quad \frac{\mu_r}{\mu_c}\frac{1}{\lambda_{min}^2 V_{min}} = \xi_{min} \qquad (10.21)$$

which allows expressing V_{min} as a function of the minimum spacing and the other parameters as:

$$V_{min} = \frac{1}{\lambda_{min}^2 \xi_{min}}\frac{\mu_r}{\mu_c} \qquad (10.22)$$

This expression can now be used in Eq. (10.1) to obtain:

$$\lambda_{min}\Delta T_{ex} = \frac{1 + \xi_{min}}{\xi_{min}}\mu_r \qquad (10.23)$$

A predictive methodology for the factor ξ_{min} does not exist at this time, but experimental measurements can be used to determine its value.

When examining the experimental data for Fe-graphite and Fe-Fe$_3$C (Wilkinson and Hellawell 1963, Deschanvres. and Dufournier 1970, Jones and Kurz 1981, Magnin and Kurz 1988, Park and Verhoeven 1996), and Al-Si (Toloui and Hellawell 1976, Elliot and Glenister 1980, Hogan and Song 1987, Magnin et al. 1991, Liu and Elliot 1993, Gunduz et al. 2003) and comparing them to the second part of Eq. 10.21 it was found that an appropriate value for ξ_{min} would be $\xi_{min} = f_\alpha$, which means that the minimum spacing is determined by the fraction of majority phase, f_α. Then:

$$\lambda_{min} \Delta T_{ex} = \frac{1 + f_\alpha}{f_\alpha} \mu_r \quad \text{or} \quad \lambda_{min} = \frac{1 + f_\alpha}{2 f_\alpha} \lambda_{ex} \tag{10.24}$$

Similarly, on the curve with constant ΔT_{ex} we also have $\left(\frac{\lambda}{V}\frac{dV}{d\lambda}\right)_{max} = \xi_{max} - 1$. Proceeding as for the minimum spacing an equation similar to Eq. 10.24 is derived, in which the subscript min is substituted by max. The maximum spacing is determined by the fraction of minority phase, f_β. Then, assuming $\xi_{max} = f_\beta$, we have:

$$\lambda_{max} \Delta T_{ex} = \frac{1 + f_\beta}{f_\beta} \mu_r \quad \text{or} \quad \lambda_{max} = \frac{1 + f_\beta}{2 f_\beta} \lambda_{ex} \tag{10.25}$$

With the minimum and maximum spacings known, the average spacing is simply:

$$\lambda_{av} = \frac{\lambda_{min} + \lambda_{max}}{2} \tag{10.26}$$

Obviously, for $\xi = 1$ both Eq. (10.24) and (10.25) express the extremum condition for eutectic growth.

The results obtained through Eq. 10.24, 10.25 and 10.26 for lamellar spacing and undercooling of Fe–graphite, Fe–Fe$_3$C, and Al–Si eutectics are in very good agreement with the experimental measurements, as illustrated in Fig. 10.13 for Fe–graphite eutectics. Application 10.2 and 10.3 present detailed calculations for the Fe-Gr and Fe-Fe$_3$C systems, respectively. From Fig. 10.42 it can be concluded that the minimum spacing is determined by the austenite, while the maximum spacing is determined by the graphite. As can be observed on Fig. 10.13a, the experimental data for the Fe–C and Fe–C–Si alloys depart from the theoretical line at velocities above 10^{-5} m/s. This is because, as noted by Magnin and Kurz, above this velocity the structure becomes degenerate and exhibits a higher spacing than that predicted by theory. It is also noted that small sulfur additions of 0.01 % increase the spacing considerably.

The correlation between the measured undercooling for three different Fe–C alloys and the model calculation is presented in Fig. 10.13b. An excellent agreement is seen for the Fe–C alloy. Increased silicon content produced higher undercooling than the calculated one. Note that the undercoolings are measure with respect to the equilibrium temperatures, which are 1154.5, 1155.2, and 1157 °C for the Fe–C, Fe–C–0.1 %Si, and Fe–C–0.5 %Si, respectively.

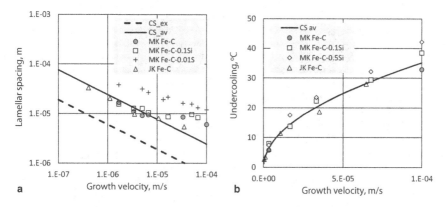

Fig. 10.13 CS model predictions vs experiments for average lamellar spacing (**a**), and growth undercooling (**b**), in the Fe–Cgr eutectic system. *CS ex* proposed modified JH model; *CS av* proposed model for irregular eutectics; *MK FeC* Magnin/Kurz 4.31 %C, 27ppmS; *MK Fe–C–0.1Si* Magnin/Kurz 4.21 %C, 0.09 %Si, 24ppmS; *MK Fe–C–0.5Si* Magnin/Kurz 4.08 %C, 0.45 %Si, 16ppmS; *MK Fe–C–0.01S* Magnin/Kurz 4.28 %C, 100ppmS; *JK Fe–C* Jones/Kurz 4.2?%C, < 5ppmS. (Experimental data from Magnin and Kurz 1988; Jones and Kurz 1981)

Fig. 10.14 Growth of the divorced spheroidal graphite–austenite eutectic: graphite spheroids and austenite

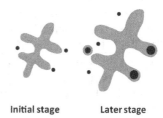

Initial stage   Later stage

10.1.5 Divorced Eutectics

In the case of divorced eutectics, the two phases of the eutectic grow independently from one another. This may happen when in a eutectic forming at the end of solidification of a two-phase system the fraction of remaining liquid is so small that the width is comparable to the eutectic spacing. In such a situation, the second phase may form as single particle or layer between the dendrites.

Another case is that of the spheroidal graphite–austenite eutectic (ductile iron). Solidification starts with primary austenite and/or primary graphite spheroids that grow in contact with the liquid. The graphite may be encapsulated in an austenite shell before having any contact with the primary austenite. The eutectic solidification involves primary/eutectic austenite and graphite spheroids. While the austenite continues to grow in contact with the liquid, the graphite is enveloped in an austenite shell and grows through solid diffusion of carbon through this shell (Fig. 10.14).

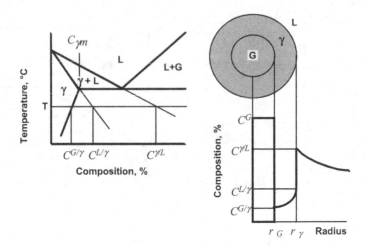

Fig. 10.15 Schematic diagram showing the concentration profile of carbon throughout the liquid and the austenite shell based on the binary phase diagram

The first analytical model to describe growth of the eutectic in spheroidal graphite (SG) iron was proposed in 1956 by Birchenall and Mead. Later, Wetterfall et al. (1972) made calculations of the diffusion-controlled steady-state growth of graphite through the austenite shell (Fig. 10.15) based on Zener's growth equation for an isolated spherical particle in a matrix of low supersaturation. The growth velocity of the γ shell was derived to be (see derivation in inset):

$$\frac{dr_\gamma}{dt} = D_C^\gamma \frac{r_G}{(r_G - r_\gamma)r_\gamma} \frac{C^{G/\gamma} - C^{L/\gamma}}{C^{\gamma/L} - C^{L/\gamma}}, \tag{10.27}$$

where D_C^γ is carbon diffusivity in austenite. This equation can be further simplified if it is assumed that, as demonstrated experimentally, the ratio between the radius of the austenite shell and that of the graphite spheroid remain constant: $r\gamma = 2.4\ r_{Gr}$. The simplified growth rate is (Svensson and Wessen 1998):

$$\frac{dr_{Gr}}{dt} = 2.87 \cdot 10^{-11} \frac{\Delta T}{r_{Gr}}. \tag{10.28}$$

Derivation of Growth Law for Solidification of SG Iron
The basic assumptions are:

- Growth controlled by carbon diffusion through the austenite shell
- Steady-state diffusion of austenite through the austenite shell

Solutal balance at the S/L interface ($r = r_\gamma$ in Fig. 10.15) gives:

$$\frac{d}{dt}\left[\frac{4}{3}\pi r_\gamma^3 \rho_\gamma \left(C^{\gamma/L} - C^{L/\gamma}\right)\right] = 4\pi r_\gamma^2 \rho_\gamma D_C^\gamma \left(\frac{\partial C}{\partial r}\right)_{r=r_\gamma} \quad \text{or}$$

$$V = \frac{dr_\gamma}{dt} = \frac{D_C^\gamma}{C^{\gamma/L} - C^{L/\gamma}}\left(\frac{\partial C}{\partial r}\right)_{r=r_\gamma}.$$

To calculate the carbon flux at the S/L interface, the steady-state diffusion equation assuming constant diffusivity, no advection and no source term is used:

$$D\nabla^2 C = 0 \quad \text{or, in spherical coordinates:} \quad D_L \frac{\partial C}{\partial r}\left(r^2 \frac{\partial C}{\partial r}\right) = 0.$$

The solution of this equation can be obtained by successive integration, as follows: $\partial C/\partial r = C_1/r^2$ and $C = -C_1/r + C_2$ with the boundary conditions:

BC1: at $r = r_G$ $C = C^{G/\gamma}$ and BC2: at $r = r_\gamma$ $C = C^{L/\gamma}$.

After evaluating the constants, the flux at the interface can be obtained from the above equation as:

$$\left(\frac{\partial C}{\partial r}\right)_{r=r_\gamma} = \frac{r_G(C^{L/\gamma} - C^{G/\gamma})}{r_\gamma(r_\gamma - r_G)}.$$

Substituting in the velocity equation we obtain the growth velocity of the austenite shell:

$$V_\gamma = \frac{dr_\gamma}{dt} = D_C^\gamma \frac{r_G}{(r_G - r_\gamma)r_\gamma} \frac{C^{G/\gamma} - C^{L/\gamma}}{C^{\gamma/L} - C^{L/\gamma}}.$$

The radius of the austenite shell is calculated from the velocity equation. Once r_γ is known, r_G is calculated from solute balance (Chang et al. 1992):

$$\rho_G \frac{4}{3}\pi \left[\left(r_G^{i+1}\right)^3 - \left(r_G^i\right)^3\right]C_G = \rho_\gamma \frac{4}{3}\pi \left[\left(r_\gamma^{i+1}\right)^3 - \left(r_\gamma^i\right)^3\right]C^{L/\gamma}. \quad (10.29)$$

This model has survived the test of time and is used today in most computational models for microstructure evolution in one form or another. This approach has also been extended to the eutectoid transformation for the austenite–ferrite transformation.

Although the mathematical description of the eutectic solidification of SG iron previously described ignores the contribution of the austenite dendrites, it can predict surprisingly well the microstructural outcome. It is believed that the mass balance equations used in these models are smoothing out the error, by attributing the dendritic austenite to the austenitic shell. Nevertheless, Lacaze et al. (1991) included calculation of the off-eutectic austenite by writing the mass balance as follows:

$$\rho_G \int_0^{r_G} r^2 dr + \rho_\gamma \int_{r_G}^{r_\gamma} r^2 dr + [\rho_L g_L + \rho_\gamma (1 - g_L)] \int_{r_\gamma}^{r_0} r^2 dr = \rho_L \frac{r_0^3}{3}.$$

Here, r_o is the final radius of the volume element. In previous models, only the first two terms on the left hand side were included.

Fredriksson et al. (2005) argued on the basis of observations on samples quenched during solidification that spheroidal graphite may be in contact with γ without being surrounded by a shell. They contended that when the nodules are small the interface kinetics produces a high growth rate. An austenite shell will not develop in this case because of lack of sufficient driving force for the plastic deformation of austenite. After the graphite reaches a certain size, the interface kinetics of graphite growth slows down, an austenite shell is formed, and γ/Gr equilibrium can be assumed. At this time, there are no mathematical models that describe this growth sequence.

10.1.6 Interface Stability of Eutectics

Contrary to what the equilibrium diagrams suggest, eutectic-like structures (composites) may be obtained even with off-eutectic compositions. Let us assume that an alloy of composition $C_o < C_E$ is grown in such a way that plane front is maintained. The distribution of solute in the liquid ahead of the interface is given by Eq. (5.5):

$$C_L = C_o + (C_L - C_o) \exp\left(-\frac{V}{D_L}x\right) = C_o \left[1 + \frac{1-k}{k} \exp\left(-\frac{V}{D_L}x\right)\right].$$

At steady state $C_L = C_o/k = C_E$. Thus, the above equation becomes: $C_E = C_o + (C_E - C_o) \exp(-(V/D_L)x)$. Substituting in the JH equation for the eutectic composition, Eq. 10.6:

$$C_E = C_o + (C_E - C_o) \exp\left(-\frac{Vx}{D_L}\right) + \sum_{n=1}^{\infty} B_n \cos\frac{2n\pi y}{\lambda} \exp\left(-\frac{2n\pi x}{\lambda}\right).$$

$$(10.30)$$

Fig. 10.16 Dendritic/eutectic
transition

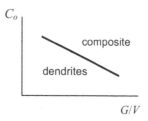

From this equation, it is seen that there are two characteristic distances in the x-direction:

- $\cong \lambda/2$: transverse transport of solute occurs within this layer
- $\cong D_L/V$: as for single-phase crystals; within this layer $C_L > C_o$

For usual solidification velocities, $\lambda/2 << D_L/V$. This means that perturbations in the y-direction dampen out much quicker than those in the x-direction. When $\lambda/2 << D_L/V$ the third RHT vanishes, and Eq. 10.30 reduces to the equation for growth of single phase alloys.

To find the influence of processing conditions on the composition of the resulting solid we will use mass balance at the eutectic front. It gives $(\partial C_L/\partial x)_{x=0} = -(C_E - \overline{C_S})(V/D)$, where $\overline{C_S}$ is the average two-phase composition. Thus, since $\partial C_L/dx = G_L/m$:

$$\overline{C_S} = C_E + (D/V)(G/m). \tag{10.31}$$

For very small G/V ratios the composition of the solid is $\overline{C_S} = C_E$. For high G/V ratios $\overline{C_S} = C_o < C_E$.

Let us try to evaluate the morphology of this solid. In the absence of convection the criterion for interface stability is:

$$G_T \geq \Delta T_o V/D = -m(C_E - C_o)V/D. \tag{10.32}$$

This is the same as Eq. 10.31 for high G/V when $\overline{C_S} = C_o$. If the stability criterion is not satisfied (small G/V), a dendritic structure will result (Fig. 10.16). Eutectic will still form at the end of solidification. If the interface is stable (high G/V), a composite structure will form, having a eutectic-like morphology.

From the eutectic phase diagram, it is apparent that a eutectic structure can be obtained only when the composition is exactly eutectic. Nevertheless, experimental observations, as well as the preceding theoretical analysis show that, depending on the growth conditions, eutectic microstructures can be obtained at off-eutectic compositions. This is possible because the eutectic grows faster than the dendrites, since diffusion-coupled growth is much faster than isolated dendritic growth. Accordingly, even in off-eutectic compositions, the eutectic may outgrow the individual dendrites, resulting in a purely eutectic microstructure. On the other hand, at high growth velocities, dendrites can be found in alloys of eutectic compositions.

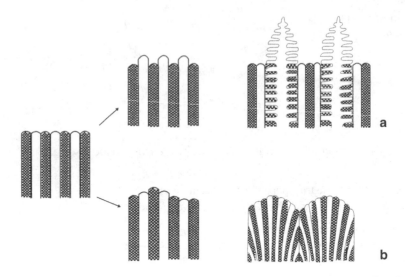

Fig. 10.17 Instability of the eutectic interface (after Kurz and Fisher 1998). **a** Instability of one phase resulting in dendrites + interdendritic eutectic. **b** Instability of both phases resulting in two-phase eutectic cells or dendrites

From the preceding analysis it is clear that, depending on the processing conditions of a two-component alloy, stable or unstable interfaces may be obtained. A third alloying element, which is partitioned between both solid phases will lead to two-phase instability. The influence of the G/V ratio and of composition on the solidification microstructure of off-eutectic alloys can be summarized as follows:

- High G/V, no third element—composite (eutectic-like structure)
- Low G/V, no third element—dendrites + planar (lamellar) eutectic (Fig. 10.17a); this is because one phase becomes heavily constitutionally undercooled
- Low G/V, some third element—two phase eutectic cells (colony; Fig. 10.17b), or dendrites + eutectic cells; this is because both phases are constitutionally undercooled
- Lower G/V, some third element—dendrites + eutectic grains; a long range diffusion boundary layer is established ahead of the S/L interface

An analysis of the possible solidification microstructure of a binary alloy can be made based on the growth velocities of the competing phases. As shown in Fig. 10.18a, the phase that will be present in the final microstructure at a given undercooling is that one which has the highest solidification velocity. Thus, this is a kinetic effect.

For regular eutectics, where the two primary dendritic phases have similar undercooling, the coupled zone is symmetric. Consider the solidification of the slightly hypereutectic alloy in Fig. 10.18. At small undercooling, the eutectic has the highest growth velocity and a planar, coupled eutectic solidifies. At higher undercooling,

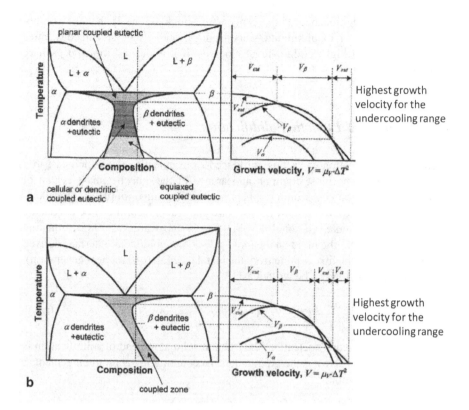

Fig. 10.18 Coupled eutectic zones. **a** Symmetric coupled zone (regular eutectic). **b** Asymmetric coupled zone (irregular eutectics)

the β phase will have higher growth velocity, and a structure made of eutectic and dendrites will result. At even higher undercooling, the eutectic velocity will become again the highest. However, because of the undercooling, a planar structure is not possible, and equiaxed coupled growth will result.

If one of the eutectic phases is faceted, the growth of this phase and consequently that of the eutectic is slowed down. Dendrites of the other phase may grow faster at a given undercooling than the eutectic, even for eutectic composition. Consequently, purely eutectic microstructures can be obtained only at hypereutectic compositions. This is exemplified in Fig. 10.18b, for the case of faceted β phase. An asymmetric coupled zone results.

The coupled zone is thus the solidification velocity dependent composition region in which the eutectic grows more rapidly, or at a lower undercooling, than the α or β dendrites. Note that the widening of the coupled zone near the eutectic temperature is observed only in directional solidification, where the thermal gradient is positive.

The practical significance of the concept of coupled zone is that the composition of cast iron or of aluminum–silicon alloys, both irregular eutectics, must be hypereutectic at high solidification velocities, if it is desired to avoid primary dendrites.

10.1.7 Equiaxed Eutectic Solidification

During the solidification of castings produced from eutectic alloys such cast iron and Al–Si alloys, the most common appearance of the eutectic phase is that of equiaxed eutectic grains. Assuming spherical eutectic grains, simple analytical calculations of the time evolution of temperature and fraction solid can be conducted (e.g., Fredriksson and Svensson 1985; Stefanescu and Kanetkar 1985). Equating the heat flow rate into the mold on the mold side at the mold/metal interface derived after Eq. 6.22 with the flux coming from the metal (see Eq. 6.10 for heat generation), the governing equation can be written as:

$$\sqrt{\frac{k_m \rho_m c_m}{\pi t}}(T - T_o)A = -v\rho c \frac{dT}{dt} + v\rho \Delta H_f \frac{d f_S}{dt}. \qquad (10.33)$$

Assuming instantaneous nucleation, the fraction solid at the end of solidification is $f_S = (4/3)\pi N r^3$, where r is the grain radius. Accounting for grain impingement at the end of solidification:

$$f_S = 1 - exp\left(-\frac{4\pi}{3}N r^3\right). \qquad (10.34)$$

The grain radius can be calculated from the growth velocity of the spherical eutectic grains as developed by Tiller (1969). The model predicted that the correlation between solidification velocity and lamellar spacing obeys the relationship $\lambda V^{1/2} = 4 \cdot 10^{-6} \text{cm}^{3/2}\text{s}^{-1/2}$. The growth velocity can be calculated as a function of the bulk undercooling with:

$$\frac{dr}{dt} = V = \mu_V \Delta T_{bulk}^2, \qquad (10.35)$$

where the growth coefficient deviates from that calculated from the JH theory. Typically, values that are calculated from experiments are (see Appendix B Table B4). Using inverse kinetic analysis of experimental data on eutectic lamellar graphite iron, Dioszegi and Svensson (2005) calculated a different relationship: $dr/dt = 48 \cdot 10^{-8} \Delta T_{bulk}^{0.66}$, in line with earlier findings by Thorgrimsson (1986).

　　Examples of calculation of the cooling curve and fraction solid evolution for the equiaxed eutectic solidification of eutectic and hypoeutectic cast irons are presented in Application 10.4 and Application 10.5, respectively.

The morphology of eutectics can be altered by small additions of elements. This process is called *modification*. Its practical purpose is to refine and compact the brittle component of the eutectic phase, when such a phase exists, or to eliminate the occurrence of a brittle eutectic. Typical examples are modification of Al–Si alloys with strontium to refine the Si phase, or the modification of lamellar graphite cast iron with Mg to spheroidize the graphite.

Modification differs from *grain refinement* in that grain refinement affects only the nucleation of the primary dendritic phase, while modification controls the undercooling of the eutectic phase. As discussed, growth of the eutectic is controlled by its undercooling. Thus, it is possible to combine different modification treatments with the same grain refinement to obtain entirely different eutectic phases, and thus entirely different casting structures. The most common example of this is cast iron. In cast iron, the graphite eutectic morphology can be controlled by a modification treatment to produce either the flake form found in gray iron, or the spheroidal form found in ductile iron. The liquid treatment of lamellar graphite iron with ferrosilicon, which is called inoculation, is in fact a modification process since it results in the elimination of the formation of the brittle iron–carbide eutectic. Modification of the eutectic Al–Si alloys with Na or Sr to refine the silicon plates is another example.

As some of the most important casting alloys, cast iron and Al–Si alloys solidify with equiaxed eutectic grains, they will be discussed in more details in Ch.19.

10.2 Peritectic Solidification

10.2.1 Classification of Peritectics

Three different types of peritectic systems can be considered (Fig. 10.19). In the first type (left), the β-solidus and the β-solvus lines have slopes of the same sign. At the peritectic invariant $\alpha + L \rightarrow \alpha + \beta$. In the second type (middle), the β-solidus and the β-solvus have opposite sign slopes. At the peritectic invariant $\alpha + L \rightarrow \beta$ solid solution. The third type (right), has a very narrow or no solubility region. At the peritectic invariant $\alpha + L \rightarrow \beta$ compound. The first two diagrams are typical for metallic alloys such as steel, while the third one is typical for rare earth permanent magnet alloys (Nd–Fe–B) and ceramic superconductors.

Depending on the G/V ratio, a planar or dendritic interface can be obtained. If the G/V ratio is sufficiently high, an $\alpha + \beta$ composite of uniform composition will solidify with planar interface even for off-peritectic compositions.

The volume fraction of each phase will be given by the lever rule if the alloy solidifies under equilibrium conditions. A simple Scheil-model has also been proposed (Flemings 1974). An example of such a calculation is given in Application 10.6 for the case of a system with cascading peritectics. Nevertheless, since kinetics and the diffusion rate in the solid phases are determining the time for reaching equilibrium, in most cases, neither the lever rule nor the Scheil model will give the correct volume fraction of the different phases.

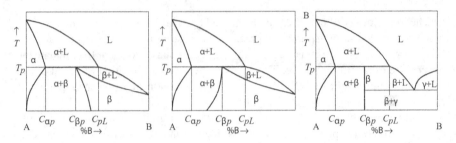

Fig. 10.19 Types of peritectics

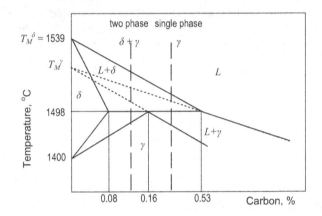

Fig. 10.20 Schematic phase diagram of the peritectic region of carbon steel

Control of the α to β transition is essential for the quality of the product in steel as well as in many other alloys. The phase diagram of the region of interest for steel is shown in Fig. 10.20. At temperatures less than 1498 °C the solidification microstructure of low-carbon steel is single phase austenite. Yet, depending on the carbon content the solidification path can be quite different. At carbon contents less than 0.16 %, at the end of peritectic solidification both δ ferrite and γ phase coexist, while over 0.16 % C liquid and γ coexist. At carbon contents higher than 0.53 % only austenite solidifies from the liquid. Also shown on the figure are the metastable extensions of the γ phase that could form at any composition directly from the liquid if nucleation of the δ phase is suppressed. Such suppression of the pro-peritectic phase has been demonstrated in a number of peritectic binary melts, including Fe–Mo, Co–Si and Al–Co alloys (Löser et al. 2004). An extensive review of peritectic solidification and its mechanism has been done by Stefanescu (2006).

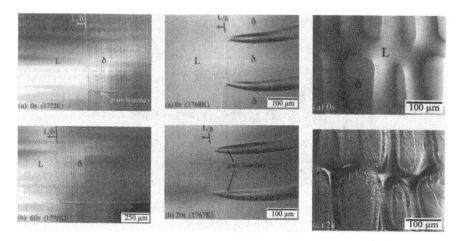

Fig. 10.21 Typical microstructures of peritectics in Fe–0.14 %C alloys. **a** Planar growth of δ crystals ($G_T = 22$ K/mm, $V = 2.5$ μm/s). **b** Cellular growth of δ crystals ($G_T = 4.3$ K/mm, $V = 2.5$ μm/s). **c** Peritectic reaction and transformation during solidification at 1768 K ($G_T = 4.3 K/mm$, $V = 38.7$ μm/s). (Shibata et al. 2000, with permission of Springer Science and Business Media)

10.2.2 Peritectic Microstructures and Phase Selection

Direct evidence of the mechanisms of peritectic solidification was provided recently through in situ dynamic observation of the progress of peritectic reactions and transformations of Fe–C alloys made with a combination of a confocal scanning laser microscope and an infrared image furnace (Shibata et al. 2000). Selected micrographs are presented in Fig. 10.21. It is observed that as the thermal gradient decreases from 22 K/mm (left column) to 4.3 K/mm (middle column) the δ solid/liquid interface becomes unstable and changes from planar to cellular. Further increase in solidification velocity from 2.5 to 19.3 μm/s showed that γ phase starts growing at the boundaries of the δ cells. Upon further increase of velocity to 38.7 μm/s, island like δ crystals appeared (Fig. 10.21, right column, a) which then underwent peritectic reaction and transformation (Fig. 10.21, right column, b). The wrinkles observed on the γ crystals that transformed from the δ crystals are thought to be due to the volume contraction of the transformation.

Similar observations were also made for the Fe–Ni system (McDonald and Sridhar 2003). The two stages of the peritectic transition involving the reaction (austenite growing along the liquid–ferrite interface) and the transformation (direct solidification of austenite from the liquid) were observed.

In general, a variety of microstructures can result from peritectic solidification, mostly depending on the temperature gradient/solidification velocity (G_T/V) ratio and nucleation conditions. The possible structures include cellular, plane-front, bands, and eutectic-like structures.

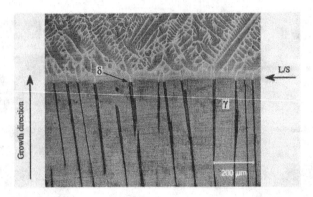

Fig. 10.22 Quenched S/L interface of simultaneous two-phase growth in peritectic Fe–Ni alloy. (Vandyoussefi et al. 2000 with permission of Elsevier)

Simultaneous growth of two phases as oriented fibers and lamellae has been observed in some peritectic alloys when the composition was on the tie-line of the two solid phases and the G_T/V ratio was close to the limit of constitutional undercooling for the stable phase having the smaller distribution coefficient (Vandyoussefi et al. 2000). An example of such a structure for a Fe–Ni alloy is presented in Fig. 10.22 Fig. 10.22.

Banded structures have been observed in peritectic alloys at low growth rates (Tokieda et al. 1999). An example is provided in Fig. 10.23a. The formation of bands is explained by nucleation and growth of the second phase during the initial transient of planar growth of the primary phase and vice versa. This occurs because the liquid at and ahead of the growing interface is constitutionally undercooled with respect to the other phase. As the second phase nucleates and grows ahead of the primary phase, the former phase cannot reach the steady state. Similarly, the primary phase nucleates again during the transient growth regime of the second phase, preventing it for reaching the steady state. Consequently, a cycle is set up leading to the layered microstructure (Trivedi 1995; Yasuda et al. 2003).

Several other structures can be obtained depending on the relative importance of nucleation diffusion and convection (Fig. 10.23 Boettinger et al. 2000). Theoretical models and experimental studies in thin samples suggest that the structures (a)–(e) can form under diffusive regime, while microstructure (f) requires the presence of oscillatory convection in the melt.

In an attempt to rationalize this plethora of peritectic microstructures, prediction of phase and microstructure selection was attempted by generating microstructure selection diagrams for peritectics. The main variables controlling microstructure evolution include interface velocity (V), thermal gradient (G_T), alloy composition (C_o), and nucleation potential.

Assuming that the leading phase that is the phase that growth at the highest interface temperature is the kinetically most stable one, Umeda et al. (1996) developed an equation that describes the transition velocity from δ dendrites to γ dendrites in

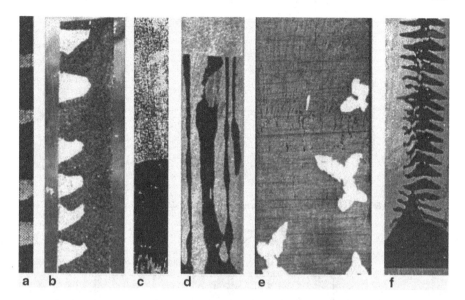

Fig. 10.23 Possible microstructures in peritectic alloys with compositions in the two-phase region. Solidification direction: upward. **a** Discrete bands of the two phases. **b** Partial bands or islands of one phase in the matrix of the other phase. **c** Single primary to peritectic phase transition. **d** Simultaneous growth of the two phases with a planar S/L interface. **e** Dispersed phases due to nucleation ahead of the interface. **f** Oscillating continuous tree-like structures of the primary phase surrounded by peritectic. (Boettinger et al. 2000 with permission from Elsevier)

directionally solidified alloys:

$$V_{tr}^{\delta-\gamma} = \frac{D_L}{4\pi^2} \left(\frac{\Delta T_m^{\delta-\gamma} - C_o \Delta m_L^{\delta-\gamma}}{\left(C_o m_L^{\delta} (k_\delta - 1) \Gamma_\delta \right)^{1/2} - \left(C_o m_L^{\gamma} (k_\gamma - 1) \Gamma_\gamma \right)^{1/2}} \right), \quad (10.36)$$

where D_L is the liquid diffusivity, $\Delta T_m^{\delta-\gamma}$ and $m_L^{\delta-\gamma}$ are the melting point difference and the liquidus slope difference between δ and γ, respectively, and k_i and Γ_i are the partition coefficient and the Gibbs–Thomson coefficient of the γ or δ phase. The numerator in the parenthesis represents the difference in liquidus temperature between the two phases (effect of phase equilibria on V_{tr}), and the denominator represents the difference in growth kinetics between the dendrites of the two phases.

Equation 10.36 was plotted in Fig. 10.24b as a function of composition for Fe–Ni alloys whose phase diagram is presented in Fig. 10.24a. It is noticed that for a given composition of 4.2 at% Ni, as the interface velocity increases, a transition from δ to γ dendrites occurs at about $8 \cdot 10^{-2}$ m/s. This model cannot explain band formation, which is apparently the result of nucleation and growth of the second phase during the initial transient of planar solidification of the primary phase and vice versa. According to Trivedi (1995), the liquid ahead of the growing interface is constitutionally undercooled with respect to the other phase. As the second phase nucleates and grows ahead of the primary phase, the former phase cannot

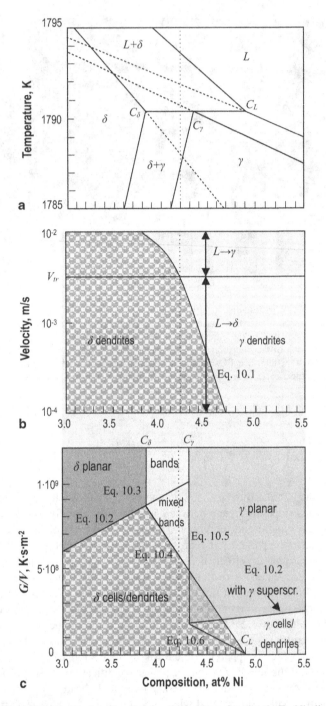

Fig. 10.24 Phase selection during directional solidification of peritectic Fe–Ni alloys. (Umeda et al. 1996; Hunziker et al. 1998)

reach steady state. Then, the primary phase nucleates ahead of the growing second phase preventing it from reaching steady state. Thus, a cycle leading to banded microstructure is set up. By combining the maximum growth temperature criterion with nucleation considerations, Hunziker et al. (1998) developed a microstructure selection diagram for peritectic alloys close to the limit of constitutional undercooling. The diagram, presented in Fig. 10.24c for Fe–Ni alloys, assumes negligible nucleation undercooling for both δ and γ phases and allows prediction of planar front, cellular, dendritic, and band solidification. The transition lines on the G/V–%Ni graph are calculated with the equations presented in the following text and plotted on Fig. 10.24c.

The transition from planar-to-cellular growth of the δ phase is given by the limit of constitutional undercooling:

$$\frac{G}{V} \geq \frac{-m_L^\delta (C_L - C_o) - m_S^\delta (C_o - C_\delta)}{D_L}, \tag{10.37}$$

where C_L is the composition of the liquid at the equilibrium peritectic temperature m_L^δ and m_S^δ are the liquidus and solidus slope of the δ phase, respectively. By substituting the superscript δ with γ, the equation can be adapted to describe the transition from planar-to-cellular growth of the γ phase.

The stability condition for planar δ with respect to γ nucleation was calculated by comparing the interface temperature with the nucleation temperature of γ, which resulted in the following equation:

$$C_o < C_\delta - \frac{m_L^\delta \Delta T_N^\gamma}{m_S^\delta (m_L^\delta - m_L^\gamma)}, \tag{10.38}$$

where ΔT_N^γ is the nucleation undercooling. Note that when the nucleation undercooling is negligible the condition reduces to $C_o < C_\delta$, as shown in Fig. 10.24c.

Similarly, by comparing the interface temperature with the nucleation temperature of γ, the stability condition for cellular δ with respect to γ nucleation was derived as:

$$\frac{G}{V} < \frac{m_L^\delta}{D_L}(C_o - C_L) + \frac{\Delta T_N^\gamma}{m_L^\delta - m_L^\gamma}. \tag{10.39}$$

The stability condition for planar γ with respect to δ is given by an equation similar to Eq. 10.38:

$$C_o > C_\gamma - \frac{m_L^\gamma \Delta T_N^\delta}{m_S^\gamma (m_L^\gamma - m_L^\delta)}. \tag{10.40}$$

When the nucleation undercooling is negligible this stability condition reduces to $C_o > C_\gamma$, as shown in Fig. 10.24c.

Additional stability conditions that will not be presented here are invoked to derive the stability condition between the two phases in the cellular regime:

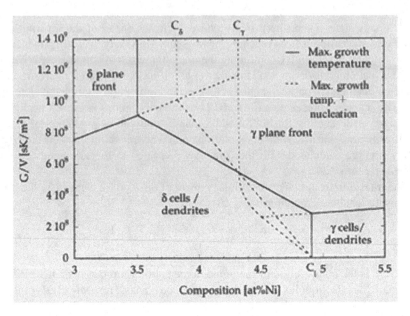

Fig. 10.25 Microstructure selection maps using the maximum temperature prediction criterion alone or in combination with nucleation considerations. (Hunziker et al. 1998 with permission of Elsevier)

$$\frac{G}{V} > \frac{\left(m_L^\delta - m_L^\gamma\right)(C_o - C_L) - \Delta T_N^\delta}{D_L \ln\left(m_L^\delta / m_L^\gamma\right)}. \tag{10.41}$$

In the region marked "bands" on Fig. 10.24, neither δ nor γ are stable at steady state, and either phase can nucleate ahead of the other's plane front. This is the condition for bands formation. In the region marked "mixed bands" alternate layers of cellular δ and planar γ are expected to form.

Nucleation undercooling can significantly affect the extent of the bands. Indeed, as the nucleation undercooling increases the stability lines defined by Eqs. 10.38 and 10.39 move to the right, while that defined by Eq. 10.40 moves to the left.

Finally, a comparison between predictions with the maximum growth temperature criterion and the combined maximum growth temperature—nucleation model is summarized in Fig. 10.25. The importance of nucleation is quite clear.

10.2.3 Mechanism of Peritectic Solidification

Two different mechanisms are involved in peritectic solidification, namely *peritectic reaction* and *peritectic transformation* (Kerr et al. 1974). These mechanisms are

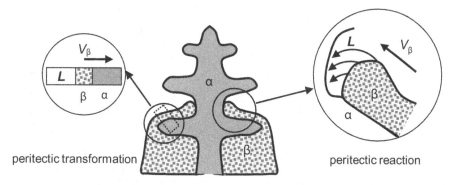

Fig. 10.26 Mechanisms of peritectic solidification. (Stefanescu 2006)

presented schematically in Fig. 10.26. The peritectic solidification starts with a *peritectic reaction* in which all three phases, α, β, and liquid are in contact with each other. The peritectic β phase will grow along the S/L interface α/L, driven by liquid supersaturation. Solute rejected by the β phase will diffuse through the liquid to the α phase contributing to its dissolution. The β phase will also thicken in the direction perpendicular to its growth, by direct growth in the liquid and at the expense of the α phase by solid state diffusion. Once the reaction is completed and all the α/L interface is covered by β, the *peritectic transformation* starts. The liquid and the primary α phase are isolated by the β phase. The transformation $\alpha \Rightarrow \beta$ takes place by long-range solid-state diffusion through the peritectic β phase. The β phase grows by direct solidification in the liquid.

The Rate of the Peritectic Reaction Depending on surface energy conditions, two different types of the peritectic reactions can occur (Fredriksson 1988):

- Nucleation and growth of the β crystals in the liquid without contact with the α crystals. Following nucleation, the secondary phase grows freely in the liquid, while the primary phase will dissolve.
- Nucleation and growth of the β crystals in contact with the primary α phase. This is the most common. Following nucleation, lateral growth of the β phase around the α phase takes place (Fig. 10.26).

The second type of peritectic reaction, which is the propagation of the triple point $L/\beta/\alpha$ along the L/α boundary of planar α crystals, consists of the dissolution of the α phase and growth of the β phase. It is controlled by the growth of β since dissolution is the fastest process. Bosze and Trivedi (1974) simplified an earlier equation developed by Trivedi to describe the relative contributions of diffusion, surface energy and interface kinetics during the growth of parabolic shape precipitates. Using their model it appears that the peritectic reaction is controlled by undercooling and liquid diffusivity according to the equations:

$$V_\gamma = \frac{9}{8\pi} \frac{D_L}{r} \frac{\Omega^2}{(1 - 2\Omega/\pi - \Omega^2/2\pi)^2} \quad \text{with} \quad \Omega = \frac{C_{L\beta} - C_{L\alpha}}{C_{L\beta} - C_{\beta L}}, \qquad (10.42)$$

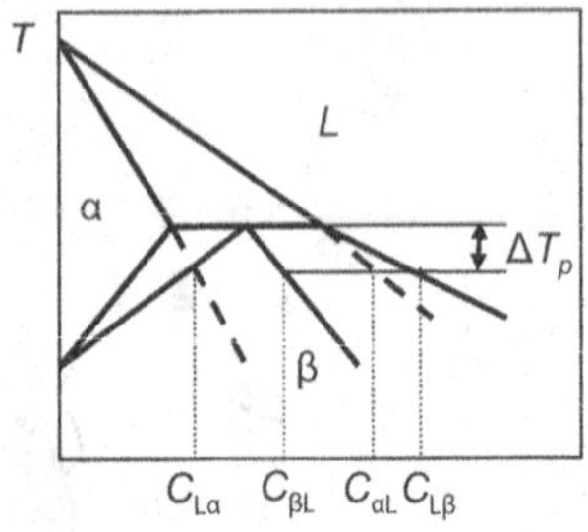

Fig. 10.27 Definition of concentration terms in Eq. 10.9

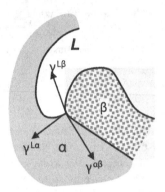

Fig. 10.28 Surface energy of phases involved in the peritectic reaction

where r is the radius of the leading edge (the plate will have a thickness of $2r$), and C_{ij} are interface concentrations (see Fig. 10.27 for definitions).

Using the maximum growth rate theory (Fredriksson and Nylen 1982; Hunziker et al. 1998) it can be shown that the thickness of β increases with lower solidification velocity and with larger surface energy difference $\Delta \gamma = \gamma^{L\beta} + \gamma^{\alpha\beta} - \gamma^{L\alpha}$ (see Fig. 10.28).

However, in situ observation in Fe–C systems (Shibata et al. 2000) showed that the experimental velocities were much higher than the ones predicted with this model. This suggests that the peritectic reaction is not controlled by carbon diffusivity in the liquid, but perhaps by either massive transformation of α into β, or direct solidification of β from the liquid. Support of theses hypotheses was brought recently by the experimental work of Dhindaw et al. (2004) who studied the peritectic reaction in medium–alloy steel (0.22 %C, 1.3 %Cr, 2.6 %Ni). Microsegregation measurements on directionally and isothermally solidified samples showed that when the segregation ratio for Ni is higher than that for Cr a peritectic reaction has occurred. However, when the segregation ratio for Cr was higher than for Ni, the liquid was transformed directly into β without undergoing a peritectic reaction. Based on the evaluation of the energy of transformation through differential thermal analysis the authors concluded that the transformation is a diffusionless transformation $\alpha \Rightarrow \beta$.

The Rate of the Peritectic Transformation The thickness of the β layer will normally increase during subsequent cooling through diffusion through the β layer,

Fig. 10.29 Definition of
quantities in Eq. 10.44

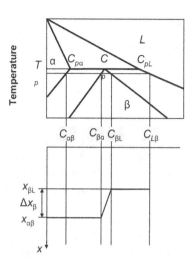

precipitation of β directly from the liquid, and precipitation of β directly from the
α phase.

Both the precipitation of β directly from the liquid and the solid, and the diffusion
process through the β layer, depend on the shape of the phase diagram and the
cooling rate. In addition, the diffusion process through the β layer depends on the
diffusion rate.

Assuming isothermal transformation, the growth of the β layer is controlled by
the diffusion rate through the layer at a temperature just below the peritectic temperature. The notations on the phase diagram and the concentration profile are shown
in Fig. 10.29. Under these assumptions, the mass balance at the interface can be
written as (Hillert 1979):

$$D\left(\frac{dC}{dx}\right)_{x=0} = -V \cdot \Delta C. \qquad (10.43)$$

Then, for the $\alpha\beta$ interface and for the βL interface we have, respectively:

$$D_\beta \left|\frac{\partial C_\beta}{\partial x}\right|_{x=x_{\alpha\beta}} = -\frac{dx_{\alpha\beta}}{dt}[C_{\alpha\beta} - C_{\beta\alpha}] \quad \text{and}$$

$$D_\beta \left|\frac{\partial C_\beta}{\partial x}\right|_{x=x_{\beta L}} = -\frac{dx_{\beta L}}{dt}[C_{\beta L} - C_{L\beta}], \qquad (10.44)$$

where D_β is the diffusion coefficient in the β phase. All other terms are concentrations that are defined in Fig. 10.29. Assuming that the concentration gradient
through the β phase is constant, it can be expressed as:

$$\frac{\partial C_\beta}{\partial x} = \frac{C_{\beta L} - C_{\beta\alpha}}{x_{\beta L} - x_{\beta\alpha}} = \frac{C_{\beta L} - C_{\beta\alpha}}{\Delta x}.$$

Substituting in the above two equations and adding the equations one obtains:

$$\frac{d(\Delta x)}{dt} = \frac{D_\beta}{C_{\beta\alpha} - C_{\alpha\beta}} \frac{C_{\beta L} - C_{\beta\alpha}}{\Delta x} + \frac{D_\beta}{C_{\beta L} - C_{L\beta}} \frac{C_{\beta L} - C_{\beta\alpha}}{\Delta x} = \frac{D_\beta}{\Delta x}[\Omega_{\alpha\beta} + \Omega_{\beta L}],$$

where $\Omega_{\alpha\beta} = \frac{C_{\beta L} - C_{\beta\alpha}}{C_{\beta\alpha} - C_{\alpha\beta}}$ and $\Omega_{\beta L} = \frac{C_{\beta L} - C_{\beta\alpha}}{C_{L\beta} - C_{\beta L}}$. Integrating:

$$\Delta x_\beta = [2D_\beta(\Omega_{\alpha\beta} + \Omega_{\beta L})t]^{1/2}. \tag{10.45}$$

This equation shows that the thickness of the β layer and the growth rate, increase with increasing undercooling. Indeed, at the peritectic temperature, $C_{\beta L} - C_{\beta\alpha} = 0$, but the difference increases with higher undercooling. This equation also shows that the growth rate is dependent on diffusivity. For substitutionally dissolved alloying elements (e.g. Fe–Ni) in face-centered cubic metals, the diffusion coefficient near the melting point is of the order of 10^{-13} m^2/s. In such a case, the growth rate will be very low and the time for the peritectic transformation will be very large. In a normal casting process, the reaction rate will be so low that the amount of β phase formed by the peritectic transformation will be negligible in comparison with the precipitation of β from the liquid.

For interstitially dissolved elements in BCC metals (e.g. Fe–C) the diffusion rates are much higher and the peritectic transformation is completed within 6–10 K of the equilibrium temperature (Chuang et al. 1975). The rate controlling phenomenon is carbon diffusion. Indeed, in situ observation in Fe–C systems (Shibata et al. 2000) demonstrated that the growth of the γ phase thickness follows a parabolic law. This supports the opinion that carbon diffusion determines growth rate. Calculations with a simple finite difference model (Ueshima et al. 1986) showed good agreement between calculated and experimental migration distances in time of the γ/δ and L/γ interfaces.

Fredriksson and Stjerndahl (1982) expanded Hillert's model to continuous cooling by assuming that the boundary conditions change during cooling and that the cooling rate is constant. From the isothermal equation they calculated that the thickness of the γ layer is given by:

$$\Delta x = ct. \cdot \left(\frac{dT}{dt}\right)^{-1} \cdot \Delta T_p. \tag{10.46}$$

This equation was used to plot the peritectic temperature range as a function of the carbon content (see Fig. 10.30). It is seen that the reaction is relatively fast and is finished at maximum 6 or 10 K below the peritectic temperature, depending on the cooling rate.

For multicomponent alloys it is necessary to use numerical models that depend on microsegregation models that describe the multiple solutal fluxes (Thuinet et al. 2003).

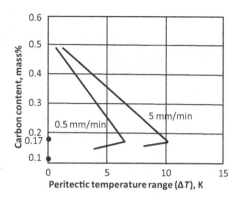

Fig. 10.30 Temperature range of peritectic reaction in Fe–C alloys as a function of carbon content and solidification velocity at a temperature gradient of 60 K cm^{-1}. (After Fredriksson and Stjerndahl 1982)

Growth of Banded (Layered) Peritectic Structure

The formation of layered structures has been observed in several peritectic systems including Sn–Cd and Zn–Cu (Boettinger 1974), Sn–Sb (Titchener and Spittle 1975), Ag–Zn (Ostrowski and Langer 1977), and Pb–Bi (Barker and Hellawell 1974; Tokieda et al. 1999). Layered structures have been observed in both hypo-peritectic and hyper-peritectic systems, but always at high G/V ratios. In principle the banded structure results when the second phase nucleates ahead of the planar primary phase. If the lateral growth of the secondary phase is higher than the normal growth of the primary phase, a planar band of secondary phase will form ahead of the planar primary phase. If lateral growth is slower than normal growth incomplete bands will result. Some typical microstructure for layered growth are given in Fig. 10.23a, 10.23b, 10.23f and in Fig. 10.31.

To determine conditions that control the volume fraction and spacing of bands Trivedi (1995) has developed a model based on the following assumptions: solute transport by liquid diffusion only; no liquid convection; negligible diffusion in solid; growth conditions are such as to produce planar S/L interface. The widths of the α and β layers were derived to be:

$$\lambda_i = \frac{D_L}{V k_i} \ln \Lambda_i, \tag{10.47}$$

where $i = \alpha$ or β, the concentrations and slopes are defined as in Fig. 10.32, and the functions Λ_i are given by:

$$\Lambda_\alpha = \left[1 - \frac{C_{\alpha p}}{C_o} \left(1 - \frac{\Delta T_N^\alpha}{C_p \left(m_L^\beta - m_L^\alpha \right)} \right) \right] \Big/ \left[1 - \frac{C_{\alpha p}}{C_o} \left(1 + \frac{\Delta T_N^\beta}{C_p \left(m_L^\beta - m_L^\alpha \right)} \right) \right]$$

and

$$\Lambda_\beta = \left[1 - \frac{C_{\beta p}}{C_o} \left(1 - \frac{\Delta T_N^\beta}{C_p \left(m_L^\beta - m_L^\alpha \right)} \right) \right] \Big/ \left[1 - \frac{C_{\beta p}}{C_o} \left(1 + \frac{\Delta T_N^\alpha}{C_p \left(m_L^\beta - m_L^\alpha \right)} \right) \right]. \tag{10.48}$$

Fig. 10.31 Banded structure in a Pb-33 at% Bi alloy grown at $G = 2.7 \cdot 10^4$ K m^{-1} and $V = 0.56$ μm s^{-1}. *Black α phase, white β phase*. (Barker and Hellawell 1974)

Fig. 10.32 Definition of quantities in Eq. 10.48

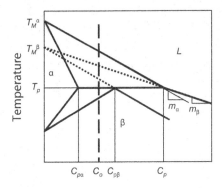

From the analysis of the last two equations it is apparent that the thickness of the layers and their periodicity scales inversely with velocity for a given composition. It is also clear that the nucleation undercooling of the two phases plays a significant role in establishing the length scale of the layers.

However, in experiments by Tokieda et al. (1999) with hyper-peritectic Pb-33 at% Bi alloy, no dependence of the band spacing on velocity was observed. They explained the morphology of the banded structure based on the interface temperature. As the temperature gradient increases, the growth rate required for a planar interface also increases, decreasing macrosegregation in the growth direction. Therefore, not low growth rate but high temperature gradient is important for the formation of banded structures.

The periodicity of the layers can then be written as:

$$\lambda = \lambda_\alpha + \lambda_\beta = \frac{D_L}{V}(\ln \Lambda_\alpha^{1/k_\alpha} + \ln \Lambda_\beta^{1/k_\beta}). \qquad (10.49)$$

10.3 Monotectic Solidification

Monotectic solidification occurs in alloys where the liquid separates into two distinct liquid phases of different composition during cooling. While these alloys have limited commercial applications, one could mention Pb containing copper alloys, and the attempt to fabricate thin microfilters, as possible applications. In this latter case,

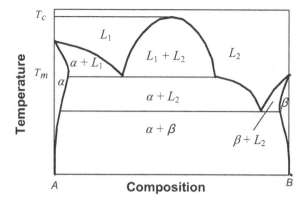

Fig. 10.33 Phase diagram showing a monotectic reaction at temperature T_m

Grugel and Hellawell (1981) have pursued directional solidification of monotectic alloys followed by the selective etching of the fibers.

A dome-shaped region within which the two liquids mix and coexist is seen on the phase diagram (Fig. 10.33). The maximum temperature of this dome, T_c, is called the critical (or consolute) temperature. At the monotectic temperature, T_m, the monotectic reaction occurs:

$$L_1 = \alpha + L_2.$$

10.3.1 Classification of Monotectics

Monotectic alloys can be classified based on the difference $T_c - T_m$ as follows:

- High-dome alloys (high $T_c - T_m$); such alloys include Al-In (206 °C) and Al-Bi (600 °C)
- Low-dome alloys (low $T_c - T_m$); typical examples include Cu-Pb (35 °C) and Cd-Ga (13 °C) alloys

They can also be classified based on the T_m/T_c ratio.

10.3.2 Mechanism of Monotectic Solidification

The morphology of the microstructure produced during directional solidification is a function of the density difference between the two liquids, $\rho_{L1} - \rho_{L2}$, and of the wetting between L_2 and α. The role of contact angle is explained in Fig. 10.34. It is seen that when L_1 wets the α phase the contact angle is 180°. The relationship between the three interface energies is governed by Young's equation:

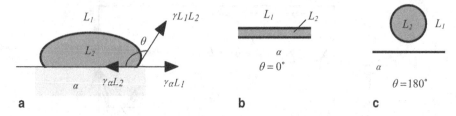

Fig. 10.34 Possible scenarios for interaction between L_1, L_2, and α. **a** General correlation. **b** Complete wetting of α by L_2. **c** Complete wetting of α by L_1

Fig. 10.35 Correlation between surface energies at various temperatures (*up*) and profile of monotectic growth front (*down*). (After Cahn 1979)

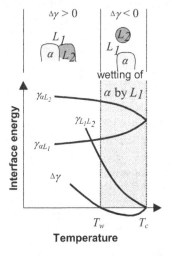

$$\gamma_{\alpha L_1} = \gamma_{\alpha L_2} + \gamma_{L_1 L_2} \cos\theta. \tag{10.50}$$

Assuming negligible density difference between the two liquid phases, microstructure evolution is controlled by the interaction between L_1 and α (Cahn 1979; Elliot 1983). At the consolute temperature, T_c, there is total mixing between the two liquids. Thus, since there is no free surface, $\gamma_{L_1 L_2} = 0$. This is shown on Fig. 10.35, assuming that $\gamma_{L_1 L_2}$ decreases with temperature.

From Young's equation, for $\gamma_{L_1 L_2} = 0$, it follows that at T_c we have $\gamma_{\alpha L_1} = \gamma_{\alpha L_2}$. This is also shown in Fig. 10.35, assuming again some variation of the two surface energies with temperature.

From the examination of the phase diagram it is concluded that, during the monotectic reaction, the interface energies are correlated by $\gamma_{\alpha L_1} + \gamma_{L_1 L_2} \rightarrow \gamma_{\alpha L_2}$. Thus, the difference in free energy during the reaction is:

$$\Delta\gamma = (\gamma_{\alpha L_1} + \gamma_{L_1 L_2}) - \gamma_{\alpha L_2}. \tag{10.51}$$

$\Delta\gamma$ can be zero at two temperatures as shown in Fig. 10.35. The first case is when the temperature is T_c, where $\gamma_{\alpha L_1} = \gamma_{\alpha L_2}$ and $\gamma_{L_1 L_2} = 0$. It can also be zero

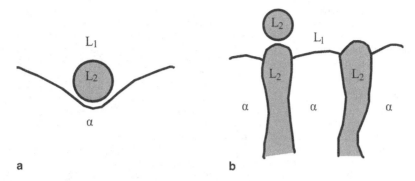

Fig. 10.36 Monotectic solidification for low-dome alloys. **a** low-growth velocity. **b** High-growth velocity

Fig. 10.37 Growth front in succinonitrile − 20 % ethanol, showing incorporation of ethanol droplets. $V = 0.27$ μm s^{-1}, $G = 4.8$ K mm^{-1} (Grugel et al. 1984). With permission of Springer Science and Business Media

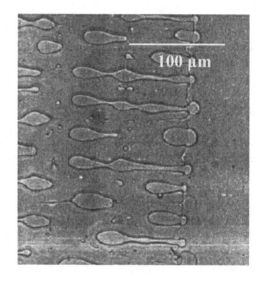

when complete wetting of α by L_1 occurs. Then the contact angle in Fig. 10.34 is 180°, and Young's equation gives $\gamma_{\alpha L_1} + \gamma_{L_1 L_2} = \gamma_{\alpha L_2}$. This temperature is the highest temperature above which wetting can occur and is defined as the wetting temperature, T_w.

Based on experimental work it was assumed (Cahn 1979) that for low-dome alloys the monotectic temperature is higher than the wetting temperature, $T_w < T_m < T_c$. Consequently, at T_m, $\Delta\gamma < 0$ and thus $\gamma_{\alpha L_2} > \gamma_{\alpha L_1} + \gamma_{L_1 L_2}$.

The system will choose the configuration with the lowest free energy, i.e., where $\gamma_{\alpha L_1}$ and $\gamma_{L_1 L_2}$ exist, that is when αL_1 and $L_1 L_2$ interfaces exist. The phases α and L_2 are separated by L_1 as shown on Fig. 10.35. At low growth rate, L_2 particles are pushed by the S/L interface (Fig. 10.36a). If the solidification velocity increases above a critical velocity, V_{cr}, L_2 is incorporated with formation of an

Fig. 10.38 Microstructure of a Cu–70 % Pb alloy solidified at $V = 778$ μm s^{-1} and $G = 12$ K mm^{-1}. Solidification direction right-to-left. (Dhindaw et al. 1988) With permission of Springer Science and Business Media

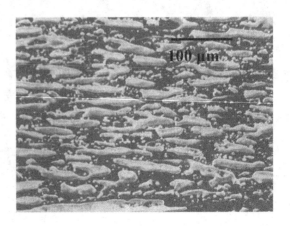

irregular fibrous composite (Fig. 10.36b). A detailed discussion on particle engulfment and pushing by solidifying interfaces is given in Section 15.1. This mechanism has been observed experimentally in a number of transparent organic alloys (e.g., succinonitrile − 20 % ethanol by Grugel et al. 1984, Fig. 10.37) as well as metallic alloys. For a Bi–50 % Ga alloy, Dhindaw et al. (1988) report that even at very high solidification velocities only a short fibrous structure was obtained (Fig. 10.38). The same authors found that, for a hyper-monotectic Cu–Pb alloy, the critical solidification velocity above which the lead droplets are engulfed decreases as the size of the droplets increases.

The range of existence of the fibrous composite is limited by the constitutional undercooling on one side and by the critical velocity of pushing-to-engulfment transition on the other (Fig. 10.39). When the solidification velocity is smaller than V_{cr} a banded structure may result. An example of such a structure is provided in Fig. 10.40. It is suggested the L_2 phase, which precipitates at the solid/liquid interface, piles up and covers the S/L interface. This produces a Pb-rich layer and increases the undercooling of the L_1/L_2 interface with respect to the monotectic temperature. Then, nucleation of the α-Cu phase occurs on the Pb-rich layer. The temperature at the growth front is also returned to the monotectic temperature. The repetition of this process will result in the banded structure.

For high-dome alloys it is assumed that the monotectic temperature is below the wetting temperature, $T_m < T_W < T_c$ and $\Delta\gamma > 0$. Thus, $\gamma_{\alpha L_2} < \gamma_{\alpha L_1} + \gamma_{L_1 L_2}$. In this case, the lowest energy exist when an α/L_2 interface exists. Consequently, α and L_2 will grow together (L_2 wets α) resulting in a regular (uniform) fibrous composite (Fig. 10.41). An assessment of the validity of the theory can be made by studying the experimental data in Table 10.1. It is seen that the high-dome ($T_m/T_c < 0.9$) alloys exhibit a fibrous structure at all velocities, while the low-dome alloys ($T_m/T_c > 0.9$) have irregular fibrous structures only at high velocities, as predicted by the theory.

Further analysis of the data in Table 10.1 reveals that the $\lambda^2 \cdot V$ relationship is about two orders of magnitude larger for irregular than for regular composites (with the exception of the Al–Bi alloy), and about one order of magnitude higher

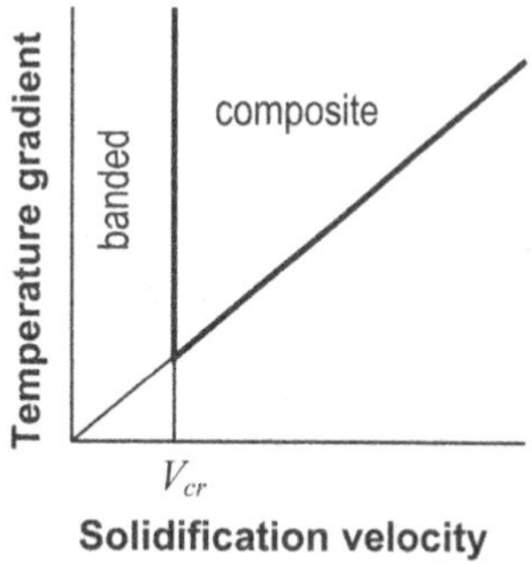

Fig. 10.39 Restriction on composite growth of low-dome alloys imposed by the critical velocity for the pushing-engulfment transition and by constitutional undercooling

Fig. 10.40 Microstructure of upward DS of a Cu-37.7 wt% Pb alloy in longitudinal section. $V = 4.4$ μm/s (Aoi et al. 2001). With permission of Elsevier

Fig. 10.41 Monotectic solidification for high-dome alloys

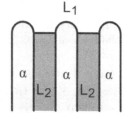

for regular monotectic composites than for regular eutectics. Indeed, for example, a Sn–Pb eutectic has $\lambda^2 \cdot V = 0.25$. This difference comes from the controlling mechanism. For irregular fibrous eutectics, this is the pushing-engulfment transition, which is a function of solidification velocity and surface energy. For regular fibrous monotectics, the spacing is controlled by surface energy, while for eutectics by diffusion.

A theoretical model based on JH's analysis of eutectic growth has been proposed by Arikawa et al. (1994). Coriell et al. (1997) extended the model to treat immiscible alloys with large density differences between phases. However, they concluded that,

Table 10.1 Dome size, interphase spacing, and structure in selected monotectic alloys. (after Grugel and Hellawell 1981)

System	$T_c - T_m$, °C	T_m/T_c, K/K	$\lambda^2 V$, m^{3}s^{-1} 10^{-16}	Structure
Cu–Pb	35	0.97	284	Irregular fibers at high V
Cd–Ga	13	0.98	120	Irregular fibers at high V
Al–In	206	0.75	4.5	Regular fibers
Al–Bi	600	0.59	250	Regular fibers
Sb–Sb$_2$S$_3$		∼ 0.5	4.8	Regular fibers

for reasons that they could not explain, there is a significant discrepancy between theory and experiment.

Ratke (2003) pointed out that an additional mode of mass transfer must be considered: the thermo-capillary effect causes Marangoni convection at the interface between the liquid L_2 phase and the molten L_1 matrix. The resulting flow affects solute transport and thus constitutional undercooling at the S/L interface. Following the JH approach with ΔT_c modified by the thermo-capillary convection, he derived analytical equations that extend the classic $V \cdot r^2 = $ const. relationship by the term $V \cdot G_T \cdot r^2$. This implies that the inter-rod distance decreases with higher gradient. It should be noted that the dependence of lamellar spacing on the temperature gradient has been derived for eutectics from purely thermal considerations.

10.4 Applications

Application 10.1
Demonstrate that the diffusion characteristic length is much smaller for the eutectic than for the primary phase. Perform the calculation for a Sn-15 % Pb alloy solidifying at $2 \cdot 10^{-6}$m/s.

Answer: For the eutectic, the diffusion characteristic length is $\delta = \lambda/2$. In turn, $\lambda = \mu_\lambda \cdot V^{-0.5} = (\mu_r/\mu_c)^{0.5} \cdot V^{-0.5}$. Using the data in Appendix B Table B1, $\delta = (0.207 \cdot 10^{-6}/5.93 \cdot 10^9)^{0.5} \cdot (2 \cdot 10^{-6})^{0.5}/2 = 2.09 \cdot 10^{-6}$. For the primary phase, $\delta = D_L/(2V) = 1.1 \cdot 10^{-9}/2/(2 \cdot 10^{-6}) = 2.75 \cdot 10^{-4}$. Thus, the diffusion characteristic length of the eutectic is two orders of magnitude smaller than that of the primary phase.

Application 10.2
Calculate the dependency of lamellar spacing on the solidification velocity for the Fe-graphite system using the JH and CS models. Assume a velocity range between 10^{-4} and 10^{-7} m/s. Compare with the experimental data from Jones and Kurz (1981) provided in Table 10.2.

Table 10.2 Experimental data for the lamellar spacing in the Fe-Graphite system

V(μm/s)	λ_{min}(μm)	λ_{av}(μm)	λ_{max}(μm)
0.41	14.75	33	90
1.11	6.1	20.3	46
3.47	4.0	9.8	24
10.7	4.1	8.0	18.6
35.2	3.1	5.4	18.4

Fig. 10.42 Correlation between the lamellar spacing and growth velocity in the Fe-graphite system: JH – Jackson-Hunt model for regular eutectics; CS – Catalina-Stefanescu model for irregular eutectics; JK – Jones-Kurz experimental data

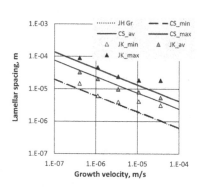

Answer: Using equations from sections 10.1.3 and 10.1.4, for the Fe-graphite system, it is calculated that $\mu_c = 1.34 \cdot 10^{11}$ and $\mu_r = 4.89 \cdot 10^{-6}$ for the JH model. For the CS model we obtain $\mu_c = 1.37 \cdot 10^{11}$ and $\mu_r = 4.89 \cdot 10^{-6}$ when using Eq. 10.19. Thus, for the extremum situation, the JH and CS model calculations are very close. Using Eqs. 10.22 to 10.26 it is seen that the CS model predictions for the minimum, average and maximum spacing are reasonable (Fig. 10.42). The CS calculation for the minimum spacing is slightly above the JH extremum calculation.

Application 10.3

Calculate the dependency of lamellar spacing on the solidification velocity for the Fe-Fe$_3$C system using the Jackson-Hunt and Catalina-Stefanescu models. Assume a velocity range between 10^{-7} and 10^{-3} m/s. Compare with experimental data provided in Table 10.3.

Answer: For the Fe-Fe$_3$C system, the following parameters were calculated with the JH Eq. $\mu_c = 5.79 \cdot 10^9$ and $\mu_r = 7.37 \cdot 10^{-7}$. The calculation results are presented in Fig. 10.43. Note that the CS irregular eutectics model predicts considerably higher average spacing than the minimum (extremum) spacing calculated with the classic JH model. At solidification velocities above 15 μm/s the experimental data deviate increasingly from the theoretical CS line.

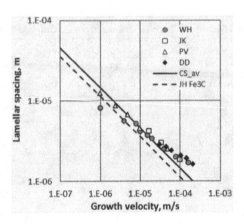

Fig. 10.43 Correlation between the lamellar spacing and growth velocity in the Fe-Fe₃C system: JH – Jackson/Hunt model for regular eutectics; CS – Catalina/Stefanescu model for irregular eutectics; WH – Wilkinson/Hellawell experimental; JK – Jones/Kurz experimental; PV – Park/Verhoeven experimental; DD – Deschanvres/Dufournier experimental

Table 10.3 Experimental data for the lamellar spacing in the Fe-Fe₃C system

Park and Verhoeven (1996)		Deschanvres and Dufournier (1970)		Wilkinson and Hellawell (1963)		Jones and Kurz (1991)	
V (μm/s)	λ (μm)	V (μm/s)	λ (μm)	V (μm/s)	λ (μm)	V (μm/s)	λ (μm)
1.0	1.24	31.5	2.84	1.0	8.20	16.8	4.2
2.49	8.76	46.0	2.66	3.98	5.55	34.2	3.0
5.0	6.72	63.0	2.39	7.94	5.16	98.8	1.84
10.0	4.55	107	2.07	10.0	4.16		
50.0	2.49	147	1.91	15.6	3.69		
		211	1.63	20.0	3.38		
				40.0	2.49		
				79.7	1.99		
				159	1.69		

Application 10.4

Calculate the final solidification time, the cooling curve and the time evolution of the solid fraction of a eutectic cast iron cube of volume 0.001 m³ solidifying with equiaxed grains. The casting is poured into a silica sand mold, from a superheating temperature of 1350 °C. Use a time-stepping analysis assuming resistance in the mold. Use transformation kinetics to calculate the temporal evolution of the eutectic solid fraction. Assume instantaneous nucleation. Compare with the solidification time obtained when using the Chvorinov's equation.

Answer: The governing equation in time-stepping format is the same as derived in Application 6.2, that is:

$$T^n = T^{n-1} - \frac{A}{\upsilon\rho c}\sqrt{\frac{k_m\rho_m c_m}{\pi t^i}}(T^{n-1} - T_o)\Delta t + \frac{\Delta H_f}{c}\Delta f_S^n, \tag{a}$$

where T^n and T^{n+1} are the temperatures at time n and $n+1$ respectively, and T_o is the ambient temperature. The solid fraction at time n is:

$$f_S^n = f_S^{n-1} + \Delta f_S^n. \tag{b}$$

Assuming instantaneous nucleation, the fraction solid at the end of solidification is given by $f_S = (4/3)\pi N r_E^3$, where r_E is the final radius of the eutectic grain and N is their number. The evolution in time of the fraction solid is then $df_S/dt = 4\pi N r_E^2 dr_E/dt$, or in discretized format, since $dr_E/dt = V_E$:

$$\Delta f_S^n = 4\pi (r_E^{n-1})^2 N V_E^n (1 - f_S^{n-1})\Delta t. \tag{c}$$

Then:

$$r_E^n = r_E^{n-1} + V_E^n \Delta t(1 - f_S^{n-1}). \tag{d}$$

Note that both the fraction solid and the radius have been decreased by the factor $(1 - f_S)$ to account for grain impingement. The solidification velocity is calculated with Eq. 10.35:

$$V_E^n = \mu_E^n (\Delta T_E^n)^2. \tag{e}$$

For eutectic the growth coefficient, μ_E, can be assumed constant (see Appendix B). The eutectic undercooling is:

$$\Delta T_E^n = T_E - T^{n-1}. \tag{f}$$

Some comments are necessary regarding the use of nucleation laws. According to the present calculations, the cooling rate before the beginning of solidification is 0.4 °C/s. Using this value and the numbers in Appendix B, it is calculated that $N_E = 3.32 \cdot 10^7 \text{m}^{-3}$. Note that the number of grains per unit area was transformed in number of grains per unit volume with the relationship $N_V = 0.87 (N_A)^{1.5}$. Since eutectic solidification occurs in a very narrow range, it is reasonable to use an instantaneous nucleation law. This algorithm is implemented in an Excel spreadsheet that is organized as shown in Table 10.4. In the second row of the table, the equations used in the various cells are shown. In addition, some calculation results at different times are also included. It is seen that the solidification time, corresponding to fraction solid 1, is 1953s. Chvorinov's equation gave 1817s.

As the fraction solid is calculated only during solidification, two IF statements must be introduced as follows:

for Δf_S: $IF(f_S^{n-1} > 0.99, 0, Eq.(c))$ $\Delta f_S = 0$ when $f_S^{n-1} > 0.99$ and

for Δf_S: $IF(f_S^{n-1} > 0.99, 1, f_S^{n-1} + \Delta f_S^n)$ $f_S^{n-1} = 1$ when $f_S^{n-1} > 0.99$.

The calculated cooling curve and fraction of solid evolution are given in Fig. 10.44. A slight recalescence is seen at the beginning of eutectic solidification.

Table 10.4 Organization of Excel spreadsheet

t	ΔT_E	V_E	r_E	Δf_S	f_S	T
Eq.	(f)	(e)	(d)	(c)	(b)	(a)
1						1350
2	0	0	0	0	0	1345
10	0	0	0	0	0	1323
286	0.04	4.0E-11	4.0E-11	2.7E-23	2.7E-23	1154
1953	19.6	1.1E-05	1.9E-03	0	1	1134

Fig. 10.44 Temperature and fraction of solid evolution for cast iron solidifying in a sand mold

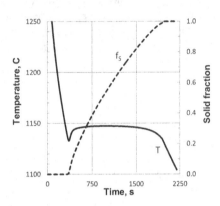

Application 10.5

Calculate the final solidification time, the cooling curve and the time evolution of the solid fraction of a 3.9 % C cast iron cube of volume 0.001 m³, poured into a silica sand mold, from a superheating temperature of 1350 °C. Use a time-stepping analysis assuming resistance in the mold. Use transformation kinetics to calculate the temporal evolution of both the primary and eutectic solid fraction. Assume instantaneous nucleation with $N_\gamma = 10^7$ and $N_E = 3.32 \cdot 10^7$ (m⁻³). Compare with the solidification time obtained when using the Chvorinov's equation.

Answer: The difference compared with Application 10.4 is that this is a hypoeutectic iron. Consequently the solidification microstructure will include both a primary phase (austenite) and eutectic. The governing equation in time-stepping format is the same as Eq. (a) in Application 10.4, while the solid fraction at time n is given by equation (b). However, because the composition is hypoeutectic, the solid fraction is the sum of the fraction of austenite dendrites and of the fraction of eutectic:

$$\Delta f_S^n = \Delta f_{S\gamma}^n + \Delta f_{SE}^n. \tag{c}$$

The fraction of austenite or eutectic is calculated with an equation similar to Eq. (c) in Application 10.4:

$$\Delta f_{S\phi}^n = 4\pi (r_\phi^{n-1})^2 N_\phi V_\phi^n (1 - f_S^{n-1}) \Delta t, \tag{d}$$

where the subscript ϕ is either the austenite, γ, or the eutectic, E. N_ϕ is the number of grains per unit volume. Then:

$$r_\phi^n = r_\phi^{n-1} + V_\phi^n \Delta t (1 - f_S^{n-1}).$$ (e)

Note that both the fraction solid and the radius have been decreased by the factor $(1 - f_S)$ to account for grain impingement. The growth velocity of the phases is:

$$V_\phi^n = \mu_\phi^n (\Delta T_\phi^n)^2.$$ (f)

For austenite, as implied in Application 9.5 Eq. d

$$\mu_\gamma^n = \frac{D_L}{2\pi^2 \Gamma m (k - 1) C_L^n}.$$ (g)

$$\Delta T_\gamma^n = T_f + m C_L^n - T^{n-1}.$$ (h)

$$C_L^n = C_o (1 - f_S^{n-1})^{k-1} \text{(the Scheil model is assumed for diffusion)}.$$ (i)

For eutectic the growth coefficient, μ_E, can be assumed constant (see Appendix B). The eutectic undercooling is:

$$\Delta T_E^n = T_E - T^{n-1}.$$ (j)

This algorithm is implemented in an Excel spreadsheet that is organized as shown in Table 10.5. In the second row of the table, the equations used in the various cells are shown. In addition, some calculation results at different times are also included. It is seen that the solidification time, corresponding to fraction solid 1 is 1954s. Chvorinov's equation gave 1817s.

The calculated cooling curve and fraction of solid evolution are given in Fig. 10.44. A slight recalescence is seen at the beginning of both primary and eutectic solidification. The fraction solid curve shows two regions: the first one when only primary austenite grains are formed and a second one where mostly eutectic solidification occurs.

Some comments are necessary regarding the use of nucleation laws. The number of eutectic nuclei was calculated as in Application 10.4 based on the cooling rate and data in Appendix B4. It is calculated that the number of grains for primary solidification is $N_\gamma = 5.03 \cdot 10^8$ m^{-3}. If this number in conjunction with instantaneous nucleation is used the model becomes unstable (the temperature oscillates back and forth around the liquidus temperature at the beginning of solidification). This is because in reality, nucleation occurs over the whole interval of primary solidification, and only a limited number of nuclei are available for growth at the beginning of solidification. Since implementation of a continuous nucleation law is more difficult on the Excel spreadsheet, an instantaneous nucleation law, with a smaller number of nuclei was proposed ($N_\gamma = 10^7$ m^{-3}) (Fig. 10.45).

Table 10.5 Organization of Excel spreadsheet

t	C_L	$\mu\gamma$	$\Delta T\gamma$	$V\gamma$	$r\gamma$	$\Delta f\gamma$	ΔT_E	V_E	r_E	Δf_E	Δf_S	f_S	T
	(i)	(g)	(h)	(f)	(e)	(d)	(h)	(f)	(e)	(d)	(c)	(b)	(a)
1	3.9												1350
2	3.9	1.97E-6	0	0	0	0	0	0	0	0	0	0	1345
10	3.9	1.97E-6	0	0	0	0	0	0	0	0	0	0	1323.0
172	3.9	1.97E-6	7.84	1.21E-4	1.21E-4	1.79E-4	0	0	0	0	1.79E-04	1.79E-04	1198.3
533	4.51	1.71E-6	0.97	1.59E-6	1.94E-3	4.51E-4	0.03	2.72E-11	2.05E-11	3.60E-24	4.51E-04	2.47E-01	1153.9
1954	40.9	1.88E-7	0	0	2.08E-3	0	21.67	1.41E-05	1.70E-03	1.69E-04	0	1	1132.2

Fig. 10.45 Temperature and fraction of solid evolution for cast iron solidifying in a sand mold

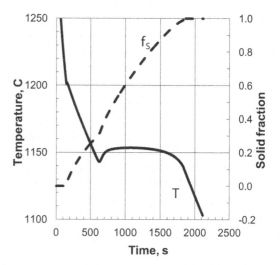

Fig. 10.46 The Ag–Sn phase diagram

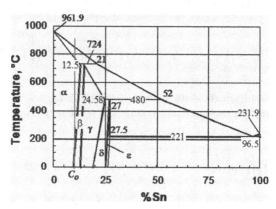

Application 10.6

Calculate the solute redistribution curve during the directional solidification of an Ag–10 % Sn alloy. The phase diagram is given in Fig. 10.46. Assume Scheil-type solidification.

Answer: From the Scheil equation the evolution of solid fraction can be calculated as: $fs = 1 - (C_S/kC_o)^{\frac{1}{k-1}}$.

For the α-phase the partition coefficient is: $k = 12.5/21 = 0.595$. Also $C_o = 10$ and $C_S = 12.5$. Then, the amount of α-phase will be: $f\alpha = 0.84$.

For the γ-phase the partition coefficient is: $k = 24.58/52 = 0.47$. Also $C_o = 21$ and $C_S = 24.58$. Then, the amount of γ-phase will be:

$$f_\gamma = (1 - f\alpha)fs = (1 - 0.84)\left[1 - \left(\frac{24.58}{0.47 \cdot 21}\right)^{\frac{1}{0.47-1}}\right] = 0.13.$$

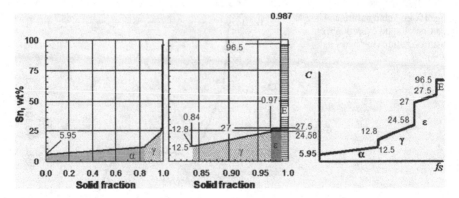

Fig. 10.47 Compositional profile during the solidification of an Ag–10% Sn alloy assuming Scheil-type solidification

For the ε-phase the partition coefficient is: $k = 27.5/96.5 = 0.28$. Also $C_o = 52$ and $C_S = 27.5$. Then, the amount of ε-phase will be:

$$f_\varepsilon = (1 - f_\alpha - f_\gamma)f_S = (1 - 0.84 - 0.13)\left[1 - \left(\frac{27.5}{0.28 \cdot 52}\right)^{\frac{1}{0.28-1}}\right] = 0.017.$$

Finally, the amount of eutectic is simply $f_E = 1 - f_\alpha - f_\gamma - f_\varepsilon = 0.013$. The solid fraction evolution as a function of composition is given in Fig. 10.47.

References

Aoi I, Ishino M, Yoshida M, Fukunaga H, Nakae H (2001) J. Crystal Growth 222:806–815
Arikawa Y, Andrews JB, Coriell SR, Mitchell WF (1994) in: Schiffman RA, Andrews JB (eds) Experimental Methods for Microgravity Material Science. TMS, Warrendale Pa, p 137
Barker NJW, Hellawell A (1974) Metal Sci. 8:353
Boettinger WJ (1974) Metall. Trans. 5:2023–31
Boettinger WJ, Coriell SR, Greer AL, Karma A, Kurz W, Rappaz M, Trivedi R (2000) Acta mater. 48:43–70
Bosze WP, Trivedi R (1974) Metall. Trans. 5:511–12
Birchenall CE, Mead HW (1956) J. of Metals 1104
Cahn JW (1979) Metall. Trans. 10A:119
Catalina AV, Stefanescu DM (2010) in: Choi JK, Hwang HY, Kim JT (eds) Proc. 8th Pacific Rim Conf. on Modeling of Casting and Solidif. Proc. Korea, p 125–132
Chang S, Shangguan D, Stefanescu DM (1992) Metall. Trans. 23A:1333
Chuang YK, Reinisch D, Schwendtfeger K (1975) Metall. Trans. A 6A:235–38
Cooper SP,. Billingham J (1980) Metal Sci. 14:225
Coriell SR, Mitchell WF, Murray BT, Andrews JB, Arikawa Y (1997) J. Crystal Growth 179:647–57
Croker MN, Fidler RS, Smith RW (1973) Proc. Roy. Soc. London, A335:15
Deschanvres A, Dufournier D (1970) Mem. Sci. Rev. Metall. 67:277–84
Dhindaw BK, Stefanescu DM, Singh AK, Curreri PA (1988) Metall. Trans. 19A: 2839

Dhindaw BK, Antonsson T, Tinoco J, Fredriksson H (2004) Metall. Mater. Trans. A, 35A:2869–287

Dioszegi A, Svensson I (2005) Int. J. Cast Metals Res. 18(1):41–46

Elliott R (1983) Eutectic Solidification Processing. Butterworths, London

Elliot R, Glenister SMD (1980) Acta Metallurgica 28:1489–94

Fisher DJ, Kurz W (1980) Acta Metallurgica 28:777–94

Flemings MC (1974) Solidification Processing. McGraw-Hill, New York

Fredriksson H (1988) Solidification of peritectics. In: Stefanescu DM (ed) Metals Handbook Ninth Edition, vol. 15 Casting. ASM International, Ohio, p.125–129

Fredriksson H, Nylen T (1982) Met. Sci., 16:283–94

Fredriksson H, Stjerndahl J (1982) Met. Sci., 16:575–85

Fredriksson H, Svensson L. (1985) in: Fredriksson H, Hillert M (eds) The Physical Metallurgy of Cast Iron. Elsevier, New York, p 273–284

Fredriksson H, Stjerndahl J, Tinoco J (2005) Mat. Sci. Eng A 413–414:363–372

Grugel RN, Hellawell A (1981) Metall. Trans. 12A:669

Grugel RN, Lagrasso TA, Hellawell A (1984) Metall. Trans. 15A:1003

Gunduz M, Kaya H, Cadirli E, Ozmen A (2003) Mat. Sci. Engr. A369:215–229

Hillert M (1979) in Solidification and Casting of Metals. The Metals Society, p 81–87

Hogan LM, Song H (1987) Metall. Trans. A 18A:707–13

Hunt JD, Jackson KA (1966) Trans. AIME 236:843

Hunziker O, Vandyoussefi M, Kurz W (1998) Acta Mater., 46:6325–633

Izumi T, Shiohara Y (2005) J. of Physics and Chemistry of Solids 66:535–545

Jackson KA, Hunt JD (1966) Trans. Met. Soc. AIME 236:1129

Jin L, Xiao F (2004) Thermochimica Acta 424:1–5

Jones H, Kurz W (1981) Z. Metallkde. 72:792–97

Kerr HW, Cissé J, Bolling GF (1974) Acta Metall. 22:667

Kurz W, Fisher DJ (1989) Fundamentals of Solidification, 3rd ed. Trans Tech Publications, Switzerland

Kurz W, Fisher DJ (1998) Fundamentals of Solidification, 4th ed. Trans Tech Publications, Switzerland

Lacaze J, Castro M, Selig C, Lesoult G (1991) in: Rappaz M et al. (eds) Modeling of Casting, Welding and Advanced Solidification Processes – V. TMS, Warrendale Pa. p 473

Löser W, Leonhardt M, Lindenkreuz HG, Arnold B (2004) Mat. Sci. Eng. A, 375–377: 534–539

Liu J, Elliot r (1993) Mat. Sci. Engr. A173:129–32

Magnin P, Kurz W (1987), Acta Metall. 35:1119

Magnin P, Kurz W (1988a) Solidification of Eutectics. In: Stefanescu DM (ed) Metals Handbook Ninth Edition, vol 15 Casting. ASM International, Ohio, p 119–125

Magnin P, Kurz W (1988) Metall. Trans. A 19A:1955–63

Magnin P, Mason JT, Trivedi R (1991) Acta metall. mater. 39(4):469–80

Magnin P, Trivedi R (1991) Acta metal.mater. 39(4):453

McDonald NJ, Sridhar S (2003) Metall. and Mat. Trans. A 34A:1931–1940

Ostrowski A, Langer EW (1977) in: Int. Conf. on Solidification and Casting, Sheffield, UK. Inst. of Metals, London, 1:139–4

Park JS, Verhoeven JD (1996) Metall. Mater. Trans. A 27A:2328–37

Ratke L (2003) Metall. Trans. 34A:449–57

Sato T, Sayama Y (1974) J. Crystal Growth 22:259–71

Seetharaman V, Trivedi R (1988) Metall. Trans. 19A:2955

Shibata H, Arai Y, Emi T (2000) Metall. and Mat. Trans. B 31B:981–991

Stefanescu DM, Kanetkar C (1985) in: Srolovitz DJ (ed) Computer Simulation of Microstructural Evolution. The Metallurgical Soc., Warrendale, Pa. p 171–188

Stefanescu DM (2006) ISIJ International, 46:786–794

Svensson IL, Wessen M (1998) in: Thomas BG, Beckermann C (eds) Modeling of Casting, Welding and Advanced Solidification Processes-VII. TMS, Warrendale Pa. p 44

Thorgrimsson (1986) PhD Dissertation, KTH, Stockholm

Thuinet L, Lesoult G, Combeau H (2003) in: Stefanescu DM et al. (eds) Modeling of Casting, Welding and Advanced Solidification Processes X. TMS, Warrendale Pa, p 237–244

Tiller WA (1958) in: Liquid Metals and Solidification. ASM, Metals Park, OH, p 276

Tiller WA (1969) in: Merchant HD (ed) Recent Research on Cast Iron. Gordon and Breach, New York, p 129

Titchener AP, Spittle JA (1975) Acta Metall. 23:497–502

Tokieda K, Yasuda H, Ohnaka I (1999) Mat. Sci. Eng. A262:238–45

Toloui B, Hellawell A (1976) Acta Metallurgica 24:565–73

Trivedi R (1995) Metall. Trans. A 26A:1583

Ueshima Y, Mizoguchi S, Matsumiya T, Kajioka H (1986) Metall. Trans. B 17B:845–59

Umeda T, Okane T, Kurz W (1996) Acta Mater. 44:4209–4216

Vandyoussefi M, Kerr HW, Kurz W (2000) Acta Mater. 48:2297–2306

Wetterfall SE, Fredriksson H, Hillert M (1972) J. Iron and Steel Inst. p 323

Wilkinson MP, Hellawell A (1963) British Cast Iron Res. Assoc. J. 11:439–50

Yasuda H, Ohnaka I, Tokieda K and Notake N (2003) in: Cantor B, O'Reilly K (eds) Solidification and Casting. Inst. of Physics Publishing, Bristol, p 160–74

Zener C (1946) Trans. AIME 167:550

Chapter 11
Solidification of Multicomponent Alloys

Most commercial casting alloys are multicomponent (three or more) and involve multiphase solidification. Processing objectives include either avoidance of microsegregation associated with the formation of detrimental minority phases, or control of the amount, the morphology, and the distribution of beneficial minority phases. Yet, understanding morphological stability and multiphase pattern formation in such systems is not trivial because of the high number of degrees of freedom and solidifying phases. For example, the eutectic reaction which is nonvariant in binary alloys becomes univariant in ternary alloys. As a consequence, the two-phase solid/liquid (S/L) interface can exhibit cellular or dendrite morphological transitions. The example in Fig. 11.1 shows multiphase patterns formed during directional solidification of ternary eutectic alloys. It is reproduced from the excellent review by Hecht et al. (2004), which was used extensively for this chapter.

11.1 Thermodynamics of Multicomponent Alloys

For binary alloys some of the required thermodynamic properties can be obtained from the phase diagrams, while others are tabulated. This becomes increasingly difficult as the number of components increases, and computational thermodynamics must be used. The CALculation of PHAse Diagram (CALPHAD) method developed by Kaufman and Bernstein (1970) succeeded in the creation of thermodynamic database through critical assessment and systematic evaluation of experimental and theoretical data that is then used by a number of software, such as Thermo-Calc (KTH, Stockholm), MTData (NPL), Chemsage (RWTH, Aachen), Pandat, for the calculation of multicomponent phase diagrams. Such databases are currently available for Fe-base, Al-base, Ti- and TiAl-base alloys, and for Ni-base superalloys.

A thermodynamic database gives a parametric description of the Gibbs free energy for all phases of the considered alloy system as function of temperature, composition, and pressure. The parametric description is based on Gibbs free energy

© Springer International Publishing Switzerland 2015   251
D. M. Stefanescu, *Science and Engineering of Casting Solidification*,
DOI 10.1007/978-3-319-15693-4_11

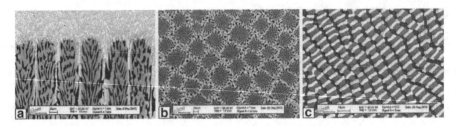

Fig. 11.1 Multiphase patterns in ternary Al–Cu–Ag alloys after unidirectional solidification: eutectic cells in longitudinal (**a**) and cross section (**b**) of Al–Cu 13.66 at.%–Ag 10.27 at.% and a three-phase eutectic pattern (**c**) in cross section of Al–Cu 13.6 at.%–Ag 16.4 at.%. (Hecht et al. 2004; with permission of Elsevier)

models that account for the specific crystal structures of the individual phases, e.g., substitutional solution phases, stoichiometric compounds, and ordered phases.

For a ternary substitutional solution phase ϕ, the molar Gibbs energy is expressed as (Saunders and Miodownik 1998):

$$G_m^\phi = G_m^{\phi,ref} + G_m^{\phi,id} + G_m^{\phi,Ex}, \qquad (11.1)$$

where the superscripts represent reference, ideal mixing, and excess contributions. The reference term corresponds to the mixture of the Gibbs energy of the constituent components:

$$G_m^{\phi,ref} = \sum_{i=1}^{n} x_i G_i^o(T). \qquad (11.2)$$

The temperature dependence of $G_i^o(T)$ is formulated as a power series of T:

$$G_i^o(T) = a + bT + cT\ln(T) + \sum d_n T^n. \qquad (11.3)$$

The ideal mixing term corresponds to the entropy of mixing for an ideal solution,

$$G_m^{\phi,id} = RT \sum_{i=1}^{n} x_i \ln(x_i). \qquad (11.4)$$

The excess term that describes deviations from ideal mixing because of specific interactions between the components can be represented by the Redlich–Kister–Muggianu polynomial function:

$$G_m^{\phi,Ex} = \sum_{i=1}^{n-1}\sum_{j=i+1}^{n} x_i x_j L_{i:j}^\phi \quad with \quad L_{i:j}^\phi = \sum_{v=0}^{k} (x_i - x_j)^v L_{i:j}^{v,\phi}, \qquad (11.5)$$

where x_i is the molar fraction of component i, L_{ij}^v are pairwise interaction parameters that can be temperature dependent according to an expression similar to Eq. 11.3.

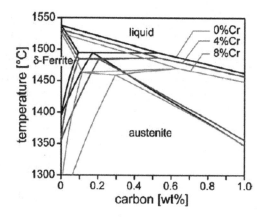

Fig. 11.2 Isopleth showing the influence of Cr on the δ-ferrite field. (Hecht et al. 2004; With permission of Elsevier)

For a ternary phase three pairwise contributions, L_{ij}^v, to the excess term are needed. The respective pairwise interaction parameters can be taken from existing binary databases or can be fitted. If experimental information relative to ternary interaction is available, then an extra term is added to Eq. 11.5 to account for the ternary interaction (Muggianu et al. 1975):

$$G_m^{\phi,Ex} = \sum_{i=1}^{n-1}\sum_{j=i+1}^{n} x_i x_j L_{i:j}^{\phi} + \sum_{i=1}^{n-2}\sum_{j=i+1}^{n-1}\sum_{k=j+1}^{n} x_i x_j x_k L_{i:j:k}^{\phi}.$$

$$(11.6)$$

A detailed example of calculations is presented by Hecht et al. (2004).

Once the thermodynamic database becomes available, numerical codes are used to calculate the total Gibbs energy minima (thermodynamic equilibria) at any composition, temperature, and pressure for complex phase equilibria.

Again, Hecht et al. (2004) give an example for the case of a steel having the following chemical composition (in wt%): 0.45 % C, 1 % Mo, 0.5 % Si, 0.5 % V, 5 % Cr, 0.5 % Ni, 0.7 % Mn, balance Fe. Due to the high amount of carbide forming elements, the peritectic reaction L+δ-ferrite → austenite is shifted to higher carbon contents as exemplified in Fig. 11.2 for Cr. To avoid the formation of δ-ferrite during solidification, the carbon content of the alloy must be adjusted. This task can be performed through thermodynamic equilibrium calculations as shown in the example in Fig. 11.3 for calculation performed with Thermo-Calc (Sundman et al. 1985). It is seen that the mole fraction of δ-ferrite in equilibrium with the liquid first increases as the temperature decreases, followed by decrease during the peritectic reaction.

Thermodynamic calculations are just a first step in the task of predicting microstructure evolution during the solidification of multicomponent alloy. Microstructure evolution models are required to complete the job, which in turn need thermophysical properties.

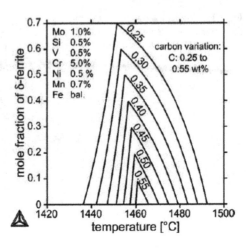

11.2 Thermophysical Properties

The essential thermophysical properties in microstructure evolution modeling are diffusion coefficients and interfacial energies. Binary diffusivities are now routinely measured. However, as in a multicomponent solution the diffusion of each component is affected by the chemical gradients of all other components. This complicates the problem.

Morphological stability is controlled by the S/L interface energy. The anisotropy of the interface energy, resulting from the crystalline anisotropy of the solid, controls the interface shape, and thus the pattern of the morphology, as discussed in Sect. 9.3.4. Generating these data is nontrivial even for single-component systems.

11.2.1 Multicomponent Diffusion

Following Onsager (1931) the fluxes J_i of the components $i = 1, \ldots, n$ are expressed as linear functions of the driving forces. The driving forces may be expressed as potential gradients $\nabla \phi_j$, where the potentials ϕ_j are functions of composition:

$$J_i = -\sum_{j=1}^{n} L_{ij} \nabla \phi_j. \tag{11.7}$$

The parameters L_{ij} relate the flux of component i to the potential gradients of all other components. They are the elements of an $n \times n$ matrix and the potential gradients constitute n interdependent driving forces for diffusion. According to Anderson and Ågren (1992), when diffusion is treated in the volume-fixed frame of reference and the driving forces are expressed via concentration gradients, rather than chemical potential gradients, the fluxes are:

$$J_i = -\sum_{j=1}^{n} D_{ij}^n \nabla C_j \tag{11.8}$$

The chemical diffusivities D_{ij}^n constitute an $(n-1) \times (n-1)$ diffusivity matrix. The superscript n denotes the choice of the solvent, which represents the dependent component.

At least two software packages are available to assess the chemical diffusivity matrix from one-dimensional (1D) multicomponent diffusion data obtained from diffusion couples (Hecht et al. 2004). First, Profiler, is based on the square root diffusivity method, developed by Brockman and Morral. It provides a solution to 1D multicomponent diffusion, for constant chemical diffusivities. Second, DICTRA, is based on the diffusion mobility formalism developed Anderson and Ågren (1992). It uses thermodynamic factors that are calculated directly from thermodynamic databases. Both methods have been extensively used to assess solid state diffusion in multicomponent alloys.

11.2.2 Interface Energy

There is no experimental method for the direct measurement of the interfacial energy. However, indirect techniques have been developed (see for example, the review by Kelton 1991).

Computational methods for the prediction of the equilibrium properties of S/L and solid/solid (S/S) interfaces, such as atomistic simulation, are promising alternatives to experiments. Molecular dynamics and Monte Carlo simulations can accurately determine even small values of the crystalline anisotropy of the S/L interface energy (Hoyt et al. 2003).

11.2.3 Microstructure

Hunziker (2001) developed analytical solutions for the solute diffusion fields during plane front and dendritic growth of multicomponent alloys taking into account the diffusive interaction between the solutes in the liquid. In principle, the composition field for each of the n solutes is given by a sum of n expressions, each corresponding to the binary solution, but where the diffusion coefficients are replaced by the eigenvalues of the diffusion matrix. The composition field for each solute i was written as:

$$C_i(Z) - C_i^\infty = \sum_{k=1}^n A_{ik} \cdot \exp\left(-\frac{Vz}{B_k}\right) \tag{11.9}$$

where A_{ik} and B_k are coefficients to be calculated. The S/L interface is at $z=0$. This equation can be solved by:

- Calculating the n eigenvalues B_k and the n corresponding unit length eigenvectors $\mathbf{N_k}$ of the diffusion matrix $\mathbf{D}$.

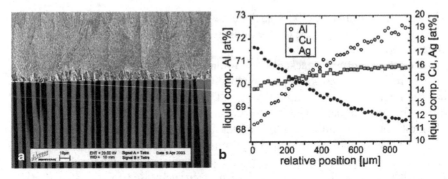

Fig. 11.4 Solutal configuration at a planar eutectic interface during steady-state growth in the ternary alloy Al–Cu 15.52 at.%–Ag 9.04 at.% after unidirectional solidification with a velocity 1.417×10^6 m/s in a temperature gradient 27×10^3 K/m. **a** Quenched interface in longitudinal section, with Al$_2$Cu (θ-phase) *black* and α(Al) *light grey*. **b** Element distribution in the liquid ahead of the quenched interface. (Hecht et al. 2004; With permission of Elsevier)

- Solving the *n*-equations system (Eq. 11.10) with A_k as variables.

$$\Delta C_i = \sum_{k=1}^{n} N_{ik} \|A_k\| \tag{11.10}$$

- Introducing the A_{ik} from the equation $A_{ik} = N_{ik} \|A_k\|$ into Eq. 11.9.

The extension to *n* solutes of the constitutional undercooling criterion for binary and ternary alloys was derived as:

$$\frac{G}{V} > \sum_{i=1}^{n} m_i \cdot \sum_{j=1}^{n} \frac{A_{ij}}{B_j} \tag{11.11}$$

Equations for interface stability, dendrite temperature tip, and interface composition were also derived.

Univariant Eutectics The coupled growth of two nonfaceted solid phases from the ternary liquid during the solidification of a univariant eutectic results in a regular lamellar or rod microstructure. They grow with an interlamellar exchange of solute species but additionally segregate the third component into the liquid, such that a long-range solute boundary layer establishes in the liquid ahead of the interface (Fig. 11.4). McCartney et al. (1980a, b) derived equations for the relationship between interface undercooling, growth velocity, and eutectic spacing, that for minimum undercooling yields $\lambda^2 V = \mathrm{ct}$. Calculation of the constant requires material properties such as thermodynamic data, diffusion, and interface energies for the ternary alloy. The relationship was confirmed for both rod-like eutectic (in Fe–Si–Mn and Fe–Si–Co) and for lamellar eutectics in Al–Cu–Ni, Al–Cu–Ag, and Ni–Al–Cr–Mo (see summary in Hecht et al. 2004).

According to the analysis by McCartney et al. (1980a), in a ternary eutectic system the possible microstructures are a function of composition as summarized

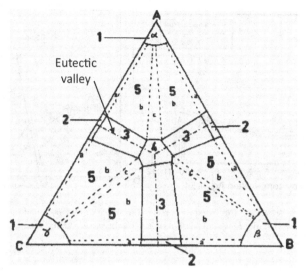

Region 1 - Near the pure component corners: planar growth of single phase;

Region 2 - Near the two component eutectic compositions: two phase eutectic growth with planar growth interface;

Region 3 - For alloys near the eutectic valley: two phase eutectic cells or dendrites with the third component appearing near the ternary eutectic temperature either as a thin layer or as a three phase eutectic layer around each cell (depending on the volume fraction of the ternary eutectic);

Region 4 - Near the ternary eutectic composition: three phase eutectic structure with planar interface and without any prima-ry single phase or two phase eutectic;

Region 5 - For alloys away from the eutectic valleys: single phase cells or dendrites are to be expected, followed by the structures described in regions 2, 3 and 4.

Fig. 11.5 Schematic representation of the composition boundaries of the various structural regions (*dotted* and *solid lines*) for a fixed velocity and temperature gradient. (McCartney et al. 1980a; With permission of Springer Science and Business Media)

in Fig. 11.5. They suggested that the lines defining the limits of the regions in three component systems can be predicted following the same concepts used in the Mullins–Sekerka stability analysis for two components. Thus, the limit of region 1 (planar single-phase growth), will be given by the constitutional undercooling condition modified for three components:

$$G > m_{j,B}\frac{V}{D_B}\left(\frac{1}{k_{j,B}} - 1\right)C_{oB} - m_{j,C}\frac{V}{D_C}\left(\frac{1}{k_{j,C}} - 1\right)C_{oC}, \qquad (11.12)$$

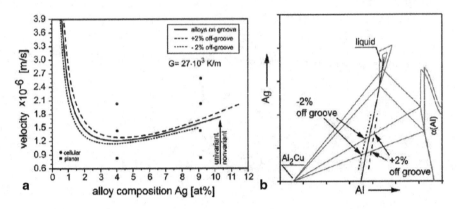

Fig. 11.6 Calculated velocities for instability onset for ternary Al–Cu–Ag alloys along the univariant eutectic groove: liquid → α(Al) + Al₂Cu. (Hecht et al. 2004; With permission of Elsevier)

where $j = \alpha$, β, or γ phases, $m_{j,B} = (\partial T/\partial C_B)_{C_C}$ and $m_{j,C} = (\partial T/\partial C_C)_{C_B}$ are the slopes of the liquidus surfaces, C_{oB} and C_{oC} are the initial concentration of components A and B, respectively, and subscript B and C refers to component B and C.

For a planar two-phase eutectic interface the line between regions 2 and 3 is given by:

$$G > m_C(1 - 1/k_C)C_{oC}(V/D_C), \tag{11.13}$$

where $k_C = (k_{\alpha C}S_\alpha)/(S_\alpha + S_\beta) + (k_{\beta C}S_\beta)/(S_\alpha + S_\beta)$ is the weighted eutectic distribution coefficient for component C and $m_C = dT/dC_C^L$ is the slope of the eutectic valley (assumed constant).

This approach of impurity driven Mullins–Sekerka instability, cannot explain morphological destabilization of a planar eutectic interface because of the interplay between the eutectic pattern and the instability.

For an alloy Al–Cu 15.52 at.%–Ag 9.04 at.% the critical velocity calculated with Eq. 11.13 was 1.67×10^{-6} m/s (Hecht et al. 2004). The same equation can also be used to calculate the velocities of instability onset for alloys with composition on and slightly off (but still on the tie lines) the groove. The calculations shown in Fig. 11.6 were done for alloys with on-groove composition and alloys that are 2 % off-grove toward the Al-rich side and the Cu-rich side, respectively (Fig. 11.6b), for a thermal gradient $G = 27 \times 10^3$ K/m. The alloys richer in Al grow with higher fractions of the Ag-rich α(Al) and thus segregation of Ag is less pronounced. Consequently they are more stable. The reverse holds for alloys that are richer in Cu. Experimentally observed transitions from planar and cellular growth structures fit quite well to the obtained transition criteria.

Since the early investigations of eutectic alloys, researchers found that in addition to the lamellae or rods, cellular morphologies, termed colonies, may also exist. The

Fig. 11.7 Eutectic colonies in the transparent organic alloy CBr_4–C_2Cl_6 obtained through directional solidification. (Hunt and Jackson 1966)

length scale of these colonies is 10–100 times the lamellar spacing. An example of such structure is given in Fig. 11.7. According to both theory and experiments, a planar lamellar eutectic front in a binary alloy is completely stable for compositions sufficiently close to the eutectic composition, and for lamellar spacings near λ_{min}. Experiments (e.g., Chilton and Winegard 1961) revealed that colonies appear only in the presence of a ternary impurity, rejected by both solid phases.

Plapp and Karma (1999) analyzed the stability of a lamellar eutectic interface in the presence of a low content of a ternary impurity. Studying the stability of an alloy of eutectic composition with a symmetric phase diagram, they concluded that the primary instability of the eutectic interface is qualitatively similar to the Mullins–Sekerka instability of a single-phase interface in binary alloys. However, for very small concentrations of the ternary impurity, the effective capillary length is significantly affected because of the surface tension of the eutectic interface consisting of an array of arcs, and because of the eutectic cross-diffusion that acts as a capillary force. As a consequence, two stabilizing terms occur in the expressions for the critical velocity and the critical wave number at onset, in addition to the classic Mullins–Sekerka.

Plapp and Karma (1999) also derived the following expression for the Gibbs–Thomson coefficient of a 2D lamellar eutectic interface for the completely symmetric case:

$$\Gamma_{eff} = \Gamma_E + 0.4965 \frac{M \Delta C Pe \lambda_o}{4\pi^2} \quad \text{with} \quad \Gamma_E = \frac{\Gamma_\alpha \Gamma_\beta \cos\theta_\alpha \cos\theta_\beta}{(1-\eta)\Gamma_\alpha \cos\theta_\alpha + \eta\Gamma_\beta \cos\theta_\beta},$$
$$(11.14)$$

where $M = 2m_\alpha m_\beta/(m_\alpha + m_\beta)$, $\Delta C = C_\alpha - C_\beta$ (C_α and C_β, are the concentrations limiting the eutectic plateau in the phase diagram), and λ_o is the unperturbed spacing. Γ_E is the "geometric part" of the effective the Gibbs-Thomson coefficient for a moving eutectic front, in which the stabilizing effect of the interlamellar diffusion has to be included. It describes the shift of the average interface temperature when

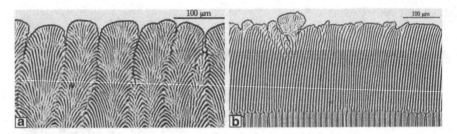

Fig. 11.8 The quasi-2D cellular pattern in DS of CBr_4–C_2Cl_6–$5 \cdot 10^{-2}$ at % naphthalene 12 mm-thick samples: (**a**) chaotic behavior for $V = 31 \cdot 10^{-6}$ m/s and $G = 110 \cdot 10^2$ K/m; (**b**) transient leading to the pattern in (**a**) for $V = 0.9 \cdot 10^{-6}$ to $31 \cdot 10^{-6}$ m/s. (Akamatsu and Faivre 2000; With permission of Elsevier)

the composite interface is curved on a scale much larger than the lamellar spacing. The undercooling at the interface is:

$$\Delta T = \Gamma_E K. \tag{11.15}$$

While exploring the formation of eutectic cells in the binary organic eutectic CBr_4–C_2Cl_6 doped with low concentrations of naphthalene (2.5×10^{-2} to 10^{-1} at.%) in thin samples. Akamatsu and Faivre (2000) found that for high impurity concentrations, travelling waves occurred prior to the formation of cells, followed by local two-phase structures (fingers). No steady-state pattern was reached on the scale of individual cells (Fig. 11.8). The initially stable lamellar eutectic interface is subject to branching, traveling waves, and formation of two-phase fingers. The coupling between the interlamellar and the cellular dynamics clearly results in an over stabilization of the lamellar front, which is only destroyed when two-phase fingers appear.

Nonvariant Eutectics As summarized by Hecht et al. (2004), in the nonvariant eutectic reaction three solid phases can simultaneously grow from the ternary liquid alloy by exchanging solute species through diffusion perpendicular to the growth direction. The composition of the phases in equilibrium at the interface and the mole fraction of the phases are given by the tie triangle corresponding to the temperature of the nonvariant eutectic point.

Except for the ideal lamellar pattern expected for a symmetric phase diagram, and observed in Cd–Sn–Pb alloys (Kerr et al. 1964; Quac Bao and Durand 1972), all experimentally observed three-phase patterns show periodic arrangements of duplex structures that exhibit quadruple points in addition to triple lines (Cooksey and Hellawell 1961). The three-phase patterns generally obey a $\lambda^2 V = ct$. relationship for each pair of phases (e.g., Rinaldi et al. 1972; Rios et al. 2002).

Himemiya and Umeda (1999) developed three-phase eutectic growth models for the ideal lamellar pattern and constructed rod + hexagon and semi-regular brick type patterns, following the procedure applied by Jackson and Hunt for binary

Fig. 11.9 Effects of fluid flow on the average spacing in quasi-binary InSb–NiSb eutectics in the ternary alloy system In–Sb–Ni. (Müller and Kyr 1984; With permission of Elsevier)

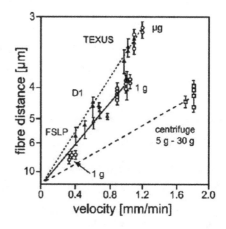

eutectics. For Sn–Pb–Cd and Al–Cu–Ag ternary eutectics the computed interface undercooling versus growth rate was lowest for the rod + hexagon pattern and highest for the ideal lamellar pattern.

Fluid Flow Effects The characteristic length during coupled growth in nonvariant eutectic and monotectic reactions is not D/V, but the lamellar or rod spacing since the controlling phenomenon is lateral diffusion. Therefore the effects of buoyancy driven convection are rather minor. In addition to buoyancy, fluid flow can be induced by density differences between the growing phases and, specifically relevant to monotectics, by thermo-capillary effects.

The effect of buoyancy driven convection on eutectic structures was investigated with the help of solidification processing under microgravity. A summary of experiments by Müller and Kyr (1984) is presented in Fig. 11.9. It is seen that the spacing increases significantly, in particular at higher V, as the gravity level increases.

Steinbach, in work quoted in Hecht et al. (2004), observed that as the fluid flow induced by rotating magnetic fields was increased, the eutectic spacing decreased. When fitting the measured spacings with a Jackson–Hunt relation $\lambda^2 V = ct.$, he noted that the constant is a linear function of flow velocity.

Ma et al. (1998) analyzed a flow field parallel to the interface having a constant velocity gradient G_u in a direction perpendicular to the interface. Performing a classical Jackson–Hunt analysis they derived an approximate analytical solution for the convection–diffusion equation of eutectic growth under the condition of weak convection:

$$\left(\frac{\lambda}{\lambda_o}\right)^2 = \frac{1}{1 - 2DG_u/V^2}, \tag{11.16}$$

where λ is the spacing with flow, and λ_o is the spacing without flow. The analysis shows that increased convection decreases the spacing. The authors attributed this effect to a slight shift of solute concentration field induced by convection, which leads to a decrease of the eutectic growth undercooling. The theory fits Steinbach's data.

For a more in-depth discussion of the effects of fluid flow on the solidification of multicomponent alloys the reader is referred to the review paper by Hecht et al. (2004).

References

Akamatsu S, Faivre G (2000) Phys. Rev. E 61:3757
Anderson JO, Ågren JJ (1992) J. Appl. Phys. 72:1350
Brockman WB, Morral JE Profiler: Diffusion Couple Software to Predict Concentration Profiles and [D] (available from J.E. Morral, Dept. of Mater. Sci. and Eng., The Ohio State University
Chilton JP, Winegard WC (1961) J. Inst. Met. 89:162
Cooksey DJS, Hellawell A (1967) J. Inst. Met. 95:183
Hecht U, Gránásy L, Pusztai T, Böttger B, Apel M, Witusiewicz V, Ratke L, De Wilde J, Froyen L, Camel D, Drevet B, Faivre G, Fries SG, Legendre B, Rex S (2004) Mater. Sci. Eng. R 46:1–49
Himemiya T, Umeda T (1999) Mater. Trans. JIM 40:665
Hoyt JJ, Asta M, Karma A (2003) Mater. Sci. Eng. 41:R121
Hunt JD, Jackson KA (1966) Trans. Metall. Soc. AIME 236:843
Hunziker O (2001) Acta mater. 49:4191–4203
Kaufman L, Bernstein H (1970) Computer Calculation of Phase Diagrams with Specific Reference to Refractory Materials. Academic Press, New York
Kerr HW, Plumtree A, Winegard WC (1964) J. Inst. Metals 93:63
Kelton KF (1991) Solid State Phys. 45:75
Ma D, Jie WQ, Li Y, Ng SC (1998) Acta Mater. 46:3203
McCartney DG, Hunt JD, Jordan RM (1980) Metall. Trans. A 11:1243
McCartney DG, Jordan RM, Hunt JD (1980a) Metall. Trans. A 11:1251
Muggianu YM, Cambino M, Bros JP (1975) J. Chim. Phys. 22:83
Müller G, Kyr P (1984) Results of Spacelab-1, 5th European Symposium Material Sciences Under Microgravity. SP 222, ESTEC Noordwijk, p 141
Onsager L (1931) Phys. Rev. 37:405
Plapp M, Karma A (1999) Phys. Rev. E 60:6865
Quac Bao HA, Durand FCL (1972) J. Cryst. Growth 15:291
Rinaldi MD, Sharp RM, Flemings MC (1972) Metall. Trans. 3:3139
Rios CT, Milenkovic S, Gama S, Caram R (2002) J. Cryst. Growth 90:237 -239
Saunders N, Miodownik AP (1998) CALPHAD. Elsevier Science, New York
Sundman B, Jansson B, Anderson JO (1985) Calphad 9

Chapter 12
Microshrinkage

During liquid cooling and solidification, a significant amount of the dissolved gas, such as hydrogen for Al alloys or nitrogen for ferrous alloys and superalloys, is rejected by the liquid and, if a critical pressure is overcome, a gas bubble forms thereby initiating porosity. If the gas bubble is formed in the liquid, it floats and failing to find an open liquid surface it interacts with the solid/liquid (S/L) interface eventually forming *gas porosity*. This is not a shrinkage defect. If the gas bubble forms in the mushy zone in the later stages of solidification, after dendrite coherency, it will be entrapped in the dendritic network and nucleate small local shrinkage cavities termed *microporosity* or *microshrinkage*.

In most cases if the macroshrinkage is not limited to the riser the casting is rejected. On the other hand, while not always a reason for scrapping the casting, microshrinkage and porosity impact negatively on mechanical properties such as ductility (Uram et al. 1958, Samuel and Samuel 1999), dynamic properties (Skallerud et al. 1993), and fatigue life. Confirming earlier work (Major 1997), Boileau and Allison (2003) have shown that fatigue life decreases in direct relation to the increase in the maximum pore size (Fig. 12.1) because of the effect of porosity on crack initiation. In turn, pore size increased as the local solidification time and the secondary dendrite arm spacing increased. The elimination of porosity through hot isostatic pressing resulted in significant increase in fatigue life.

12.1 Defect Size and Shape

The pores found in Al alloys can range in size from micrometer to hundreds of micrometer and can have shapes from almost spherical (Fig. 12.2a, b) to irregular, following the geometry of the interdendritic region (Figs. 12.2c, d, 12.3). In well-degassed melts, microshrinkage takes the shape of the interdendritic liquid that remains just before eutectic solidification (Fig. 12.2e), and the stress concentration factors resulting from these shapes are much higher than for spherical pores (Berry 1995). Both the size and the shape of the microshrinkage are important. As shown in Fig. 12.4, with the increase in solidification time the size of the porosity increases

© Springer International Publishing Switzerland 2015   263
D. M. Stefanescu, *Science and Engineering of Casting Solidification*,
DOI 10.1007/978-3-319-15693-4_12

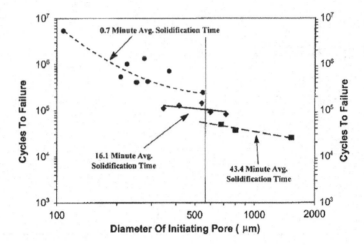

Fig. 12.1 The relationship between pore size and fatigue life for aluminum alloy W319-T7 (alternating stress of 96.5 MPa) (Boileau and Allison 2003). With permission of Springer Science and Business Media

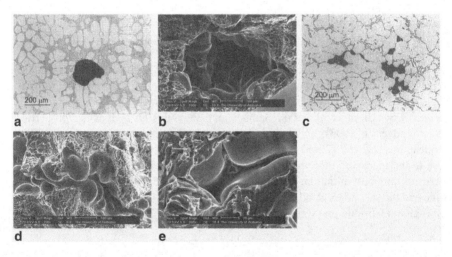

Fig. 12.2 Microporosity in aluminum alloys. **a** Optical micrograph; Na-modified Al–Si alloy (Fuoco et al. 1994). Copyright 1994 American Foundry Soc., used with permission. **b** SEM image; unmodified Al–Si alloy (compliments of R. Ruxnda). **c** Optical micrograph of interdendritic microshrinkage in Sb refined Al–Si alloy (Fuoco et al. 1994). Copyright 1994 American Foundry Soc., used with permission. **d** SEM image of interdendritic microshrinkage in Al–Si alloy (compliments of R. Ruxanda). **e** SEM image of microshrinkage between the eutectic grains (compliments of R. Ruxanda)

together with the SDAS. In addition, the fatigue strength also decreases (Boileau and Allison 2003).

Overall, there is a direct relationship between the amount of gas initially dissolved in the melt and the amount of porosity, as shown in Fig. 12.5.

Fig. 12.3 Microshrinkage in ductile iron. The figure at the *right* is an enlargement of the one on the *left* (Ruxanda et al. 2001). Copyright 2001 American Foundry Soc., used with permission

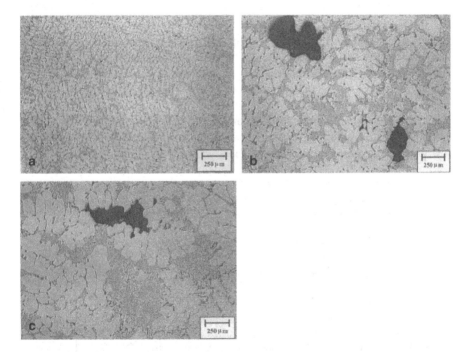

Fig. 12.4 The effect of solidification time on the microstructure of an Al7.4Si3.3Cu alloy (Boileau and Allison 2003). With permission of Springer Science and Business Media. **a** Average solidification time 0.7 min (average SDAS 23 μm). **b** Average solidification time 16 min (average SDAS 70 μm). **c** Average solidification time 43 min (average SDAS 100 μm)

The shrinkage defects at different length scale (macro and micro) interact with one another and must be understood together. A certain correlation seems to exist between the total amount of porosity, open shrinkage cavity, and caved surface. Awano and Morimoto (2004) who investigated the shrinkage behavior of Al–Si alloys with various silicon and gas (vacuum degassed, nontreated, and gas enriched)

Fig. 12.5 Amount of porosity as function of hydrogen
content in the aluminum melt
from various sources (after
Shih et al. 2005)

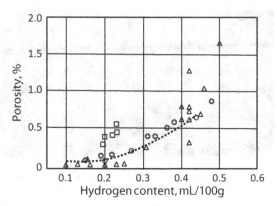

contents, concluded that the total amount of shrinkage is constant at the same silicon
level, but varies with the amount of gas in the melt. As summarized in Fig. 12.6, the
total shrinkage depends little on the amount of gas, but decreases with higher silicon
content. Microporosity, on the other hand is a direct function of the gas content, as
further evident from Fig. 12.7.

Awano and Morimoto further found that for alloys with large solidification interval (mushy solidification) the amount of pipe is constant in the low porosity region
(low to moderate gas content), but decreases with increasing porosity in the high

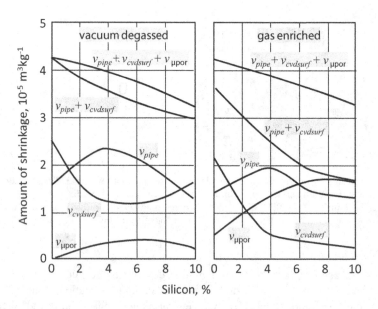

Fig. 12.6 Effect of silicon content on the amount of shrinkage defects in Al–Si alloys. v_{pipe} is the
volume of pipe, $v_{cvdsurf}$ is the volume of caved surface, and $v_{\mu por}$ is the volume of microporosity
(after Awano and Morimoto 2004)

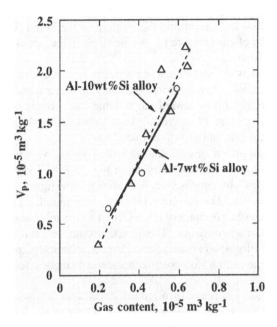

Fig. 12.7 Correlation between microporosity and hydrogen content in two Al–Si alloys (Awano and Morimoto 2004). With permission of Maney

porosity region (high gas content). For alloys that solidify with small solidification interval (skin solidification), the amount of caved surface is constant in all porosity regions, while the amount of pipe decreases with increased porosity as the pore generation during the early stage of solidification compensates shrinkage.

This analysis reveals the complexity of the problem. Computational modeling of microshrinkage formation must describe phenomena such as pore nucleation and growth, elastic and plastic deformation of the solidification shell, and interdendritic flow.

12.2 The Physics of Shrinkage Porosity Formation

The picture that emerges from the analysis of all these experimental data is that microporosity is a complex phenomenon in which the hierarchy of the numerous process and material parameters that affect the outcome is difficult to assess. The present understanding of microporosity formation is that metal flows toward the region where shrinkage is occurring until flow is blocked, either by solid metal or by a solid or gaseous inclusion. The early theory by Walther et al. (1956) assumed that feeding ceases simply because the cross sectional area of the feeding channel continuously decreases during solidification. When the section of the channel has decreased too much, the pressure drop ruptures the liquid in the channel, forming a pore. However, pure liquids have high tensile strengths capable of collapsing the surrounding solid, and preventing the fracture of the liquid (Campbell 1991). Thus, in the absence of gas pressure the tensile stress in the liquid will prevent any discontinuity formation.

More recent models assume that when a gas pore appears in the mushy zone during late solidification, after dendrite coherency, it is entrapped in the dendritic network. When the metal flow toward the solidification front is blocked, the pore becomes the starting point of microshrinkage. Thus, microshrinkage formation depends on the nucleation and growth of micropores.

The physics of gas pore nucleation and growth in the mushy zone of a solidifying alloy is presented in the conceptual drawing in Fig. 12.8. The gas solubility in the liquid C_L^{max} decreases with temperature, or as shown in the figure, from the dendrite tips to the dendrite roots. The gas concentration in the liquid, C_L, is higher at the solid/liquid (S/L) interface because of gas rejection by the solid, and decreases away from the interface through diffusion. The two curves intersect at the critical fraction solid for the onset of gas supersaturation. Thermodynamically, pore formation is possible at this fraction solid. However, nucleation and kinetic effects may delay the pore formation.

The growth of the gas pores in the mushy zone can be understood from the analysis of the local pressure summarized in Fig. 12.8 for the case of ductile iron. Mathematically, this is expressed through a pressure balance equation stating that the pressure exerted by gas evolution, P_G, must be higher than the sum of the local pressure in the mushy zone, P_{mush}, and the pressure induced by the surface tension on the pore, P_γ:

$$P_G > P_{mush} + P_\gamma \quad \text{where } P_{mush} = P_{appl} + P_{st} + P_{exp} - P_{shr} \quad (12.1)$$

where, P_{appl} is the applied pressure on the mold (e.g., atmospheric pressure), P_{st} is the metallostatic pressure, P_{exp} is the expansion pressure because of phase transformation, and P_{shr} is the negative pressure from resistance to shrinkage induced flow through the fixed dendrite network. The equation can be rearranged to highlight the driving force on the left hand side:

$$P_G + P_{shr} > P_{appl} + P_{st} + P_{exp} + P_\gamma \quad (12.2)$$

In the case of graphitic cast iron, eutectic graphite produces a positive expansion pressure, P_{exp}, which counteracts the shrinkage and gas pressure. If the expansion pressure equals or exceeds the shrinkage pressure, porosity is completely avoided. However, this requires a completely rigid mold which is rarely the case in practice.

This equation shows that the driving forces for pore occurrence are the gas and shrinkage pressure. If the metal is completely degassed the shrinkage pressure must reach the level of the shear stress of liquid metal for a vacuum pore to occur.

Once a pore is formed, its interaction with the S/L interface is complex. Work on Al and Al–25 %Au (Catalina et al. 2004) has shown that at a certain interface velocity, a planar S/L interface may engulf the gas pore. The pore continues to grow during its engulfment. The real-time measurements of pore growth in pure Al have revealed that when the pore is relatively far from the S/L interface, the mechanism of

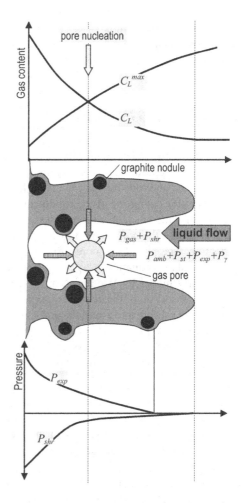

Fig. 12.8 Pressure and gas content along the mushy zone of ductile iron (Stefanescu 2005). With permission of Maney

pore growth is the hydrogen diffusion through the liquid phase. A sudden increase of the pore growth rate occurs when the solutal (i.e., hydrogen) field ahead of the S/L interface begins interacting with the pore. If the interface is nonplanar, the pore is entrapped by the solid. Its size may be of the order of micrometer to millimeter. While the terminology used to describe these pores is different (e.g., shrinkage pore versus gas pore), the mechanism of their occurrence is the same.

Dendrite coherency occurs at about 0.2–0.3 for most alloys. It follows that at low initial gas contents, the gas pore can continue to grow in the liquid with no immediate neighbors before being surrounded by solid and thus acquire a quasi-spherical shape. If the initial gas content is high, then gas pores will continue to form a long time after the onset of dendrite coherency and will grow in the interdendritic spaces, acquiring an irregular shape. A transition from shrinkage pores to micropores occurs with the increase of the hydrogen content in the melt.

To calculate the volume of gas porosity and the pore size at the end of solidification, Eq. 12.1 must be solved, i.e., the pressure in the gas pore and the pressure in the mushy zone must be formulated.

12.2.1 Pressure in the Mushy Zone

The assumption used by many models in the calculation of the pressure in the mushy zone P_{mush} is that the length of the mushy zone is established at the beginning of solidification and thus can be calculated from the cooling rate and the temperature gradient at the beginning of the solidification.

The early models tackle more or less comprehensively the whole transport problem but ignore the contribution of gas rejected by the solidifying melt to porosity formation. The physics of pore formation can be described through the simplified pressure balance equation: $P_{shr} > P_{amb} + P_{st}$.

An early model by Walther et al. (1956) assumed that void pores are formed because the section of the channel along which feed metal travels continuously narrows during solidification until the pressure drop ultimately ruptures the liquid in the channel. Further assuming shrinkage driven mass flow and conduction heat transport, an expression for the pressure drop along the channel, P_{shr}, was derived. It is given here in the format modified by Piwonka and Flemings (1966):

$$P_{st} - P_{shr} = \frac{32\mu\beta\varsigma^2 L^2}{(1-\beta)r^4} \tag{12.3}$$

where μ is the dynamic viscosity of the melt, β is the shrinkage ratio (Eq. 7.1), L is the length of liquid zone in the casting, r is the radius of liquid channel (central cylinder) in the partially solid casting, and $\varsigma = k_{mold}(T_f - T_o)/(\rho_s \Delta H_f \sqrt{\pi \alpha_{mold}})$, where k_{mold} is the thermal conductivity of the mold, T_f is the melting point of the metal, and α_{mold} is the thermal diffusivity of the mold. Assuming further that the first pore forms when $P_{shr} = 1$ atm and that once formed, the pore occupies the space previously occupied by the melt, the radius of the pore can be calculated from Eq. 12.3. Reasonable agreement with experiments was obtained.

Piwonka and Flemings (1966) were the first to use Darcy's law for the flow in a porous medium to describe the liquid flow through the coherent network of dendrites in the mushy zone. They used a tortuosity factor, $\xi \geq 1$, to account for the fact that the liquid flow channels are not straight and smooth. After some manipulations an equation that shows dependency on the primary dendrite arm spacing (PDAS), λ_I, is obtained:

$$P_{st} - P_{shr} = \frac{32\pi\mu\beta\varsigma^2 L^2}{(1-\beta)}\left(\frac{\xi\lambda_I}{r^2\pi r_{feed}^2}\right)^2 \tag{12.4}$$

where r_{feed} is the radius of the cylindrical feeding channel, L is the length of the feed (mushy) zone, and r is the radius of the feeding capillary, and ς is the thermal function in Eq. 12.5. In general $\xi^2/\left(\pi r_{feed}^2 n\right) \ll 1$. Thus by comparison with Eq. 12.3, pores in mushy freezing alloys will be much finer than those at the centerline of pure metals. This equation was used in their model that included gas pressure on the pore (see Sect. 12.2.2).

Niyama et al. (1982) further elaborated on Pellini's idea in their development of a criterion for low-carbon steel castings. They used Darcy's law in cylindrical coordinates and expressed the pressure drop in the mushy zone as an inverse function of the ratio $G_T/\sqrt{dT/dt}$. Shrinkage defects form in the region where the ratio is smaller than a critical value, to be determined experimentally. While this criterion works well for low-carbon steel, its application by many nonferrous foundries is questionable (see for example Spittle et al. 1995).

Attempting to improve this model, Huang et al. (1993) and then Suri et al. (1994) performed a 1-D analysis of the conservation of mass and momentum in the two-phase interdendritic region. It was found that the only significant term in the momentum equation responsible for loss of liquid pressure is that of the friction drag. Assuming flow along channels in the mushy zone, the nondimensional frictional drag on the feeding fluid can be expressed as: $F_{drag} = C_o V'_L$. Here, V_L' is the nondimensional liquid velocity and C_o is the "feeding resistance number" that controls the feeding in the mushy region expressed as:

$$C_o = \frac{N\mu\Delta T_{SL}}{\rho_L V_S \beta G_T d^2} \tag{12.5}$$

where N is a numerical constant (16π for columnar dendrites and 216 for equiaxed dendrites), and d is the characteristic length scale of the solid phase, i.e., either the primary dendrite arm spacing or the equiaxed grain diameter. C_o can be used as a criterion for microporosity formation. A high value indicates a high resistance to feeding and thus higher potential for pore formation. Indeed, experimental verification of this criterion showed excellent correlation between C_o and the percent porosity in an equiaxed A356 alloy.

In another approach by Stefanescu and Catalina (2011), assuming Darcy flow in the mushy zone, the liquid velocity in the mushy zone (shrinkage velocity) for horizontal flow was calculated as:

$$V_L = -\frac{K_{mush}}{f_L \mu}\frac{dP_{shr}}{dx} \tag{12.6}$$

where K_{mush} is the permeability, f_L is the fraction liquid, and μ is the viscosity. Using $V_L = \beta V_S$ where β is the shrinkage coefficient and V_S is the solidification velocity (after Carlson and Beckermann 2009), and approximating the width of the mushy zone as $L = (T_L - T_S)/G_T$, dx is $dx = \Delta T/G_T$, and the change in shrinkage pressure becomes:

$$\Delta P_{shr} = -\frac{f_L \mu \, \beta T \, \Delta T}{K_{mush} \, G_T^2} \tag{12.7}$$

The permeability can then be calculated with the Blake-Kozeny model (Sect. 7.4.2).

12.2.2 Gas Pressure in Pore

More advanced models include in the analysis the contribution of rejected gas to microporosity formation. The full pressure balance Eq. 12.2 is used.

Pore nucleation can occur when the gas dissolved in the liquid, C_L, exceeds the maximum solubility in the liquid, C_L^{max} (Fig. 12.8). The stability of such a pore is controlled by the surface energy pressure on the gas pore given by:

$$P_\gamma = 2\gamma_{LG}/r_P \tag{12.8}$$

where γ_{LG} is the gas–liquid surface energy and r_P is the pore radius. This equation suggests that homogeneous nucleation of a pore in the liquid is unlikely since $P\gamma$ is immense at the very small initial radius of the pore.

Piwonka and Flemings (1966) were the first to consider the role of the dissolved gas pressure. Eq. 12.2 was written as $P_G > P_{st} - P_{shr} + P\gamma$. The gas pressure term for a diatomic gas such as hydrogen was formulated as:

$$P_G = [v_i/(K_S(1 - f_L) + K_L f_L)]^2 \tag{12.9}$$

where v_i is the initial gas volume in the melt, and K_S and K_L are the equilibrium constants of the gas dissolution reaction for the solid and liquid, respectively.

> **Derivation of the gas pressure term in Eq. 12.9 (Piwonka and Flemings 1966)**
>
> A diatomic gas, G_2, dissolves in liquid metal according to the reaction: $(1/2)G_{2 \, gas} \leftrightarrow \underline{G}$. Assuming that the gas behaves ideally, so that the activity, a_G, is proportional to the partial pressure, P_{G2}, the equilibrium constant of this reaction is $K_L = a_G/\sqrt{P_{G2}}$. Further assuming that the activity of the liquid follows Raoult's law (activity coefficient = 1), the weight percent gas in the liquid is given by $wt\%\underline{G} = K_L\sqrt{P_{G2}}$. This is Sievert's law. It can be written in terms of volumes of gas for both liquid and solid, as follows: $v_L = K_L\sqrt{P_G}$ and $v_S = K_S\sqrt{P_G}$
>
> Mass balance dictates that $v_S(1 - f_L) + v_L f_L = v_i$, where v_i is the initial volume of gas in the liquid. Combining these equations, we obtain Eq. 12.9.

To verify the square dependency of the pressure drop on the primary dendrite arm spacing (PDAS) suggested by Eq. 12.4, Piwonka and Flemings conducted some experiments and demonstrated that this is true at $f_L < 0.3$. The surface tension pressure on the gas pore was calculated with Eq. 12.8. It was assumed that there was no barrier to the nucleation of pores, which seems to be a good assumption for metal casting, where pores nucleate on inclusions and second phases.

Another approach (Stefanescu and Catalina 2011) was to rewrite Eq. 12.1 as:

$$P_G = P_{mush} + 2\gamma / r_P \tag{12.10}$$

Assuming that the gas pores and the liquid are in thermodynamic equilibrium, Sievert's law gives:

$$C_L^{eq} = \overline{K}\left(P_{mush} + \frac{2\gamma}{r_P}\right)^{1/2} \quad \text{with} \quad \overline{K} = \frac{K_{eq}}{101325^{1/2} f_H} \tag{12.11}$$

where $101325^{1/2}$ is a unit transformation factor, f_H is the fraction of hydrogen, and P_{mush} is in Pascal. As pressure equilibrium at the G/L interface is valid at any time:

$$P_{mush} + 2\gamma / r_P = nRT / v_P \tag{12.12}$$

where v_P is the pore volume, n is the number of moles of gas, and R is the gas constant. Assuming spherical pores and N_P the number of pores in Δv the equation becomes:

$$\left(P_{mush} + \frac{2\gamma}{r_P}\right) N_P \frac{4\pi r_P^3}{3} = \frac{m_P}{M_G} RT \tag{12.13}$$

where m_P is the total mass of gas in all pores, and M_G is the molecular mass of gas.

12.2.3 Gas Evolution in Liquid

According to Sievert's law, the equilibrium reaction of a gas in equilibrium with a melt can be written as:

$$1/2 G_2 = [G] \tag{12.14}$$

The equilibrium constant of this reaction is:

$$K_{eq} = f_G [\%G](P_G)^{-1/2} \tag{12.15}$$

where f_G is the activity coefficient, $[\%G]$ is the amount of gas in solution and P_G is the partial pressure of the gas (in atmosphere). The activity coefficient accounts for the effect of alloying elements on the solubility of the gas.

The equilibrium constant is a function of temperature:

$$ln K_{eq} = -a/T - b \qquad (12.16)$$

where T is the temperature a and b are numerical coefficients. Then, the amount of gas in equilibrium with the melt can be expressed as:

$$C_L^{eq} = [\%G] = \frac{(P_G)^{1/2}}{f_G} exp(-a/T - b) \qquad (12.17)$$

This equation indicates that the amount of gas in equilibrium with the melt decreases as the melt cools. If at any time the amount of gas dissolved in the liquid exceeds the equilibrium content for the current melt temperature, molecular gas will form, and the gas will bubble out of solution. When solidification starts, the melt is continuously enriched in the dissolved gas because of the gas rejection at the S/L interface. Assuming equilibrium solidification (fast diffusion of gas in liquid and solid), the gas dissolved in the liquid C_L^G, resulting from the rejection at the S/L interface, can be calculated as (Stefanescu and Catalina 2011):

$$C_L^G = \frac{\rho_S + (\rho_L - \rho_S)f_L}{k_H \, \rho_S + (\rho_L - k_H \, \rho_S)f_L} C_L^{in} \qquad (12.18)$$

where C_L^{in} is the initial gas content in the liquid (i.e., at $f_L = 1$). This equation indicates that as the S/L mixture cools/solidifies, the gas concentration in the remaining liquid continues to increase. Alternatively, the Scheil equation can be used to calculate C_L^G.

12.2.4 Pore Nucleation

Experimental evidence suggests that gas pores nucleate heterogeneously on inclusions that are present in the melt (Rooy 1993, Roy et al. 1996, Mohanty et al. 1995). It appears that for "oxide free" aluminum alloys a rather high hydrogen concentration of 0.3 cc/100 g is required for pore formation (Fisher and Roy 1968). This theory is supported by the fact that inoculation seems to play an important role in the morphology of the gas pores. It may change both the morphology and the distribution of the pores (Dinnis et al. 2005) and the total amount of porosity. At low gas content (0.13 cc/100 g), the increase in porosity with modification is negligible, except at the slowest cooling rates. At a higher gas content (0.25 cc/100 g), porosity increased significantly as Sr was added (Sigworth 2009).

A typical case is the increased tendency to porosity formation in aluminum alloys modified with Sr. Fig. 12.9 reveals that in the specimen with Sr addition the pores are larger than in the nonmodified specimen. Also, most of the new pores emerge above the liquidus, while in the non-modified sample specimen most of new pores form around the liquidus. Liao et al. (2013) attribute this effect to the pore nucleating effect of Sr. Sr in the melt diffuses to the oxide inclusions, which enhances the

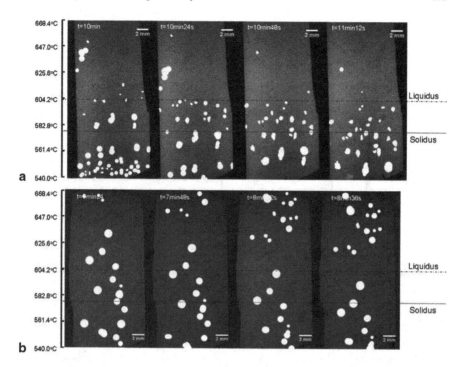

Fig. 12.9 X-ray images captured by XIDS showing evolution of pores during directional solidification at solidification velocity of 0.1 mm/s and temperature gradient of 4.28 ° C/mm in a Al–7Si–0.4Mg alloy. **a** No additions, **b** modified with 0.02 %Sr. (Liao et al. 2013). With permission of Maney

efficiency of the oxide inclusions as nucleation sites for pore, leading to an increase in the nucleation temperature of pores.

Depending on the amount of gas dissolved in the liquid, pore nucleation can occur in the liquid or in the mushy zone. As shown in Fig. 12.9, as solidification proceeds the pores continue to grow. Quenching experiments on a Al–4 %Si alloy demonstrated that most of the pores are formed immediately after the start of primary solidification (Fig. 12.10). According to Fig. 12.9 this is true for the nonmodified melt, but not for the Sr modified alloy.

Heterogeneous nucleation on inclusions is consistent with the observation that filtering molten Al castings reduces the porosity.

In general, but not always, the amount of microshrinkage in Al–Si alloys decreases with the amount of hydrogen dissolved in the melt. As shown in Fig. 12.11, work on small tapered plates and end-chilled plates of Al 7Si 0.4Mg found that at short solidification times (small DAS) the pore density rose with increasing hydrogen content. However, at long solidification times the pore density fell as hydrogen content increased (Tynelius et al. 1993).

According to Campbell (2003), this apparently confusing effect can be explained if it is accepted that pore nucleation occurs inside oxide films as thin as 20 nm

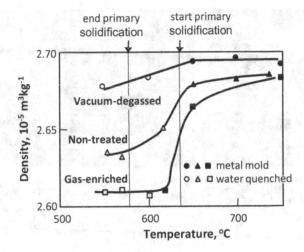

Fig. 12.10 Effect of quenching temperature on the density of Al–4Si alloys (after Awano and Morimoto 2004)

Fig. 12.11 The effect of gas content and DAS and on the number of pores per square cm (after Tynelius et al. 1993)

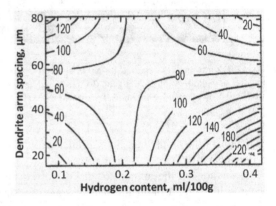

formed in liquid aluminum during mold filling. Oxide films may fold and produce bifilms (Fig. 12.12). If the liquid is assumed full of bifilms (i.e., cracks) the liquid has the potential to initiate pores with negligible difficulty. The bifilms simply open by the separation of their unbonded halves. Such a pore initiation mechanism can explain the data in Fig. 12.11 as follows. At low gas contents and short solidification times the bifilms remain folded and porosity is minimal. As the solidification time (DAS) increases, at the same gas level, the bifilms start opening and gas porosity increases. This is also true for high gas contents and low solidification times. At high gas contents and solidification times, large gas pores are formed, which decreases the area density of pores.

Nucleation on bifilms is now widely accepted as one of the mechanisms for the nucleation of pores. However, gas bubbling in liquids supersaturated with gas occurs

Fig. 12.12 Schematic view of surface turbulence acting to fold an oxide film and bubbles (after Campbell 2003)

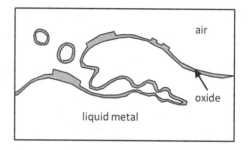

when the pressure is decreased (try opening a coke can). Clearly, there are no bifilms to facilitate nucleation. It is reasonable to conclude that the nucleation of gas bubbles before the beginning of solidification occurs at the mold walls, on impurities in the liquid (oxides, bifilms, etc.), or on microscopic size defects. Furthermore, micrometer sized gas bubbles can be assumed to exist in the liquid following their incorporation from turbulent flow during pouring and mold filling.

As discussed, the surface tension pressure, $P\gamma$, is extremely high at the beginning of pore formation when $r \sim 0$ which all but precludes homogeneous nucleation of pores. Thus, some additional assumptions regarding pore nucleation are necessary. Kubo and Pehlke (1985) argued that the pores nucleate at the root of the secondary dendrite arms. The free energy change of pore formation is:

$$\Delta G = v_p(P_G - P_{mush}) + A_{SG} \cdot \gamma_{SG} + A_{LG} \cdot \gamma_{LG} - A_{SG} \cdot \gamma_{SL} \qquad (12.19)$$

where v_P is the volume of the pore, and A is the area between the phases interfaces. If homogeneous nucleation is assumed, the first right-hand-term, which represents the free energy change during the liquid-to-gas phase transformation, must have a very large negative value to overcome the effect of $P\gamma$. However, in the Kubo and Pehlke model, since the effect of surface energy is reduced by $A_{SG} \cdot \gamma_{SL}$ a large negative pressure is not required. The initial pore radius was thus assumed half of the secondary arm spacing.

Other researchers worked with significantly smaller numbers, such as 3 μm (Felicelli et al. 2009) or even 1 μm (Fang and Granger 1989). It appears that SDAS/2 is too large an initial radius, as experimental measurements show that the final radius is on the order of SDAS/2. The number of nuclei is even more uncertain. Felicelli et al. (2009) worked with $N = 2 \cdot 10^{11} m^{-3}$, which gives a pore concentration of ~ 5 ppm. They based their selection of N on the assumption that most of the pore nucleation occurs on Al_2O_3 particles or films and on research by Simensen and Berg (1980) showing the smallest alumina particles in Al alloys to range from 0.2–10 μm, with the oxide concentration in the range of 6–16 ppm.

The change in gas concentration after the nucleation of a pore can be calculated assuming that the gas composition in the liquid has reached supersaturation and there is no nucleation barrier (Stefanescu and Catalina 2011). If a certain quantity of gas leaves the solution to form N_P gas pores in the volume element Δv, then the melt concentration will decrease to a value C_L^n that can be calculated through the following procedure. The amount of gas in the gas pore is given by the difference between the initial mass (superscript "o") of gas in the volume Δv and the mass left in the melt after the pore formation (superscript "n"):

$$m_P = m_L^o - m_L^n = \frac{\rho_L f_L}{100} \Delta v \left(C_L^o - C_L^n \right) \tag{12.20}$$

Combining this equation with Eq. 12.13, and writing the volumetric pore density (m^{-3}) as $N = N_P/\Delta v$ the gas concentration after the nucleation of a pore of radius r_P can now be calculated as:

$$C_L^n = C_L^o - \frac{100 \, N \, M_G}{\rho_L \, f_L R \, T} \left(P_{mush} + \frac{2\gamma}{r_P} \right) \frac{4\pi r_P^3}{3} \tag{12.21}$$

12.2.5 Pore Growth in the Mushy Zone

Pore growth can be calculated using a *diffusion model* or assuming *equilibrium solidification*. For the diffusion model, the main assumption of mass transport calculation is that the volume element is closed (no gas transport in or out of the volume element). Starting with Eq. 12.12, after a number of manipulations and assuming pores of spherical shape, the mass rate entering the pore by diffusion from the liquid can be calculated with:

$$\frac{dm_P}{dt} = \pi r_P^2 \rho_L \left(C_L^G - C_L^n \right) \left(D_L^G / t \right)^{1/2} \tag{12.22}$$

where D_L^G is the liquid diffusivity of gas, C_L^G is the gas concentration in the bulk liquid, C_L^n is the gas solubility at the local pressure and temperature, and t is the time measured from pore nucleation.

An alternative calculation of pore growth can be performed on the assumption of equilibrium solidification, as proposed by Catalina et al. (2007). As solidification proceeds, the gas dissolved initially only in the liquid will be redistributed between the new solid and the remaining liquid. The new pore radius can be calculated with:

$$\left(r_P^n \right)^3 = \frac{S_H^o \, C_L^o - S_H^n \, C_L^n + (100/\Delta v) m_P^o}{P_{mush}^n + 2\gamma/r_P^n} \frac{3RT}{400\pi \, M_G \, N_P/\Delta v} \tag{12.23}$$

where $S_H^i = \rho_S k_G + (\rho_L - \rho_S k_H) f_L^i$ with i being either o (old) or n (new), k_G is the equilibrium partition coefficient of gas between the solid and liquid phases, m_P is the mass of gas in the pore phase, and M_G is the molecular mass of gas. This equation can be solved numerically for the new pore radius r_P^n as all the other quantities are known either from the previous time step or from temperature and liquid fraction.

The model was implemented on the Excel spreadsheet for the case of aluminum alloy A356. The time–temperature data, including the local cooling rate and the temperature gradient, were computed with a one-dimensional finite difference method

Fig. 12.13 Effect of cooling rate on the final pore radius and comparison with literature data from Fang and Granger (1989). $N = 10^9$ (Stefanescu and Catalina 2011). With permission of Maney

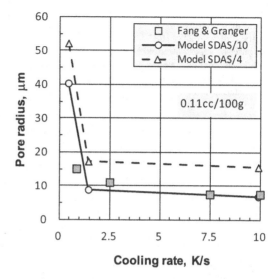

Fig. 12.14 Evolution of pore radius r_p^n, hydrogen dissolved in the liquid C_L^G, and equilibrium hydrogen solubility in the liquid C_L^{eq-P}, with the fraction solid. Equilibrium model (Stefanescu and Catalina 2011). With permission of Maney

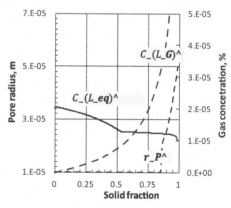

program. The initial pore radius was assumed SDAS/10 in most cases, although some calculations with SDAS/4 were run for comparison. The number of nuclei was taken as $10^9 \mathrm{m}^{-3}$ in most cases. Some other values were used for comparison. The predicted dependency of the final radius on cooling rate is exemplified in Fig. 12.13. An initial radius of SDAS/4 results in over estimation of the final radius for both models.

A typical output of the model is presented in Fig. 12.14. It is seen that, when the amount of dissolved hydrogen exceeds the equilibrium value (supersaturation) pore nucleation occurs. The pore grows rapidly to the end of solidification. The fraction solid at which supersaturation occurs is a strong function of the initial hydrogen content of the melt (Fig. 12.15). As the initial hydrogen content increases the onset of supersaturation moves to lower fraction solids. This allows more time for pore growth and the result is larger pore at the end of solidification (Fig. 12.15). It is also seen that as the initial hydrogen concentration decreases under 0.05 cc/100 g, the

Fig. 12.15 Effect of the
initial dissolved hydrogen
on the critical fraction for
the onset of supersaturation
and on the final pore radius.
Diffusion model (Stefanescu
and Catalina 2011). With
permission of Maney

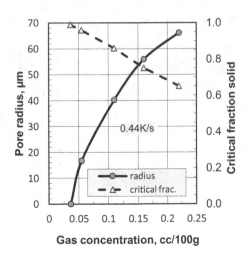

final pore radius approaches zero very fast. Calculations demonstrate that the final pore radius and porosity volume decrease dramatically with lower initial hydrogen concentration in the melt, higher applied pressure, and lower level of impurities (pore nuclei). Sufficiently low levels of initial hydrogen content and impurities may be conducive to minimal (if any) porosity.

A more complete mathematical description of microporosity formation on gas pores or on oxide films requires at least two-dimensional numerical treatment of the problem. It will be discussed in more detail in Chap. 18.

References

Awano Y, Morimoto K (2004) Inter. J. of Cast Metals Res. 17(2):107
Berry JT (1995) AFS Trans. 103:837
Boileau JM, Allison JE (2003) Met. Mater. Trans. 34A:1807
Campbell J (1991) Casting. Butterworth Heinemann, Oxford, UK
Campbell J (2003) in: Stefanescu DM, Warren J, Jolly M, Krane M (eds) Modeling of Casting, Welding and Adv. Solidif. Processes X. TMS, Warrendale PA, p 209
Carlson KD, Beckermann C (2009) Metall. Mater. Trans. A 40A:163
Catalina AV, Stefanescu DM, Sen S, Kaukler W (2004) Metall. and Mater. Trans. 35A:1525
Catalina AV, Leon-Torres JF, Stefanescu DM, Johnson ML (2007) in: Proc. 5th Decennial Int. Conf. on Solidification Processing, Sheffield. The University of Sheffield, p 699–703
Dinnis CM, Dahle AK, Taylor JA (2005) AFS Trans. 113:112
Felicelli SD, Wang L, Pita CM, Escobar de Obaldia E (2009) Metall. Mater. Trans. B 40B:169
Fang QT, Granger DA (1989) AFS Trans. 97:989–1000
Fisher EF, Roy EL (1968) AFS Trans. 76:237–290
Fuoco R, Goldenstein H, Gruzleski JE (1994) AFS Trans. 102:297
Huang H, Suri VK, EL-Kaddah N, Berry JT (1993) in: Piwonka TS, Voller V, Katgerman L (eds) Modeling of Casting, Welding and Advanced Solidification Processes VI. TMS, Warrendale, Pa, p 219
Kubo K, Pehlke RD (1985) Met. Trans. 16B:359
Liao HC, Song W, Wang QG, Zhao L, Fan R, Jia F (2013) Int. J. Cast Metals Res. 26 (4):201

Major JF (1997) AFS Trans. 105:901

Mohanty PS, Samuel FH, Gruzleski JE (1995) AFS Trans. 103:555

Niyama E, Uchida T, Morikawa M, Saito S (1982) AFS Cast Metals Res. J. 7:52

Piwonka TS, Flemings MC (1966) Trans. AIME 236:1157

Rooy EL (1993) AFS Trans. 101:961

Roy N, Samuel AM, Samuel FH (1996) Met. Mater. Trans. 27A:415

Ruxanda R, Beltran-Sanchez L, Massone J, Stefanescu DM (2001) Trans. AFS 109:1037

Samuel AM, Samuel FH (1999) Met. Mater. Trans. 26A:2359

Shih TS, Huang LW, Chen YJ (2005) Int. J. Cast Met. Res. 18:301–308

Sigworth G (2009) Mod. Cast. 99(5): 41–45

Simensen CJ, Berg G (1980) Aluminium, 56(5): 335–40

Skallerud B, Iveland T, Harkegard G (1993) Eng. Fract. Mech. 44:857

Spittle JA, Almeshhedani M, Brown SGR (1995) Cast Metals 7:51

Stefanescu DM (2005) Int. J. of Cast Metals Res. 18(3):129–143

Stefanescu DM, Catalina AV (2011) Int. J. Cast Met. Res. 24 (3/4):144

Suri VK, Paul AJ, EL-Kaddah N, Berry JT (1994) Trans. AFS 138:861

Tynelius K, Major JF, Apelian D. (1993) Trans. AFS 101:401

Uram SZ, Flemings MC, Taylor HF (1958) AFS Trans. 66:129

Walther WD, Adams CM, Taylor HF (1956) AFS Trans. 64:658

Chapter 13
Rapid Solidification and Amorphous Alloys

As discussed in some detail in Sect. 2.3, an increase in the rate of heat extraction (cooling rate) results in gradual departure from equilibrium up to global and interface nonequilibrium. As the rate of heat extraction increases, the microstructure length scale of solidified alloys decreases. Eventually, solidification without crystallization may occur.

Crystallization must occur via a process of nucleation and growth. If nucleation is suppressed (for example through the imposition of high undercooling on the system), the liquid will solidify without crystallization as an amorphous material also termed as glass. Thus, condensed matter can be classified into three categories, as listed in Table 13.1. Liquids are stable above the fusion temperature T_f, crystalline solids are stable under the solidification temperature T_S, and glasses are stable at a temperature lower than the glass transition temperature T_g.

The glass transition temperature (Tg) is defined as the temperature at which the material exhibits a sudden change in the derivative thermodynamic properties, such as heat capacity and expansion coefficient, from crystal-like to liquid-like values. This is illustrated in Fig. 13.1. When the cooling rate is low, phase transformation occurs at T_f with formation of a crystalline solid. When the cooling rate is high, the property changes continuously and the extrapolation method (broken lines) is used to determine T_g.

Thus, depending on the solidification velocity and the degree of interface nonequilibrium, rapidly solidified materials may exhibit a crystalline or amorphous microstructure.

13.1 Rapidly Solidified Crystalline Alloys

The term rapid solidification is normally applied to casting processes in which the liquid cooling rate exceeds 100 K/s (Boettinger 1974). This definition may be outdated in light of recent work on bulk metallic glasses. While different alloys respond differently to high cooling rates, some microstructures observed in rapidly solidified alloys can be achieved by slow cooling when large liquid undercooling is achieved prior to nucleation.

© Springer International Publishing Switzerland 2015
D. M. Stefanescu, *Science and Engineering of Casting Solidification,*
DOI 10.1007/978-3-319-15693-4_13

Table 13.1 Classification of condensed matter

Condensed matter	Temperature range	Thermal condition	Atomic configuration
Liquid	$T > T_f$	Stable	Disordered
Solid—Crystal	$T < T_S$	Stable	Ordered
Solid—Amorphous (glass)	$T < T_g$	Metastable	Disordered

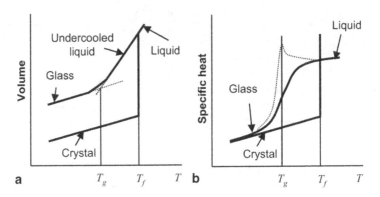

Fig. 13.1 Schematic illustration of the change in volume (**a**) and specific heat (**b**) as the liquid is cooled. *Solid lines* are experimental results. *Broken lines* illustrate extrapolations for determining T_g. (After Yonezawa 1991)

The following techniques are used to produce rapidly solidified alloys:

- Melt spinning, strip casting, or melt extraction, which produce thin ($\sim$25–100 μm) ribbon, tape, sheet, or fiber
- Atomization, which produces powder ($\sim$10–200 μm)
- Surface melting and resolidification, which produce thin surface layers

Some examples of such techniques are presented in Fig. 13.2. These methods may be considered as casting techniques where at least one physical dimension of the final product is small. Consolidation is used to yield large products from rapidly solidified alloys. This consolidation often alters the solidification microstructure of the final products. Yet, many features of the solidification structure can remain in the final product.

At "normal' rates of cooling the tip radius of the dendrite decreases as the solidification velocity increases (Fig. 13.3). However, as the cooling rate increases in the rapid solidification range the tip radius increases. This is accompanied by a decrease in branching. The equiaxed dendrite becomes globular/cellular. Typical examples of the evolution of the microstructure as a function of the solidification velocity are given in Fig. 13.4. Figure 13.4a shows a transverse section of a fine cellular structure of the Ag-rich phase in Ag–15 % Cu alloy. Most of the intercellular regions are filled with the Cu-rich phase, not the Ag–Cu eutectic as the Gulliver–Scheil

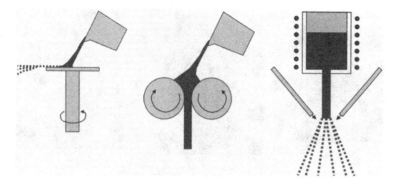

Fig. 13.2 Examples of techniques for producing rapidly solidifying alloys

Fig. 13.3 Schematic correlation between solidification
velocity and dendrite tip
radius

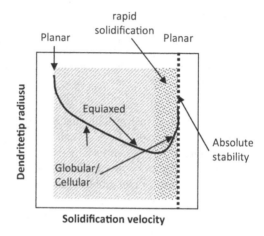

equation would predict. Figure 13.4b shows a longitudinal view of a cellular solidification structure. In Fig. 13.4c, the alloy has solidified with a planar interface to
produce a microsegregation-free alloy. The fine particles are the result of a solid
state precipitation.

A common occurrence in some rapidly solidified alloys is a change in the identity
of the primary solidification phase from that observed for slow solidification. Many
examples are found in hypereutectic Al alloys containing transition elements such
as Fe, Mn, or Cr. If the alloy is hypereutectic, slowly cooled castings will contain
intermetallics such as Al_3Fe or Al_6Mn as the primary (or first) phase to solidify.
However, under rapid solidification conditions the primary phase in these alloys is
the Al solid solution, usually found in a cellular structure with an intermetallic in
the intercellular regions. This transition from an intermetallic to an Al solid solution
as the primary phase can be understood by a careful examination of the kinetics of
the competitive nucleation and growth of the intermetallic and α-Al solid solution
(Hughes and Jones 1976).

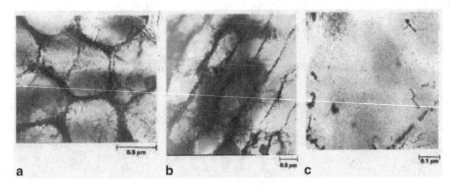

Fig. 13.4 Microstructures of Ag–Cu alloys electron beam melted and resolidified at different velocities. Thin foil transmission electron micrograph prepared by ion milling. **a** Ag–15 %Cu alloy resolidified at 0.025 m/s. Globular microsegregation pattern. Magnification: 32000 × . **b** Ag–15 %Cu alloy resolidified at 0.3 m/s. Cellular microsegregation pattern revealed by dislocation networks along cell walls. Magnification: 18000 × . **c** Ag–15 %Cu alloy resolidified at approximately 0.6 m/s. No cellular structure. The solid produced is uniform in composition except for fine Cu precipitates formed during solid state cooling. Magnification: 87000 × . (Boettinger et al. 1984). With permission of Springer Science and Business Media

In some cases, an intermetallic that is not given on the equilibrium phase diagram may compete with α-Al. In Al–Fe alloys, a metastable phase, Al_6Fe, rather than the stable phase, Al_3Fe, can form under some rapid solidification conditions. This situation is analogous to the appearance of cementite rather than graphite in some cast irons. The use of metastable phase diagrams to assist in the interpretation of rapidly solidified microstructures is described by Perepezko and Boettinger (1983).

As explained in Sect. 2.3, other rapidly solidified alloys have microsegregation-free structures formed by a liquid–solid transformation similar to a massive solid–solid transformation (partitionless or diffusionless transformation). The liquid transforms to solid without a change in composition. The ratio of the solid composition at the interface to the liquid composition is $C_S/C_L = 1$, rather than the equilibrium partition coefficient. Velocities required to produce partitionless solidification must exceed 5 m/s. The phase diagrams do not apply in this situation.

Rapidly solidified alloy powders exhibit a broad spectrum of solidification structures, depending on alloy composition and solidification conditions. Figure 13.5 shows single powder particles of stainless steel with dendritic or cellular structure. The size of the particles is less than 25 μm. Different degrees of undercooling prior to nucleation for particles of almost the same size determine the type of structure. The dendritic structure radiates from a point on the surface where nucleation has occurred. The scale of the structure is relatively uniform across the powder particle.

Other rapidly solidified powders often show significant microstructural variations across individual powder particles. Initial growth of the solid may occur very

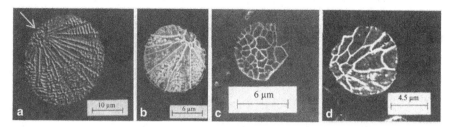

Fig. 13.5 SEM micrographs of atomized droplets of martensitic stainless steel. **a** and **b**—dendritic structures; **c** and **d**—cellular structures. (Pryds and Pedersen 2002). With permission of Springer Science and Business Media

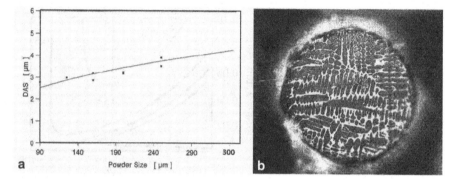

Fig. 13.6 Dendritic microstructure of stainless steel particles produced by the PREP process. (Wosch et al. 1995)

rapidly in a partitionless manner. The interface velocity decreases as the S/L interface crosses the particle, because of the release of the latent heat of fusion and warming of the powder particle, and the solidification front becomes cellular.

More sophisticated processing techniques, such as plasma rotating electrode process (PREP) are reported to produce nearly perfect stainless steel spherical form particles with relatively uniform structure (Wosch et al. 1995). PREP is in principle an atomization process based on the pulverization of metal bars in contact with an Ar/N$_2$ plasma arc. The particle diameter, d, can be approximated from force balance through the equation:

$$d = ct. \cdot (\sigma/D\rho)^{1/2} \cdot (2\pi n)^{-1},$$

where D is the bar diameter, σ is the surface tension, ρ is the density, and n is the rotational speed. The constant includes other process variables such as melting rate, dynamic viscosity, and the composition of the melted material. The SDAS of the particles was a direct function of the particle size (Fig. 13.6).

The growth velocity of rapidly solidified dendrites can be calculated with Eqs. 9.33 and 9.35. For the eutectic solidification the simplifications used in the Jackson–Hunt model (low Péclet number, interface composition in the liquid close to the eutectic composition, constant partition coefficient) cannot be used anymore.

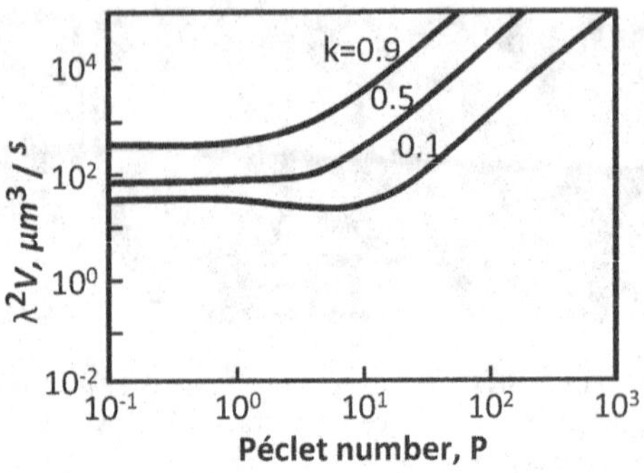

Fig. 13.7 Eutectic growth at high Péclet numbers. (After Kurz and Fisher 1998)

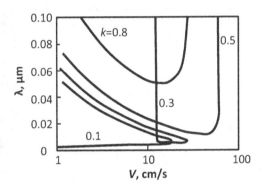

Fig. 13.8 Eutectic spacing–growth velocity relationship (After Kurz and Fisher 1998)

The constant in the $\lambda^2 V = ct = \mu_\lambda$ relationship becomes a function of the Péclet number at $Pe > \sim 10$, as shown in Fig. 13.7. The partition coefficient has a significant effect on both the growth parameter μ_λ and the velocity at which absolute stability is reached. Indeed, as seen in Fig. 13.8, for relatively high k, λ decreases with V before increasing again upon approaching the limit of stability. However, for small k values, the behavior is more complicated because of the increase in eutectic undercooling, which decreases diffusivity and bends the curve back before increasing it again.

13.2 Metallic Glasses

While liquid polymers and silicates can be easily converted into amorphous solid forms at cooling rates as low as 1–10 K/s, metallic glasses are relatively new products of the scientific ingenuity. The first reported metallic glass, $Au_{75}Si_{25}$, was obtained by quenching the liquid metal at very high rates of 10^5–10^6 K/s (Klement et al. 1960). By the late 1970s the continuous casting processes for commercial manufacture of metallic glasses ribbons, lines, and sheets (Kavesh 1978) were developed. These materials have an interesting combination of properties such as high mechanical strength, good thermal stability, large supercooled liquid region, and potential for easy forming.

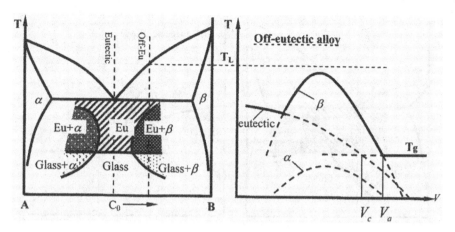

Fig. 13.9 Schematic phase diagrams for symmetric coupled zone eutectics. (Li 2005)

Turnbull and Fisher (1949) predicted that a ratio between the glass transition temperature, T_g, and the melting or liquidus temperature of an alloy, T_f, referred to as the reduced glass transition temperature $T_{rg} = T_g/T_f$, can be used as a criterion for determining the glass-forming ability (GFA) of an alloy. According to Turnbull's criterion (Turnbull 1969), a liquid with $T_g/T_f = 2/3$ can only crystallize within a very narrow temperature range. Such liquids can thus be easily undercooled at relatively low cooling rates to solidify as glasses.

The concepts of eutectic coupled zone introduced in Sect. 10.1.6 are also useful in understanding glass formation. For the symmetric coupled zone (regular) eutectics, glass will form preferentially to eutectic when the growth velocity is higher than the critical growth velocity for glass formation, V_a (Fig. 13.9). For eutectic composition fully amorphous phase will form, while for off-eutectic compositions glass $+ \beta$ dendrites composites will form at velocities between V_a and V_c, where V_c is the critical growth velocity for composite formation. Experimental work seems to indicate that the complete glass region is rather narrow.

For the asymmetric coupled zone (irregular) eutectics, eutectic will be replaced with glass in a larger region once the growth velocity is above V_a' (Fig. 13.10). Thus, it appears that the best glass forming alloy should be at an off-eutectic composition, on the faceted size of the eutectic.

For all practical purposes, it is not necessary to completely eliminate nucleation in order to produce a glass. If the crystallized volume fraction is below the detection limit, the material is practically amorphous. The cooling rate necessary to achieve this depends on the nucleation and growth rates and their temperature dependence. In the absence of convection or some adiabatic cooling mechanism, the cooling rate is limited by thermal conduction in the liquid, which scales with the square of the smallest sample dimension (Peker and Johnson 1993). As a result, a critical cooling rate for glass formation corresponds to a critical thickness for glass formation (see for example Fig. 13.11). This critical thickness for glass formation can be used to

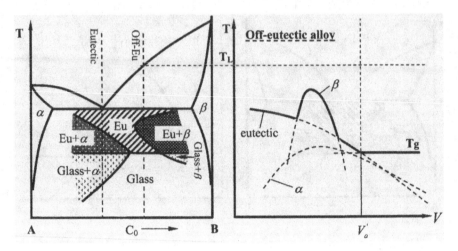

Fig. 13.10 Schematic phase diagrams for asymmetric coupled zone eutectics. (Li 2005). Copyright 2005 by The Minerals, Metals & Materials Society. Reprinted with permission

quantify the concept of GFA. It is directly related to the critical cooling rate for glass formation, but easier to determine experimentally.

The Turnbull criterion for the suppression of crystallization in undercooled melts has played a key role in the development of various metallic glasses including the new materials termed bulk metallic glasses (BMGs). If, as shown in Fig. 13.11, one arbitrarily defines the millimeter scale as "bulk" (Wang et al. 2004), the first bulk metallic glass was the ternary Pd–Cu–Si alloy prepared in the form of millimeter–diameter rods through a suction-casting method in 1974 (Chen 1974). The cooling rate of 10^3 K/s was significantly lower than reported before for the production of metallic glasses. Recently, amorphous Fe-based alloys with high carbon content termed as "structural amorphous steels" by the researchers were cast in rods of up to 12 mm diameter. A typical composition is, for example, $Fe_{44.3}Cr_5Co_5Mo_{12.8}Mn_{11.2}C_{15.8}B_{5.9}$ (Lu et al. 2004). BMGs can now be produced at cooling rates as low as 0.1 K/s which brings them in the range of castable materials.

According to Inoue (2000) there are three empirical rules for BMG formation: (1) multicomponent systems consisting of more than three elements; (2) significant difference in atomic size ratios above ~12 % among the three main constituent elements; and (3) negative heats of mixing among the three main constituent elements.

In general, the GFA in BMGs increases as more components are added to the alloy (the "confusion principle") (Greer 1993). This is construed to mean that larger number of components in an alloy system destabilizes competing crystalline phases that may form during cooling. Thus the crystallization tendency of the melt increases. It is claimed that the alloys satisfying the three empirical rules have atomic configurations in the liquid state that are significantly different from those

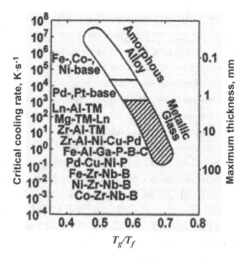

Fig. 13.11 Relationship between the critical cooling rate for glass formation, maximum sample thickness for glass formation, and reduced glass transition temperature (T_g/T_f) for bulk amorphous alloys. The data of the ordinary amorphous alloys, which require high cooling rates for glass formation, are also shown for comparison. (Inoue 2000). With permission of Elsevier

of the corresponding crystalline phases and that favor glass formation in terms of thermodynamics (free energy), kinetics (atom mobility), and microstructure evolution.

If the liquid-to-glass transformation is assumed to be the result of crystallization suppression within the supercooled liquid because of lack of nucleation, the steady state nucleation equation can be used to identify both the thermodynamic and the kinetic controlling parameters. The nucleation rate can be written as:

$$I = AD_{ef} \exp\left(-\frac{16\pi\gamma^3}{3k_B T \Delta G_{LS}^2}\right), \tag{13.1}$$

where A is a constant, k_B is the Boltzmann's constant, T is the absolute temperature, D_{ef} is the effective diffusivity, γ is the solid/liquid interface energy, and ΔG_{LS} is the energy difference between the liquid state and the crystalline state. The derivation of this equation is discussed in Sect. 3.1. ΔG_{LS} is the driving force for crystallization. Note that diffusivity can be related to viscosity through the Stokes–Einstein equation, $D = k_B T/6\pi\eta r$ where, r is the atomic radius. Based on the above considerations, the driving force (thermodynamic factor), diffusivity or viscosity (kinetic factor), and configuration (structural factor) are crucial parameters for understanding the glass formation in multicomponent alloys.

Further expanding on thermodynamic considerations, from Eq. 13.1 it follows that high γ and small ΔG_{LS} are conducive to low nucleation rates and thus favors high GFA. In turn, ΔG_{LS} can be calculated as:

$$\Delta G_{LS}(T) = \Delta H_f - T_f \Delta S_f - \int_T^{T_f} \Delta c_p^{LS}(T)\, dT + \int_T^{T_f} \frac{\Delta c_p^{LS}(T)}{T}\, dT, \tag{13.2}$$

where Δc_p^{LS} is the specific heat difference between the liquid and solid states. Low ΔH_f and high ΔS_f will thus decrease the nucleation rate. As ΔS_f is proportional to the number of microscopic states (Inoue 1995), a large ΔS_f is expected to be associated with multicomponent alloys. Therefore, a higher number of alloy components leads to the increase in ΔS_f and causes the increase in the degree of dense random packing in the liquid state. This is favorable to the decrease in ΔH_f and the S/L interfacial energy. The concept is consistent with the "confusion principle" and Inoue's first empirical rule (Wang et al. 2004).

The strong liquid behavior implies high viscosity and sluggish kinetics in the supercooled liquid state. The nucleation and growth of the thermodynamically favored phases is inhibited by the poor atom mobility resulting in high GFA and thermal stability of the supercooled liquid, as illustrated in Fig. 13.12.

The second empirical criterion for BMG formation suggested by Inoue requires large difference in the size of the component atoms. This is thought to lead to complex structures that experience difficulties in crystallization. Density measurements show that the density difference between BMG and fully crystallized state is in the range 0.3–1.0 % (Inoue et al. 1998; Wang et al. 2000), which is much smaller than the previously reported range of about 2 % for ordinary amorphous alloys. Such small differences in values indicate that the BMGs have higher dense randomly packed atomic configurations.

While the number of quantitative criteria proposed to evaluate the GFA of liquid alloys is large (at least 12 by last count), their applicability appears to be limited in most cases to a few alloy systems. The criterion that seems to be applicable to a majority of alloy systems is the γ parameter proposed by Lu and Liu (2002):

$$\gamma = T_x / \left(T_g + T_L\right), \tag{13.3}$$

where T_x is the temperature of onset of crystallization, T_g is the glass transition temperature, and T_L is the liquidus temperature.

As the novel BMG-forming liquids can be studied on significantly larger time and temperature scales, the opportunities for studying nucleation and growth in undercooled liquids and the glass transition have been highly improved. It is now possible to produce time–temperature-transformation (TTT) diagrams that describe the competition between the increased driving force for crystallization produced by increasing undercooling and the deceleration of atom movement (effective diffusivity). The TTT diagram for Vitalloy 1 in Fig. 13.13, shows the typical "C" curve and a minimum crystallization time of 60 s at 895 K and a critical cooling rate of about 1 K/s. For older glass-forming alloys, the times were of the order of 10^3 s.

In summary, the BMG-forming liquids are typically dense liquids with small free volumes and high viscosities, which are several orders of magnitude higher than those of previously known amorphous metals and alloys. Their microstructure configurations are significantly different from those of conventional amorphous metals. Thermodynamically, these melts are energetically closer to the crystalline state than other metallic melts. They have high packing density in conjunction with a tendency to develop short-range order (Wang et al. 2004).

Extensive work by numerous researchers has resulted in the development of BMGs with critical diameters larger than 20 mm in alloy systems based on Cu

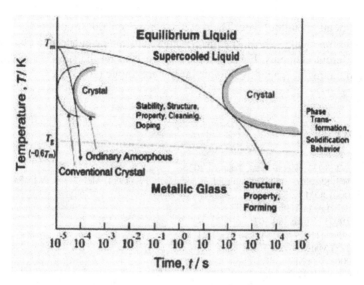

Fig. 13.12 Schematic diagram showing the high stability of the BMG forming supercooled liquid for long periods reaching several thousands of second. (Inoue and Takeuchi 2002)

Fig. 13.13 A TTT diagram for the primary crystallization of Vitalloy 1. Data obtained by electrostatic levitation (●) and processing in high-purity carbon crucibles (▲) are included. (After Busch 2000)

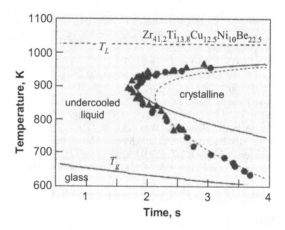

(Zhang and Inoue 2004), Ln (Li et al. 2007), Mg (Zheng et al. 2007), Ni (Zeng et al. 2009), Pd (Inoue et al. 1997), Pt (Schroers and Johnson 2004) and, Zr (Inoue and Zhang 1996; Yokoyama et al. 2007).

While the material systems coming under the attention of researchers is immense and growing, there is particular interest in the synthesis and characterization of Fe-based BMGs, because they exhibit characteristics which have not been obtained in conventional Fe-based crystalline alloys. These properties qualify them for practical use as soft magnetic and structural materials. While BMGs can be produced in both the metal–metalloid and metal–metal varieties, all the current Fe-based BMGs are of the metal–metalloid type. The metallic component constitutes ~80 % and the metalloid component ~ 20 %. The metal component can be either only Fe or a mixture of different elements. For a complete discussion on the subject the reader is referred to the review paper by Suryanarayana and Inoue (2013).

References

Boettinger WJ (1974) Metall. Mat. Trans. 5:2026
Boettinger WJ, Shechtman D, Schaefer RJ, Biancaniello FS (1984) Met. Trans. A 15:55
Busch R (2000) JOM 52:39
Chen HS (1974) Acta Metall. 22:1505
Greer AL (1993) Nature 366:303
Hughes IR, Jones H (1976) J. Mater. Sci. 11:1781
Inoue A (1995) Mater. Trans. JIM 36:866
Inoue A (2000) Acta Mater. 48:279
Inoue A, Zhang T (1996) Mater. Trans. JIM 37:185–187
Inoue A, Nishiyama N, Kimura HM (1997) Mater. Trans. JIM 38:179–183
Inoue A, Negishi T, Kimura HM, Zhang T, Yavari AR (1998) Mater. Trans. JIM 39:318
Inoue A, Takeuchi A (2002) Mater. Trans. 43(8):1892–1906
Kavesh S (1978) Chapter 2. In: Gillman JJ, Leamy HL (eds) Metallic Glasses. ASM Int., Metals Park, OH
Klement W, Willens RH, Duwez P (1960) Nature 187:869
Kurz W, Fisher DJ (1998) Fundamentals of Solidification, 4th edition. Trans tech Publ., Switzerland
Li Y (2005) JOM March:60–63
Lu Z.P. and C.T. Liu, 2002, Acta Mater. 50:3501–3512
Lu ZP, Liu CT, Thompson JR, Porter WD (2004) Phys. Rev. Lett. 92:245503
Li R, Pang S, Ma C, Zhang T (2007) Acta Mater. 55:3719–3726
Peker A, Johnson WL (1993) Appl. Phys. Lett., 63:2342
Perepezko JH, Boettinger WJ (1983) in: Mat. Res. Soc. Symp. Proc. 19:223
Pryds NH, Pedersen AS (2002) Metall. Mat. Trans. A 33A:3755–3761
Schroers J, Johnson WL (2004) Appl. Phys. Lett. 84:3666–3668
Suryanarayana C, Inoue A (2013) Int. Mater. Rev. 58(3):131
Turnbull D (1969) Contemp. Phys. 10:437
Turnbull D, Fisher JC (1949) J. Chem. Phys. 17:71
Wang WH, Wang RJ, Zhao DQ, Pan MX, Yao YS (2000) Phys. Rev. B 62:11292
Wang D, Li Y, Sun BB, Sui ML, Lu K, Ma E (2004) Appl. Phys. Lett. 84:4029
Wosch E, Feldhaus S, Gammal T (1995) ISIJ Int. 35(6):764–770
Yokoyama Y, Mund E, Inoue A, Schultz L (2007) Mater. Trans. 48:3190–3192
Yonezawa F (1991) in: Ehrenreich E, Turnbull D (eds) Solid State Physics - Advances in Research and Applications. Aca-demic Press, Boston p 179–254
Zeng YQ, Nishiyama N, Yamamoto T, Inoue A (2009) Mater. Trans. 50:2441–2445
Zheng Q, Xu J, Ma E (2007) J. Appl. Phys. 102:113519-1–113519-5
Zhang W, Inoue A (2004) Mater. Trans. 45:1210–1213

Chapter 14
Semisolid Processing

For an incompressible and isotropic Newtonian fluid, the viscous stress is related to the strain rate by the simple equation:

$$\tau = \mu \frac{dV}{dy} \tag{14.1}$$

where τ is the shear stress ("drag") in the fluid, μ is a scalar constant of proportionality (the shear viscosity of the fluid), and dV/dy is the derivative of the velocity component that is parallel to the direction of shear, relative to displacement in the perpendicular direction. Assuming that the liquid remains Newtonian after the incorporation of the particles, the viscosity of the slurry can be approximated with Eq. 7.23 through the concept of *relative viscosity*. Yet, while studying the rheology of semisolid metals, Spencer et al. (1972) found that a semisolid slurry with a fraction solid higher than 0.2 behaves like a non-Newtonian fluid with a much lower relative viscosity than that of the static slurry. This discovery triggered the development of semisolid processing.

14.1 Phenomenology

During mushy-type solidification in the early solidification stages the grains can move freely and can settle. At some critical fraction of solid, the grains form a network, dendrite coherency occurs, and mass feeding is replaced by interdendritic feeding. This critical fraction of solid, f_S^{cr}, depends on dendrite morphology, size, and number. It is typically in the range of 0.1–0.2 (Flemings 1991). Above 0.2 f_S^{cr}, the shear stress required to isothermally shear a batch of solidifying alloy steadily increases (Fig. 14.1). The alloy in Fig. 14.1 was partially solidified before the shear was applied. In well grain-refined alloys, strength did not begin to develop until 0.4 fraction solid.

Deformation at fractions solid up to 0.9 is primarily by grain-boundary sliding. At sufficiently high strain, continuous fissures open and stress falls rapidly. This type

© Springer International Publishing Switzerland 2015
D. M. Stefanescu, *Science and Engineering of Casting Solidification,*
DOI 10.1007/978-3-319-15693-4_14

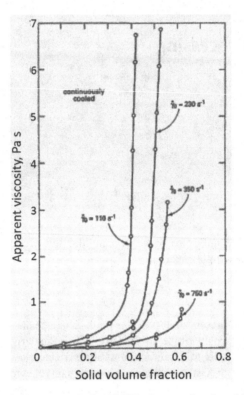

Fig. 14.1 Viscosity and shear stress of a Sn–15 %Pb alloy as a function of fraction solid and test arrangement (Joly and Mehrabian 1976). With permission of Springer Science and Business Media

of strain and the resulting fluid flow determine the existence of localized regions with macrosegregations in castings.

When shear was applied before the beginning of solidification, and continued through solidification, shear stress increased only slowly as the temperature was decreased below liquidus. For example, while at $f_S = 0.4$, for the dendritic material the stress was 200 kPa, for the nondendritic material the stress was only 0.2 kPa. The grain structure is nondendritic. Initially this process was called "rheocasting." Later on it was dubbed "semisolid metal (SSM) forming."

The mechanism of structure evolution during rheocasting is shown in Fig. 14.2. Without stirring, the solid grows as dendrites. The stronger the stirring and the longer the time, the greater the tendency to form equiaxed dendrites, rosettes, and finally spheroids. An example of the effect of time and shear rate on structure evolution is given in Fig. 14.3.

Semisolid slurries possess "pseudoplasticity" which can be described by the empirical relation:

$$\mu = K\,\dot{\gamma}^{n-1} \tag{14.2}$$

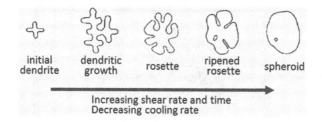

initial
dendrite

dendritic
growth

rosette

ripened
rosette

spheroid

Increasing shear rate and time
Decreasing cooling rate

Fig. 14.2 Schematic illustration of evolution of structure during solidification with vigorous agitation. (After Flemings 1991)

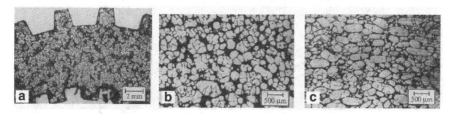

Fig. 14.3 Sn–15 %Pb alloy cooled at 0.006 K/s under various shear rates and water quenched. **a** Low shear rate, low fraction solid ($f_S = 0.35$). **b** Low shear rate, $f_S = 0.5$. **c** High shear rate, $f_s = 0.5$. (Flemings 1991). With permission of Springer Science and Business Media

where μ is the apparent viscosity (defined as shear stress/shear rate), $\dot{\gamma}$ is the shear rate, n is the power law index, and K is the "consistency." The viscosity of a SSM slurry measured during continuous cooling decreases with higher shear rate (Fig. 14.4) and cooling rate. This may be because increasing shear rate and decreasing cooling rate result in denser, more rounded particles that move easier past one another.

The combined influence of shear rate on viscosity and structure at constant fraction solid is demonstrated in Fig. 14.4. For the continuously cooled alloy, viscosity decreases with increased shear rate as agglomeration decreases and grain shape changes from dendritic to rosette. For the isothermally held alloy (steady-state curve), the grains are spheroidal, viscosity is lower overall because of the spheroidal grain shape, and decreases with higher shear rate because of lower agglomeration.

Approximately reversible pseudoplastic behavior is typical for steady-state experiments. There is, however, a time dependency (*thixotropy*), so that when shear rate is abruptly changed, the new steady-state viscosity is attained only after some time at that shear rate (lower curve on Fig. 14.4). The viscosity is less than the steady-state value because the structure did not have time to adjust to that of the new shear rate. With time agglomerates build and the viscosity at a given shear rate approaches the steady-state value. The difference between this curve of "instantaneous viscosity" and the steady-state curve is a measure of the thixotropy of the slurry.

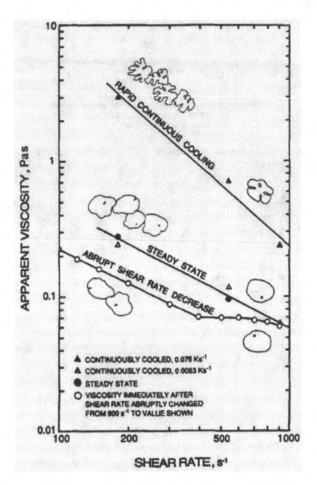

Fig. 14.4 Influence of shear rate on viscosity and structure for Al–6.5 %Si alloy at 0.4 fraction solid (Flemings 1991). With permission of Springer Science and Business Media

The constitutive relations for these materials might be written (Brown 1990) as:

$$\mu = \mu(\dot{\gamma}, f_S, m, s) \tag{14.3}$$

$$ds/dt = s(\dot{\gamma}, f_S, m, s) \tag{14.4}$$

where m is a measure of particle morphology and s is a measure of degree of particle agglomeration.

A comparison of viscosities of various materials and of the shear rates of some processes is given in Table 14.1. For additional details the reader is referred to the extensive review paper by Fan (2002).

Table 14.1 Some typical viscosities and shear rates of familiar processes. With permission of Springer Science and Business Media

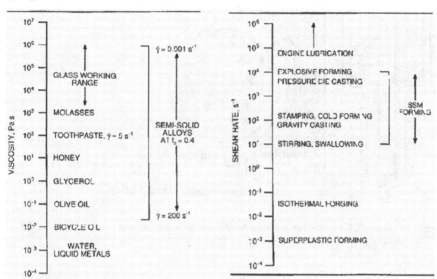

14.2 Typical Process Routes

14.2.1 Semisolid Slurry Processing

The original Massachusetts Inst. of Technology method for producing nondendritic slurries relied on batch stirring, which then evolved into continuous stirring and cooling into rod– cylindrical crucible assembly (Fig. 14.5a). The scaling of this system to industrial levels was unsuccessful because of the erosion of the ceramic stirrer, dross, and gas entrapment and limitation to maximum 0.4 fraction solid. An alternative to the mechanical stirring is the magnetohydrodynamic (MHD) stirring process (Kenney et al. 1988) in which local shear is generated by rotating electromagnetic fields. This process produces a material with uniform fine grains (30 μm) and little contamination.

Another approach is the shearing cooling roll (SCR) process (Kiuchi and Sugiyama 1995) in which the molten alloy is poured into a roll-shoe gap (Fig. 14.5b) where the solidifying melt is sheared to produce a fine slurry. The process has been tested on Al alloys, cast iron, and steel, and has produced grains in the range of 30–50 μm.

The simplest process to produce semisolid slurries is the cooling plate technique developed by Muumbo et al. (2003) described in Fig. 14.6. It consists in pouring the

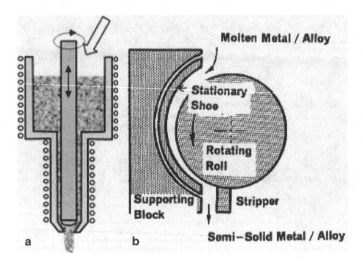

Fig. 14.5 Methods for producing nondendritic slurries. **a** Continuous mixer (after Flemings 1991). **b** The shearing cooling roll (*SCR*) process (Kiuchi and Sugiyama 1995)

Fig. 14.6 Schematic representation of the cooling plate process for producing semisolid slurries (Muumbo et al. 2003). With permission of Maney

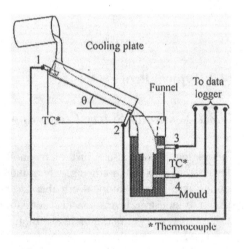

molten metal of interest over an inclined water-cooled copper plate into a mold, and allowing the metal to solidify.

Spray casting processes, such as the Osprey process (Leatham 1989) in which a stream of liquid metal is atomized by a gas jet and deposited in the semisolid state onto a cold target, have also been reported to produce nondendritic structures with grain size as small as 20 μm. Upon reheating the deposited alloy back to semisolid, a thixo-forgeable material is generated.

14.2.2 Forming of the Semisolid Slurry

Once the slurry is obtained, a number of forming processes are available. The first process, *rheocasting*, which consists in solidifying the slurry directly into amold, can be done through continuous casting, centrifuging the slurry into a mold cavity, or through injection molding. The second process, *thixoforming*, consists in pouring the slurry into an ingot, reheating the ingot, and then forming it by die casting or by closed-die forging. Finally, *compocasting* involves processing of metal matrix composites.

14.3 Material Models/Systems

A large number of materials are amenable to semisolid processing. Brabazon et al. (2002) studied the effects of controlled stirring during solidification on the microstructure and mechanical properties of aluminum alloys Al–4Si and Al356, in comparison to conventionally gravity chill cast material. A more globular primary phase was achieved at low values of f_S, but this was not the optimum morphology for mechanical properties. The properties were found to be at their maximum for a low degree of primary phase structural breakdown, which occurred at high f_S and low $\dot{\gamma}$ and shear time. In all cases, improved mechanical properties and reduced porosity were obtained in the stir cast alloy compared with conventionally cast alloy. Comparison with alloy commercially rheocast via electromagnetic stirring, however, showed that the latter had superior mechanical properties. It was proposed that the mechanical stir casting process be considered as an alternative to gravity die casting in cases where very simple and thick-walled shapes are required.

Xing et al. (2012) developed a method which they called the self-inoculation method for the preparation of semisolid slurry of magnesium alloy Mg–5.9Al–0.3Mn–0.2Zn (AM60). The process consists in mixing the molten alloy with particles of solid alloy (self-inoculants) and subsequently pouring the liquid mixture through a two-stream steel static mixer. The process produces high nucleation and survival rates of the primary α-Mg phase at superheating as high as 85 °C. To obtain high-quality semisolid slurries, a melt treatment temperature range of 680–700 °C and self-inoculant addition of 5–7 % are suggested.

Kotadia et al. (2011) studied the effect of the processing temperature on the microstructural and mechanical properties of hypoeutectic Al–Si (Al–9.4Si) alloys solidified from intensively sheared liquid metal. Intensive shearing produced a significant refinement of the primary α-Al grain size and improved the distribution of the α-Al(Mn, Fe)Si intermetallic phase with a narrow size distribution. The average α-Al(Mn, Fe)Si intermetallic particle size was reduced from 8 to 5 μm. Defect bands were observed in both sheared and nonsheared samples. However, intensive shearing distributes externally solidified crystals, which are believed to form in the shot sleeve, more uniformly, provides an ideal condition to nucleate primary α-Al

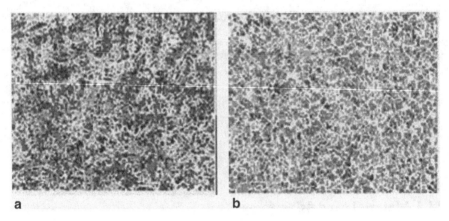

Fig. 14.7 Microstructure of hypoeutectic gray iron. **a** Sand mold. **b** Cooling plate and sand mold (Muumbo et al. 2003). With permission of Maney

that is spherical in shape, and significantly reduces the defect band size and porosity. The reduction in porosity was explained through the collapse and/or dispersion of gas bubbles that exist in the liquid melt into a smaller size by the application of high (intensive) shearing.

Semisolid forming (thixoforming) was used by Sen et al. (2012) for near-net shape forming of hard and wear-resistant tool steels (crucible particle metallurgy steel Fe-3.4C-5.25Cr-14.5V-1.3Mo-0.5Mn-0.9Si) that are very difficult to machine conventionally. The novel structures obtained resulted in improved hardness.

Muumbo et al. (2003) applied the cooling plate process to hypoeutectic gray cast iron of carbon equivalent 3.93–4 %. When the metal was delivered from the cooling plate to a sand mold, they obtained globular grains with about 50 μm diameter (Fig. 14.7).

Similar experiments were conducted on high speed steels with high V and C contents (Shirasaki et al. 2008). It was reported that nucleation increased through cooling plate semisolid processing. A large amount of smaller spherical MC carbides crystallized.

Present commercial interest is in high-integrity aluminum components, especially for the automotive industry because near-net shape parts are produced at lower cost than forging or machining. In addition, they exhibit higher quality than die casting because of the higher viscosity of thixocast material resulting in mold filling with "solid-front fill" as compared with metal spraying and air entrapment. Also, improved integrity is achieved because of the lower solidification shrinkage of semisolid alloys. Examples of applications include: master brake cylinders (in Europe) and electrical connectors for military aerospace applications.

References

Brabazon D, Browne DJ, Carr AJ (2002) Mat. Sci. Eng. A326:370–381

Brown SB (1990) in: Rappaz M (ed) Proc. 5th Int. Conf. on the Modeling of Casting, Welding, and Solidification Processing. TMS Warrendale PA

Fan Z (2002) Int. Mater. Reviews 47(2):49

Flemings MC (1991) Metall. Trans. 22A:957

Joly PA, Mehrabian R (1976) J. Mater. Sci. 11:1393–1418

Kenney MP, Courtois JA, Evans RD, Farrior GM, Kyonka CP, Koch AA, Young KP (1988) in: Stefanescu DM (ed) Metals Handbook 9th edition, vol. 15. ASM International, Metals Park OH p327–338

Kiuchi M, Sugiyama S (1995) ISIJ International 35(6):790–797

Kotadia HR, Babu NH, Zhang H, Arumuganathar S, Fan Z (2011) Metall. Mater. Trans. 42A:1117

Leatham A, Ogilvy A, Chesney P, Wood JV (1989) Met. Mater. 5(3):140–143

Muumbo A, Nomura H, Takita M (2003) Int. J. Cast Metals Res. 16(1–3):359–364

Sen I, Jirkova H, Masek B, Bohme M, and Wagner MF-X (2012) Metall. Mat Trans. 43A:3034

Shirasaki K, Takita M, Nomura H (2008) Int. J. Cast Metals Res. 21(1–4):45–48

Spencer DB, Meharbian R, Flemings MC (1972) Metall. Trans. 3:1926–1932

Xing B, Li YD, Ma Y, T. Chen J, Hao T (2012) Int. J. Cast Metals Res. 25(4):232

Chapter 15
Solidification of Metal Matrix Composites

Composite materials are engineered materials in which two or more materials are combined to form a new material with specific properties, in order to maximize some properties (stiffness, strength-to-weight ratio, tensile strength, etc.) and minimize others (weight, cost). A metal matrix composite (MMC) is a material with at least two constituent parts, one of which is a metal. The other material may be a different metal or another material, such as a ceramic or an organic compound. The majority phase (the matrix), which in MMCs is the metal, transfers external load to the strengthening phase. The minority phase is the strengthening phase (reinforcement).

Both the amount and size of the reinforcements affect strengthening, mostly through their effect on the dislocation that increases with higher volume fraction and smaller particle size (Fig. 15.1). According to Arsenault et al. (1991), with certain assumptions, the increase in the yield strength of the composite over that of the matrix can be estimated with:

$$\Delta\sigma_Y = \alpha\mu b\, \rho^{1/2} \tag{15.1}$$

where α is a constant (1.25 for Al), μ is the shear modulus of the matrix (2.64×10^4 MPa for Al), b is the Burgess vector (2.86×10^{-10} m), and ρ is the dislocation density.

MMCs produced by solidification/casting techniques fall into two broad categories: ex-situ and in-situ. Manufacture of ex-situ composites is based on one of the following processes:

- Mixing of the reinforcements (particles or short fibers) in the liquid matrix and subsequent solidification (stir casting)
- Pressure infiltration of reinforcement preforms by the liquid matrix and solidification
- Spray deposition of streams of molten metal and ceramic into a mold or on a substrate

A classification of MMCs based on the shape of the reinforcement is given in Table 15.1.

© Springer International Publishing Switzerland 2015
D. M. Stefanescu, *Science and Engineering of Casting Solidification,*
DOI 10.1007/978-3-319-15693-4_15

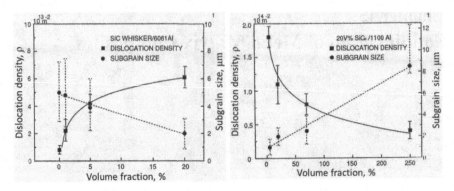

Fig. 15.1 Effect of volume fraction and particle size on the dislocation density (SiC whiskers length/diameter ratio = 2, diameter 0.5 μm) (Arsenault et al. 1991). With permission of Elsevier

Table 15.1 Shape and size of typical reinforcements for MMCs

Type	Strengthener	Processing	Size	Volume, %
Dispersion strengthened	Particles	In-situ	0.001–0.1 μm	1–15
Particle reinforced	Particles Platelets	Liquid mixing and casting, spray casting	>1 μm	1–50
Fiber reinforced	Fibers (continuous or discontinuous)	Infiltration	0.1 μm–0.1 mm dia. 0.01 mm–continuous	2–80 2–80

Typical reinforcements used for particle and fiber MMCs are as follows (Stefanescu 1993):

- Carbides: SiC, B_4C (e.g., Kalaiselvan et al. 2011), WC, TiC (e.g., Liu et al. 2011), TaC, ThC_2, ZrC, Cr_7C_3, HfC, MoC
- Nitrides: AlN, BN, TaN, TiN, HfN, Si_3N_4 (e.g., Ma et al. 1996), ZrN, ThN
- Borides: TaB_2, MoB, HfB, TiB_2 (e.g., Wang and Arsenault 1991; Chen et. al. 2014), CrB_2, ZrB_2, WB
- Oxides: ZrO_2, Al_2O_3 (e.g., Singh and Alpas 1995), HfO_2, ThO_2, etc.
- Metals: Nb, Mo, Ta, W, Be
- Graphite

Of particular interest are graphite (continuous fiber)/Cu or Al, SiC (whisker and particles)/Al, Al_2O_3 or SiC (particles)/Mg, and reinforced-ordered intermetallic composites, such as titanium and nickel aluminides, and the berylides. An extensive compilation of possible reinforcement and their thermodynamic and physical properties was produced by El-Mahallawy and Taha (1993) and Stefanescu (1993).

The fundamental science issues pertinent to processing of MMCs include the reinforcement/liquid and reinforcement/solid interaction and the calculation of the pressure required for the infiltration of the preform. These topics are discussed in the following two sections.

Fig. 15.2 Microstructure of
investment cast A356 Al–Si
alloy with 20 % SiC particles
(Kennedy 1991)

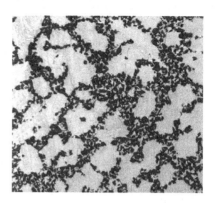

15.1 Solidification in the Presence of Freely Moving Particles

The phenomenon of interaction of particles that can move freely in the liquid with solid/liquid (S/L) interfaces has been studied since mid-1960s. While the original interest stemmed from geology applications (frost heaving in soil), researchers soon realized that understanding particle behavior at solidifying interfaces might yield practical benefits in other fields, including metallurgy. The issue is the location of particles with respect to grain boundaries at the end of solidification. Considerable amount of experimental and theoretical research was lately focused on applications to MMCs produced by casting (Stefanescu et al. 1988; Kennedy and Clyne 1991) or spray forming techniques (Wu et al. 1994; Lawrynowicz et al. 1997). In the most common cast MMCs, Al–Si alloy with SiC particles, the particles are normally distributed at the grain boundaries (Fig. 15.2). This results in decreased plastic properties. Similar issues are pertinent to inclusion management in steel (Shibata et al. 1998). The particle S/L interface interaction was also found to play an important role in the solidification of ternary eutectics (Hecht and Rex 2001) as well as in the formation of microporosity during solidification (Mohanty et al. 1995).

Another application of particle—S/L interface interaction is in the growing of $Y_1Ba_2Cu_3O_5$ (123) superconductor crystals from an undercooled liquid (Endo et al. 1996; Shiohara et al. 1997). The oxide melt contains $Y_2Ba_1Cu_1O_5$ (211) precipitates, which act as flux pinning sites. Other applications include phagocytosis (literally "cell-eating," i.e., large particles are enveloped by the cell membrane of a larger cell and internalized to form a food vacuole) (Torza and Mason 1969), and particle chromatography (separation of various solids) (Kuo and Wilcox 1973). For directional solidification, the large body of experimental data on various organic and inorganic systems (e.g., Uhlmann et al. 1964; Zubko et al. 1973; Omenyi and Neumann 1976; Körber et al. 1985; Stefanescu et al. 1988; Shibata et al. 1998), demonstrates that there exist a critical velocity of the planar S/L interface below which particles are pushed ahead of the advancing interface, and above which particle engulfment occurs. As shown in Fig. 15.3a under certain conditions the

Fig. 15.3 Pushing and engulfment of 10–150 μm SiC
particles during DS (from *left*
to *right*) of an Al–2 %Mg
alloy. V_{SL} = 8μm/s. The
transition from the equiaxed
zone at the *left* to the columnar zone marks the beginning
of DS **a** particles engulfed at
G_T = 117 K/s, **b** particles
pushed at G_T = 74 K/s, then
engulfed (Stefanescu et al.
1988). With permission of
Springer Science and Business Media

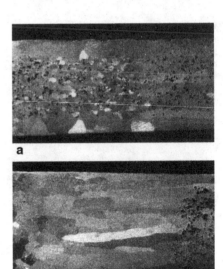

particles are engulfed by the columnar front. In Fig. 15.3b particles are pushed during directional solidification and then engulfed as the particle volume fraction builds
up.

However, the problem is more complicated because in most commercial alloys
dendritic interfaces must be considered. Indeed, most data available on metallic
alloys, as summarized by Juretzko et al. (1998), are on dendritic structures. At high-
solidification velocity (V_{SL}) or low-temperature gradient (G_T) cellular/dendritic or
even equiaxed interfaces will develop. The tips of cells or dendrites may *engulf* the
particles. Alternatively, solute trapping in the particle/interface gap decreases interface curvature to the point that it changes sign, resulting in tip splitting followed
by engulfment. However, because convection will move the particles in the interdendritic regions, *entrapment* between the dendrites arms will be more common. In
general, for cellular/dendritic and equiaxed interfaces the probability of engulfment
is much smaller than that of entrapment. Thus, most particles will be distributed at
the grain boundaries (Stefanescu et al. 2000). The physics of these two phenomena,
engulfment and entrapment, is quite different.

15.1.1 Particle Interaction with a Planar Interface

In liquids solidifying with planar interfaces, it has been observed that the interface
can instantaneously engulf the particles, the interface may continuously push the
particle ahead of it, or the interface may push the particle up to a certain distance
before engulfing it (Omenyi and Neumann 1976). Most experimental findings point

Fig. 15.4 Correlation between the critical velocity for PET and particle radius for water–nylon particles system (after Azouni et al. 1990)

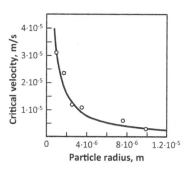

out to the existence of a critical velocity, V_{cr}, above which particles are engulfed (the pushing–engulfment transition—PET), and to an inverse relationship between V_{cr} and particle radius, r_P. An example is given in Fig. 15.4 for water doped with nylon particles. The V_{cr}—r_P relationship can be described by a power function:

$$V_{cr} = M \cdot r_P^{-m} \tag{15.2}$$

where M is a material constant and m is an exponent that can have different values. Some examples of the experimental values of these parameters for a number of systems are given in Table 15.2.

As the particle approaches the S/L interface, the difference in thermal conductivity between the particle, k_P, and the matrix (liquid), k_L, will impose an interface curvature (Fig. 15.5). If the thermal conductivity ratio $k^* = k_P/k_L$ is larger than unity, the interface will form a trough. In the opposite case, a bump will grow on the interface. However, for engulfment to occur a trough should form on the bump. These effects have been demonstrated on transparent organic metal analog materials (TOMA) as well as on aluminum matrices through X-ray transmission microscopy (Omenyi and Neumann 1976; Uhlmann et al. 1964; Sen et al. 1997).

However, as demonstrated by Hadji (1999a) the interface shape in pure substances is not determined solely by the thermal conductivity ratio, but also by the local pressure. In alloys, the solutal field will also affect interface shape.

Table 15.2 Values of parameters in Eq. 15.2 obtained from experimental data on particle pushing by a planar interface

Matrix/particle system	m	M, $\mu m^2/s$	Reference
Biphenyl/acetal	0.90	1132	Omenyi and Neumann (1976)
Biphenyl/nylon	0.64	199	Omenyi and Neumann (1976)
Naphthalene/acetal	0.30	195	Omenyi and Neumann (1976)
Naphthalene/nylon	0.46	142	Omenyi and Neumann (1976)
Succinonitrile/polystyrene	1.0	12.1	Pang et al. (1993)
Steel/silica–alumina (liquid)	1.0	24	Shibata et al. (1998)
Aluminum/zirconia	1.0	250	Juretzko et al. (1998)

Fig. 15.5 Interface shape
during particle engulfment

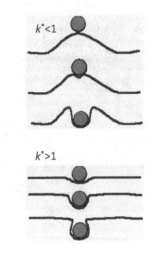

Fig. 15.6 Schematic representation of the forces acting
on a particle in the vicinity of
the S/L interface

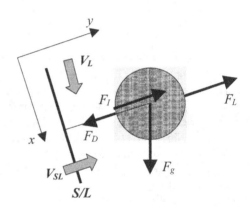

The PET is also affected by the flow of the liquid at the S/L interface. As summarized in Fig. 15.6, two distinct liquid flow patterns exist: flow into the interface
generated by solidification, and flow parallel to the interface at velocity V_L induced
by natural convection. Four main forces are identified:

- The interaction force between the particle and the S/L interface F_I, (typically a
  repulsive force)
- The drag force exercised by the solidification induced liquid flow around the
  particle into the interface F_D, (which pushes the particle into the interface)
- The lift force produced by the liquid flow parallel to the interface, F_L, (which
  pushes the particle away from the interface)
- The gravity force, F_g

Table 15.3 Regimes of particle-interface interaction

Melt convection	Interface velocity	Governing phenomenon	Outcome
Zero or small	Low	Interface interaction	Pushing
Zero or small	High	Interface interaction	Engulfment
Large	Any	Fluid flow velocity	No particle-interface interaction (pushing)

There is an agreement on the fact that at low convection regime in the melt particle/interface interaction governs the PET, while under conditions of high convection there is no particle/interface interaction and particles are continuously pushed. Considering the two velocities, V_{SL} and V_L, it can be rationalized that the behavior of the particle is governed by either the interaction with the interface, or by the fluid flow velocity between the particle and the interface, as summarized in Table 15.3.

To explain the experimental findings several theories have been proposed. They fall largely into two categories: (i) theories that attribute the PET to select material properties such as surface energy or thermal properties and (ii) theories that acknowledge the role of solidification and particle kinetics. Numerous models have been suggested based on these approaches.

15.1.2 Material Properties Models

The net change in free energy of a single spherical particle that moves from the liquid to the solid is:

$$\Delta\gamma_o = \gamma_{PS} - \gamma_{PL} < 0 \tag{15.3}$$

where γ_{PS} and γ_{PL} are the particle/solid and particle/liquid interface energy, respectively. Omenyi and Neumann (1976) have postulated that if $\Delta\gamma_o < 0$, engulfment is to be expected; while for $\Delta\gamma_o > 0$, pushing should result.

Since the thermal properties of the particle and the liquid affect the shape of the interface, some researchers assumed that the ratio between these properties determines particle behavior. Zubko et al. (1973) suggested that the value of k^* determines the outcome. If $k^* > 1$, the trough that forms at the S/L interface will engulf the particle. On the contrary, if $k^* < 1$, a bump forms and the particle will be continuously pushed. Following a similar line of thinking, Surappa and Rohatgi (1981) proposed an empirical heat diffusivity criterion; engulfment is postulated when $\left(k_P c_P \rho_P / k_L c_L \rho_L\right)^{1/2} > 1$, where c and ρ are the specific heat and density, respectively.

15.1.3 Kinetic Models

Kinetic models are complex attempts at describing the physics of PET and include energy and mass transport (diffusive or convective), particle motion, and the interaction force between the particle and the S/L interface. This problem does not yet have a complete solution.

The physics of PET is best described in terms of force balance as shown in Fig. 15.6. The governing equation is the equation of particle motion (Catalina et al. 2000):

$$F_I + F_L - F_D - F_g - F_o = m\frac{dV_P}{dt} \qquad (15.4)$$

where F_o is the force required to accelerate the fluid that adheres to the particle, m is the particle mass, and dV_P/dt is the particle acceleration. An additional force, the thermal force was suggested by Hadji (2001). Formulating the various forces *dynamic models* can be developed. Assuming steady state, the equation simplifies to:

$$F_I + F_L - F_D - F_g = 0 \qquad (15.5)$$

Further simplification is possible by ignoring the gravitational acceleration (microgravity experiments):

$$F_I - F_D = 0 \qquad (15.6)$$

These last two equations have been used in developing *steady-state models* that calculate an *equilibrium velocity*.

Let us now address the problem of formulating the various forces in the governing equation.

Formulation of Forces **The interface force.** The formulation of the interface force depends on the source considered for this force. In the earliest model proposed for PET, Uhlmann et al. (1964) assumed that the interfacial repulsive force results from the variation of the surface free energy with the distance d_{gap} from the interface given as:

$$\Delta\gamma = \Delta\gamma_o \cdot (d_o/d_{gap})^n \quad \text{where} \quad \Delta\gamma_o = \gamma_{PS} - \gamma_{PL} - \gamma_{LS} \qquad (15.7)$$

Here, d_o is the minimum separation distance between particle and solid, and the subscripts P, S, L stand for particle, solid, and liquid, respectively. Note that as $d_{gap} \to \infty$, $\Delta\gamma \to \Delta\gamma_o$. The value of the exponent n was assumed to be between 4 and 5. Further assuming $n = 2$ and non-retarded van der Waals interaction the interaction force was derived (Stefanescu et al. 1988; Pötschke and Rogge 1989; Shangguan et al. 1992) to be:

$$F_I = 2\pi \, \Delta\gamma_o d_o^2 \frac{r_P}{d_{gap}^2} \xi \qquad (15.8)$$

where ξ is a correction factor for the curved interface. For example, in the SAS model (Shangguan et al. 1992), for $d_{gap} \ll r_P$, the repulsive interface force was derived as:

$$F_I = 2\pi r_P \Delta\gamma_o \left(\frac{a_o}{a_o + d}\right)^2 \frac{r_I}{r_I - r_P} = 2\pi r_P \Delta\gamma_o \left(\frac{a_o}{a_o + d}\right)^2 k^* \quad (15.9)$$

where r_I is the radius of the S/L interface, a_o is the atomic diameter of the matrix material, and d_{gap} is the equilibrium distance between the interface and the particle.

The main problem with this formulation is the calculation of $\Delta\gamma_o$. In spite of extensive efforts and sometimes contentious debate (Kaptay 1999, 2000) it is clear that because of the uncertainties in the evaluation of the various interface energies, $\Delta\gamma_o$ is at best a fitting parameter.

Attempting to avoid the complexities of solid/solid interface energy, Ode et al. (2000) assumed a sinusoidal interface and proposed the equation:

$$F_I = \pi^2 \gamma_{SL} \sqrt{(r_P + d_{gap})l} \left(\frac{d_o}{d_o + d_{gap}}\right)^2 \quad (15.10)$$

where l is the amplitude of the sin function.

Another approach to the estimation of the interface force was proposed by Chernov et al. (1976, 1977). They assumed that as the interface approaches the particles, the difference in chemical potential between the bulk liquid, μ_L^∞, and that in the liquid film between the particle and the interface, μ_L, produces a disjoining pressure given by:

$$\Pi = \frac{\mu_L^\infty - \mu_L}{v_o} = \frac{B_n}{d^n} \quad (15.11)$$

where v_o is the molecular volume of the liquid and B_n is a constant. The disjoining pressure (which is in fact a volume energy expressed in J/m³) is assumed positive if the films thicken, and negative in the opposite case. The disjoining pressure introduces a local undercooling, ΔT_P. Further assuming that the disjoining pressure is the source of the interface force, the following equation was derived:

$$F_I = \pi B_3 \frac{r_P}{d_{gap}^2} \xi \quad (15.12)$$

where B_3 is a constant suggested to be 10^{-21} J. However, just as $\Delta\gamma_o$, B_3 becomes a fitting parameter. The correlation between B_3, the Hamaker constant and surface energy is discussed by Stefanescu (2002) and Asthana and Tewari (1993).

Experimental measurements of the interface (repulsive) force were attempted by Smith et al. (1993). By observing particles being pushed up an incline by an advancing S/L interface, they measured the repulsive force between the particle and the front in three systems where the energy of particle adhesion to the solidification front was known. The equation of their force of adhesion is identical with that for $\Delta\gamma_o$ given in Eq. 15.7. A linear relationship was found between $\Delta\gamma_o$ and F_I, as predicted by the theory. A summary of their experimental data is given in Table 15.4.

Table 15.4 The interface (repulsive) force between acetal particles in three matrix materials (Smith et al. 1993)	Matrix	$2\,r_P$, μm	F_I, nN	d_o, nm
	Salol	20	0.0010	22
	Benzophenone	20	0.0051	18
	Biphenyl	20	0.0145	19

Note that the minimum separation distance calculated from these experiments is constant at about 20 nm.

The Drag Force To maintain a stable liquid film between the particle and the S/L interface during particle pushing, liquid must continue to flow into the gap. The pressure gradient behind the particle is the driving force for the fluid flow. It induces the drag force on the particle. In its most general form the drag force on a particle of radius r_P can be expressed as:

$$F_D = \frac{1}{2}C_D\rho_L V_P^2\pi r_P^2 \tag{15.13}$$

where C_D is the drag coefficient. The simplest expression for the drag force was derived by Stokes for a spherical particle in an unbounded fluid, $F_D = 6\mu\eta V_P r_P$, where η is the dynamic viscosity. It is only valid for Reynolds numbers (Re = $V_P r_P \rho / \eta$) smaller than 0.5.

When the particle is very close to the interface, the assumption of unbounded flow does not hold anymore. The drag force can be calculated from the lubrication theory assuming the process is controlled by fluid flow in the particle/interface gap. For a gap of width d, much smaller than the particle radius, the first approximation of the drag force as calculated by Leal (1992) is:

$$\text{for a circle: } F_D = 6\pi\eta V_P\frac{r_P^2}{d} \quad \text{for a cylinder: } F_D = 3\sqrt{2}\pi\eta V_P\left(\frac{r_P}{d}\right)^{1.5} \tag{15.14}$$

where η is the viscosity. Note that these equations are approximations obtained by integrating the pressure distribution only within the gap between the particle and the interface, and not all around the particle.

This equation has been modified by different investigators to account for the interface curvature (e.g., Pötschke and Rogge 1989; Shangguan et al. 1992). For locally deformed interface, it was derived that the drag force becomes:

$$F_D = 6\pi\eta V_P\frac{r_P^2}{d}\left(\frac{r_I}{r_I - r_P}\right)^2 = 6\pi\eta V_P\frac{r_P^2}{d}k^{*2} \tag{15.15}$$

Due to the limitations introduced by the assumptions used in the derivation, this equation is valid for trough formation in the interface ($k^* < 1$). For $k^* \geq 1$, calculated results match better the experimental data when k^* is assumed equal to one in this equation as well as in Eq. 15.9.

Using numerical modeling (Catalina et al. 2000) it was demonstrated that the drag force is higher than given by the preceding equations. For a cylinder it is:

$$F_D = \sqrt{3}\pi\eta V_P(r_P/d)^{1.92} \tag{15.16}$$

However, numerical work by Garvin and Udaykumar (2004) seems to confirm Leal's equation. Indeed, they calculated that:

$$F_D = 10\eta V_P (r_P/d)^{1.53} \tag{15.17}$$

Regardless of which equation is more accurate, the main problem in the calculation of the drag force rests with the value used for viscosity. As the gap width is of the atomic distance order, it is to be expected that the value of the viscosity in the gap will significantly differ from that in the bulk liquid.

The Gravity Force The net gravity force acting on the particle is:

$$F_g = \frac{4}{3}\pi r_P^3 (\rho_L - \rho_P)g \tag{15.18}$$

Lift Forces The general expression for the lift forces is Eq. 15.13 where the drag coefficient is replaced by the lift coefficient, C_L. Using the method of matched asymptotic expansions, Saffman (1965) derived the following equation for the lift force on a rigid sphere translating parallel the streamlines of a unidirectional linear shear flow field:

$$F_S = 6.46\eta V_{rel} r_P^2 \sqrt{\frac{\rho_L}{\eta}\left(\frac{dV_{Lx}}{dy}\right)_{avg}} \tag{15.19}$$

where V_{rel} is the velocity of the particle relative to the liquid, and $(dV_{Lx}/dy)_{avg}$ is the average liquid velocity gradient over the particle diameter. This equation holds only for small Reynolds numbers.

An additional lift force comes from particle rotation—the Magnus force, F_M. For a particle rotating with an angular velocity, ω, and translating with a velocity, V_{rel}, relative to the liquid, the lift coefficient for the Magnus force was calculated (Rubinow and Keller 1961)) to be:

$$C_{LM} = 2r_P\omega/V_{rel} \tag{15.20}$$

Kinetic Steady-State Models A large number of analytical steady-state models for PET were developed over the years. The first kinetic steady-state model was proposed by Uhlmann et al. (1964). It assumed that mass transport in the particle/solid gap is by mass diffusion alone. In most cases, this model underestimates the critical velocity.

Numerous kinetic models were developed based on Eq. 15.6. By equating the repulsive and the drag force, these models calculate an equilibrium velocity at which the particle is continuously pushed. The first such model was published in by Stefanescu et al. (1988). Assuming that the equilibrium velocity is the critical velocity they suggested that:

$$V_{cr} = \frac{\Delta\gamma_o d_o}{6\eta r_P}\left(2 - \frac{k_P}{k_L}\right) \tag{15.21}$$

One year later, Pötschke and Rogge (1989) obtained an equation for V_{cr} by solving numerically an analytical solution that included the effect of the solutal field. Many other models followed. In most of these models the equation is of the form $V_e \propto 1/d_e$ and states that at any interface velocity the particle will find an interface distance (equilibrium distance) at which steady state exists. An additional criterion must be imposed to obtain the critical velocity. The criteria imposed by various researchers ranged from complex mathematical exercises (e.g., Chernov et al. 1976—CTM model, Bolling and Cissé 1971), to additional hypotheses on the mechanism of engulfment including maximization of equilibrium velocity (Pötschke and Rogge 1989; Shangguan et al. 1992) or on gap thickness (Sen et al. 1997; Stefanescu et al. 1998). For example, in the Shangguan et al. (1992) (SAS) model, when equating Eqs. 15.9 and 15.15 the following equilibrium velocity is obtained:

$$V_e = \frac{\Delta \gamma_o}{3\eta \, k^*} \frac{d}{r_P} \left(\frac{a_o}{a_o + d} \right)^2 \tag{15.22}$$

Maximizing this equation with respect to d (i.e., $dV_e/dd = 0$) gives the minimum separation distance (critical distance): $d_{cr} = a_o$. Substituting the critical distance in the equilibrium distance gives the critical velocity:

$$V_{cr} = \frac{a_o \Delta \gamma_o}{12\eta k^* r_P} \tag{15.23}$$

In the CTM model the equilibrium velocity is obtained by equating Eqs. 15.12 and 15.15: $V_e = B_3/(6\eta dr_P)$. It becomes the critical velocity if d is substituted by the minimum separation distance d_o. However, in this case, both the velocity and the distance are unknown. Additional manipulations are required to derive the critical velocity without imposing a critical distance.

Pötschke and Roge (1989) (PR model) assumed that the interaction force could be treated as the van der Waals force between two spheres. Their final analytical solution for the critical velocity was solved numerically, to yield:

$$V_{cr} = \frac{1.3\Delta \gamma_o}{\mu} \left[16 \left(\frac{r_P}{a_o} \right)^2 k^* \left(15 \, k^* + x \right) + x^2 \right]^{-1/2} \quad \text{with} \quad x = \frac{C_o \, |m_L| \, \Delta \gamma_o}{k G_L \eta D_L} \tag{15.24}$$

For pure metals $x = 0$, and the above equation becomes:

$$V_{cr} = \frac{0.084 a_o \Delta \gamma_o}{\mu \, k^* r_P} \tag{15.25}$$

Note that this equation is identical to Eq. 15.23. Both the SAS and PR model overestimate the critical velocity because they use a critical distance $d_{cr} = a_o$, which is too small.

Rempel and Worster (2001) derived a similar equation to the CTM equation for $r_p, < 500\mu$m. They elaborated on the role of intermolecular forces other than those dominated by non-retarded van der Waals interactions, such as long-range electrical

Table 15.5 Selected steady-state models for PET

Model/Reference	Critical velocity, $V_{cr} =$	Exponent m in Eq. 15.2	Assumption for d_{cr}
Uhlmann/Chalmers/ Jackson(UCJ)	$\frac{n+1}{2}\left(\frac{\Delta H_f a_o v_a D}{k_B T r_P^2}\right)$	2	
Bolling/Cissé (BC)	$\left(\frac{1.36 k_B T a_o \gamma_{SL}}{\pi}\right)^{1/2}\frac{1}{3\eta r_P^{3/2}}$	3/2	Maximum force
Chernov/Temkin/ Melnikova (CTM)	$r_P > 500\ \mu m\ \frac{B^{3/4}(\Delta S_f G_T)^{1/4}}{24\eta r_P(k^*)^{3/4}}$	1	
	$r_P < 500\ \mu m\ \frac{0.14 B^{2/3}\gamma_{SL}^{1/3}}{\eta r_P^{4/3}}$	4/3	
Pötschke/Rogge (PR)	$\frac{0.084 a_o \Delta \gamma_o}{\eta k^* r_P}$	1	$dV_e/dd = 0$
Sen et al. (1997) (modified SAS)	$\frac{\Delta \gamma_o a_o}{156\eta k^* r_P}$	1	$d_{cr} = 50 a_o$
Stefanescu et al. (1998)(SC)	$\left(\frac{\Delta \gamma_o a_o^2}{3\eta k^* r_P}\right)^{1/2}$	1/2	$dV_e/dd = -1$
Kim and Rohatgi (1998) (KR)	$\frac{a_o \Delta \gamma_o}{18\eta}\left[\frac{G a_o}{\Gamma}\left(\frac{k^*-1}{3}+\frac{1}{r_P}\right)\right]$	1	$d_{cr} = a_o$
Hadji (1999b)	$\lvert A\rvert\, G_L\left[\frac{36\eta\left[\Delta T_L + (1-k^*)\Delta T_P\right]}{\pi r_P + 6\pi\eta(1-k^*)}\frac{12(3\pi)^{1/3}\gamma_{SL}v_a}{\lvert A\rvert^{1/3}\Delta H_f}r_P^{2/3}\right]^{-1}$ A: *Hamaker ct.*, ΔT_L, ΔT_P: temperature change across the liquid and particle		Limit of vanishing disjoining pressure

interactions and retarded van der Waals interaction. Using linear stability analysis, Hadji (2003) demonstrated that presence of a particle in the melt modifies the threshold value of the thermal gradient for the inset of morphological instability. With the exception of the UCJ model derived on the assumption that the controlling mechanism is mass diffusion in the P-S gap, all others assume flow in the P-S gap as the controlling mechanism.

The equations for most of the models have been summarized in some references including Stefanescu (2002) and Youssef et al. (2005). A summary of the steady-state equations for critical velocity derived by different researchers is provided in Table 15.5.

A comparison between calculation with some of these models for the systems biphenyl/glass particles and succinonitrile/polystyrene particles on one hand and

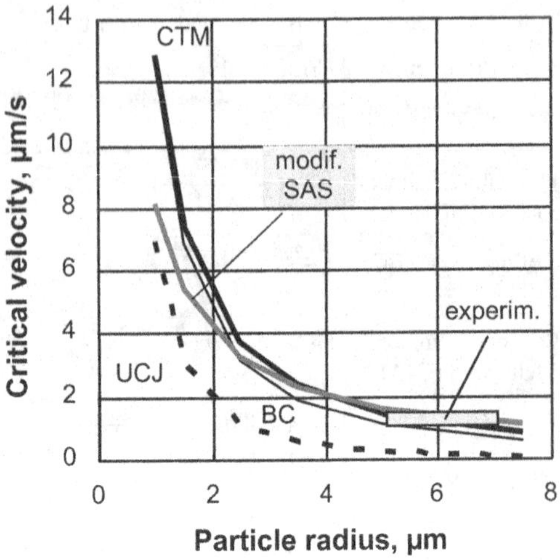

Fig. 15.7 Experimental and calculated critical velocities for the biphenyl/glass system (Sen et al. 1997a). With permission of Elsevier

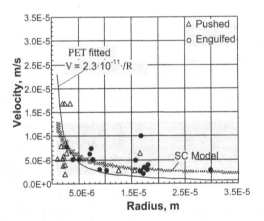

Fig. 15.8 Experimental (Shibata et al. 1998) and calculated (Stefanescu and Catalina 1998) correlation between the critical velocity for the pushing/engulfment transition of globular inclusions in 0.01 wt% C steel

experimental results on the other is presented in Fig. 15.7. The BC, CTM, and modified SAS models calculate critical velocities in the same range. The UCJ model underestimates V_{cr}, while PR overestimates it by several orders of magnitude.

Interesting results were obtained when theoretical predictions with the SC model were compared with the experimental results of Shibata et al. (1998) on inclusions in steel. The experimental data are given in Fig. 15.8 for alumina agglomerates and for complex globular inclusions in steel. The lower limit for the PET of alumina clusters is fitted to the experimental data as the $V = 1.9 \cdot 10^{-11}/r_P$ curve. Most clusters that have been pushed are located under this lower limit. For the globular inclusions, a fitted PET critical velocity is superimposed on the experimental points. The curve corresponds to the equation $V = 2.3 \cdot 10^{-11}/r_P$, as suggested by Shibata et al. It is seen that in both cases predictions with the SC model are reasonable.

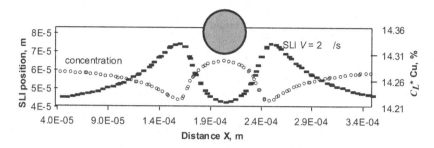

Fig. 15.9 S/L interface shape and liquid interface Cu concentration (Al-2 wt% Cu alloy, ZrO_2 *particle r_P* $= 22.5\mu m$, $V_{SL} = 2\mu m/s$) (Catalina and Stefanescu 1999)

Kinetic Interface Shape Models With the development of numerical techniques, it became possible to attempt to calculate the change in interface shape at least in two dimensions. In most cases, the particle was assumed to be at rest, while the S/L interface advanced toward the particle. The first work on this subject seems to be by Sasikumar et al. (1989, 1991) who developed a steady state heat flow numerical model that considered curvature and pressure undercooling and accounted for the role of solutal field. Fluid flow around the particle was ignored. The critical velocity for PET was calculated from the maximization of the equation of particle velocity. While they demonstrated for the first time that for $k^* < 1$ a trough will occur on the bump under certain conditions, they were unable to produce interface shape plots because of the complications of their numerical method.

Casses and Azouni (1993) used numerical simulation, including disjoining pressure and curvature effects, to demonstrate the change in the shape of a pure substance S/L interface approaching a particle that has a different thermal conductivity than that of the liquid (bump and trough formation). Kim and Rohatgi (1998) studied the effect of diffusion field in the gap, but did not present any calculated interfaces.

A more complete analysis was performed by Catalina and Stefanescu (1999). They developed a diffusive model for interface shape based on a 2D numerical model for interface tracking. The interface temperature is controlled by solute concentration and curvature. Fluid flow around the particle was ignored. Some typical results are shown in Fig. 15.9. Note the higher concentration in the gap. Later (Catalina et al. 2004), the model was validated for the system Au–H_2 pores, and used to explain the mechanism of comet tail-shape segregation observed behind the particles in many experiments (Fig. 15.10).

This model demonstrated elegantly the formation of a trough-in-bump for alloys, because of solute accumulation in the gap. Thus, engulfment of particles for the case of $k^* < 1$ was explained. However, it remained to unambiguously prove that a trough-in-bump can form even for the case of pure matter. Otherwise, engulfment in pure matter for $k^* < 1$ would not be possible.

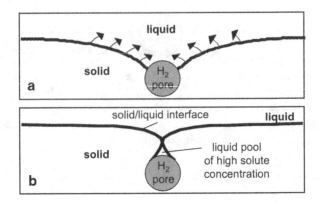

Fig. 15.10 The mechanism of comet-tail-shaped segregation region behind an H_2 pore: **a** solute diffusion toward the sample centerline before complete engulfment, **b** solute trapping behind the pore at the time of complete engulfment (Catalina et al. 2004). With permission of Springer Science and Business Media

Fig. 15.11 The influence of the Gibbs–Thomson and disjoining pressure effects on the shape of the S/L interface (SCN/polystyrene system, r_P=5 μm, $d_{gap} = 11 \cdot a_o$, G_T =6 K/mm except for curve 3' for which G_T =8 K/mm) (Catalina et al. 2003). Copyright 2003 by The Minerals, Metals & Materials Society. Reprinted with permission

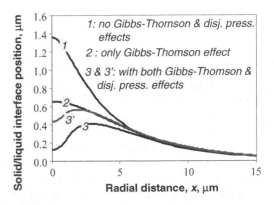

After Hadji (1999a) used perturbation analysis to demonstrate that the interface shape is affected not only by the thermal conductivity but also by the disjoining pressure, Catalina et al. (2003) developed a semi-analytical model for the interface shape that accounts for the Gibbs–Thomson and disjoining pressure effects. Calculations were performed for the SCN/polystyrene system for which $k_P < k_L$. It was found that the shape of the S/L interface is fundamentally different from the situation when the Gibbs–Thomson and disjoining pressure effects are neglected (Fig. 15.11). For systems characterized by $k_P < k_L$ the disjoining pressure causes the sign change of the interface curvature near the particle. The increase of the temperature gradient in the liquid diminishes the effect of the disjoining pressure.

Kinetic Dynamic Models Dynamic models consider the nonsteady-state nature of PET, meaning that a particle, initially at rest, must have an accelerated motion in

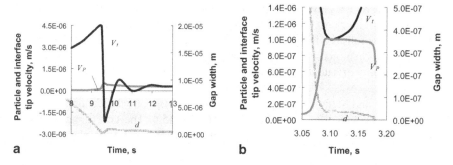

Fig. 15.12 Calculations for the Al–ZrO$_2$ particle system (Al viscosity was taken as 2 mPa s, and $r_P = 250$ μm) **a** Particle velocity (V_p) and S/L interface tip velocity (V_t) versus time, for a subcritical solidification velocity $V_{SL} = 0.3$ μm/s, **b** Time evolution of particle velocity and S/L interface tip velocity for $V_{SL} = V_{cr} = 1$ μm/s (Catalina and Stefanescu 1999; Catalina et al. 2000). With permission of Springer Science and Business Media

order to reach the steady-state velocity, which is the solidification velocity. The governing equation is the equation of particle motion, introduced earlier as Eq. 15.4.

The first dynamic models were presented at the same conference by Schvezov (1999) and by Catalina and Stefanescu (1999). In the latter model, Eq. 15.8 was used for the interface force with $\Delta\gamma_0$ given by Eq. 15.3. The drag force for a sphere was obtained from manipulations of Eqs. 15.14 and 15.16. Lift and gravity forces were ignored. The force required to accelerate the fluid that adheres to the particle was included in the formulation. Model calculations result for a constant solidification velocity smaller than V_{cr} are shown in Fig. 15.12a. It is seen that the particle velocity, V_P, increases from zero to the velocity of the interface as the particle/interface gap, d, decreases. The velocity of the interface under the particle, called tip velocity, V_t, behaves in an oscillatory manner, and so does the particle/interface distance. Eventually, when steady state is reached, $V_P = V_t$.

The evolution of V_P and V_t for the case when the solidification velocity is equal to the critical value, is presented in Fig. 15.12b. As solidification proceeds, both V_P and V_t are continuously increasing, as in the previous case. In the second stage, when V_t decreases, it is seen that it only decreases to a value close to V_{SL}, without going below this value. At the same time, V_P increases close to V_{SL} but without exceeding it. At the critical velocity, the drag force becomes higher than the pushing force and when the particle comes close enough to the S/L interface, it is eventually caught by its tip and then engulfed into the solid. Clearly this model describes well the observed experiments. It demonstrates that the interaction is essentially non-steady state and that the steady state eventually occurs only when solidification is conducted at subcritical velocities.

Kinetic Models Considering Fluid Flow In all the models discussed so far fluid flow parallel to the S/L interface has been ignored. Han and Hunt (1995) introduced in their analysis the force acting on a particle because of the difference in flow velocity in the region between the particle and the interface and in the region on the opposite side of the particle ($F_L \neq 0$). They also proposed a different mechanism for engulfment, as follows: the particle near the solidification front rolls/slides on the interface and is captured by the front because of its roughness. However, because of the unavailability of data regarding fluid velocity and friction conditions, this model is impractical.

Mukherjee and Stefanescu (2000, 2004) further developed the Catalina and Stefanescu (1999a) model to include all terms in Eq. 15.4. The governing equation, which is the equation for particle acceleration, was written in its complete form as:

$$\frac{4\pi}{3}\rho_P r_P{}^3 \frac{dV_P}{dt} = 2\pi\, r_P \Delta\gamma_o \left(\frac{d_o}{d}\right)^2 k^* - 6\pi\,\eta\, V_P \frac{r_P{}^2}{d}\left(\frac{r_P}{d}\right)^{0.423} k^{*2} -$$

$$C_A \frac{4\pi}{3}\rho_L r_P{}^3 \frac{dV_P}{dt} + 6.46\eta\, V_{rel} r_P{}^2 \sqrt{\frac{1}{\nu}\left(\frac{dV_{Lx}}{dy}\right)_{avg}} + r_P{}^3 \rho_L \omega V_{rel}$$

$$(15.26)$$

where the LHT is the force on the particle, the first RHT is the interface force, the second RHT is the drag force, the third RHT is the force required to accelerate the virtual "added" mass, the fourth RHT is the Safmann force, and the fifth RHT is the Magnus force. ω is the angular velocity and V_{rel} is the velocity of the particle relative to the fluid. Then, the particle acceleration was expressed in terms of the Reynolds and Weber numbers. The effect of particle radius and the Reynolds number of flow, modified by changing the gravity level, on the critical velocity of engulfment is shown in Fig. 15.13. It is seen that the critical velocity increases as the flow Reynolds number increases. No effect of the level of convection on the critical velocity of engulfment was observed in the low-convection regime. In the moderate-convection regime ($0.1 < \mathrm{Re} < 0.6$), the critical velocity is significantly increased with the increase in the Reynolds number. When $\mathrm{Re} > 0.6$, i.e., when a high-convection regime is established, no interaction occurs between the interface and the particle. For this regime, the convection velocities are so large as to cause the particles to be swept away from the interface. This was calculated for $g = 1$. In a subsequent paper, Mukherjee et al. (2004) verified the validity of the analytical equations proposed for the drag and lift forces through a 2D numerical calculation.

15.1.4 Microstructure Visualization Models

A phase filed model by Ode et al. (2000) applied to a Fe–C alloy with an alumina particle was able to reproduce the interface movement during particle pushing and engulfment, and to estimate the critical velocity for PET. Phase field simulation was also used to describe the growth of an L_2 droplet in front of a solid planar front in a monotectic system (Nestler et al. 2000). The computed pictures are compared well with the results of experimental work on monotectics with transparent organic metal analogues.

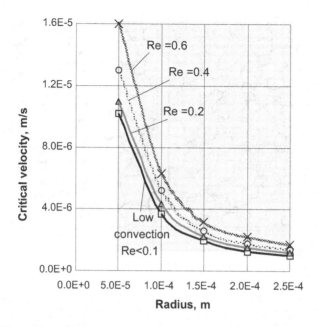

Fig. 15.13 Effect of Reynolds number of flow on the critical velocity for engulfment as a function of particle radius for the aluminum–zirconia particle system (Mukherjee and Stefanescu 2004). With permission of Springer Science and Business Media

Shelton and Dunand (1996) carried out 2D cellular automaton computer simulations to model the geometric interactions between mobile, equiaxed particles, and growing matrix grains. The model allows the study of particle pushing by growing grains, resulting in particle accumulation and clustering at the grain boundaries. It was found that certain parameters such as particle area fraction, particle settling speed, particle cluster mobility, and grain nucleation rate strongly affect the spatial distribution of particles. An example of computed microstructures is given in Fig. 15.14.

15.1.5 Mechanism of Engulfment (Planar S/L Interface)

Most steady-state models do not provide a clear mechanism for engulfment. In principle, the majority of them attempt to calculate the breakdown of steady state and consider that engulfment occurs at that time. Some of them suggest that when the particle/interface gap becomes so small as to prevent mass transport by either diffusion or fluid flow, the particle "sticks" to the interface and is engulfed. Note that only in two-dimensions the particle/interface contact will prevent mass transport. In three dimensions, a point contact between a sphere and a surface will not be sufficient to stop mass transport.

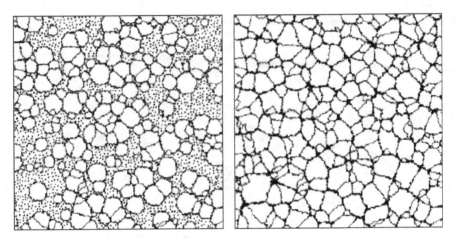

Fig. 15.14 Particles segregation at grain boundaries after 13 time-steps (*left*) and after 24 time-steps (*right*) (Shelton and Dunand 1996). Reprinted with permission from Elsevier

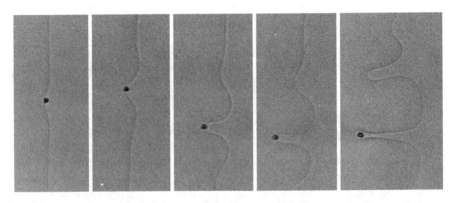

Fig. 15.15 Engulfment of a silicon carbide particle by the S/L interface in succinonitrile. Velocity was gradually increased from 9.2 to 21 μm/s (Stefanescu et al. 1998)

Stefanescu et al. (1995) explained particle engulfment through the "sinking" of the particle in the S/L interface because of the combined effect of the thermal and solutal field. As shown in Fig. 15.5, local perturbation of the interface allows the solid to grow around the particle and engulf it. In other words, there is no requirement of particle "sticking" to the interface for engulfment to occur.

There is reliable experimental evidence to support such a mechanism for alloys. An example is provided from work on transparent organic materials in Fig. 15.15. It is seen that a SiC particle initiates destabilization of the interface and is eventually engulfed by the solid. As the interface approaches the particle at a distance smaller than D/V, the particle obstructs solute diffusion. The composition gradient at the interface decreases and the interface under the particle is decelerated, while the rest of the interface moves at the velocity imposed by the thermal field. A trough is formed on the interface. This trough appears in the bump of systems with $k^* > 1$, or increases the trough of systems with $k^* < 1$. This mechanism has been confirmed through numerical calculations.

In the case of pure materials, trough formation in systems with $k^* < 1$ explains engulfment. A trough can also occur for systems where $k^* > 1$ (Sasikumar et al. 1989), because of the undercooling produced by the disjoining pressure term. This effect is demonstrated in Fig. 15.11.

15.1.6 Particle Interaction with a Cellular/Dendritic Interface

The information on particles interaction with dendritic or cellular interfaces is scarce. Although, some information for metallic alloys has been published (e.g., Stefanescu et al. 1988; Premkumar and Chu 1993; Yaohui et al. 1993; Kennedy and Clyne 1991; Fasoyinu and Schvezov 1990; Hecht and Rex 1997), the details of the interaction have not been clearly described because the analysis has to rely on quenched samples. A better understanding can be obtained at this stage from experiments with transparent organic materials. When conducting directional solid-ification experiments with succinonitrile (99.5 % SCN, balance water) and a variety of particles (nickel, alumina, cobalt), Sekhar and Trivedi (1990) observed that engulfment occurred in cells and dendrites, as for planar interfaces. Particle engulf-ment into the dendrite tip caused tip splitting. Particle entrapment was observed between the secondary dendrites.

Stefanescu et al. (2000) observed a variety of particle behaviors during micro-gravity experiments with succinonitrile—polystyrene particles systems solidifying with a cellular/dendritic interface. In some instances, particles were engulfed as for planar interfaces with local deformation of the interface (Fig. 15.16a). In some other cases, solute trapping in the particle/interface gap decreased interface curva-ture to the point that it changed sign. This means that the tip of the cell has split (Fig. 15.16b). Dendrite/cell tip splitting results in engulfment. Thus, in dendritic solidification engulfment is still possible. However, because in a 1-g environment convection will move the particles, the residency time of a particle on a particular dendrite tip is short. The particle will most likely be moved in the interdendritic regions where it will be entrapped. In μg the tip of the cell/dendrite may push the particle, which is then entrapped in the intercellular space (Fig. 15.16c). Con-sequently, for cellular/dendritic interfaces the probability of engulfment is much smaller than that of entrapment. Thus, a large number of particles will be distributed at the grain boundaries. In the case of columnar solidification, this will result in particle alignment.

Perturbation of the solutal field by the particle changes the dendrite tip radius and its temperature. When the particle approaches the interface the solute gradient will

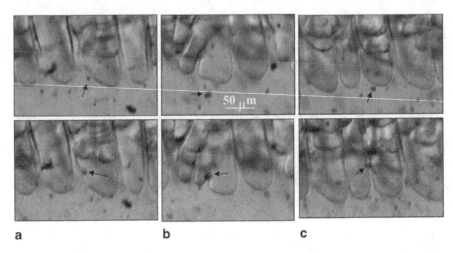

Fig. 15.16 Particle interface interaction during cellular/dendritic solidification at 10 μm/s of SCN-polystyrene particles systems in microgravity **a** engulfment, **b** cell tip splitting, **c** pushing and alignment (Stefanescu et al. 2000)

decrease and the tip radius will increase since the tip radius, r_t, depends on the liquidus temperature gradient, G_L, produced by the solute. This effect is obvious from Eq. (9.12) written here as $r_t = 2\pi\sqrt{\Gamma/(G_L - G_T)}$. Thus, a dendritic to cellular transition may occur. This effect is stronger as the number of particles ahead of the interface increases (Sekhar and Trivedi 1990).

Dutta and Surappa (1998) have attempted a theoretical analysis of particle interaction with a dendritic S/L interface assuming no convection. They concluded that a higher growth velocity is required to engulf a particle during dendritic solidification than during planar solidification. They claim however that if the growth velocity is sufficiently small, entrapment may result. Clearly, the convection level in the liquid will play a particularly significant role in the PET of dendritic interface since it may prevent any significant interaction with the dendrite tips. Thus, the outcome cannot be predicted only on the basis of energy and solute transport.

For a chart of the influence of experimental conditions on the outcome of particle/interface interaction for dendritic interface the reader is referred to Stefanescu (2002, p. 240).

15.2 Solidification in the Presence of Stationary Reinforcements; the Infiltration Pressure

Another class of MMC is that produced through infiltration of a preform by applying pressure on the liquid metal. The main concern related to this topic is the physical/chemical interaction (wetting) between the reinforcement and the melt, which in large extent determines the required infiltration pressure.

Fig. 15.17 Surface energy
balance for a liquid droplet on
its solid

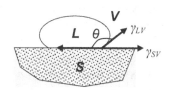

15.2.1 Surface Energy Considerations

The pressure difference required to infiltrate a fiber preform: is described by the following pressure balance:

$$\Delta P = \Delta P_\mu + \Delta P_\gamma = P_o - P_a \qquad (15.27)$$

where ΔP_μ is the pressure drop due to fluid flow, ΔP_γ is the pressure drop due to capillarity, P_o is the metal pressure at the entrance of the preform, and P_a is the atmospheric pressure. In this section, we address the capillary pressure drop.

Young's equation describes the surface energy balance at the trijunction between a liquid droplet (L), its vapor phase (V), and a solid substrate (S) (Fig. 15.17):

$$\gamma_{SV} - \gamma_{SL} = \gamma_{LV} \cos\theta \qquad (15.28)$$

where γ is the surface energy between pairs of the three components.

Based on Young's equation it can be demonstrated that the pressure-drop along the channel (resistance to flow) because of metal-reinforcement interaction that is the resistance to flow through a channel of radius r, can be calculated with:

$$\Delta P_\gamma = -K \, \gamma_{LA} \cos\theta \qquad (15.29)$$

where γ_{LA} is the liquid/atmosphere interface energy and $K = 2/r$ is the curvature of the interface, calculated as the curvature at the narrowest constriction between the fibers, This equation can be used to describe the pressure drop during flow through capillary channels.

When the melt wets the mold (i.e., $\theta < 90°$), $\Delta P_\gamma < 0$, which means that flow occurs spontaneously (no outside pressure is required). On the contrary, if the melt does not wet the mold (i.e., $\theta > 90°$), $\Delta P_\gamma > 0$, and outside pressure is required for flowing. As metals do not wet ceramics and most reinforcements are ceramics, it follows that most MMCs have to contend with the problem of the resistance to infiltration resulting from a positive capillary pressure. It is particularly difficult to infiltrate the wedges between the fibers (Fig. 15.18).

A simple approach to the calculation of the capillary pressure required for the infiltration of fiber preforms was suggested by Mortensen and Cornie (1987). The following assumptions were introduced: (i) infiltration takes place reversibly, i.e., no friction forces; (ii) gravity is ignored; (iii) venting of the preform causes no

Fig. 15.18 Voids at fiber-to-fiber contacts in Al–4.5 wt%Cu/SiC composites infiltrated at 1000 psi (Mortensen and Cornie 1987). With permission of Springer Science and Business Media

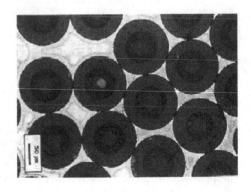

problems. Then the energy for the process of replacing the unmixed metal + fibers + atmosphere with the composite, per unit volume, is:

$$W = (\gamma_{SL} - \gamma_{SA}) S_F \tag{15.30}$$

where S_F is the fiber surface area per unit volume. Since $W = -P \cdot \Delta v$, where Δv is the volume over which metal is displaced, for unit volume of metal matrix:

$$\Delta P_\gamma = (\gamma_{SL} - \gamma_{SA}) S_F = W_i S_F \tag{15.31}$$

where W_i is the work of immersion, which is the critical parameter that governs the wettability of the reinforcement. With the appropriate values for the W_i, this equation gives the lower bound for the pressure required for infiltration. For a volume fraction of reinforcement v_F:

$$S_F = n \frac{v_F}{1 - v_F} \frac{1}{d_F} \tag{15.32}$$

where $n = 4$ for fibers, $n = 6$ for spherical particles, and d_F is the diameter of the reinforcement. If γ_{LA} is known ($\gamma_{SL} - \gamma_{SA}$) can be calculated from the work of adhesion as $W_a = \gamma_{SA} - (\gamma_{SL} - \gamma_{SA})$, or from the contact angle (Eq. 15.28).

The problem can be further complicated by assuming the occurrence of interfacial reactions. For limited interfacial reactions Laurent et al. (1991) have suggested that the immersion work can be calculated with:

$$\gamma_{SA} - \gamma_{SL} = (\gamma_{SA} - \gamma_{SL})_o - \Delta\gamma_\tau - \Delta G_\tau \tag{15.33}$$

where $(\gamma_{SA} - \gamma_{SL})_o$ signifies wetting without reaction, $\Delta\gamma_\tau$ is the change in interfacial energy because of the replacement of the unreacted S/L interface with one new interface after reaction, and ΔG_τ is the free energy released at the S/L/atmosphere triple contact. This last term is system dependent and its calculation is not trivial.

Another major limitation of Eq. 15.31 is that it ignores the irreversible energy losses associated with the wetting of a porous medium. A more detailed discussion on this issue was provided by Mortensen and Jin (1992).

Heat and solute transport must then be analyzed at the scale of the same volume element.

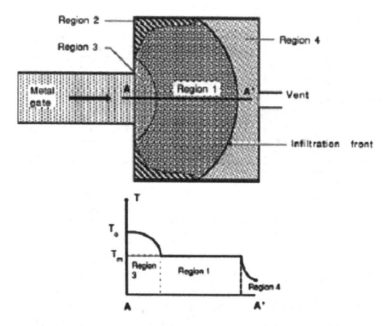

Fig. 15.19 Infiltration of a preforms: *region 1*—fibers, solid metal and flowing liquid metal; *region 2*—solid metal; *region 3*—flowing liquid metal; *region 4*—un-infiltrated preform (Mortensen et al. 1989). With permission of Springer Science and Business Media

15.2.2 Transport Phenomena Considerations

A schematic representation of the infiltration of a preform by liquid metal is presented in Fig. 15.19. The transport equations governing the infiltration process, according to Mortensen et al. (1989), are as follows:

- Assuming flow through a porous medium (Darcy's law, see also Eq. 7.11):

$$V_o = -(\mathbf{K}/\mu) \cdot (\nabla P - \rho_m g) \qquad (15.34)$$

where V_o is the superficial velocity, $\mathbf{K}$ is the permeability tensor of the preform, μ is the viscosity of the metal, ∇P is the pressure drop, ρ_m is metal density, and g is the gravitational acceleration.

- The continuity equation:

$$\nabla V_o = 0 \qquad (15.35)$$

- The heat transport equation:

$$\rho_c c_c \frac{\partial T}{\partial t} + \rho_m c_m V_o \cdot \nabla T = \nabla \cdot (k_c \nabla T) \qquad (15.36)$$

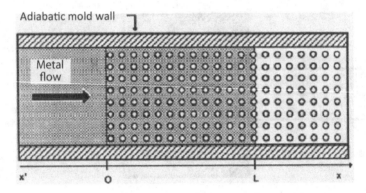

Fig. 15.20 Illustration of unidirectional adiabatic infiltration perpendicular to fiber axis (Mortensen et al. 1989). With permission of Springer Science and Business Media

where the subscripts c and m stand for composite and metal, respectively.

A complete solution of these equations requires numerical methods. However, simpler analytical solutions can be derived (Mortensen et al. 1989). The calculation of the infiltration length of aligned fiber composites assumes unidirectional infiltration of a porous preform of constant reinforcement geometry, size, and volume fraction v_F. The process is driven by a constant pressure differential $\Delta P_\gamma = P_o - P_g$, where P_o is the pressure in the un-reinforced metal in contact with the preform at $x = 0$ and $t = 0$ (Fig. 15.20).

With these simplifying assumptions, the preceding three equations become:

$$V_o = -(K/\mu) \cdot (dP/dx) \tag{15.37}$$

$$\partial V_o / \partial x = 0 \tag{15.38}$$

$$\rho_c c_c \frac{\partial T}{\partial t} + \rho_m c_m V_o \frac{\partial T}{\partial x} = k_c \frac{\partial^2 T}{\partial x^2} \tag{15.39}$$

Using similarity solutions, it can be demonstrated that for unidirectional adiabatic infiltration without solidification the position of the infiltration front can be calculated with:

$$L = \left(\frac{2K \Delta P_\mu}{\mu (1 - v_f)} t \right) \tag{15.40}$$

where L is the infiltration length, $\Delta P_\mu = \Delta P_T - \Delta P_\gamma$ is the pressure drop between 0 and L, and v_F is the volume of fibers. For flow parallel to the fiber axes the permeability can be calculated with:

$$K = \frac{0.427 \, r_f^2}{v_f} \left[1 - \sqrt{\frac{2v_f}{\pi}} \right]^4 \cdot \left[1 + 0.473 \left(\sqrt{\frac{\pi}{2vf}} - 1 \right) \right] \tag{15.41}$$

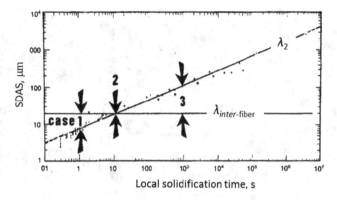

Fig. 15.21 The average secondary arm spacing (λ_2) as a function of the solidification time for Al–4.5 wt% Cu. The average interfiber spacing ($\lambda_{\text{inter-fiber}}$) is about 20 μm with 55 vol% fibers (Mortensen 1986). Copyright 1986 by The Minerals, Metals & Materials Society. Reprinted with permission

where r_f is the radius of the fibers. For flow perpendicular to the fiber axes the permeability is:

$$K = \frac{2\sqrt{2}\, r_f^2}{9 v_f} \left[1 - \sqrt{\frac{4 v f}{\pi}} \right]^{5/2} \tag{15.42}$$

A similar but more complicated equation has been derived for the case of unidirectional adiabatic infiltration with solidification.

For a more complete analysis and experimental results the reader is referred to the reference Mortensen and Michaud (1990).

15.2.3 Microstructure Effects

The reinforcing phase affects the solidification of the matrix. This precludes unrestricted application of solidification rules developed for metals and alloys. The complexities of the effects of the reinforcement on nucleation and S/L interface stability are too complex to be addressed here, and the reader is referred, once again to the review by Mortensen and Jin (1992). We will confine this discussion to the analysis of a simple example.

Consider the case of an Al–4.5 %Cu/SiC fiber composite. The average interfiber spacing is about 20 μm with 55 vol% fibers. The secondary arm spacing of the un-reinforced alloy is described by the equation $\lambda_2 = 7.5 \cdot 10^{-6}\, t_f^{0.39}$ (SI units). Depending on the cooling rate of the system and the volume of fibers, three situations may ensue, as summarized in Fig. 15.21:

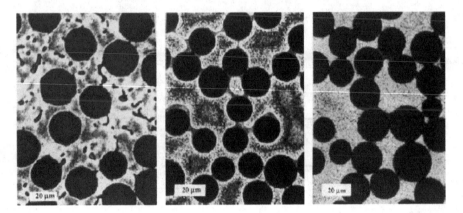

Fig. 15.22 Microstructures corresponding to the three cases in Fig. 15.21 (Mortensen 1986). Copyright 1986 by The Minerals, Metals & Materials Society. Reprinted with permission

1. Unperturbed structure: at high cooling rate (Fig. 15.22a) and/or v_f the matrix will solidify in an unperturbed fashion (case 1 in Fig. 15.21)
2. Microsegregation between fibers (Fig. 15.22b): at moderate cooling rate (cooling rate is sufficiently low for SDAS in the un-reinforced alloy to be commensurate with the fiber spacing), and no nucleation catalysis on the fiber. The fibers affect the final structure (case 2 in Fig. 15.21). The isoconcentrates are parallel to the fiber surface. The second phase (Al₂Cu) is predominantly at the fiber/metal interface. The grain size is large compared to the interfiber spacing.
3. No microsegregation: at low cooling rate (Fig. 15.22c) and/or nucleation catalysis on the fiber structures free of, or with very limited, microsegregation can be obtained (case 3 in Fig. 15.21).

15.3 Processing of Ex-Situ MMCs by Solidification Techniques

While there are many industrial routes to produce MMCs, the casting techniques offer a number of significant advantages including, low cost, flexibility, complex to near-net shape, and adaptability to mass production. The ex-situ processes fall into three broad categories: stir casting, infiltration, and spray casting.

15.3.1 Stir Casting

In the stir casting process, reinforcement particles are added into the liquid or the semi-solid alloy through mechanical or electromagnetic mixing that produces a vortex. A summary of the stir casting process is provided in Table 15.6.

Table 15.6 Typical ex-situ processes for the manufacture of cast MMCs

Liquid processing	Casting processing
• Addition of particles during mechanical mixing of liquid or semi-solid alloy	• Gravity sand or die casting
• Addition of composite briquettes in the melt followed by mild mixing	• Low pressure casting
• Injection of particulates in the melt with an inert gas	• High pressure die casting
• Addition of particles to ultrasonically irradiated melt (Tsunekawa et al. 1994)	• Squeeze casting
• Addition of particles to electromagnetically stirred melt	• Centrifugal casting
• Centrifugal dispersion of particles	
• Single roll strip casting (Gupta et al. 2007)	

One of the main problems is the incorporation of the particles in the liquid during mixing, and in the solid during solidification.

Significant development work for enhancement of the wettability of reinforcements by metals was done. It can be classified into three categories: (a) reinforcement pretreatment, (b) alloying modifications of the matrix, (c) reinforcement coating (see Mortensen and Jin 1992 for an in-depth analysis).

It was found that by heat treating reinforcements such as Al_2O_3, SiC, or graphite to promote desorption of gaseous species from the surface, the γ_{SA} of the reinforcements is increased. This decreases W_i and therefore improves their incorporation into aluminum melts (e.g., Krishnan et al. 1981).

Another approach to wettability improvement is alloying of the metal matrix to promote reactions between the reinforcement and the matrix. Typical examples include Li in Al for wetting of Al_2O_3 fibers or SiC (e.g., Delannay et al. 1987), Ti in Al–Sn alloys for wetting of SiC (e.g., Kobashi and Choh 1990), Cu, Mg, or Ni in Al for SiC wetting (e.g., Stefanescu et al. 1990), carbide formers in Al for wetting of carbon fibers, oxide coatings for carbon fibers in Mg, or SiC in Al. All of these processes lower the wetting angle. Alternatively, coatings that are designed to react with the oxide layer covering molten Al can be used.

Yet another method to improve wettability consists in coating of the reinforcement. Typical examples include various processes involving fiber treatment by molten Na for infiltration of carbon or Al_2O_3 fibers by Al or Mg, the use of nitrogen in increasing wettability of SiC and Al_2O_3 particles by Mg, and Ni coating of SiC for Al–SiC composites.

As discussed in Chap. 14 , the amount of solid in the slurry and the shear rate will affect the behavior of the viscosity of the system. At increasing fractions of solid, departure from the Newtonian behavior is expected (Fig. 15.23). Moon et al. (1993) found that the rheology of fully molten Al–6.5 %Si/SiC composite slurries indicates the existence of a low shear rate Newtonian region, an intermediate pseudoplastic region and a high-shear rate Newtonian region. Unexpectedly, the semisolid composite slurries with 10 and 20 % volume of SiC exhibited lower viscosity than the

Fig. 15.23 Variation of apparent viscosity of semisolid composite slurries during continuous cooling experiments (Moon et al. 1993)

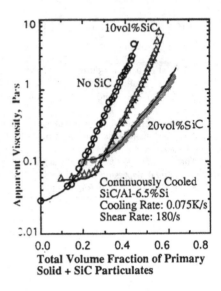

un-reinforced matrix. This is probably because the SiC particles concentrate in the regions between the dendrites and prevent contact and welding of adjacent dendrites.

Many instances of successful industrial stir casting are reported for SiC reinforced aluminum alloys (e.g., Carity 1989; Kennedy and Church 1991; Cox et al. 1994). The main concerns during processing include the influence of solidification velocity, particle settling, and melt fluidity. Most stir-cast MMCs exhibit density mismatch between the reinforcement and the metallic matrix that often leads to settling or floating, inhomogeneous reinforcement distribution, and formation of particle enriched and particle denuded zones in composite castings. Sobczak et al. (2013) developed the mathematical frame for practical assessment of the sedimentation and flotation behaviors of particulate reinforcement in remelting and shape casting for composites such as A359/SiC particles, A359/Al$_2$O$_3$ particles, and A359/graphite particles.

More recently researchers became interested in using nanoparticles as reinforcements. Ghanaraja et al. (2014) investigated the effects of coarse and fine particles on the microstructure and properties of aluminum-based composites. High energy milling of MnO$_2$ particles with excess of aluminum powder was used to produce MnO$_2$ or alumina nanoparticles surrounded by aluminum particles. When the milled powder was added to molten aluminum the excess aluminum particles were melted leaving behind separate oxide nanoparticles without significant agglomeration. The slurry was cast into composites, which also contains coarser (μm size) alumina particles formed by internal oxidation of the melt during processing. The microstructure exhibited good distribution of particles without significant clustering. The oxide particles were primarily γ-alumina in a matrix of Al–Mg–Mn alloy containing some Fe picked up from the stirrer. These composites failed during tensile tests by ductile fracture due to debonding of coarser particles. The presence of nanoparticles along

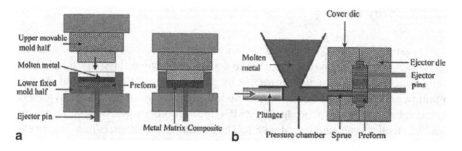

Fig. 15.24 Squeeze casting processes for MMCs **a** direct squeeze casting **b** indirect squeeze casting

with coarser particles improved both strength and ductility. The work did not optimize the relative amounts of the different sized particles for achieving maximum ductility.

15.3.2 Infiltration of Reinforcements

Infiltration is a liquid state method for the fabrication of MMCs in which a preformed dispersed phase (ceramic particles, fibers, woven) is soaked in a molten matrix metal, which fills the space between the dispersed phase inclusions. When the driving force for infiltration is the capillary force (when the metal wets the reinforcement) *spontaneous infiltration* occurs. If the system metal/reinforcement is non-wetting, an external pressure (gaseous, vacuum, mechanical, electromagnetic, centrifugal, or ultrasonic) must be applied to the liquid metal to produce *forced infiltration*.

In the *gas pressure infiltration* process, pressurized gas forces the molten metal to penetrate into a preformed dispersed reinforcement. Infiltration results in low damage to the fibers, in contrast to the methods using mechanical force.

In the *squeeze casting infiltration* process the metal is forced into the cavities of a short or long fibers preform, through mechanical pressure applied via a ram. Two variants of the process exist, direct squeeze casting and indirect-squeeze casting (a hybrid process between pressure die casting and squeeze casting) (Fig. 15.24). A variety of material combinations are used (e.g., B/Al, graphite/Al, graphite/Mg, Al_2O_3/Al, SiC/Al, Borsic/Ti). The process has high productivity and can be used to infiltrate fibers that are not wetted by the molten metal. However, the process is not suitable for complicated shapes and requires particularly strict control of the process variables, including temperature of molten metal and of fibers, infiltration velocity, and final squeeze pressure (Fukunaga, 1988).

15.3.3 Spray Casting

Spray casting is a method of casting near net shape metal components via the deposition of semisolid sprayed droplets onto a shaped substrate. In spray forming an alloy is melted in an induction furnace or by gas melt arc. The molten metal exits as a thin free-falling stream and is broken-up into droplets by an annular array of gas jets. In the variant used for producing MMCs, the reinforcement is blown by a gas jet into the metal stream, and mixed with it. The mixture metal droplets/reinforcements impact a water-cooled substrate while the droplets are still in the semisolid condition. Deposition continues, gradually building up a metal layer on the substrate. Any metal, including refractory metals, and any reinforcement can be used in this process. Examples of material systems include Al matrix with SiC fibers (1–6 mm) or whiskers (0.1–1 μm). Because of the rapid solidification against the cooled substrate the microstructure exhibits fine grains and reduced segregation, but is often nonuniform.

15.3.4 Ultrasonic Cavitation

As the stir casting process cannot disperse reinforcing nanoparticles uniformly into the melt due to their large surface-to-volume ratio and poor wettability, researchers have attempted to use ultrasonic vibration as a method of dispersing nanoparticles into the melt. Ultrasonic cavitation can create small-size transient domains that could reach very high temperatures and pressures. The shock force occurring during ultrasonic cavitation coupled with local high temperatures could break nanoparticle clusters and clean the surface of the particles (Lan et al. 2004; Cao et al. 2008). In addition, there is experimental evidence that ultrasonic vibration can improve the wettability of the nanoparticles. Indeed, Tsunekawa et al. (1994) demonstrated that the contact angle of a water droplet on a paraffin-coated substrate vibrated ultrasonically, decreased from a non-wetting system of 1.82 rad to a wetting one of 1.01 rad (Fig. 15.25). The apparent wettability was improved because of the hysteresis between the advancing and receding contact angle and the ultrasonic pressure. Full infiltration in the Al_2O_3 particle/molten Al system was achieved even without pressure application (Fig. 15.26).

Liu et al. (2014) used the ultrasonic technology to produce Al_2O_3 and SiC nanoparticles reinforced A356 matrix composite castings. Nanoparticles (spherical β-SiC of about 50 nm and Al_2O_3 of about 20 nm) were blown into the melt by Ar gas and dispersed by ultrasonic cavitation and acoustic streaming. The microstructures were greatly refined (grain size and SDAS decreased, columnar-to-equiaxed-transition was completely eliminated) and with the addition of nanoparticles, tensile strength, yield strength, and elongation it increased significantly (Table 15.7).

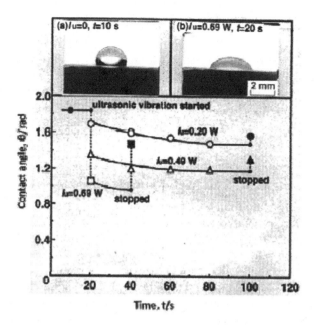

Fig. 15.25 Influence of holding time on the contact angle of water droplets on paraffin-coated substrates as a function of the ultrasonic power (Tsunekawa et al. 1994). Copyright 1994 American Foundry Soc., used with permission

Fig. 15.26 Optical micrographs showing infiltrated area without and with ultrasonic vibration (Tsunekawa et al. 1994). Copyright 1994 American Foundry Soc., used with permission

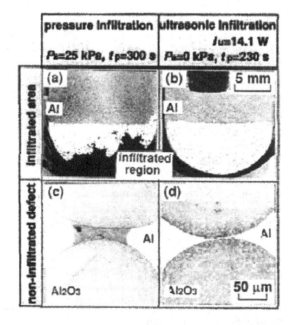

Table 15.7 Effect of 5 min. ultrasonic treatment on the mechanical properties of a A356 alloy (Liu et al. (2014)

Samples	Tensile strength(MPa)	Yield strength(MPa)	Elongation (%)
Standard A356	228 ± 4	180 ± 3	4.0 ± 0.3
Ultrasound + 1 % Al_2O_3	263 ± 8	196 ± 5	6.3 ± 0.5
Ultrasound + 1 % SiC	283 ± 4	196 ± 2	11.2 ± 0.9

15.4 Processing of In-Situ Metal Matrix Composites

Ex-situ MMC processing is afflicted by problems such as surface contamination of the reinforcements, poor bonding, particulate agglomeration, and formation of interfacial reaction products. In-situ techniques promise to overcome some of these deficiencies.

In-situ MMC processing is reactive processing in which a chemical reaction occurs between the matrix and an added chemical to produce dispersed reinforcements. For Al-based composites, a number of in-situ fabrication methods have been developed including self-propagating high-temperature synthesis (Merzhanov et al. 1981), exothermic dispersion (Ma et al. 1994), liquid metallurgy using master alloys (Ramesh et al. 2011), and the halide salt route (Wood et al. 1993). This last technique uses K_2TiF_6 and KBF_4 salts as the starting Ti and B sources to form a dispersion of in situ TiB_2 particles in the Al matrix. The mechanical properties of the Al/TiB_2 composites produced by the halide salt route differ appreciably, since the microstructure of the final composite is highly sensitive to the processing parameters employed in the preparation. Efforts to obtain better and consistent mechanical properties of in-situ Al/TiB_2 composites were directed to the manipulation of timing of additions, holding, and stirring (Chen et al. 2014), resulting in a Al–5TiB_2 composite with 140 % improvement in tensile strength without decreasing ductility with respect to the Al matrix.

In-situ processing is not restricted to aluminum-based alloys. For example, short alumina fiber and in situ Mg_2Si particle reinforced hybrid composites were fabricated by infiltration with the molten Mg alloy into the preforms consisting of the fibers having Si particles attached to their surfaces (Asano and Yoneda 2008). All of Si particles reacted with the Mg alloy to form Mg_2Si particles. As the melting temperature decreased or the cooling rate after the infiltration increased, the Mg_2Si particles became finer. The introduction of P or CaF_2 further promoted the refinement of the Mg_2Si particles. When the composite was fabricated by squeeze casting in a permanent mold, fine Mg_2Si particles with a size of ~ 5 μm were formed even without the refiners because of rapid solidification. The strength of the hybrid composites was higher than that of the conventional fiber reinforced composite at both room temperature and high temperature.

High-temperature composites can also be manufactured by this technique. For example, Sen et al. (1994) investigated TiC-reinforced Ni₃Al composites, fabricated by melting together an 87.3Ni-12.6Al-0.1B (wt%) alloy and various additions of TiC powder in a water cooled copper crucible under argon atmosphere. The TiC is dissolved during melting and re-precipitated during cooling. These precipitates are thermodynamically stable in the Ni_3Al at 1000°C up to 50 h. Room- and high-temperature mechanical testing showed significant improvement in modulus, yield strength, and ultimate tensile strength, while still maintaining a 2.5 % elongation at 0.1 vol. fraction TiC and 500°C. The volumetric density of precipitates (number/m^3) was strongly dependent on the fraction of TiC, and on the temperature and the holding time at that temperature. The critical nucleation temperature for these precipitates was determined to be at 1475°C. Experimentally, evaluated growth rate constants suggested a diffusion-controlled coarsening mechanism along with coalescence and multiparticle interaction at higher volume fractions of TiC.

References

Arsenault RJ, Wang L, Feng CR (1991) Acta Metall. 39:47
Asano K, Yoneda H (2008) Int. J. Cast Metals Res. 21(1–4):239–245
Asthana R, Tewari SN (1993) J.Mater. Sci. 28:5414
Azouni MA, Kalita W, Yemmou M (1990) J. Crystal Growth 99:201
Bolling GF, Cissé J (1971), J. Crystal Growth 10:56
Cao G, Choi H, Konishi H, Kou S, Lakes R, Li X (2008) J Mater Sci 16:5521–6
Carity RE (1989) Trans. AFS 97:743–746
Casses P, Azouni MA (1993) J. Crystal Growth 130:13
Catalina AV, Stefanescu DM (1999) in: Hong CP, Choi JK, Kim DH (eds) Proc. of Modeling of Casting and Solidification Processes IV. Center for Computer-Aided Mat. Proc., Seoul, p 3
Catalina AV, Stefanescu DM (1999a) in: Hofmeister WH, Rogers JR, Singh NB, Marsh SP, Vorhees PW (eds) Solidification 99. TMS, Warrendale, PA p 273
Catalina AV, Mukherjee S, Stefanescu DM (2000) Metall. Mater. Trans. 31A: 2559
Catalina AV, Stefanescu DM, Sen S (2003) in: Stefanescu DM et al. (eds) Modeling of Casting, Welding and Advanced Solidification Processes X. TMS, Warrendale PA p 125
Catalina AV, Stefanescu DM, Sen S, Kaukler W (2004) Metall. and Mater. Trans. 35A:1525
Chen ZN, Wang TM, Zheng YP, Zhao YF, Kang HJ, Gao L (2014) Mat. Sc. Eng. 605:301–309
Chernov AA, Temkin DE, Mel'nikova AM (1976) Sov. Phys. Crystallogr. 21:369
Chernov AA, Temkin DE, Mel'nikova AM (1977) Sov. Phys. Crystallogr. 22:656
Cox BM, Doutre D, Enright P, Provencher R (1994) in: Stefanescu DM, Sen S (eds) Second International Conf. on Cast Metal Matrix Composites. AFS, Des Plaines IL p 88
Delannay F, Froyen L, Deryttere A (1987) J. Mater. Sci. 22:1–16
Dutta B, Surappa MK (1998) Metall. and Mater. Trans. 29A:1319
El-Mahallawy NA, Taha MA (1993) Key. Eng. Mater. 79–80:1–14
Endo A, Chauhan HS, Egi T, Shiohara Y (1996) J. Mater. Res. 11:795
Fasoyinu Y, Schvezov CE (1990) in: Proceedings F. Weinberg Intl. Symposium on Solidification Processing. Ontario, Pergamon Press p 243
Fukunaga H (1988) in: Cast Reinforced Metal Composites. ASM International pp 101–107
Garvin JW, Udaykumar HS (2004) J. Crystal Growth, 267:
Ghanaraja S, Nath SK, Ray S (2014) Metall. Mater. Trans. 45A:3467
Gupta RK, Mehrotrab SP, Gupta SP (2007) Mater. Sci. Eng. A 465:116–123

Hadji L (1999a) in: Hofmeister WH, Rogers JR, Singh NB, Marsh SP, Vorhees PW (eds) Solidification 99. TMS, Warrendale, PA p 26
Hadji L (1999b) Phys. Rev. E 60:6180
Hadji L., 2001, Physical Rev. E, 64:051502
Hadji L., 2003, Scripta Materialia 48:665
Han Q. and J.D. Hunt, 1995, ISIJ International. 35:693
Hecht U, Rex S (1997) Met. Trans. 28A:867
Hecht U, Rex S (2001) in: Stefanescu DM, Ruxanda R, Tierean M, Serban C (eds) The Sci. of Casting and Solidification. Editura Lux Libris, Brasov, Romania p 53
Juretzko FR, Dhindaw BK, Stefanescu DM, Sen S, Curreri P (1998) Metall. Mater. Trans. 29A:1691
Kalaiselvan K, Murugan N, Parameswaran S (2011) Mater. Des. 32:4004–4009
Kaptay G (1999) Metall. and Mater. Trans. 30A:1887
Kaptay G (2000) Metall. and Mater. Trans. 31A:1695
Kennedy DO, Church JC (1991) Trans. AFS 99:729–735
Kennedy DO (1991) Advanced Mater. & Proc. 6:42
Kennedy AR, Clyne TW (1991) Cast Metals 4,3:160
Kim JK, Rohatgi PK (1998) Metall. and Mater. Trans. 29A:351
Kobashi M, Choh T (1990) Mater. Trans. JIM 31:1101–1107
Körber C, Rau R, Cosman MD, Cravalho EG (1985) J. Crystal Growth 72:649
Krishnan BP, Surappa MK, Rohatgi PK (1981) J. Mater. Sci. 16:1209–1216
Kuo VHS, Wilcox WR (1973) Sep. Sci. 8:375
Lan J, Yang Y, Li XC (2004) Mater Sci Eng A 386:284 –90
Laurent V, Chatain D, Eustathopoulos N (1991) Mater. Sci. Eng. A135:89–94
Lawrynowicz DE, Li B, Lavernia HJ (1997) Met. Trans. 28B:877
Leal LG (1992) Laminar Flow and Convective Transport Processes: Scaling Principles and Asymptotic Analysis. Butterworth-Heinemann
Liu ZW, Rakita M, Han Q, Li JG (2011) Mater. Res. Bull. 46:1674–1678
Liu X, Jia S, Nastac L (2014) Int. J. Metalcasting 8(3):51–58
Ma ZY, Li JH, Luo M, Ning XG, Lu YX, Bi J, Zhang YZ (1994) Scr. Metall. Mater. 31:635–639
Ma ZY, Li TL, Liang Y, Zheng F., Bi J., Tjong SC (1996) Mater. Sci. Eng. 219:229–231
Merzhanov AG, Karyuk GG, Borovinskaya IP, Sharivker SY, Moshkovskii EI, Prokudina VK, Dyad'ko EG (1981) Sov. Powder Metall. Met. Ceram. 20:709–713
Mohanty PS., Samuel FH, Gruzleski JE (1995) AFS Trans. 103:555
Moon KK, Ito Y, Cornie JA, Flemings MC (1993) Key Eng. Mater. 79–80:105–116
Mortensen A et al. (1986) JOM p 30
Mortensen A, Cornie JA (1987) Met. Trans. 18A:1160
Mortensen A, Masur IJ, Cornie JA, Flemings MC (1989) Metall. Trans. 20A:2535–2547
Mortensen A, Michaud V (1990) Met. Trans. A 21A:2059–2072
Mortensen A, Jin I (1992) Int. Mater. Rev. 37(3):101–128
Mukherjee S, Stefanescu DM (2000) in: Rohatgi PK (ed) State of the Art in Cast Metal Matrix Composites in the Next Millennium. The Minerals, Metals & Materials Society, Warrendale PA p 89
Mukherjee S, Stefanescu DM (2004) Metall. and Mater. Trans. 35A:613
Mukherjee S, Sharif MAR, Stefanescu DM (2004) Metall. and Mater. Trans. 35A:623
Nestler B, Wheeler AA, Ratke L, Stöcker C (2000) Physica D 141:133
Ode M, Lee JS, Kim SG, Kim WT, Suzuki T (2000) ISIJ International 40:153
Omenyi SN, Neumann AW (1976) J. Applied Physics 47:3956
Pang H, Stefanescu DM, Dhindaw BK (1993) in: Stefanescu DM, Sen S (eds) Proceedings of the 2nd International Conference on Cast Metal Matrix Composites. American Foundrymen's Soc. p 57
Pötschke J, Rogge V (1989) J. Crystal Growth 94:726
Premkumar MK, Chu MG (1993) Met. Trans. 24A:2358
Ramesh CS, Pramod S, Keshavamurthy R (2011) Mater. Sci. Eng. A528:4125–4132

Rempel AW, Worster MG (2001) J. Crystal Growth 223:420

Rubinow SI, Keller JB (1961) J. Fluid Mechanics. 11:447

Saffman PG (1965) J. Fluid Mechanics. 22:385

Sasikumar R, Ramamohan TR, Pai BC (1989) Acta Metall. 37:2085

Sasikumar R, Ramamohan TR (1991) Acta Metall. 39:517

Schvezov C (1999) in: Hofmeister WH, Rogers JR, Singh NB, Marsh SP, Vorhees PW (eds) Solidification 99. TMS, Warrendale, PA p 251

Sekhar JA, Trivedi R (1990) in: Rohatgi P (ed) Solidification of Metal Matrix Composites. TMS, Warrendale, PA p 39

Sen S, Stefanescu DM, Dhindaw BK (1994) Metall. Trans. 25A:2525–541

Sen S, Kaukler WF, Curreri P, and Stefanescu DM (1997) Metall. Mater. Trans 28A:2129

Sen S, Dhindaw BK, Stefanescu DM, Catalina AV, Curreri P (1997a) J. Crystal Growth 173:574

Shangguan DK, Ahuja S, Stefanescu DM (1992) Metall. Trans. 23A:669

Shelton RK, Dunand DC (1996) Acta mater. 44:4571

Shibata H, Yin H, Yoshinaga S, Emi T, Suzuki M (1998) ISIJ Int. 38:149

Shiohara Y, Endo E, Watanabe Y, Nomoto H, Umeda T (1997) in: Beech J, Jones H (eds) Solidification Processing 97. Univ. of Sheffield, UK p 456

Singh J, Alpas AT (1995) Scr. Metall. Mater. 32:1099–1105

Smith RP, Li D, Francis W, Chappuis J, Neumann AW (1993) J. Colloid Interf. Sci. 157:478

Sobczak JJ, Drenchev L, Asthana R (2013) Int. J. Cast Metals Res. 26(2) 122

Stefanescu DM (1993) Key Eng. Mater. 79-80:75–90

Stefanescu DM (2002) Science and Engineering of Casting Solidification, Kluwer Academic/Plenum Publishers, New York

Stefanescu DM, Dhindaw BK, Kacar SA, Moitra A (1988) Met. Trans. 19A:2847

Stefanescu DM, Moitra A, Kacar SA, Dhindaw BK (1990) Met. Trans. 21A:231

Stefanescu DM, Phalnikar RV, Pang H, Ahuja S, Dhindaw BK (1995) ISIJ International 35:700

Stefanescu DM, Catalina AV (1998) ISIJ International 38:503

Stefanescu DM, Juretzko FR, Dhindaw BK, Catalina A, Sen S, Curreri PA (1998) Metall. Mater. Trans. 29A:1697

Stefanescu DM, Catalina AV, Juretzko FR, Mukherjee S, Sen S, Dhindaw BK (2000) in: Microgravity Research and Applications in Physical Sciences and Biotechnology, SP-454 vol. 1. Sorrento, European Space Agency p 621

Surappa MK, Rohatgi PK (1981) J. Mater. Sci. Lett. 16:765

Tsunekawa Y, Nakanishi H, Okumiya M, Mohri N (1994) in: Stefanescu DM, Sen S (eds) Second International Conf. on Cast Metal Matrix Composites. AFS, Des Plaines IL p 70

Torza S, Mason SG (1969) Science 162:813

Uhlmann DR, Chalmers B, Jackson KA (1964) J. Appl. Phys. 35:2986

Wang L, Arsenault R (1991) Metall. Mater. Trans. 22:3013–3018

Wood JV, Davies P, Kellie JLF (1993), Mater. Sci. Technol. 9:833–840

Wu Y, Liu H, Lavernia EJ (1994) Acta Metall. Mater. 42:825

Yaohui L, Zhenming H, Shufan L, Zhanchao Y (1993) J. Mat. Sci. 12:254

Youssef YM, Dashwood RJ, Lee PD (2005) Composites A36:747

Zubko AM, Lobanov VG, Nikonova VV (1973) Sov. Phys. Crystallogr. 18:239

Chapter 16
Multiscale Modeling of Solidification

Computational modeling is now a well-established technique in solidification science. It has increasingly gained ground as a form of numerical experimentation as well as a predictive tool in casting processing. Modeling is not a stand-alone theoretical technique, as effective modeling is directly related to theory and physical experimentation.

A hierarchy of multiscale simulation techniques currently used in solidification science plotted in the time scale–length scale space is presented in Fig. 16.1. The areas of overlap permit "mapping" or "zooming" from one scale to the next, which often are required for parametric coupling of higher scale methods, or for finer scale resolution of selected parts of the higher scale system.

Multiscale modeling means the application of simulation techniques, which can be dissimilar in their theoretical character, at two or more different length and time scales (Elliott 2011). A distinction is made between the hierarchical approach (Berendsen 2005), which involves running separate models with parametric coupling, and the hybrid approach, in which models are run concurrently over different length scales. A discussion of hybrid finite-element/molecular-dynamics/electronic-density-functional approach to materials simulations on parallel computers is provided by Ogata et al. (2001). Multiscale atomistic/continuum coupling methods are discussed by Miller and Tadmor (2009). For a more in-depth treatment of the problem, the reader is referred to the work of Raabe (1998) and Elliott (2011).

Multiscale modeling is facilitated by the existence of many commercial and academic modeling software packages such as ABAQUS (a finite element analysis code with full multiphysics capability), FLUENT (a computational fluid dynamics code), Car–Parrinello molecular dynamics (an ab-initio molecular dynamics code), AMBER (a code for molecular dynamics simulations on large molecules), and others (see for example http://www.hpcvl.org/). Some commercial software packages, such as ProCAST, have integrated Finite Elements/Cellular Automaton software and are successful in calculation and visualization of solidification microstructures. A discussion of the main principles of numerical modeling and typical examples are provided in the following two sections.

© Springer International Publishing Switzerland 2015   343
D. M. Stefanescu, *Science and Engineering of Casting Solidification,*
DOI 10.1007/978-3-319-15693-4_16

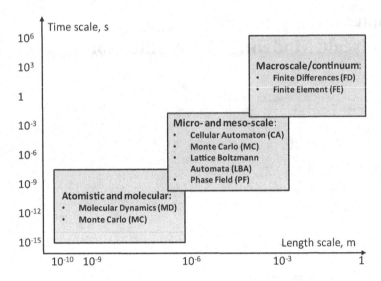

Fig. 16.1 The hierarchy of multiscale modeling techniques used for solidification. (After Elliott 2011)

References

Berendsen HJC (2005) Simulating the physical world: hierarchical modeling from quantum mechanics to fluid dynamics. Cambridge University Press, Cambridge

Elliott JA (2011) Int. Mater. Rev. 56 (4):207

Miller RE, Tadmor EB (2009) Model. Simul. Mater. Sci. Eng. 17(5)

Ogata S, Lidorikis E, Shimojo F, Nakano A, Vashishta P, Kalia RK (2001) Comput. Phys. Commun., 138(2):143–154

Raabe D (1998) Computational materials science. Wiley-VCH, Weinheim

Chapter 17
Numerical Macroscale Modeling of Solidification

From the analysis of solidification based on the energy transport equation presented in the earlier chapter, it was seen that analytical solutions of this equation are not always available. Significant simplifying assumptions must be used, assumptions that are many times debilitating to the point that the solution is of little engineering interest. Fortunately, with the development of numerical methods and their application to the solution of partial differential equations, the most complicated equations can be solved numerically. Numerical solutions rely on replacing the continuous information contained in the exact solution of the differential equation with discrete values. Discretization equations are derived from the governing differential equation.

Process modeling has become possible in a much larger extent than allowed by the use of analytical solutions. Process modeling has emerged as a practical industrial tool for the design of manufacturing processes, troubleshooting, and identifying the dependent and independent variables of the process.

Computer simulation of solidification is based on numerical solutions of energy, mass, and momentum transport. Its main computational purpose is calculation of the evolution of the thermal and compositional field throughout the casting. To produce a solidification model the following steps are necessary:

- Problem formulation
- Discretization of governing equations
- Solving of the system of algebraic equations

17.1 Problem Formulation

Heat transfer (HT) modeling for a given casting—mold combination requires solving of the energy conservation equation for heat conduction with heat generation. Ignoring for the time being the convective term of the energy transport equation, the governing equation is:

© Springer International Publishing Switzerland 2015
D. M. Stefanescu, *Science and Engineering of Casting Solidification,*
DOI 10.1007/978-3-319-15693-4_17

$$\frac{\partial T}{\partial t} = \alpha \nabla^2 T + \frac{\dot{Q}_{gen}}{\rho c}. \tag{17.1}$$

The source term associated with the phase change, which describes the rate of latent heat evolution during the liquid/solid transformation given by Eq. 6.10 is:

$$\dot{Q}_{gen} = \rho \Delta H_f \frac{\partial f_S(x,t)}{\partial t}. \tag{17.2}$$

To solve Eq. 17.1 an appropriate expression for $f_S(x,t)$ must be found.

In one approach, the solution of the discretized energy transport equation at all the nodes or elements of the computational domain, is found by prescribing a solidification path, i.e., by assuming a relationship between f_S and T. The fraction of solid is rewritten as $\partial f_S/\partial t = (\partial f_S/\partial T)(\partial T/\partial t)$. Then, some functional dependency of the fraction of solid on temperature is assumed. Such typical assumptions include linear dependency for eutectics, and equilibrium or Scheil equation for dendritic alloys. For example, assuming that the composition field is governed by Scheil-type diffusion an equation for the fraction solid evolution can be derived as follows. The interface temperature depends on composition according to the relationships $T_f - T_L = -mC_o$ or $T_f - T^* = -mC_L^*$. The liquid composition is given by the Scheil equation: $C_L^* = C_o(1 - f_S)^{1-k}$. Then, $f_S = 1 - (C_L^*/C_o)^{1/(1-k)}$ and finally:

$$f_S = 1 - \left(\frac{T_f - T^*}{T_f - T_L}\right)^{1/(k-1)}. \tag{17.3}$$

With these assumptions, Eq. 17.1 can be rewritten in several ways and solved by numerical techniques.

17.1.1 The Enthalpy Method

In the enthalpy method (Pham 1986) for 1D Eq. 17.1 is rewritten as:

$$k\frac{\partial^2 T}{\partial x^2} = \left(\rho c - \rho \Delta H_f \frac{\partial f_S}{\partial T}\right)\frac{\partial T}{\partial x}. \tag{17.4}$$

Defining the enthalpy as: $H(T) = \int_0^T \rho\, c dT + \rho \Delta H_f[1 - f_S(T)]$. Substituting in the previous equation, we obtain:

$$k(\partial^2 T/\partial x^2) = \rho \partial H/\partial t . \tag{17.5}$$

The enthalpy discontinuity at the solidification temperature is usually circumvented by the arbitrary selection of a solidification interval.

Fig. 17.1 Schematic illustration of the temperature recovery method

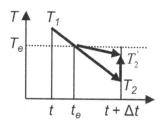

17.1.2 The Specific Heat Method

In the specific heat method, an effective specific heat, c^*, is defined from Eq. 17.4 as:

$$c^*(T) = \frac{\partial H}{\partial T} = \rho\, c(T) - \rho\, \Delta H_f \frac{\partial f_S}{\partial t}\frac{\partial t}{\partial T},$$
(17.6)

which, when reintroduced in the same equation gives:

$$k\frac{\partial^2 T}{\partial x^2} = c^*(T)\frac{\partial T}{\partial x}.$$
(17.7)

The specific heat method has several drawbacks. Firstly, because of the strong variation of the specific heat close to the liquidus or eutectic temperature, it is difficult to assure energy conservation while solving the equations numerically. Secondly, it requires that $(\partial f_S/\partial t)(\partial t/\partial T) \le 0$ so that $c^* \ge c$. This requirement precludes the possibility that $\partial f_S/\partial t$ takes both positive and negative values required to predict recalescence.

17.1.3 The Temperature Recovery Method

For molds having high Biot numbers, that is higher cooling rates, the previously described methods may omit the effect of the latent heat evolved during solidification. The temperature recovery method described graphically in Fig. 17.1 correctly recovers the latent heat. In this approach the temperature T_2 at time step $t + \Delta t$ is calculated from Eq. 17.1 without including the latent heat generation term. Then the increased temperature because of latent heat generation, T_2', is calculated from the integral energy balance for the volume element:

$$\int_{T_e}^{T_2'} \rho c\, dT - \int_{T_e}^{T_2'} \rho \Delta H_f \frac{\partial f_S}{\partial t}\frac{\partial T}{\partial t}\, dT = \int_{T_1}^{T_2} \rho c\, dT - \int_{T_1}^{T_e} \rho c\, dT.$$
(17.8)

Finally, the temperature in the volume element at time $t + \Delta t$ is updated to T_2'.

A more accurate approach can be used by calculating the fraction solid evolution as a function of time from solidification kinetics, thus relaxing the previous assumptions on the fraction solid. This technique will be discussed in detail in Chap. 18.

17.2 Discretization of Governing Equations

The continuous variables may be represented by a number of discrete values associated with the volume elements (cells) or volume vertices (nodes) of a computational grid. This process is called discretization. The partial differential equations are approximated by a set of algebraic equations. For a discussion on computational grids the reader is referred to Winterscheidt and Huang (2002).

The discretization equations for the differential equations can be derived in several ways:

- Finite difference method (FDM) based on Taylor series
- Variational formulation, based on calculus of variations
- Boundary element method (BEM)
- Method of weighted residuals—control-volume method (CVM), or finite elements method (FEM)

In the following sections the FDM, as well as the CVM, will be introduced. The application of finite differences to solidification modeling will also be discussed.

17.2.1 The Finite Difference Method: Explicit formulation

The finite difference calculus is a technique for differentiating functions by employing only arithmetic operations. The method is largely known as the Euler method. However, Taylor's series can also be used to derive the relevant equations. While a good understanding of Taylor's series is fundamental to FDM applications, a simpler approach can be used to derive some of the basic FDM equations.

Let us produce a finite difference discretization of the heat conduction equation. If, for the time being, we ignore heat generation, the 1D format is:

$$\frac{\partial T}{\partial t} = \alpha \frac{\partial^2 T}{\partial x^2}.$$
(17.9)

The first step is to divide the physical domain over which the calculation will be performed, in a number of volume elements or nodes. Higher the number of nodes, higher is the accuracy of the solution. A graphic representation of the problem, where the computational space has been divided in nodes, is given in Fig. 17.2. Let us write the discretization equation. The time derivative can be approximated as:

$$\frac{\partial T}{\partial t} \cong \frac{T_i^{n+1} - T_i^n}{t^{n+1} - t^n} = \frac{T_i^{n+1} - T_i^n}{\Delta t},$$
(17.10)

where the superscript n refers to the time level. In terms of Taylor's series this is a *forward difference representation*.

The gradient at time n between nodes i and $i + 1$, or between nodes i and $i - 1$ is, respectively:

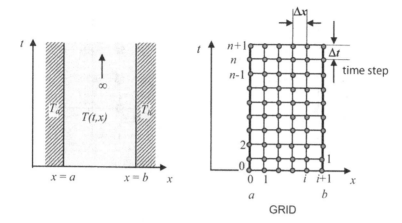

Fig. 17.2 Graphic representation of the problem described by Eq. 17.9 and temporal and spatial grid. Boundary conditions are: at $x = a$, $T = T_a$, and at $x = b$, $T = T_b$

$$\left(\frac{\partial T}{\partial x}\right)_{right} \cong \frac{T_{i+1}^n - T_i^n}{\Delta x} \quad \left(\frac{\partial T}{\partial x}\right)_{left} \cong \frac{T_i^n - T_{i-1}^n}{\Delta x}, \quad (17.11)$$

where the subscript i refers to the spatial level. The second derivative in Eq. 17.9 can now be written as:

$$\frac{\partial}{\partial x}\left(\frac{\partial T}{\partial x}\right) \cong \frac{1}{\Delta x}\left[\left(\frac{\partial T}{\partial x}\right)_{right} - \left(\frac{\partial T}{\partial x}\right)_{left}\right] \quad \text{or} \quad \frac{\partial^2 T}{\partial x^2} = \frac{T_{i+1}^n - 2T_i^n + T_{i-1}^n}{(\Delta x)^2}.$$

$$(17.12)$$

This is a *central difference representation* in terms of Taylor's series. Substituting the two difference representations in Eq. 17.9 and solving for T_i^{n+1} yields:

$$T_i^{n+1} = T_i^n + \frac{\alpha(\Delta t)}{(\Delta x)^2}\left(T_{i+1}^n - 2T_i^n + T_{i-1}^n\right). \quad (17.13)$$

Note that at time step $n+1$ all values are known for step n. This is known as an *explicit solution*. The temperature in a node i can be found by "marching forward in time."

Let us now apply these principles to a 2D problem. The graphic representation is given in Fig. 17.3. Alternative notations commonly used are i–j and N-S-E-W, as illustrated in the figure. Also for time *old* and *new* instead of n and $n+1$.

The governing equation is:

$$\frac{1}{\alpha}\frac{\partial T}{\partial t} = \frac{\partial^2 T}{\partial x^2} + \frac{\partial^2 T}{\partial y^2}. \quad (17.14)$$

Central difference representations are:

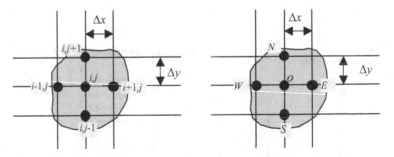

Fig. 17.3 Representation of the computational grid for 2D calculations

$$\frac{\partial^2 T}{\partial x^2} = \frac{T_{i+1,j}^n - 2T_{i,j}^n + T_{i-1,j}^n}{(\Delta x)^2} \quad \text{and} \quad \frac{\partial^2 T}{\partial y^2} = \frac{T_{i,j+1}^n - 2T_{i,j}^n + T_{i,j-1}^n}{(\Delta y)^2}$$

$$(17.15a)$$

or, alternatively:

$$\frac{\partial^2 T}{\partial x^2} = \frac{T_E^{old} - 2T_o^{old} + T_W^{old}}{(\Delta x)^2} \quad \text{and} \quad \frac{\partial^2 T}{\partial y^2} = \frac{T_N^{old} - 2T_o^{old} + T_S^{old}}{(\Delta y)^2}. \quad (17.15b)$$

The forward difference representation for $\partial T/\partial t$ is the same as for 1D that is Eq. 17.10. Then, taking $\Delta x = \Delta y = \Delta s$, and combining with Eq. 17.15, the 2D discretization equation is:

$$T_{i,j}^{n+1} = \frac{\alpha(\Delta t)}{(\Delta s)^2}\left(T_{i-1,j}^n + T_{i+1,j}^n + T_{i,j-1}^n + T_{i,j+1}^n\right) + \left[1 - 4\frac{\alpha(\Delta t)}{(\Delta s)^2}\right]T_{i,j}^n$$

$$(17.16a)$$

or, using the notations *old* for n and *new* for $n+1$:

$$T^{new} = \frac{\alpha(\Delta t)}{(\Delta s)^2}\left(T_W^{old} + T_E^{old} + T_S^{old} + T_N^{old}\right) + \left[1 - 4\frac{\alpha(\Delta t)}{(\Delta s)^2}\right]T^{old}. \quad (17.16b)$$

From this point on we will only use this last type of notations. Defining the Fourier number as $Fo = \alpha(\Delta t)/(\Delta x)^2$, the 1-, 2-, and 3-dimensional explicit discretization equations for uniform and equal spatial increments are respectively:

$$T^{new} = Fo\left(T_W^{old} + T_E^{old}\right) + [1 - 2Fo]T^{old} \qquad (17.17a)$$

$$T^{new} = Fo\left(T_W^{old} + T_E^{old} + T_S^{old} + T_N^{old}\right) + [1 - 4Fo]T^{old} \qquad (17.17b)$$

$$T^{new} = Fo\left(T_W^{old} + T_E^{old} + T_S^{old} + T_N^{old} + T_B^{old} + T_T^{old}\right) + [1 - 6Fo]T^{old}$$

$$(17.17c)$$

Fig. 17.4 Computational grid for derivation of the discretization equation for the prescribed flux boundary condition

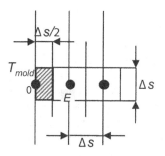

Here the subscripts T and B in the last equation stand for top and bottom (the z-direction), respectively.

The stability of the numerical solution during iterations for successive time steps requires that the coefficients of T^{old} be non-negative. A negative coefficient would yield fluctuations in the numerical solution. Thus, the stability criteria are:

- For 1D $Fo \leq 1/2$
- For 2D $Fo \leq 1/4$
- For 3D $Fo \leq 1/6$

To meet these criteria Δs and Δt must be appropriately selected.

To solve Eq. 17.17 two boundary conditions (BC) and an initial condition (IC) are necessary. The BCs shown in Fig. 17.2, *prescribed temperature* and *constant initial temperature*, are written as:

- BC1 $T(a, t) = T_a$
- BC2 $T(b, t) = T_b$
- IC $T(x, 0) = T_{initial}$

A more useful boundary condition is *prescribed flux at the interface* (Fig. 17.4). Heat conservation on the shaded area having unit depth gives:

$$hA(T - T_{mold}) = -v\rho c \frac{\partial T}{\partial t} + kA \frac{\partial T}{\partial x} \qquad \text{or, in discretized format:}$$

$$\rho c \frac{(\Delta s)^2}{2} \left(\frac{T^{new} - T^{old}}{\Delta t} \right) = k \Delta s \left(\frac{T_E^{old} - T^{old}}{\Delta s} \right) + h \Delta s (T_{mold} - T^{old}).$$

After rearranging, we have:

$$T^{new} = 2Fo \left(T_E^{old} + \frac{h \Delta s}{k} T_{mold} \right) + \left[1 - 2Fo \left(\frac{h \Delta s}{k} + 1 \right) \right] T^{old}. \qquad (17.18)$$

For an exterior nodal point subject to convective boundary conditions in a 2D system, a similar derivation results in:

$$T^{new} = Fo \left(T_S^{old} + 2T_E^{old} + T_N^{old} + 2 \frac{h \Delta s}{k} T_{mold} \right) + \left[1 - 2Fo \left(\frac{h \Delta s}{k} + 2 \right) \right] T^{old}. \qquad (17.19)$$

The stability criteria for convective boundary conditions are:

$$\text{for 1D:}\quad Fo \le \frac{1}{2}\left(\frac{h\,\Delta s}{k}+1\right)^{-1} \qquad\qquad \text{for 2D:}\quad Fo \le \frac{1}{2}\left(\frac{h\,\Delta s}{k}+2\right)^{-1}$$

The most stringent of the stability requirements applicable in a given problem dictates the maximum time step that can be used for calculation.

Let us now consider the 1D case with heat generation. Using the enthalpy method, the governing equation is Eq. 17.4. Using a central difference representation for $\partial^2 T/\partial t^2$, Eq. 17.12, and a forward difference representation for $\partial T/\partial t$, Eq. 17.10, and then rearranging yields the sought off discretization equation:

$$T^{new} = T^{old} + \frac{Fo}{1 - \frac{\Delta H_f}{c}\frac{\partial f_S}{\partial T}}\left(T_E^{old} - 2T^{old} + T_W^{old}\right). \tag{17.20}$$

In order to solve this equation it is necessary to know $f_S(T)$. The thermophysical parameters can be calculated with the mixture-theory relationships, e.g., Eq. 4.7.

An example of the implementation of a 1D FDM scheme for the solidification of a steel casting in conjunction with the enthalpy method is presented in Application 17.1. A 2D FDM problem is solved in Application 17.2.

In the preceding discussion, the FDM method based on forward, backward, and central differences has been introduced and used to derive discretization equations for the energy transport equation. This explicit formulation that only requires information obtained in a preceding time step is also called the *Euler method*. More precise numerical approximation methods can be used, such as the Runge–Kutta method, which is a fourth order (truncation error of the order h^5). These are all one-step methods, in which each step uses only values obtained in the preceding step. There are also multistep methods (e.g., Adams–Moulton method) in which more than one preceding step is used.

17.2.2 The Finite Difference Method: Implicit Formulation

Equation 17.1 will be discretized using the following representations:

- Central difference representation for $\partial^2 T/\partial x^2$ and $\partial^2 T/\partial y^2$:

$$\frac{\partial^2 T}{\partial x^2} = \frac{T_E^{new} - 2T^{new} + T_W^{new}}{(\Delta x)^2} \quad\text{and}\quad \frac{\partial^2 T}{\partial y^2} = \frac{T_N^{new} - 2T^{new} + T_S^{new}}{(\Delta y)^2}$$

- Backward difference representation for $\partial T/\partial t$:

$$\partial T/\partial t = (T^{new} - T^{old})/\Delta t$$

For $\Delta x = \Delta y = \Delta s$, the discretization equation is:

$$(1 + 4Fo)T^{new} = Fo\left(T_W^{new} + T_E^{new} + T_S^{new} + T_N^{new}\right) + T^{old}.$$

Note that the future temperature of node n depends on the future temperatures of the adjacent nodes and on its present temperature. Consequently, a set of simultaneously algebraic equations must be solved for each time step. There are as many equations as many nodes. Because all coefficients are positive, there is no stability problem. This is called an *implicit solution* and also the *Crank–Nicholson method*. For details see, for example, Poirier and Geiger (1994).

17.2.3 The Finite Difference Method: General Implicit and Explicit Formulation

Generalization of the discretization equations previously derived allows easy formulation of boundary conditions for any number of spatial dimensions:

- Explicit $\quad T_n^{new} = T_n^{old} + \Delta t \left[\sum_m \frac{T_m^{old} - T_n^{old}}{R_{mn}C_n}\right]$
- Implicit $\quad T_n^{new} = T_n^{old} + \Delta t \left[\sum_m \frac{T_m^{new} - T_n^{new}}{R_{mn}C_n}\right],$

where m are the nodes adjacent to node n, $R_{mn} = \Delta s/(k\, A_{k,mn})$ for conduction, $R_{mn} = 1/(h_{mn}\, A_{h,mn})$ for convection, and $C_n = \rho\, c\, v_n$. Here, $A_{k,mn}$ and $A_{h,mn}$ are the areas for conductive and convective HT between nodes m and n, respectively; and v_n is the volume element determined by the value of Δs at node n. Boundary nodes are included by proper formulation of R_{mn} and C_n.

17.2.4 Control-Volume Formulation

The fundamental concept used in the CVM is that the governing partial differential equation is constrained to a finite control volume over which the specific phase quantity (mass, enthalpy, momentum) must be conserved. As an example, the control-volume statement will be developed for the transient heat conduction equation with heat generation:

$$\rho c \frac{\partial T}{\partial t} = \nabla \cdot (k\nabla T) + \dot{Q}_{gen}.$$

Integrating over the control-volume element v, and using the divergence theorem, which states that $\int_v \nabla \cdot (k\nabla T)dv = \int_A (k\nabla T) \cdot \mathbf{n}\, dA$, we have:

$v\rho c \frac{\partial \bar{T}}{\partial t} = \int_A (k\nabla T) \cdot \mathbf{n}\, dA + v\bar{\dot{Q}}_{gen}$ and, since $\Delta T \cdot \mathbf{n} = \partial T/\partial n$:

$$v\rho c \frac{\partial \bar{T}}{\partial t} = \int_A k\frac{\partial T}{\partial n}dA + v\bar{\dot{Q}}_{gen}, \qquad (17.21)$$

where $\mathbf{n}$ is the unit outward normal vector, $\partial/\partial n$ is its derivative along the outward normal to the surface of the control volume, and $\bar{T}$ and $\bar{\dot{Q}}_{gen}$ are suitable averages of the temperature and the energy generation rate over the control volume. The advantage of this formulation is that the numerical solution of Eq. 17.21 is fully conserving, since it is derived on the basis of global conservation, unlike finite difference equations that are derived based on interface flux conservation. This equation is used to derive discretization equations for various problems. For details of these derivations as well as for the formulation of various terms in the diffusion equations, the reader is referred to the specialized texts by Patankar (1980) and Özisic (1994).

17.3 Solution of the Discretized Equations

The explicit method introduced in a previous section is very simple computationally. However, it has the disadvantage that the time step is restricted by stability considerations, which may result in unacceptable high computational time.

The implicit formulation has the advantage that it is unconditionally stable. However, a large number of equations may have to be solved simultaneously requiring a large computer memory. Typical methods of solving large sets of algebraic equations include direct methods such as Gauss elimination or the Thomas algorithm, or iterative methods such as the Gauss–Seidel iteration. Additional complications may result if the equation to be solved becomes nonlinear because of the source term or because of time-dependent thermophysical properties. Linearization methods must then be used. For a detailed discussion of these methods, the reader is referred to Özisic (1994).

For a realistic description of the physical phenomenon, all transport equations must be coupled and solved simultaneously. The main difficulty in solving these equations is the unknown pressure field required in the computation of the velocity field. A number of numerical methods have been developed to tackle this problem, e.g., Semi-Implicit Method for Pressure-Linked Equations (SIMPLE, Patankar 1980), SIMPLE Revised Consistent (SIMPLEC, Van Doormaal and Raithby 1984), volume of fluid (VOF, Hirt and Nichols 1981). However, complications that still linger are not disclosed by producers of commercial codes, as pointed out in the review of fluid flow modeling by Ohnaka (1993).

The possible errors in solving the linear and nonlinear algebraic equations by numerical techniques must be considered in addition to the errors introduced by the various assumptions. First, discretization techniques introduce discretization errors. Then, as iterative methods rather than direct methods are often used to save computing time, iteration errors are introduced. The iteration error is the difference between the exact and iterative solution of the discretized algebraic equation system. A convergence criterion that stops the iteration must be selected. Even if high number of iterations is used, the exact solution of the discretized equation is not obtained.

17.4 Macrosegregation Modeling

Occurrence of macroscale local chemical differences termed macrosegregations, continues to be an issue in casting production. Macrosegregation effects may range from local differences in the mechanical properties of the part to rejection of the casting, as in the case of single crystal superalloy turbine blades scrapped because of freckles. In the first model for macrosegregation by Flemings and Nereo (see Sect. 7.6) the importance of convection occurring in the mushy zone was revealed. The complexities of the physics of macrosegregation that requires considering both macroscale phenomena (energy, solute, mass and momentum transport, solid movement and deformation) and microscale (nucleation and grain growth, microsegregation, phase equilibria) phenomena, continues to be an active filed of research. Many numerical macrosegregation models have been proposed. To list a few, Benon and Incropera (1987), Voller et al. (1989), Fellicelli et al. (1991), Diao and Tsai (1993), Vannier et al. (1993), Schneider and Beckermann (1995), Chang and Stefanescu (1996). Most of the early models are single domain models, based on either the mixture theory (Fig. 4.1a) or the volume averaging techniques (Fig. 4.1b).

While alloy specific models and multiscale models are discussed in Chap. 19, as an example of mixture-theory models, the Chang-Stefanescu (1996) model will be reviewed in some detail in the following section.

17.4.1 A Mixture-Theory Model

The transport equations previously introduced were simplified in order to describe mathematically phenomena of engineering interest. The following basic assumptions were used:

- Only liquid and solid phases are present (i.e., L, S, $L + S$, no pores formation).
- The density of the liquid phase is different than that of the solid.
- The properties of the liquid and solid phases are homogeneous and isotropic.
- Complete diffusion of solute in liquid, no diffusion in solid; since the solid diffusion coefficient is four orders of magnitude smaller than the liquid diffusion coefficient, this is a reasonable assumption.

There are two stages during the solidification process:

- Stage I: equiaxed grains move freely with the liquid, i.e., $V = V_S = V_L$, the viscosity of the mixture is described by a relative viscosity; Darcy flow is not important.
- Stage II: a rigid dendritic skeleton is established (dendrite coherency is reached) and $V_S = 0$, the viscosity value returns to liquid viscosity; Darcy flow becomes significant.

Liquid flow is driven by thermal and solutal buoyancy as well as by solidification contraction. To solve the problem at hand it is now necessary to develop the governing transport equations with the appropriate source terms.

Conservation of mass (continuity) is described by Eq. 4.2. The energy transport equation for a two-phase system was derived based on Eq. 4.3. A detailed discussion of the energy transport equation is provided in Sect. 6.1. Conservation of species and momentum will be discussed in the following paragraphs.

Conservation of Species The governing equation is Eq. 4.4, where the source term is:

$$S_C = \nabla \cdot [\rho D \nabla (C_L - C)] - \nabla \cdot [\rho f_S (\mathbf{V} - \mathbf{V}_S)(C_L - C_S)].$$

In the first stage, $V = V_S = V_L$, and equation Eq. 4.4 can be simplified to:

$$\frac{\partial}{\partial t}(\rho C) + \nabla \cdot (\rho C \mathbf{V}) = \nabla \cdot (\rho D \nabla C_L). \tag{17.22}$$

In the second stage the relative velocity is $V_r = V_L - V_S = V_L$. Then, Eq. (4.4) becomes:

$$\frac{\partial}{\partial t}(\rho C) + \nabla \cdot (\rho C_L \mathbf{V}) = \nabla \cdot (\rho D \nabla C_L). \tag{17.23}$$

The only difference between these two equations is in the advective term. In the first stage, the average of solid and liquid concentrations is used in Eq. 17.22 because solid and liquid move together, whereas liquid concentration is used in Eq. 17.23 because of zero solid velocity. After further manipulations (see Chang and Stefanescu 1996 or Stefanescu 1993) the final equation for the second stage can be written as:

$$\rho f_L \frac{\partial C}{\partial t} + \rho \mathbf{V} \nabla \cdot C_L = \nabla \cdot (\rho D \nabla C_L) + \rho C_L (1 - k)\frac{\partial f_S}{\partial t} - (C_L - C)\nabla \cdot (\rho \mathbf{V}). \tag{17.24}$$

This equation indicates that the variation of solute in the liquid phase in a given volume element should be equal to the net loss or gain of solute due to convection, diffusion, interfacial reaction, and solidification contraction. Quite interestingly, this equation can be reduced to the local solute redistribution equation when the assumptions made by Flemings and Nereo (1967) are used. In addition, under the assumption of equal solid and liquid density it reduces to the Scheil equation. Thus, it is apparent that Eq. 17.24 can be used as a general equation for macrosegregation calculation without many simplifying assumptions.

Conservation of Momentum The governing equation is Eq. 4.8. Since two-phase flow has to be described, the source term must include an additional term. Assuming flow through the mushy zone to be Darcy type flow, the pressure drop through the mushy zone is calculated from Eq. 7.11 as:

$$\nabla P = -\mu g_L V_L / K \tag{17.25}$$

Table 17.1 Velocity, viscosity, and permeability in the Chang–Stefanescu model

State	f	V	Viscosity	Permeability
Liquid	$f_L = 1$	$V = V_L$	μ_L	∞
Mushy	$f_L + f_S = 1$	$V_L = V_S$	μ^*	∞
Mushy	$f_L + f_S = 1$	$V_S = 0$	μ_L	K

Then, ignoring again the additional viscous term (interfacial interaction term) the source term is:

in the x direction: $\quad S = -(\mu^*/K)(\rho/\rho_L)\left(V^x - V_S^x\right),$

in the y direction: $\quad S = -\frac{\mu^*}{K}\frac{\rho}{\rho_L}\left(V^y - V_S^y\right) + \rho g\left[\beta_T(T - T_o) + \beta_C(C_L - C_{L,o})\right].$

In these equations, the RH1 term is the drag force term (assuming Darcy flow), and the RH2 term is the buoyancy term describing natural convection based on the Boussinesq approximation.

The treatment of the viscosity and of the permeability in the two-phase system requires further discussion. We can divide the computation in three stages. For each of these stages, the viscosity and the permeability must satisfy the conditions summarized in Table 17.1.

Many models were derived for liquid flowing into a permeable medium, and, in most cases, the local viscosity was considered to be the weighted average of the liquid and solid viscosity. However, when equiaxed solidification is considered, in the early stage of solidification when the solid fraction is small, the Darcy flow is not important. This is because the equiaxed grains will move with the liquid. Darcy flow becomes significant when dendrite coherency is reached, that is when a rigid dendritic skeleton is established. As suggested by Bennon and Incropera (1987), an effective viscosity should be used instead. A hybrid model was developed by Oldenburg and Spera (1992) to account for this behavior. A switching function was defined to control the transition. This switching function should handle different solidification morphologies, such as eutectic and dendritic. In dendritic solidification, the critical switching value will be relatively small, compared to eutectic solidification.

The variation of the viscosity of the mixture, μ^*, as a function of solid fraction can be described with equations of the form of Eq. 7.23. In the Chang–Stefanescu model the empirical equation proposed by Metzner (1988) modified to include a *switching function* was used:

$$\mu^* = \mu_L[1 - F_\mu g_S/A]^{-2}, \tag{17.26}$$

where A is a crystal constant, depending on the aspect ratio and surface roughness of the crystal, and F_μ is the switching function for viscosity. The switching function was introduced to smooth the transition from the first to the second stage. The form of the switching function is a modification of that suggested by Oldenburg and Spera (1992), as follows:

$$F_\mu(g_S) = 0.5 - \frac{1}{\pi}tan^{-1}\left[s\left(g_S - g_S^{cr}\right)\right], \tag{17.27}$$

where s is a constant (typical value 100), and g_S^{cr} is the critical solid fraction. The critical solid fraction is, for example, 0.27 for the aluminum alloy A201 (Bäckerud et al. 1990). The value of A is equal to 0.68 for smooth spherical crystals (aspect ratio = 1). For equiaxed dendritic crystals (aspect ratio from 1 to 10), the value of A was calculated to be 0.3. The viscosity of the mixture is employed in the first stage of solidification, and then turned off in the second stage (i.e., the viscosity becomes again that of the liquid).

For equiaxed grains, the permeability can be regarded as isotropic. Thus, a *permeability function* can be obtained based on the theory of flow through a porous medium. The Carman–Kozeny equation (Carman 1937), similar to the previously discussed Blake–Kozeny equation, was used:

$$K = C_2 \frac{(1 - g_S)^3}{g_S^2 F_K} = \frac{(1 - g_S)^3 d^2}{180 g_S^2 F_K}, \qquad (17.28)$$

where F_K is another switching function $F_K(g_S) = 1 - F_\mu(g_S)$, C_2 is a function of grain size, and d is the grain size (e.g., 50 μm). The switching function shuts off permeability in the first stage and turns it on in the second stage.

To obtain a solution of the derived transport equations a numerical model was developed (for details see Chang 1994; Chang and Stefanescu 1996). Results of calculation of the concentration along a directionally solidified casting are shown in Fig. 17.5a. It is noticed that *positive segregation* is formed near the chill followed by *negative segregation*. This behavior was demonstrated theoretically by Kato and Cahoon (1985). The occurrence of positive segregation can be explained as follows. The liquid ahead of the solid/liquid (S/L) interface is rich in solute because of solute rejection at the S/L interface. Solidification shrinkage induces back flow of enriched liquid into the solidifying region. This creates a region of positive segregation next to the chill. The back-flowing liquid is replaced by liquid that is depleted in solute. This produces a negative segregation zone. Positive segregation occurs only in the regions where the convection induced liquid velocity is very small, and has little effect on solute redistribution. A comparison of model prediction and experimental data is presented in Fig. 17.5b.

Figure 17.6a shows the calculated final solute distribution for flow driven only by thermosolutal convection. While the symmetry of the pattern is altered because of the flow pattern, the highest segregation is in the middle of the casting, as expected. The last region to solidify is the region richest in solute.

Figure 17.6b shows the solute redistribution for the case of liquid flow driven by thermosolutal buoyancy and solidification contraction. Solidification contraction imposes a back flow of enriched liquid in the area adjacent to early-solidified regions, which results in the occurrence of isolated, highly segregated regions aligned almost parallel to the right-side wall where solidification starts.

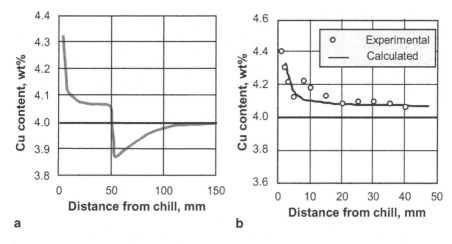

Fig. 17.5 Copper concentration in a directionally solidified Al–4 % Cu casting. **a** Calculated evolution of the average concentration at a given time after the beginning of solidification (after Chang and Stefanescu 1996). **b** Comparison of the calculated final solute distribution (Chang and Stefanescu 1996) and experimental data (Kato and Cahoon 1985). (With permission of Springer Science and Business Media)

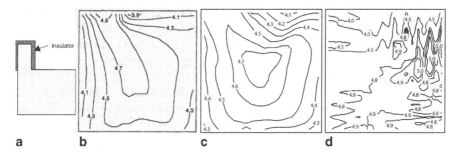

Fig. 17.6 Calculated macrosegregation maps for an Al–4 % Cu alloy (Chang and Stefanescu 1996); the numbers are Cu concentration in mass%. **a** Casting. **b** Flow driven by thermal buoyancy. **c** Flow driven by thermosolutal convection. **d** Flow driven by thermosolutal convection and solidification contraction. (With permission of Springer Science and Business Media)

17.4.2 Effect of Solid Deformation

An important parameter that affects macrosegregation is the deformation of the solid under the weight of the metal or because of stress, encountered in most shaped castings and in particular in continuous casting. Thus, the strains and deformations in the solid shell must be taken into account. To solve this complicated problem, Lesoult and Sella (1990) and Raihle and Fredriksson (1994) modified the local solute redistribution equation (Eq. 7.20) to account for changes in the volume of the mush element because of the deformation of the dendritic skeleton. This allowed them to explain the formation of positive centerline segregations (type V). To fully include the effects of solid deformation (shell and mush) numerical multiscale models that include stress analysis are required. Significant progress has been made in this direction, but there is much more work to do until these models will become industrial tools.

17.5 Macroshrinkage Modeling

The prediction of the location and size of shrinkage-related defects is a difficult task. Numerous attempts at answering the problem through complex numerical 3D models that solve the transport equations are on record. Depending on the assumptions on the physics of the problem and the mathematical apparatus used the different approaches to macroshrinkage defect prediction can be summarized as follows (see also reviews in Piwonka 2000; Lee et al. 1992; Stefanescu 1994):

- Thermal models: Solve energy transport equations to identify the last region to solidify or regions where feeding becomes restricted.
- Thermal/volume calculation models: Solve energy transport equations and mass conservation to predict the position of the free surface and of the last region to solidify.
- Thermal/fluid flow models: Solve mass and energy transport equations to predict the position of the free surface and of the last region to solidify.
- Transport/stress analysis models.
- Nucleation and growth of gas pores models: Compute pore nucleation and growth when the dendrites have formed a coherent network; liquid flow is described as flow through a porous medium.

Many of these models attempt to predict not only macroshrinkage, but also porosity formation, while ignoring the direct contribution of gas evolution. The last type of models is used to predict microshrinkage and will be discussed in detail in Sect. 18.2.4.

17.5.1 Thermal Models

These models include in the analysis only heat transport and ignore fluid flow and the role of gas. After the pioneering work of Henzel and Keverian (1965) that used a finite difference (FD) mesh to represent the irregular shape of casting and adapted a general purpose transient HT program to solve for the temperature field, it became possible to identify the last region to solidify (hot spots) in a casting by mapping the isotherms. Assuming further that shrinkage cavities are located in the last region to solidify, prediction of position but not of the size of shrinkage cavities could be mapped. This technique has been widely used and its simplicity made it possible to implement it even on Excel sheets (see Application 17.2). However, the assumption that the mold is full at the beginning of solidification and that the liquid has uniform temperature, and ignoring the role of gravity on thermo-solutal convection are serious sources of errors.

Simpler thermal models based on the derivation of some analytical equation used as a criterion function were also proposed. The best known is the Chvorinov (1940) rule, which recognizes that for adequate feeding the riser must solidify after the casting, and consists in comparing the final solidification time, t_f, of the riser and the casting. It relates t_f to the modulus of the casting (volume to cooling surface area ratio, v/A), based on 1D heat transport across the mold/metal interface, Eq. 6.24.

In complex castings, where modulus calculations are cumbersome, numerical discretization schemes have been used to calculate casting and riser moduli. Incorporation of such calculations in casting simulation models has allowed a first approximate prediction of macroshrinkage defects (Upadhya and Paul 1992; Suri and Yu 2005).

The observation that occurrence of porosity can be avoided or at least minimized by maintaining a minimum thermal gradient, G_T, in the casting is at the basis of the Pellini (1953) criterion. It states that a shrinkage defect may occur in a region where the thermal gradient is smaller than a critical value, $G_T < G_T^{cr}$, at the end of solidification (when the fraction solid is $f_S = 1$, or when a critical fraction solid is reached ($f_S < f_S^{cr}$). This critical fraction solid may be that when flow is interrupted or dramatically decreased (for example, when dendrites reach coherency).

Sigworth and Wang (1993) have proposed a "geometric" model in which the critical thermal gradient required to avoid microshrinkage is a function of the angle of the inner feeding channel inside the casting, θ. Microshrinkage will not occur if:

$$G_T^{cr} \geq \left(1 + \frac{2\Delta T_{SL} k_S}{\rho_S \Delta H_f} \frac{t_c}{l^2}\right) \cdot \frac{\rho_S \Delta H_f l}{4 k_S t_c} \cdot \tan\theta,$$

where ΔT_{SL} is the solidification interval, t_c is the solidification time for the center of the plate, k_S is the thermal conductivity of the solid, ρ_S is the solid density, l is half the thickness of the plate, and ΔH_f is the heat of fusion. The model seems to work fine for narrow freezing range alloys, including steel. They also point out that surface energy effects may prevent feeding of the casting once dendrite coherency is reached in the riser.

Lee et al. (1985) developed a criterion function for wide freezing range alloys. The critical value under which feeding becomes difficult and porosity occurs was derived to be $G_T \cdot t_f^{2/3}/V_S$, where V_S is the solidification velocity.

17.5.2 Thermal/Volume Calculation Models

These models solve the heat transport problem and attempt to predict defect locations through simple change in volume calculations based on mass conservation, thus avoiding rigorous flow analysis of the molten metal during solidification. Imafuku and Chijiiwa (1983) were the first to propose such a model for prediction of the shape of macroshrinkage in steel sand castings. The main assumptions of the model are: (i) gravity feeding occurs instantly (liquid metal moves only under the effect of gravity, solidification velocity is much smaller than flow velocity); (ii) liquid metal free surface is flat and normal to the gravity vector; (iii) the volume of shrinkage cavity is equal to the volume contraction of the metal; and (iv) macroscopic fluid flow exists as long as the fraction solid, f_S, is less than a critical fraction of 0.67. The net change in volume because of shrinkage is calculated with $\Delta v = \beta \, v \, \Delta f_L$, where v is the initial volume, $\beta = (v_S - v_L)/v_L = (\rho_S - \rho_L)/\rho_L$ is the shrinkage ratio, ρ is the density, and the subscripts L and S stand for liquid and solid, respectively. If the volume loss is on the surface, e.g., on top of the riser, it is compensated by lowering the level of the liquid metal. If it is inside the casting it is compensated by introducing a void. In either case the shape of the void is governed by the solidification sequence and by the value chosen for the critical fraction solid. This model can predict the position and size of pipe shrinkage and of macroporosity. By introducing a shrinkage ratio that is a function of temperature (see, for example, Hummer 1988), Suri and Paul (1996) extended the previous model to ductile iron.

A similar approach was also used by Beech et al. (1998). Heat conduction in the casting and mold was coupled with calculation of the volume change because of solidification contraction for each isolated liquid region of the casting at any given time. The volume change was calculated as $v = \beta \, v_o \, g_S \, F$, where v_o is the initial volume of the element, g_S is the volume fraction of solid in the volume element, and F is the fluid fraction. The fluid fraction has been originally defined by Hirt and Nichols (1981), when they developed the VOF method. If the volume element is empty $F = 0$, while $F = 1$ if the volume element is full. A partially occupied volume has a value of F between 0 and 1. This volume is subtracted from the top of the liquid region. The top of the liquid region is defined by the direction of the gravity vector. A feeding criterion based on a drag force coefficient, K, is introduced to describe the feeding process. K is a function of the local solid fraction, as follows:

$$K = \begin{cases} 0 & if \quad f_S \le f_S^{coh} \\ \infty & if \quad f_S < f_S^{coh} \end{cases}.$$

For pure metals it was assumed that the solid fraction for coherency is unity. When the drag force coefficient K is infinite no feeding occurs; otherwise the shrinkage is fully fed. The method was validated by simulating the solidification of T-shaped castings (Fig. 17.7).

Jiarong et al. (1995) developed a 3D model for calculation of pipe shrinkage and macroporosity (closed shrinkage) in hypereutectic ductile iron based on the variation of the total volume of the casting over a time step:

Fig. 17.7 Simulation (**a** and **b**) and experimental (**c**) results for cylindrical castings (Beech et al. 1998). **a** During solidification. **b** End of solidification. **c** End of solidification. (Copyright 1998 by The Minerals, Metals & Materials Society; reprinted with permission)

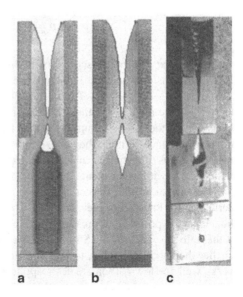

$$\Delta v = \sum \Delta v_L + \sum \Delta v_{Gp} + \sum \Delta v_{Ge} + \sum \Delta v_\gamma + \Delta v_{mold},$$

where Δv_L is the liquid contraction, Δv_{Gp} is the primary graphite expansion, Δv_{Ge} is the eutectic graphite expansion, Δv_γ is the eutectic austenite contraction, and Δv_{mold} is the volume change due to mold wall movement. A similar equation can be written for hypoeutectic iron. While the equations for the volume changes in the metal are fully given, no explanation is provided on how mold expansion was treated. No details were given on the implementation of this equation in the model.

17.5.3 Thermal/Fluid Flow Models

These models tackle more or less comprehensively the whole transport problem but ignore the contribution of gas rejected by the solidifying melt to porosity formation. Thus, the physics of pore formation can be described through the simplified pressure balance equation: $P_{shr} > P_{amb} + P_{st}$, where P_{shr}, P_{amb}, and P_{st} are the shrinkage, ambient and metallostatic pressures, respectively.

An early model by Walther et al. (1956) assumed that void pores form because the section of the channel along which feed metal travels continuously narrows during solidification until the pressure drop ultimately ruptures the liquid in the channel. Further assuming shrinkage driven mass flow and conduction heat transport, an expression for the pressure drop along the channel, P_{shr}, was derived. It is given here in the format modified by Piwonka and Flemings (1966):

$$P_{st} - P_{shr} = \frac{32\mu\beta\varsigma^2 L^2}{(1-\beta)r^4},$$ (17.29)

where μ is the dynamic viscosity of the melt, L is the length of liquid zone in the casting, r is the radius of liquid channel (central cylinder) in the partially solid casting, and $\varsigma = k_{mold}(T_f - T_o)/\rho_S \Delta H_f \sqrt{\pi \alpha_{mold}}$, where k_{mold} is the thermal conductivity of the mold, T_f is the melting point of the metal, and α_{mold} is the thermal diffusivity of the mold. Assuming further that the first pore forms when $P_{shr} = 1$ atm and that once formed, the pore occupies the space previously occupied by the melt, the radius of the pore can be calculated from Eq. 17.29. Reasonable agreement was obtained with experiments.

Niyama et al. (1982) further elaborated on Pellini's idea in their development of a criterion for low-carbon steel castings. They used Darcy's law in cylindrical coordinates and expressed the pressure drop in the mushy zone as an inverse function of the ratio $G_T/(dT/dt)^{1/2}$. Shrinkage defects form in the region where the ratio is smaller than a critical value, to be determined experimentally. While this criterion works well for low-carbon steel, its application by many nonferrous foundries is questionable (see, for example, Spittle et al. 1995).

Attempting to improve this model, Huang et al. (1993) and then Suri et al. (2002) performed a 1D analysis of the conservation of mass and momentum in the two-phase interdendritic region. It was found that the only significant term in the momentum equation responsible for loss of liquid pressure is that of the friction drag. Assuming flow along channels in the mushy zone, the nondimensional frictional drag on the feeding fluid can be expressed as: $F_{drag} = C_o V_L'$. Here, V_L' is the nondimensional liquid velocity and C_o is the "feeding resistance number" that controls the feeding in the mushy region expressed as:

$$C_o = \frac{N\mu\Delta T_{SL}}{\rho_L V_S \beta G_T d^2},$$

where N is a numerical constant (16π for columnar dendrites and 216 for equiaxed dendrites), and d is the characteristic length scale of the solid phase, i.e., either the primary dendrite arm spacing or the equiaxed grain diameter. C_o can be used as a criterion for microporosity formation. A high value indicates a high resistance to feeding and thus higher potential for pore formation. Indeed, experimental verification of this criterion showed excellent correlation between C_o and the percent porosity in an equiaxed A356 alloy.

Another proposed criterion is given by Hansen and Sahm (1998):

$$G_T \bigg/ \left[(dT/dt)^{1/4} \cdot \left(V_L^{mush} \right) \right]^{1/2},$$

where V_L^{mush} is the flow velocity through the fixed dendrite skeleton. As opposed to previously described criteria, this criterion is scale and shape independent.

Bounds et al. (2000) developed a 3D code that predicts pipe shrinkage, macroporosity, and misruns in shaped castings. They reduced the multiphase (S, L, G) problem to single phase by modifying the standard transport equations as proposed by the mixture-theory model. The conservation of mass was rewritten as $\nabla \cdot \mathbf{V} = S_{met} + S_G$, where $\mathbf{V}$ is the mixture velocity, and S_{met} and S_G are the source terms corresponding to the metal shrinkage and gas evolution respectively. S_{met} was evaluated numerically with the iterative scheme:

$$S_{met}^{new} = S_{met}^{old} - \frac{1}{\rho_{met}} \left(\frac{\partial f_{met} \rho_{met}}{\partial t} + \nabla \cdot (\rho_{met} \mathbf{V}_{met}) \right),$$

where $\mathbf{V}_{met}$ is the metal component of the mixture velocity, and the other symbols are as before.

The mixture-theory relationships were used to evaluate the density, fraction, and velocity of the metal. The equivalent fluid component of velocity is given by $\rho \mathbf{V} = \rho_{met} \mathbf{V}_G + \rho_G \mathbf{V}_G$ where no subscript indicates mixture quantities, and the subscript G stands for gas. The mixture density is given by $\rho = f_{met} \rho_{met} + (1 - f_{met}) \rho_G$, where the metal fraction is tracked using a scalar advection equation $\partial f_{met}/\partial t + \nabla \cdot (f_{met} \mathbf{V}_{met}) = S_{met}$. The momentum conservation equation was modified to account for the presence of fluid and solid phases by the introduction of an effective viscosity, μ_{eff}. A Darcy source term, S_D, was added as a momentum sink to describe flow through the fixed dendritic network after coherency:

$$\frac{\partial}{\partial t}(\rho \mathbf{V}) + \nabla \cdot (\rho \mathbf{V} \cdot \mathbf{V}) = \nabla \cdot (\mu_{eff} \cdot \nabla \mathbf{V}) - \nabla P + \rho \mathbf{g} + S_D,$$

where $\rho \mathbf{g}$ is the body force. The fraction solid was obtained from the energy conservation equation.

By extending the Chang-Stefanescu model (see Sect. 17.4) to dendritic-eutectic and columnar-equiaxed morphologies, Bounds et al. were able to show that the pressure drop increases by five orders of magnitude as the flow changes from semisolid feeding to interdendritic feeding.

The model does not deal with the complexities of pore nucleation and growth but rather assumes that below a specific pressure, arbitrarily taken as a value below atmospheric, the volume vacated by metal shrinkage is filled by gas. In this way macroporosity is predicted based only on the pressure drop in the liquid when the feed path is obstructed.

Surface-connected macroporosity occurs under the same circumstances as internal porosity with the additional requirement that the pressure drop in the liquid must be sufficient to draw air into the casting through the permeable mold. This condition is modeled through the boundary condition $Q_{air} = K^{-1}(1 - f_{met})(P_{amb} - P)$, where Q_{air} is the flow rate and K is the permeability. For $P \geq P_{amb}$, f_{met} takes the value calculated for the metal and for $P < P_{amb}$, it is taken as zero. This prevents liquid metal permeating the mold when $P \geq P_{amb}$.

Numerical predictions with this model are presented in Fig. 17.8 for the case of short- and long-freezing-range alloys. For short freezing range alloys a solid skin forms at the mold/metal interface. Once the feeding path becomes obstructed, the pressure drop in the solidifying region triggers macroporosity formation. For long

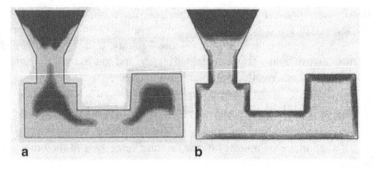

Fig. 17.8 Predicted macroshrinkage in different type alloys (Bounds et al. 2000). **a** Pipe shrinkage and closed shrinkage in short-freezing-range alloy. **b** Surface shrinkage in long-freezing-range alloy. (With permission of Springer Science and Business Media)

freezing range alloys, the mushy zone extends to the surface of the casting. The air can be drawn into the casting through the semisolid surface, and shrinkage may form at the hot spots on the surface.

17.6 Impact of Macromodeling of Solidification on the Metal Casting Industry

The use of solidification models to help production of better castings is today accepted in the metal casting industry. Foundries that are not actively using one of the many commercial models are aware of them, and most accept that soon they will find it necessary to use them to remain competitive. The development of solidification simulation models has even become a matter of national pride: each major casting producing country has developed at least one (and often more than one) model for its foundrymen to use. With rapid prototyping and concurrent engineering applications growing, modeling is clearly an established tool for metal-casters. But models of the casting process can and must include more than merely the solidification event.

Modeling implementation into the foundry industry has been led by casting solid-ification models. The first models did not incorporate fluid flow, and it was simply assumed that once the mold is full, the temperature is uniform across the casting. However, as the technology matured it was soon noticed that this assumption is a source of significant error, in particular for casting with relatively large variations in section size. Subsequently, mold-filling models were developed and implemented. Today, any casting solidification package that claims state of the art level includes both mold filling and solidification modeling.

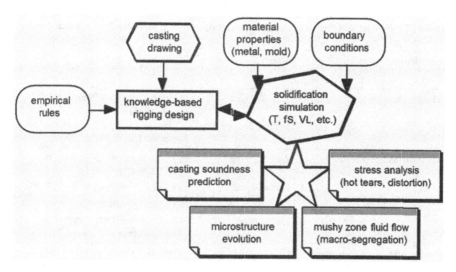

Fig. 17.9 Architecture of a comprehensive casting modeling system (Stefanescu and Piwonka 2002). (Copyright 1996 American Foundry Soc., used with permission)

More ambitious goals have also been set for casting models. By combining mold filling and solidification models with knowledge-based systems for gating and risering, a completely computerized solution for casting design becomes possible (Upadhya and Paul 1992). The architecture of such a package is summarized in Fig. 17.9. Note the addition of stress analysis models that are not discussed in this text.

The objectives of a complete computer based casting design are to design and evaluate the process, and to evaluate process output, which includes casting integrity, surface quality, microstructure, and mechanical properties. It can also be used as a process control tool. Knowledge-based systems attempt to design the process (design of casting rigging), while mold filling and solidification models strive to do the rest. Process modeling has penetrated the metal casting industry and casting markets, and has become integral part of advanced manufacturing processes.

It must be made clear that what the foundryman needs and what the models deliver is not the same thing. As far as the foundryman is concerned, the objective of the modeling is to provide information on casting quality. However, after intricate calculations the models can only deliver some physical quantities such as temperature, solidification time, composition, pressure, fluid velocity, etc. A summary of "promises and realities" deriving from objectives is given in Table 17.2. Matching the foundryman's objectives with models deliverables is still very much an area of research and continuous development.

Some of the quantities calculated by the models can be used directly in the form of property maps (e.g., temperature, solidification time, fraction of solid, composition, velocity, etc.). Composition mapping can be used to predict such features as

Table 17.2 Summary of foundry requirements and mold filling and solidification macro-models deliverables

Foundry objectives	Model deliverables	Calculations based on
Casting soundness, surface quality		
Macroshrinkage	$T, t_f, f_S, G, \dot{T}$	Criteria functions (Ohnaka 1986; Sahm 1991)
	Volume deficit (Δv)	Diffusive energy and convective mass transport (Jiarong et al. 1995)
Microshrinkage/porosity	$T, t_f, f_S, G, \dot{T}, V_L, P$	Criteria functions (Niyama et al. 1982; Lee et al. 1985)
	Pressure map	Pressure balance (Bounds et al. 2000)
Missruns, cold shuts	$G, \dot{T}, T_S > T_{min}$	Criteria functions
	Filling time	Convective–diffusive, energy and mass transport
Casting appearance—surface quality	Penetration index	Pressure balance at metal/mold interface and interfacial reactions (Stefanescu et al. 1996)
Casting composition		
Macrosegregation	Composition map	Convective– diffusive energy + mass transport (Mehrabian et al. 1998; Poirier et al. 1998; Schneider and Beckermann 1995; Chang and Stefanescu 1996)
Casting dimensions—distortions	Stress map	Residual stress
Casting microstructure and mechanical properties		
Fraction phase	$C^{macro}, T, t_f, f_S, G, \dot{T}$	Criteria functions
SL interface stability	$G/V \geq \Delta T_{SL}/D$	Criteria functions
Phase transition	$\dot{T}$	Criteria functions
Columnar-to-equiaxed transition	$G_L < G_{min}$	Criteria functions
Dendrite arm spacing	$\lambda_I = ct. \cdot \dot{T}^n$, $\lambda_{II} = ct. \cdot t_f^{1/3}$ t_f: local solidif. time	Criteria functions
Gray-to-white transition in cast iron	$V < V_{max}$ or $\dot{T} < \dot{T}_{cr}$	Criteria functions
Mechanical properties of casting	$C^{macro}, \dot{T}$	Criteria function based on composition and cooling rate
	$C^{micro}, f\alpha, N, \lambda$	Criteria fct. based on microstructure
Nomenclature		

C^{macro}: macroscale composition	P: pressure
C^{micro}: microscale composition	t_f: final solidification time
f_α: fraction of phase	ΔT_{SL}: solidification interval
G: temperature gradient	Δv: total volume variation
N: number of grains per unit volume	λ: phase spacing

micro- and macrosegregation. Examples of macrosegregation mapping have been given in Fig. 17.6.

As is probably clear to the reader, many features of interest to the foundryman, such as microshrinkage, surface quality, mechanical properties, etc. cannot or are very difficult to obtain through direct calculation. The mathematical complexity of the numerical models for shrinkage prediction and the lack of reliable database have led a number of investigators to develop simpler analytical equations termed "criterion functions" to predict when and where there is a high probability of defect formation in a casting. Criterion functions are simple empirical rules that relate the local conditions (e.g., cooling rate, solidification velocity, thermal gradient, etc.) to the shrinkage defect susceptibility. Some of the criteria functions used to predict casting soundness have been given earlier in this section. Others and microstructure prediction criteria are summarized in Table 17.2.

Most of the criteria discussed here are size (scale) and shape dependent. Hansen et al. (1993) have developed a number of scaling relationships that can be used to develop nondimensional criteria or criteria that are scale and shape independent.

Most criteria are shape dependent. That is, if the shape of the casting is changed the constants involved are changed. The classic example is that of the Chvorinov criterion, $t_f = ct. \cdot M^2$, with $M = v/A$, where M is the casting modulus. This criterion is size independent but shape dependent.

The Niyama criterion can be transformed into a dimensionless criterion if written as $G_S \cdot v^{-1/2} (\dot{T}_S \cdot \Delta T_{SL})^{-1/2}$, where v is the kinematic viscosity (in m^2 s) and ΔT_{SL} is the solidification interval.

17.7 Analysis of Shrinkage Porosity Models and Defect Prevention

A number of criteria-based models for shrinkage defect formation in the last solid to form were presented in the previous section. Although the concepts used are of the macroscale variety, the real intent of these models is to predict microporosity formation. A full treatment of this subject will be provided in Sect. 18.2.4. Because microshrinkage requires the presence of gas and inclusions to nucleate pores and purely thermal analyses ignore both, criterion functions have not been overwhelmingly successful. Laurent and Rigaut (1970) concluded that they could account for only about 75 % of the porosity that exists in castings. The effect of gas content on the use of criteria functions can be dealt with in several ways. The first way is to simply redefine that the proper use of criteria functions depends on having gas and inclusion-free melts (difficult to attain in practice). Alternatively, one can develop a customized function for a particular foundry that includes the density of a reduced pressure test sample, determined empirically (Chiesa et al. 1998; Chiesa and Mammen 1999). Yet another approach is to use statistic models that include not only the amount of gas but also the interaction between the amount of gas and some solidification parameters such as solidus velocity and final solidification time (Tynelius et al. 1993).

While criterion functions are not reliable in predicting microshrinkage in aluminum alloys, they are quite acceptable for ferrous castings (with the exception of ductile iron), where oxide films do not form, and gas evolution is not a problem, and for some copper based alloys. Indeed, the classical feeding distance rules for iron (Wallace and Evans 1958), steel (Bishop et al. 1955; Caine 1950), and copper-base alloys (Weins et al. 1964) are essentially thermal criterion functions recast in geometrical terms.

Numerical models are more complete in the scope of the underlying physics, but have to resort to numerous simplifying assumptions because of lack of data or mathematical complexity. They have many fitting parameters. Typically, only one experimental fact is used in validation—final percentage of porosity. While the influence of alloy composition on hydrogen solubility is computed in some models, in most cases its effect on microstructure is not considered. However, there are indications that, because of the change in the phase morphology, variations in composition can affect the permeability values. There is also experimental evidence that minor additions such as carbon and hafnium in superalloys (Chen et al. 2004) reduce porosity.

Concerning pore nucleation, Poirier (1991) has shown that it is improbable for simple oxides to act as nuclei for pores. Nevertheless, oxide clusters or folded oxide films probably do.

From the previous discussion it appears that, unless the gas content of the liquid is reduced to a value below the solid solubility, the only way to prevent microporosity from forming is to apply pressure during solidification. Indeed, for magnesium alloys it was found that application of pressure greater than 40 MPa completely suppressed microporosity (Yong and Clegg 2004). Possible methods to reduce the size and number of pores include melt degassing, filtering of the metal to remove nuclei and oxide films, and counter-gravity casting to decrease turbulence and to carry the remaining inclusions out of the casting cavity. A postmortem solution is the use of hot isostatic pressing, which may completely eliminate microshrinkage (Lei et al. 2001; Boileau and Allison 2003; Bor et al. 2004).

Finally, all models predict that microporosity occurs unless gas contents are below the solid solubility limit in the liquid. Since the most advanced models rely on pore nucleation and growth to calculate microporosity, it follows that in the absence of dissolved gas in the melt, there will be no porosity. It may be argued that when the feeding channel is totally closed, the pressure drop may fracture the liquid and produce voids. On the other hand, burst feeding may then release the pressure, resulting in surface shrinkage but no microshrinkage. Unfortunately, the author could not find any report on experiments attempting to verify the theory that, in the absence of gas, there is no microshrinkage.

Fig. 17.10 Problem discretization

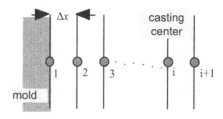

17.8 Applications

Application 17.1 Consider a cubic casting, having a volume of 0.001 m³, poured in a sand mold. The casting material is eutectic cast iron and the pouring temperature is 1350 °C. Assume resistance at the interface on two of the opposite faces of the cube and adiabatic conditions on the other four. Consider also that the solidification interval is 10 °C. Calculate the solidification time of this casting using a 1D FDM scheme. Plot the corresponding cooling curves at different positions throughout the casting. Compare calculation with the Chvorinov equation. Required data are found in Appendix B.

Answer A 1D FDM explicit formulation will be used. To decrease the computational time, since the casting is symmetric, the calculation will be performed only for half of the casting having the length $l = 0.5 \, (v)^{1/3}$. A number of i nodes will be used for the casting, as shown in Fig. 17.10. The node $i + 1$ is used for the second boundary condition at the center of the casting.

The governing equation is Eq. 17.4 in its discretized form Eq. 17.20. Using the same procedure as in Application 6.2 to discretize the fraction of solid, the time evolution of the fraction solid is $\partial f_S / \partial T = -1/\Delta T_o$, where ΔT_o is the solidification interval. Thus, the discretized governing equation is:

$$T^{new} = T^{old} + \frac{Fo}{a} \left(T_E^{old} - 2T^{old} + T_W^{old} \right), \quad \text{where} \quad a = 1 + \frac{\Delta H_f}{c \cdot \Delta T_o}.$$
(17.30)

Note that ΔH_f is expressed in J/kg. To solve this equation we need an IC and two boundary conditions. The IC is $T_{init} = 1350 \,°C$. The first BC in node 1 is a convective boundary condition. Its basic discretized form is given by Eq. 17.18. However, this equation must be modified to include heat generation:

$$hA(T - T_{mold}) = v \, \rho \Delta H_f \frac{\partial f_S}{\partial t} - v \, \rho c \frac{\partial T}{\partial t} + kA \frac{\partial T}{\partial x} \quad \text{or, in discretized format:}$$

$$T^{new} = \frac{2Fo}{a} \left(T_E^{old} + b \cdot T_{mold} \right) + \left[1 - \frac{2Fo}{a}(b + 1) \right] T^{old}, \quad \text{where} \quad b = \frac{h \Delta x}{k}.$$
(17.31)

The second BC in node i is a zero flux condition. That is: $\partial T / \partial x = 0$. An additional node, $i + 1$, past the symmetry line is added. In discretized form BC2 is:

Table 17.3 Spreadsheet structure for the 1D heat transport problem

	A	B	C	D		J	K	L
	Time	i1	i2	i3		i9	i10	i11
	Eq. 17.31	Eq. 17.30	Eq. 17.30	Eq. 17.30		Eq. 17.30	Eq. 17.30	Eq. 17.32
1	0.00	1350	1350	1350		1350	1350	1350
2	2.60	1287	1350	1350		1350	1350	1350
3	5.20	1274	1326	1350		1350	1350	1350
n	704.6	797	844	890		1114	1144	1144

Fig. 17.11 Cooling curves at different positions in the eutectic iron casting

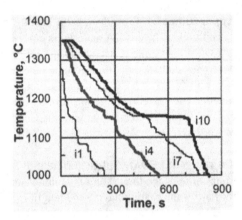

$$T_{i+1} = T_i. \qquad (17.32)$$

The time step is obtained from the stability criterion, $\Delta t \leq (\Delta x^2/2\alpha)(h\Delta x/k + 1)^{-1}$. The grid size is $\Delta x = l/(i - 1)$.

The spreadsheet is organized as shown in Table 17.3. This is a temporal (columns) and spatial (rows) grid. The times in column A result from the summation of the time step. The first row shows the nodes. Node 1 is next to the mold and node 10 is in the center of the casting. Node 11 is the additional node past the symmetry line. The numbers in columns B through C are temperature. Row 1 is for the IC, $T_{init} = 1350\,°C$. Columns B and L are for the boundary conditions. Equation 17.30 is used in all the interior nodes.

Since the constant a becomes 1 before and after solidification, a in Eq. 17.30 is substituted with the following IF statement: IF(OR($T > T_{eut}$, $T < (T_{eut}-\Delta T)$), 1, a). Equations 17.31 and 17.32 are used for the boundary cells, as shown in the table.

The computational results are shown in Fig. 17.11. Several bumps are seen on the cooling curves of nodes 1, 4, and 7. This is because these nodes "feel" the influence of eutectic solidification of adjacent nodes. A finer mesh or an implicit method should be used to avoid this problem.

Fig. 17.12 Casting dimensions and grid

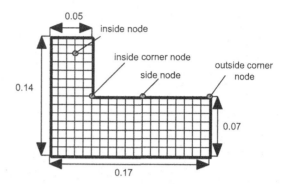

Another concern is that the same eutectic arrest temperature is shown for all nodes. This is unrealistic, since higher cooling rate should result in lower eutectic arrest. This is a consequence of the simplifying assumption that the fraction solid is only function of temperature.

As seen from the table, the solidification time (the time at which the center of the casting is solid) is of 704 s. When Chvorinov's equation (Eq. 6.24) is used for calculation with $\Delta T_{super} = 1350 - 1154 = 196\,°C$, the solidification time is 1626 s. This is considerably larger than that obtained from the 1D FDM calculation. This means that the choice of the HT coefficient is not correct. If h is decreased, the solidification time is increased.

Application 17.2 Consider an L-shaped mold cavity with the dimensions given in Fig. 17.12. A 0.6 % C steel is poured in this mold at an initial temperature of 1550 °C. Calculate the solidification time of this casting using a 2D FDM scheme and the enthalpy method.

Answer A 2D grid is superimposed on the casting, as shown in Fig. 17.12. Four different types of nodes are observed. Discretized equations must be written for each node, as follows:

- Inside node: from Eq. 17.4 written in 2D format, and using the enthalpy method as in previous examples we obtain:

$$T^{new} = \frac{Fo}{a} \left(T_E^{old} + T_W^{old} + T_N^{old} + T_S^{old} \right) + \left(1 - \frac{4Fo}{a} \right) T^{old},$$

$$\text{where} \quad a = 1 + \frac{\Delta H_f}{c \cdot \Delta T}.$$

- Side node (left side):

$$T^{new} = \frac{Fo}{a} \left(T_N^{old} + T_S^{old} + 2T_E^{old} + 2bT_{mold} \right) + \left[1 - \frac{2Fo}{a}(b+2) \right] T^{old},$$

$$\text{where} \quad b = \frac{h \cdot \Delta s}{k}.$$

Table 17.4 Iteration grid

1386	1468	1468	1468	1386										
1468	1550	1550	1550	1468										
1468	1550	1550	1550	1468										
1468	1550	1550	1550	1468										
1468	1550	1550	1550	1468										
1468	1550	1550	1550	1468										
1468	1550	1550	1550	1495	1468	1468	1468	1468	1468	1468	1468	1468	1468	1386
1468	1550	1550	1550	1550	1550	1550	1550	1550	1550	1550	1550	1550	1550	1468
1468	1550	1550	1550	1550	1550	1550	1550	1550	1550	1550	1550	1550	1550	1468
1468	1550	1550	1550	1550	1550	1550	1550	1550	1550	1550	1550	1550	1550	1468
1468	1550	1550	1550	1550	1550	1550	1550	1550	1550	1550	1550	1550	1550	1468
1468	1550	1550	1550	1550	1550	1550	1550	1550	1550	1550	1550	1550	1550	1468
1468	1550	1550	1550	1550	1550	1550	1550	1550	1550	1550	1550	1550	1550	1468
1386	1468	1468	1468	1468	1468	1468	1468	1468	1468	1468	1468	1468	1468	1386

- Outside corner node (top left):

$$T^{new} = \frac{2Fo}{a}\left(T_S^{old} + T_E^{old} + 2bT_{mold}\right) + \left[1 - \frac{4Fo}{a}(b+1)\right]T^{old}.$$

- Inside corner node:

$$T^{new} = \frac{2}{3}\frac{Fo}{a}\left(T_N^{old} + T_E^{old} + 2T_W^{old} + 2T_S^{old} + 2bT_{mold}\right)$$
$$+ \left[1 - \frac{4}{3}\frac{Fo}{a}(b+3)\right]T^{old}.$$

The time step is selected based on the stability criterion presented in Application 16.1. Then, a number of cells are selected on the Excel sheet such that each cell corresponds to a node in Fig. 17.12. We will call this grid 1. The initial temperature is written in each cell. A second group of cells is selected in a similar way, and the equations corresponding to each node are written in the cells. We will call this grid 2. Then, a macro is written using the following sequence:

- COPY grid 2 (cells with equations)
- PASTE SPECIAL, VALUE in grid 1

In this way an iteration in time has been created. The macro is saved, a short cut key is assigned to it and the calculation can be started. By holding the short cut key down, continuous iterations are obtained. Before starting the iterations, grid 2 will look as shown in Table 17.4. A similar iteration can be created between two other

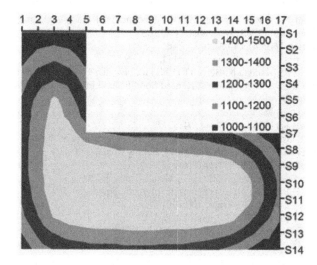

Fig. 17.13 Isotherms after 116 s. Initial temperature was 1550 °C

cells to record the time corresponding to the temperature in the cells. The data can be output using Excel surface chart type. An example of the graphic output after 116 s is given in Fig. 17.13. The central area of the casting is still in a mushy state.

References

Bäckerud L, Chai G, Tamminen J (1990) Solidification Characteristics of Aluminum Alloys: Volume 2, Foundry Alloys. AFS/Skanaluminimu, Des Plaines, Illinois

Beech J, Barkhudarov M, Chang K, Chin SB (1998) in: Thomas BG, Beckermann C (eds) Modeling of Casting Welding and Advanced Solidification Processes VIII. The Minerals, Metals and Materials Soc., Warrendale. PA, p 1071

Bennon WD, Incropera FP (1987) Int. J. Heat Mass Transfer 30:2161,2171

Bishop HF, Myskowski ET, Pellini WS (1955) AFS Trans. 63:271

Boileau JM, Allison JE (2003) Met. Mater. Trans. 34A:1807

Bor HY, Hsu C, We CN, (2004) Materials Chemistry and Physics 84:284

Bounds S, Moran G, Pericleous K, Cross M, Croft TN (2000) Metall. Mater. Trans. B 31B:515

Caine JB (1950) AFS Trans. 58:261

Carman PC (1937) Trans. Inst. Chem. Eng. 15:150

Chang S (1994) Numerical Modeling of Micro- and Macro-Segregation in Casting Alloys. PhD Disseratation, The Univ. of Alabama, Tuscaloosa

Chang S, Stefanescu DM (1996) Metall. Mater. Trans. 27A:2708

Chen QZ, Kong YH, Jones CN, Knowles DM (2004) Scripta Materialia 51:155

Chiesa F, Mammen M, Smiley LE (1998) AFS Trans. 106:149

Chiesa F, Mammen J (1999) AFS Trans. 107:103

Chvorinov N (1940) Giesserei 27:201

Diao QZ, Tsai HL (1993) Metall. Trans. 24A:963

Fellicelli SD, Heinrich JC, Poirier DR (1991) Metall. Trans. 22B:847

Flemings MC, Nereo GE (1967) Trans. AIME 239:1449

Hansen PN, Sahm PR, and Flender E (1993) Trans. AFS 101:443

Hansen PN, Sahm PR (1998) in: Giamei AF, Abbaschian GJ (eds) Modeling of Casting and Welding Processes IV. TMS, Warrendale, PA, p 33

Henzel JG, Keverian J (1965) J. of Metals 17:561

Hirt CW, Nichols BD (1981) J. Computational Physics 39:201

Huang H, Suri VK, EL-Kaddah N, Berry JT (1993) in: Piwonka TS, Voller V, Katgerman L (eds) Modeling of Casting, Welding and Advanced Solidification Processes VI. TMS, Warrendale, Pa, p 219

Hummer R (1988) Cast Metals 1:62

Imafuku I, Chijiiwa K (1983) AFS Trans. 91:527

Jiarong LI, Liu B, Xiang H, Tong H, Xie Y (1995), in: Proceedings of the 61st World Foundry Congress, International Academic Publishers, Beijing China, p 41

Kato H, Cahoon JR (1985) Metall. Trans. 16A:579

Mehrabian R, Keane M, Flemings MC (1970) Metall. Trans. 1:1209

Metzner AB (1985) Rheology of suspensions in polymeric liquids, 29:739

Laurent V, Rigaut C (1992) AFS Trans. 100:399

Lee PD, Chirazi A, See D (2001) J. Light Metals, 1:15

Lee YW, Chang E, Chieu CF (1990) Met. Trans. B 21B:715

Lei CS, Lee EW, Frazier WE (1998) in: Tiryakioglu M, Campbell J (eds) Advances in Aluminum Casting Technology. ASM, Materials Park, OH p 113

Lesoult G, Sella S (1988) Solid State Phenom. 3–4:167–178

Niyama E, Uchida T, Morikawa M, Saito S (1982) AFS Cast Metals Research J. 7:52

Ohnaka I (1986) in: Fredriksson H (ed) State of the Art of Computer Simulation of Casting and Solidification Processes. Les Editions de Physique, Les Ulis, France p 211

Ohnaka I (1993) in: Piwonka TS et al. (eds) Modeling of Casting, Welding and Advanced Solidification Processes VI. TMS, Warrendale Pa p 337

Oldenburg CM, Spera FJ (1992) Numer. Heat Transfer B 21:217

Özisic MN (1994) Finite Difference Methods in Heat Transfer. CRC Press

Patankar SV (1980) Numerical Heat Transfer and Fluid Flow. Hemisphere Publ. Corp., New York

Pellini WS (1953) Trans. AFS 61:61

Pham QT (1986) Int. J. of Heat & Mass Transf. 29:285

Piwonka TS (2000) in: Abbaschian R, Brody H, Mortensen A (eds) Proc. Merton C. Flemings Symposium on Solidification and Materials Processing. TMS, Warrendale Pa., p 363

Piwonka TS, Flemings MC (1966) Trans. AIME, 236:1157

Poirier DR, Nandapurkar PJ, Ganesan S (1991) Metall. Trans. 22B:1129

Poirier DR, Geiger GH (1994) Transport Phenomena in Materials Processing. TMS Minerals Metals Materials, Warrendale Pa. pp 571–598

Poirier DR (1998) in: Thomas BG, Beckermann C (eds) Modeling of Casting, Welding and Advanced Solidification Processes VII. TMS, Warrendale, PA p 837

Raihle CM, Fredriksson H (1994) Metall. Mater. Trans. B 25B:123–133.

Sahm PR (1991) in: Kim C, Kim CW (eds) Numerical Simulation of Casting Solidification in Automotive Applications. TMS, Warrendale PA p 45

Schneider MC, Beckermann C (1995) Metall. Trans. 26A:2373

Sigworth GK, Wang C (1993) Met. Trans. B, 24B:365

Spittle JA, Almeshhedani M, Brown SGR (1995) Cast Metals 7:51

Suri VK, Paul AJ (1993) Trans. AFS 144:949

Suri VK, Paul AJ, EL-Kaddah N, Berry JT (1994) Trans. AFS 138:861

Suri VK, Yu KO (2002) in: Yu KO (ed) Modeling for Casting and Solidification Processing. Marcel Dekker, NY, p 95

Stefanescu DM, Giese SR, Piwonka TS, Lane A (1996) AFS Trans. 104:1233

Stefanescu DM, Piwonka TS (1996) in: Applications of Computers, Robotics and Automation to the Foundry Industry. Proc. Technical Forum, 62nd World Foundry Congress, Philadelphia, PA, CIATF, American Foundrymen's Soc. p 62

Stefanescu DM (2002) Science and Engineering of Casting Solidification. Kluwer Academic/Plenum Publishers, NY

Stefanescu DM (2005) Int. J. Cast Metals Res. 18(3):129–143

Tynelius K, Major JF, Apelian D (1993) Trans. AFS 101:401

Upadhya GK, Paul AJ (1992) Trans. AFS 100:925

Van Doormaal JP, Raithby GD (1984) Numer. Heat Transfer, 7:147

Vannier I, Combeau H, Lesoult G (1993) Mater. Sci. Eng. A173:317–321

Voller VR, Brent AD, Prakash C (1989) Int. J. Heat Mass Transf. 32:1718–1731

Wallace JF, Evans EB (1958) AFS Trans. 66:49

Walther WD, Adams CM, Taylor HF (1956) Trans. AFS 64:658

Weins MJ, Bottom JLS, Flinn RA (1964) Trans. AFS 72:832

Winterscheidt DL, Huang GX (2002) in: Yu KO (ed) Modeling for Casting and Solidification Processing. Marcel Dekker, NY pp 17–54

Yong MS, Clegg AJ (2004) J. of Materials Processing Technology 145:134

Chapter 18
Numerical Microscale Modeling of Solidification

In the previous chapter, the basic elements required to build a macro-solidification model were introduced. Until recently, casting properties were evaluated based on empirical correlations between the quantity of interest and some significant output parameter of the macro-model. Most commonly used are temperature and cooling rate, $\dot{T}$. However, the revolutionary development of numerical analysis and computational technology opened the door for prediction of microstructure evolution during solidification and subsequent cooling to room temperature through models that coupled macro-transport (MT) and transformation kinetics (TK) models. This in turn made prediction of mechanical properties based on direct microstructure-property correlations possible.

The age of computational modeling of microstructure evolution was started by the brilliancy of a scientist, W. Oldfield (1966), who developed a computer model that could calculate the cooling curves of lamellar graphite iron. To the best of our knowledge this is the first attempt to predict solidification microstructure through computational modeling, and the first attempt to validate such a model against cooling curves.

There are many different approaches to the description of MT, including multi-phase and solid movement, as well as to the description of TK. In principle, the methods used for the computational modeling of TK belong to deterministic or probabilistic/stochastic methods (see Fig. 18.1). Deterministic modeling of TK is based on the solution of the continuum equations over some volume element. They are mostly concerned with the description of the solid fraction evolution over time. Probabilistic (stochastic) models are based on the local description of the material combined with some evolutionary rules rather than solving the local integral and differential equations. They can output a graphical description of the microstructure (the computer becomes a dynamic microscope).

© Springer International Publishing Switzerland 2015
D. M. Stefanescu, *Science and Engineering of Casting Solidification*,
DOI 10.1007/978-3-319-15693-4_18

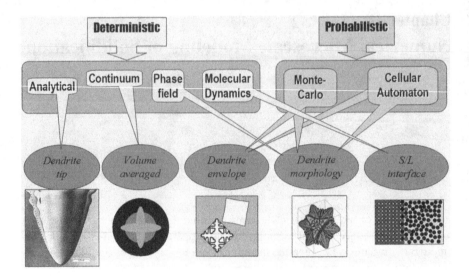

Fig. 18.1 Classification of computational models for microstructure evolution

18.1 Heterogeneous Nucleation Models

Regardless of the model used, a common need for meso- or microscale solidification models for casting solidification is a nucleation model. Looking for a fundamental solution is a difficult problem. At this time, the nucleation models used in numerical modeling of solidification are by and large empirical models which need experimental data to be functional. Fitting parameters are used to match the equations with the experiments.

Without thermosolutal convection to transport dendrite fragments from the mushy zone to the bulk liquid, it is reasonable to assume that nucleation of equiaxed grains is based on heterogeneous nucleation mechanisms. While at least two significant methods based on the heterogeneous nucleation theory have been developed, they are empirical in essence and rely heavily on metal- and process-specific experimental data. The *continuous nucleation model* (Fig. 18.2a) assumes a continuous dependency of N on temperature. Some mathematical relationship is then provided to correlate nucleation velocity, $\partial N/\partial t$, with undercooling, ΔT, cooling rate, or temperature. A summation procedure is carried on to determine the final number of nuclei.

The *instantaneous nucleation model* assumes site saturation, that is, all nuclei are generated at the nucleation temperature, T_N (Fig. 18.2b). Again, an empirical relationship must be provided to correlate the final number of nuclei (grains) in a volume element with ΔT or $\dot{T}$.

A summary of the basic equations and of the parameters that must be assumed or experimentally evaluated is given in Table 18.1. It is seen that all models require either two or three fitting parameters.

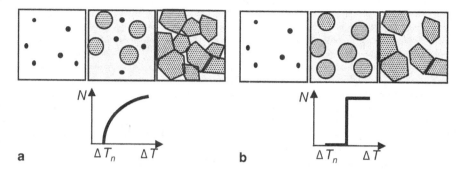

Fig. 18.2 Schematic comparison between assumptions of instantaneous and continuous nucleation models. **a** Continuous nucleation. **b** Instantaneous nucleation (site saturation). (Stefanescu 1973)

Table 18.1 Summary of nucleation models. (Stefanescu 1973)

Model	Type	Basic equation	Fitting parameters	Eq.
Oldfield (1966)	Continuous	$\frac{dN}{dt} = -n\mu_1(\Delta T)^{n-1}\frac{dT}{dt}$	n, μ_1	(18.1)
Maxwell/ Hellawell (1975)	Continuous	$\frac{dN}{dt} =$ $(N_s - N_i)\,\mu_2 \exp\left[-\frac{f(\theta)}{\Delta T^2(T_p-\Delta T)}\right]$	N_S, θ	(18.2)
Thévoz et al. (1992)	Continuous (statistical)	$\frac{\partial N}{\partial(\Delta T)} =$ $\frac{N_s}{\sqrt{2\pi}\,\Delta T_\sigma} \exp\left[\frac{(\Delta T-\Delta T_N)^2}{2(\Delta T_\sigma)^2}\right]$	$N_S, \Delta T_N, \Delta T_\sigma$	(18.3)
Goettsch/ Dantzig (1994)	Continuous (statistical)	$N(r) = \frac{3N_s}{(R_{max}-R_{min})^3}(R_{max} - r)^2$	N_S, R_{max}, R_{min}	(18.4)
Stefanescu et al. (1986)	Instantaneous	$N = a + b \cdot \dot{T}$	a, b	(18.5)

In Oldfield's (1966) continuous model, a power law function, $N = \mu_1(\Delta T)^n$, fitting the experimental data on cast iron, was used to evaluate the final number of grains. A typical experimental and fitted N-ΔT dependency is given in Fig. 18.3. From this, an equation for nucleation velocity is derived (Eq. (18.1) in Table 18.1). The exponent n has typically values between 1 and 2. The coefficient μ_1 depends on the alloy and includes inoculation effects.

Equation (18.1) was later modified (Lacaze et al. 1989) to include the residual volume fraction of liquid. The following equation was obtained:

$$dN = -n\mu_1(\Delta T)^{n-1}f_L dT. \tag{18.6}$$

However, it appears that for equiaxed solidification, this correction is negligible because the nucleation process will cease before a significant portion of the metal has solidified. Calculations for the peritectic Al–Ti system (Maxwell and Hellawell

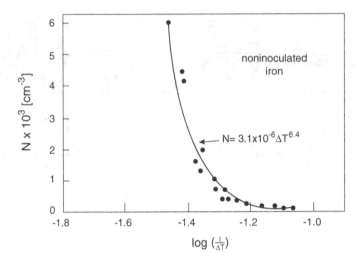

Fig. 18.3 Correlation between undercooling and volumetric nucleation density of eutectic grains in eutectic cast iron. (Mampey 1988)

1975) showed that when nucleation was completed the fraction of solid was a mere 10^{-4}. Nevertheless, it is claimed that such an equation gives a better description of nucleation during directional solidification (Lesoult 1991).

In the more fundamental model developed by Maxwell and Hellawell (1975), Eq. (18.2) in Table 18.1, N_S is the number of heterogeneous substrates, N_i is number of particles that have nucleated at time i, T_p is the peritectic temperature, and $f(\theta)$ is the classic function of contact angle.

Other continuous nucleation models introduced some statistical functions to help describe the rather large size distribution of grains sometimes encountered in castings. A Gaussian (normal) distribution of the number of nuclei with undercooling was introduced by Thévoz et al. (1989). In Eq. (18.3) in Table 18.1, ΔT_N is the average nucleation undercooling and ΔT_σ is the standard deviation. The same distribution was used by Mampey (1991) to model spheroidal graphite iron solidification. However, rather than applying this distribution to the number of nuclei, he applied it to the size of nuclei. To avoid the complication of having to specify θ for heterogeneous nucleation, the width of the substrate was used as a function of undercooling ($K_2/\Delta T$).

The main assumption used in the instantaneous nucleation model (Stefanescu et al. 1986) is that all nuclei are generated at the nucleation temperature. The fundamentals of the model are based on Hunt's (1984) equation for heterogeneous nucleation 3.17 rewritten here:

$$\frac{dN}{dt} = (N_s - N_i)\,\mu_3\,exp\left(\frac{\mu_4}{\Delta T^2}\right). \tag{18.7}$$

Fig. 18.4 Variation of volumetric eutectic grain density as a function of cooling rate for cast iron. (Upadhya et al. 1999, Basdogan et al. 1982, Tian and Stefanescu 1992)

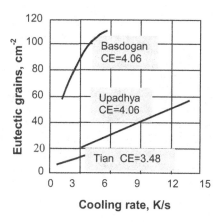

Calculations for eutectic cast iron showed that all substrates became nuclei over a very short time. Therefore, this equation can be substituted by $\partial N / \partial t = N_s \, \delta(T - T_N)$, where δ is the Dirac delta function. Integration gives a total number of nuclei of N_S at T_N. While μ_3 and μ_4 affect the nucleation rate, only N_S will determine the final grain density.

Thus, the dependency between cooling rate and grain density reflects a direct correlation between cooling rate and the number of substrates. This is illustrated by Eq. (18.5) in Table 18.1. In this equation, a and b are experimental constants. This is the most common form of the instantaneous nucleation law. Some typical numbers for cast iron of various carbon equivalents (CE) are given in Fig. 18.4. The numbers are highly dependent on the sulfur content and degree of inoculation.

The question is now which nucleation model works best? In principle, they all work since they are based on fitting experimental data. Thus, the issue is which one fits experimental data better. This is debatable. The main difference between the Oldfield and the Thévoz et al. models is the use of second or third-order polynomials, respectively, to fit the data. In other words, they are using two and three adjustable parameters, respectively. Two adjustable parameters seem to be sufficient in most cases.

In general, for alloys that solidify with narrow solidification interval the instantaneous nucleation model is recommended since it saves computational time. The use of the continuous nucleation model runs into computation complications, related to the definition of the dimensions of the micro-volume element (diffusion distance), when applied to equiaxed dendritic solidification, unless complete solute diffusion is assumed in the liquid.

Experimental evaluation of heterogeneous nucleation laws has been traditionally oversimplified. Typically, the final number of grains at the end of solidification is used to compute a nucleation law. However, as demonstrated through liquid-quenching experiments (Tian and Stefanescu 1992), the evaluation of a nucleation law from the final grain density may result in inaccurate data, since grain coalescence plays a significant role as seen in Fig. 18.5. Indeed, the final eutectic grain

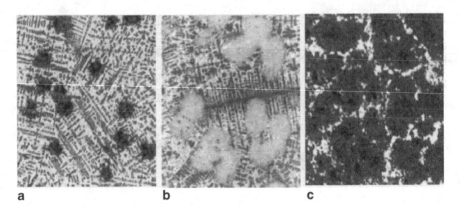

Fig. 18.5 Nucleation and coalescence of eutectic grains in cast iron. **a** Early solidification. **b** Late solidification. **c** After solidification (room temperature). (Tian 1999)

density in cast iron was found to be smaller by up to 27 % than the maximum number of grains developed during solidification.

Greer et al. (2000) introduced a numerical model for the prediction of grain size in inoculated aluminum castings. They make the assumption that, within the operating conditions of slow growth in a near-isothermal melt, only the curvature and solutal undercooling are significant. For a spherical crystal of radius r, the curvature undercooling is given by $\Delta T_r = 2\gamma/(\Delta S_v r)$ where S_v is the volumetric entropy of fusion. The solutal undercooling is calculated with $\Delta T_c = m\,(C_o - C_L^*)$. Crystal growth velocity is calculated according to the approach suggested by Maxwell and Hellawell (1975). Early-stage growth is assumed spherical. In an isothermal melt, the radius of a growing spherical crystal is given by:

$$r = \lambda_S(D_S t)^{1/2}. \tag{18.8}$$

The parameter λ_S is obtained using the invariant-size approximation:

$$\lambda_S = \left(\frac{-S}{2\pi^{1/2}}\right) + \left(\frac{S^2}{4\pi} - S\right)^{1/2}, \quad \text{with} \quad S = \frac{2(C_L^* - C_o)}{C_S^* - C_L^*}. \tag{18.9}$$

As the total undercooling is $\Delta T = \Delta T_r + \Delta T_C$, S becomes:

$$S = \frac{2(\Delta T - \Delta T_c)/m}{(k-1)[(\Delta T - \Delta T_c)/m + C_o]} \tag{18.10}$$

The growth rate of the spherical crystal can now be obtained by differentiating Eq. 18.8 with respect to time:

$$V = \lambda_S^2\, D_S/(2r). \tag{18.11}$$

This equation was solved numerically and tested against measured grain sizes obtained in standard grain-refining tests on aluminum alloys. It was concluded that for potent nucleants, the nucleation stage itself is not the controlling factor. The number of grains is determined by a free-growth condition in which a grain grows from an inoculant particle at an undercooling inversely proportional to the particle diameter. With measured particle size distributions as input, the model made quantitatively correct predictions for grain size and its variation with inoculant addition level, cooling rate, and melt composition.

18.2 Continuum and Volume-Averaged Models

18.2.1 Problem Formulation

A good description of the time dependency of the cooling rate throughout the casting can be obtained from the energy transport equation coupled with mass and momentum transport. The most important governing equation is derived from Eqs. 6.9 and 6.10:

$$\frac{\partial T}{\partial t} + \nabla \cdot (\mathbf{V}T) = \nabla \cdot (\alpha \nabla T) + \frac{\Delta H_f}{\rho c} \frac{\partial f_S}{\partial t} \qquad (18.12)$$

As discussed previously, this equation can be solved if an appropriate description of f_S as a function of time or temperature is available. The heat evolution during solidification described in the source term depends strongly on both the macro-transport (MT) of energy from the casting to the environment, and on the transformation kinetics (TK).

If it is assumed that TK does not influence MT, the two computations can be performed uncoupled. Typically, $\dot{T}$ is evaluated with an MT code, and then the microstructure length scale that includes phase spacing, λ, and volumetric grain density, N, are calculated based on empirical equations as a function of $\dot{T}$. This methodology is presently used by classic MT models that solve the mass, energy, momentum, and species macroscopic conservation equations, as discussed in the preceding chapter. They are inherently inaccurate in their attempt to predict microstructure because of the weakness of the uncoupled MT-TK assumption. Indeed, since, in effect, MT and TK are coupled during solidification, accurate prediction of microstructure evolution revolves around modeling both MT and TK, and then coupling them appropriately.

Consequently, the problem is to describe $f_S(x,t)$ in terms of TK and to select appropriate boundary conditions. The governing equation couples MT and TK through f_S, if f_S is calculated from TK.

While Oldfield's model for microstructural evolution included a MT computer model for heat flow across a cylinder similar to finite difference method (FDM), some early models used analytical modeling for MT and time-stepping procedures to calculate the cooling curve and the fraction solid evolution (e.g., Stefanescu 1973, Stefanescu and Trufinescu 1985; Fras 1975; Aizawa 1978).

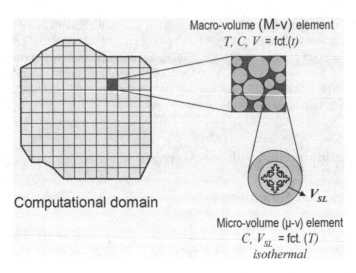

Fig. 18.6 Division of the computational space for deterministic modeling of coupled macro-transport and transformation kinetics

For complete numerical formulations the first step in building an MT-TK solidification model is to divide the computational space of the casting in macro-volume elements within which the temperature is assumed uniform (Fig. 18.6).

In the first numerical models (e.g., Su et al. 1990; Fredriksson and Svensson 1985, 1988; Stefanescu and Kanetkar 1995, 2001; Thévozet et al. 1992), the basic simplifying assumption was that the solid phase has zero velocity (*one velocity models*), meaning that once nucleated, the grains remain in fixed positions. Grain coalescence and dissolution are ignored. The macro-volume element was assumed closed to mass and momentum transport but open to energy transport. Then, each of these elements was further subdivided in micro-volume elements, typically spherical, based on some nucleation law (Fig. 18.6). Within each of these micro-volume elements only one spherical grain is growing at a velocity *V*, dictated by kinetic considerations. The isothermal micro-volume elements are considered open to species transport.

Assuming, for the time being, that the macro-volume element is closed to mass and momentum transport, and that the solid is fixed ($\mathbf{V} = 0$), the governing equation becomes:

$$\frac{\partial T}{\partial t} = \nabla(\alpha \nabla T) + \frac{\Delta H_f}{\rho c} \frac{\partial f_S}{\partial t} \qquad (18.13)$$

It must be noted that the assumption $V = 0$ is always valid once grain coherency is reached, which typically happens at about 0.2 to 0.4 fraction of solid. Thus, for many cases this is a reasonable approximation.

In more advanced models (e.g., Wang and Beckermann 1994), the macro-volume element is open to energy, mass, and momentum transport. The solid may move freely with the liquid. Within each macro-volume element, solid grains are allowed to nucleate based on some empirical nucleation laws. Since the macro-volume element is open, solid grains can be transported in or out of the volume element. The volume-averaging technique is used to manage the various phase quantities, temperature, species concentration, and velocity, as well as some microstructure quantities such as number of grains. $\mathbf{V}$ in Eq. 18.12 is obtained by coupling with the momentum equations.

Consider now an isothermal macro-volume element. The fraction of solid in this element is made of all the grains that have been nucleated and grown until that time, N, multiplied by the grain volume, v. We will introduce the following simplifying assumptions:

- Instantaneous nucleation; this means that at a given time all grains will have the same volume
- Spherical grains, that is the grain volume is given by $v(t) = (4/3)\,\pi r(t)^3$
- No grains are advected in the volume element, that is, $\mathbf{V} = 0$

Then, the fraction solid is $f_S = 4/3(N\pi r^3)$ and $df_S = 4N\pi r^2 dr$, which gives the following equation for the time evolution of the fraction solid:

$$df_S/dt = 4N\pi r^2 dr/dt = 4N\pi r^2 V, \tag{18.14}$$

where V is the solidification velocity of the grain, typically a function of undercooling ($V = \mu \Delta T^2$). Toward the end of solidification the grains come in contact with one another and grain impingement occurs. To account for slowing of grain growth after grain impingement occurs, a correction factor $(1 - f_S)$ is added to the equation:

$$df_S/dt = 4N\pi r^2 V(1 - f_S) \tag{18.15}$$

This correction factor was originally derived by Kolmogorov (1937), Avrami (1939), and Johnson and Mehl (1939). A modified expression was proposed by Hillert (1955) as $(1 - f_S)^i$, where i is greater than unity if the transformed phase is aggregated and less than unity if it is random. Since equiaxed solid grains are typically random, Chang et al. (1992) used $(1 - f_S)^{f_S}$. Note that integration of Eq. 18.15 results in $f_S = 1 - exp[-(4/3)\,\pi\,N\,r^3]$.

In a general format, the evolution of the fraction of solid in time can be written as:

$$\frac{\partial f_S}{\partial t} = \frac{\partial}{\partial t}(Nv) = \frac{\partial N}{\partial t}v + N\frac{\partial v}{\partial t}, \tag{18.16}$$

where N is the volumetric grain (nuclei) density and v is the grain volume. Here, the RHT1 describes the contribution from new nuclei (e.g., heterogeneous nucleation) and the RHT2 that of increased grains volume (grain growth).

The following conservation equation can be written to describe the contribution from new nuclei (Ni and Beckermann 1991):

$$\frac{\partial N}{\partial t} = \dot{N} = \frac{dN}{dt} + \nabla \cdot (\mathbf{V}_S\,N), \tag{18.17}$$

where $\dot{N}$ is the net nucleation rate accounting for the various mechanisms of nucleation and $dN(t)/dt$ is the local nucleation rate (rate of formation of grains per unit volume). The RH2 is the flux of grains due to a finite solid velocity.

The change in grain volume can be calculated as $\partial v/\partial t = A\,V$, where V is the growth velocity of the grain and A is its surface area.

In a simplified analysis, it can be assumed that the grains are of spherical shape. Then Eq. 18.16 describing the evolution of fraction of solid becomes (Stefanescu 2001):

$$\frac{df_S}{dt} = 4\pi \left(\frac{dN}{dt} \frac{r^3}{r} + Nr^2 \frac{dr}{dt} \right)(1 - f_S), \qquad (18.18)$$

where $r(x,t)$ is the grain radius.

If instantaneous nucleation is assumed and if $\mathbf{V}_S = 0$, $dN/dt = 0$ and the evolution of solid fraction simplifies to Eq. 18.15. The problem is then to formulate dN/dt and V (or $r(x,t)$). These calculations are based on specific nucleation kinetics and interface dynamics (to be discussed). If dN/dt is known, df_S/dt can be calculated and used in Eq. 18.13 to couple MT with TK. An enthalpy formulation can then be used.

18.2.2 Coupling of Macro-transport and Transformation-Kinetics Codes

The coupling of the macro- and micro-calculations can be done through the enthalpy method. However, much smaller time steps than obtained from the Fourier number are required to avoid massive calculated release of latent heat which will raise the temperature above the equilibrium temperature. This will result in an oscillatory behavior of the cooling curve around the equilibrium temperature.

To improve accuracy and/or reduce computational time, several coupling techniques between MT and TK have been developed. They include the Latent Heat Method (LHM) by Stefanescu and Kanetkar (2001) and Kanetkar et al. (1988a, b), the Micro-Enthalpy Method (MEM) by Rappaz and Thévoz (1993) and the Micro-Latent Heat Method (MLHM) by Nastac and Stefanescu (1992).

When the LHM is used, the governing equation is obtained by combining equations Eqs. 17.1 and 17.2:

$$\frac{\partial T}{\partial t} = \alpha \frac{\partial^2 T}{\partial x^2} + \frac{\Delta H_f}{c} \frac{\partial f_S}{\partial t}$$

Discretization of this equation gives:

$$T^{\text{new}} = T^{\text{old}} + Fo(T_E^{\text{old}} - 2T^{\text{old}} + T_W^{\text{old}}) + \frac{\Delta H_f}{c} \Delta f_S \qquad (18.19)$$

Here, Δf_S is calculated directly from nucleation and growth laws, e.g., Eq. 18.18. The LHM fully couples MT and TK and is the most accurate. However, because the time step increment necessary to solve the heat flow equation is limited by the microscopic phenomena, it has to be much smaller than the recalescence period in order to properly describe the microscopic solidification. Thus, much longer computational times are required as compared with codes that do not include TK.

The MEM and the MLHM compute heat flow and microstructure evolution at two different scales (Nastac 2004). At the macro-level the heat flow is calculated through Eq. 6.1 or 6.3 without a source term using a large time step. Thus, the macro-enthalpy change, ΔH_{macro}, or the macro-temperature change, ΔT_{macro}, are obtained. Assuming constant rate of heat extraction during the liquid/solid (L/S) transformation, the micro-enthalpy (ΔH_{micro}) or micro-temperature (ΔT_{micro}) change is calculated using a much smaller time step by including the source term in Eq. 3.1 or 3.3. Thus, the fraction solid evolution can be calculated from nucleation and growth kinetics. These two methods are partially coupled methods. While not as accurate as LHM, they substantially decrease the central processing unit (CPU) time (Nastac and Stefanescu 1992). As for accuracy, in the case of eutectic gray iron, for cooling rates between 1 and 5 °C/s, a maximum error of 0.3 % was calculated in the prediction of the recalescence and solidus temperatures with MLHM as compared to the LHM. For further details regarding the coupling schemes and stability criteria, the reader is referred to Nastac (2004).

18.2.3 Dendrite Growth Models

The main techniques used to simulate the evolution of dendrites morphology during solidification based on continuum mechanics concepts include front tracking (FT), continuum (mixture-theory), and volume-averaging techniques.

Front-Tracking Models FT models (Juric and Tryggvason 1996; Jacot and Rappaz 2002; Gandin 2000; Zhao et al. 2014) attempt to simulate time-dependent dendritic growth. They follow the dynamics of the sharp solid/liquid (S/L) interface by solving the heat and species conservation equations with appropriate interface conditions. The FT method explicitly provides the location and shape of the S/L interface at all times by means of a set of extra marker nodes, independent of the mesh, that are defined at every time and move according to the interface conditions. FT models can handle the discontinuous properties at the interface, interfacial anisotropy and topology changes. However, the algorithm for explicitly tracking the position of the interface involves complex numerical calculations, particularly when three-dimensional (3D) computations are considered (Udaykumar et al. 1993). Thus, this method seems inadequate to simulate multi-dendrite microstructures with well-developed side branches realistically because of the solving difficulties caused by the explicit nature of the algorithm.

Continuum Models The most complete continuum models are those developed by Ni and Incropera (1995) in which, by relaxing some assumptions used in the original continuum model, it was possible to account for the effect of solutal undercooling, solidification shrinkage, and solid movement. The model consists of a set of macroscopic continuum equations for conservation of mass, momentum, energy, and species in a binary mixture, expressed in terms of the mixture density ρ_m, velocity $\mathbf{V}_m$, enthalpy h_m, and species concentration f_m^x (see Eq. 4.7). Constitutive equations for the viscous stress and the thermal and mass diffusive fluxes were included and supplemental equations for calculation of phase interaction quantities provided.

Solving the basic energy transport equation used in most continuum models requires calculation of fraction solid. Because of the geometrical complexities of the solidifying grains, simplifications are used. For equiaxed eutectic, a spherical approximation of the eutectic grain works reasonably well. Describing the complexities of dendrite growth requires some simplifications.

A summary of the geometry of the equiaxed grains assumed by some of the models that will be discussed is shown in Fig. 18.7. The dendrite envelope is defined as the surface that touches all the tips of the primary and secondary dendrite arms. It includes the solid and the intra-dendritic liquid. It is the interface separating the intra-dendritic and extra-dendritic liquid phases. The equivalent dendrite envelope is a sphere that has the same volume as the dendrite envelope (r_E in Fig. 18.7a). The equivalent dendrite volume is the sphere having the same volume as the solid dendrite.

Two main types of dendrites are considered: "condensed" and "extended" dendrites. The condensed dendrites have the equivalent dendrite volume of the same order as the volume of the dendrite envelope ($r_S \cong r_E$). They form typically at high undercooling. In the case of the extended dendrite, the equivalent dendrite envelope is significantly higher than the equivalent dendrite volume ($r_E \gg r_S$). They are common at low undercooling when primary arms develop fast with little secondary and higher-order arm growth. The final radius of the grain, r_f, is in fact the micro-volume element. The main problem to solve is to formulate the radius of the equiaxed grain, or the grain growth velocity.

To calculate the radius of the equiaxed grain, it is necessary to develop a model that describes the growth of a simplified dendrite. From Fig. 18.7 a number of solid fractions can be defined as follows:

$$f_S = v_S/v_f = (r_S/r_f)^3 \qquad \text{fraction of solid in the volume element} \qquad (18.20a)$$

$$f_E = v_E/v_f = (r_E/r_f)^3 \qquad \text{volume fraction of dendrite envelope} \qquad (18.20b)$$

$$f_i = f_S/f_E \qquad \text{internal fraction of solid} \qquad (18.20c)$$

These three relationships fully define the growth and morphology of the dendrite. Spherical envelopes must not necessarily be assumed. Stereological relationships such as shape factors and interface concentrations are used for nonspherical envelopes. The time derivative of these equations defines the growth velocity of the

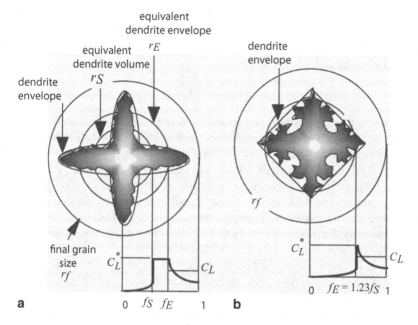

Fig. 18.7 Assumed morphologies and associated concentration profiles of solidifying equiaxed dendrites. **a** "Extended" dendrite. **b** "Condensed" dendrite

dendrite. Present volume average models assume that the movement of the dendrite envelope is governed by the growth model for the dendrite tip. Thus:

$$r_E = \int_0^t \mathbf{V}_E \, dt \text{ until } r_E = r_f = (4\pi N/3)^{-1/3}. \tag{18.21}$$

Since r_f is considered known, one more equation is necessary to solve the system of Eq. 18.20. However, additional simplifications used for the condensed dendrite allow the implementation of simple models in the HT codes (see example in Applications 9.5 and 10.4).

A first model (Dustin and Kurz 1986) was proposed for a compacted dendrite in which the r_E sphere was not completely filled by the network-type dendrite. It was assumed, based on some experimental data, that the internal fraction of solid was $f_i = 0.3$. Both thermal and solutal undercoolings were considered. The solutal field was calculated only at the tip of the dendrite, which implies the assumption of equilibrium solidification (complete diffusion in solid and in liquid), over the total volume of the grain.

Rappaz and Thévoz (1993) developed a 1D radial symmetry model for the case in Fig. 18.7a. The solutal field ahead of an equivalent sphere having the same mass as the dendrite, r_E, growing within the micro-volume element, r_f, was calculated assuming no diffusion in the solid. It was demonstrated that $f_i = \Omega \, f(P_c)$. Uniform composition in the mushy zone was assumed. Solutal balance over the elemental volume of the grain was used. The growth velocity of the sphere of radius r_E was calculated based on the growth velocity of the dendrite tip and a mass balance. Thermal undercooling was neglected. A modified form of the classic equation for the hemispherical dendrite tip was used:

$$V_E = \mu \, \Delta T_c^2,$$

where the growth constant can be calculated from an analytical model for dendrite tip growth. ΔT_c is the constitutional undercooling taken as the temperature difference between the interface and the grain boundary, $T^* - T_{rf}$. Once the dendrite tip reached r_f, which for certain alloys can occur as early as when $f_S = 0.1$, the Scheil equation was used to calculate f_S for the remaining solidification path.

The numerical implementation of this model was done by assuming that the solid phase has zero velocity. This concept was discussed earlier in this text and is schematically described in Fig. 18.6. The evolution of fraction of solid can be calculated with equations such as Eqs. 18.15 or 18.18.

To avoid the early use of the Scheil equation in calculations necessary because of the excessively high growth rate for large grains (e.g., $r_f = 100 \, \mu$m), Kanetkar and Stefanescu (1988) modified the Rappaz–Thévoz model. They used an average composition in the liquid $\langle C_L \rangle$ as the driving force for dendrite growth rather than the composition at r_f. The interface undercooling (Eq. 2.30) becomes $\Delta T = T_f + m \langle C_L \rangle - T_{bulk}$. This correction was later adopted also by Wang and Beckermann (1998).

In the Nastac and Stefanescu (1996) model, it was assumed that nuclei grow as spheres until the radius of the sphere becomes larger than the radius of the minimum instability. Then, the sphere degenerates into a dendrite. Growth of the dendrite is also related to morphological stability and is calculated as a function of the thermal and solutal undercooling of the melt, which is controlled by the bulk temperature and the average composition in the liquid. The complex shape of the equiaxed dendrite, which has several levels of instabilities (i.e., primary, secondary, etc., arms), is converted into a sphere of equivalent volume (Fig. 18.7a, b). This sphere has the same number of instabilities as the dendrite. Since the composition of the liquid, and thus the driving force for the growth of the dendritic instabilities, is position and time dependent, the volume average of the liquid composition is used to compute the growth of instabilities on the solid sphere. The number of instabilities is given by the ratio between the surface area of the solid sphere and that of one instability. Thus, the evolution of the fraction of solid is directly related to the radius of the solid phase, the number of instabilities, and the tip growth velocity. Growth is calculated until the average composition becomes the eutectic composition, or until the fraction of solid becomes one.

The basic elements of the mathematical formulation are as follows. From Eq. 18.20 we have $f_S = f_i \cdot f_E$. The time derivative of this equation is:

$$\frac{\partial f_S}{\partial t} = \frac{\partial f_E}{\partial t} f_i + \frac{\partial f_i}{\partial t} f_E \tag{18.22}$$

After calculation of shape factors, specific interfacial areas, and interfacial area concentrations, it was demonstrated that the temporal evolution of the solid fraction is:

$$\frac{\partial f_S}{\partial t} = \psi f_S \left(\frac{\mathbf{V}_E}{r_E} \frac{1}{\chi_E} + \frac{\mathbf{V}_S}{r_S^i} \frac{1}{\chi_S^i} \right)(1 - f_S) \tag{18.23}$$

with $\chi_E = 4\pi r_E^2 / A_E$ and $\chi_S^i = 4\pi \left(r_S^i \right)^2 / A_S^i$. Here, $\mathbf{V}_E$ and $\mathbf{V}_S$ are the average normal velocities of the dendrite envelope and the S/L interface, respectively; r_S^i is the radius of the instability; χ_E and χ_S^i are the shape factor of the envelope and of the instability, respectively; and A_E and A_S^i are the interfacial areas of the envelope and of the instability, respectively. ψ is a geometrical factor that is 3 for equiaxed grains and 2 for columnar grains. The factor $(1 - f_S)$ on the RH side accounts for grain impingement. The first term in this equation involves calculations at the dendritic length scale, while the second term includes calculations at the instability length scale, describing both formation and coarsening of instabilities. Here coarsening is understood as a process in which larger instabilities grow at the expense of smaller ones through diffusion. An iterative method must be used to solve this equation since the solid fraction is used implicitly.

Application of such a model to casting modeling is restricted by the lack of information regarding the shape factors and the interfacial areas. Some simplification can be introduced, as follows.

If the instabilities are assumed to be spherical, $\chi_S^i = 1$. Assuming further that the condensed dendrite has an envelope that preserves a cubic shape (Fig. 18.7b), $\chi_E = \pi/6$. Thus, Eq. 18.23 simplifies to:

$$\frac{\partial f_S}{\partial t} = \psi f_S \left(\frac{6}{\pi} \frac{\mathbf{V}_E}{r_E} + \frac{\mathbf{V}_S}{r_S^i} \right)(1 - f_S) \tag{18.24}$$

Calculation of $\mathbf{V}_S$ and r_S^i is not trivial. However, if coarsening is neglected since, as shown earlier, when the instability grows at the limit of stability, $r_S^i = 4\pi^2 \Gamma/\Delta T$:

$$\frac{\mathbf{V}_S}{r_S^i} = \frac{1}{r_S^i} \frac{\partial r_S^i}{\partial t} = -\frac{1}{\Delta T} \frac{\partial \Delta T}{\partial t}.$$

Note that from this relationship the internal fraction of solid can be calculated as: $\partial f_i/\partial t \cong f_i \dot{T}/\Delta T$.

If coarsening is considered, it was shown (Nastac and Stefanescu 1996) that:

$$\mathbf{V}_S = \frac{0.75}{\left(r_S^i\right)^2} \frac{D_L \Gamma}{m(k-1)C_d} \left(\frac{1}{f_i^{1/3}} - 1 \right)^{-1} \tag{18.25}$$

where C_d is the volume-averaged intra-dendritic concentration. f_i and r_S^i are either obtained from the previous time step or calculated implicitly.

For the particular case of the condensed dendrite, $f_i = f_S/f_E = (\pi/6)^{1/3} = 0.806$ (or $f_i \approx 1$), $\partial f_i/\partial t = 0$, and the RHT2 in Eq. 18.22 disappears. Also, the shape factor is $\chi = 4\pi r_E^2 / [6(2r_E)^2] = \pi/6$. Thus, under the assumption of constant f_i throughout solidification, Eq. 18.24 simplifies to:

$$\frac{\partial f_S}{\partial t} = \psi \frac{6}{\pi} f_S \frac{V_E}{r_E} (1 - f_S) \qquad (18.26)$$

In all these equations the movement of the dendrite envelope is directly related to dendrite tip velocity; V_E is calculated with some dendrite tip velocity model for either equiaxed or columnar dendrites. Nastac (2004) used for the velocity of columnar dendrites the following equation:

$$V_{den} = \frac{D_L}{\pi^2 \Gamma k \Delta T_o} (\Delta T^*)^2 \text{ with } \Delta T^* = d \frac{\dot{T}}{V},$$

where ΔT^* is the S/L interface undercooling, d is the mesh size, and $\dot{T}$ is the local cooling rate.

This model has the capability to calculate dendrite coherency. Coherency is reached when $f_E = 1$, that is, when the equivalent dendrite envelope reaches the final grain radius. However, the calculated coherency fraction solid for a condensed dendrite, which is about 0.55, is significantly larger than the experimental one, which is typically 0.2–0.3 for most alloys.

Catalina et al. (2007) used a different approach for both calculation of the solid fraction and coupling of the macro- and micro-calculations. They calculated the evolution of f_S directly from the evolution of temperature and solutes fields (without the use of dendrite tip models) that are coupled at the S/L interface through the temperature interface equation written as:

$$T^* = T_f + \sum_{i=1}^{n} m_i C_{Li} + \Gamma K \qquad (18.27)$$

where T_f is the melting temperature of the base element, m_i is the liquidus slope of element i, C_{Li} is the liquid concentration of i at the S/L interface, n is the number of alloying elements, Γ is the Gibbs–Thomson coefficient, and K is the curvature of the S/L interface. For spherical grains $K = -2/r_g$, where r_g is the instantaneous grain radius.

In a solidifying volume element the average concentration, $\bar{C}_i$, of element i is given by:

$$\bar{C}_i = f_s \cdot \bar{C}_{Si} + (1 - f_s)\bar{C}_{Li} \qquad (18.28)$$

where $\bar{C}_{Si}$ and $\bar{C}_{Li}$ are the average concentration of element i in the solid and liquid phase, respectively. The actual distribution of the concentration within each phase depends on the diffusion rate. For the particular case of steel, no diffusion in the solid for the substitutional elements (Mn, Si, Ni, Cr, Mo, P, and S) and complete mixing of the interstitial elements (H, N, C) was assumed (see also Sung et al. 2002). It was further assumed that complete mixing in the liquid phase, within a volume element, occurs for all the solutes, i.e., $\bar{C}_{Li} = C_{Li}$. The effect of interstitial gases on solid fraction evolution was neglected. Consequently, $\bar{C}_{Si}$ can be expressed as:

$$\bar{C}_{Si} = \frac{1}{fs} \int_0^{fs} C_S(\xi) d\xi \tag{18.29}$$

where ξ is the coordinate variable. The change in time is given by:

$$d\overline{C}_{Si}/dt = [(C_{Si}^* - \overline{C}_{Si})/fs] \, df_S/dt \tag{18.30}$$

where C_{Si}^* is the solid concentration of i at the S/L interface and is related to the liquid concentration through $C_{Si}^* = k_{pi}C_{Li}$, where k_{pi} is the partition coefficient of i. Then, using the previous equation, the change of the average concentration in the volume element, $\bar{C}_L$, can be calculated as:

$$\frac{d\bar{C}_i}{dt} = (1 - f_s)\frac{dC_{Li}}{dt} - (1 - k_{pi})C_{Li}\frac{df_s}{dt} \tag{18.31}$$

Multiplying this equation by the liquidus slope, m_i, performing the summation over all the alloying elements, and then using Eq. 18.27 the following finite difference expression can be obtained:

$$\Delta f_s = \frac{1 - f_S^o}{\sum\limits_{i=1}^{n} (1 - k_{pi})m_i C_{Li}^o} \cdot T^n$$

$$- \frac{(1 - f_S^o) \cdot \left(T_m + \Gamma K^n + \sum\limits_{i=1}^{n} m_i C_{Li}^o\right) + \sum\limits_{i=1}^{n} m_i(\bar{C}_i^n - \bar{C}_i^o)}{\sum\limits_{i=1}^{n} (1 - k_{pi})m_i C_{Li}^o} \tag{18.32}$$

where Δf_S is the change of the solid fraction during the time step, and the superscripts n and o denote quantities at the end and beginning of the time step, respectively. After the energy equation is solved for T^n, this equation is used to calculate Δf_S for the respective time step. Note that $\overline{C}_i^n$ should first be calculated from the species conservation equation by means of an explicit procedure. The curvature, K^n, is taken from the previous time step. Once Δf_s is known, the solid and liquid concentrations can be updated by means of Eqs. 18.28 and 18.30. The concentrations of the fast diffusing interstitial elements should be updated by means of the relationship:

$$\bar{C}_i = [1 - (1 - k_{pi})f_S]C_{Li} \tag{18.33}$$

This model was incorporated in Caterpillar's proprietary model, CAPS.

Volume-Averaged Models Ni and Beckermann (1991) developed a two-phase model in which separate volume-averaged mass, momentum, energy, and species conservation equations were derived for the solid and liquid phases. Microscopic features

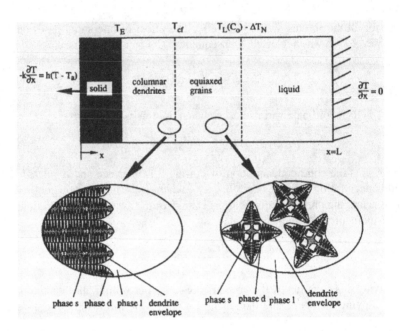

Fig. 18.8 Illustration of the physical problem in the multiphase approach. (Wang and Beckermann 1993b; with permission of Springer Science and Business Media)

were later included through the interfacial transfer terms, nucleation models, and stereological formulations which account for the geometry of the dendritic and eutectic grains. The model was later advanced to a multiphase solute diffusion model for dendritic alloy solidification (Wang and Beckermann 1993a, b, 1994, 1996). A control volume, containing columnar or equiaxed dendrites, is considered to consist of three phases: solid, interdendritic liquid (between the dendrite arms), and extra-dendritic liquid (between the dendrites) (Fig. 18.8).

The two liquid phases are associated with different interfacial length scales and have different transport behaviors. Melt convection and solid transport were included in the last version of the model. Macroscopic conservation equations were derived for each phase, using a volume-averaging technique. The model can incorporate coarsening and was used to predict the columnar-to-equiaxed transition (CET).

The governing equations for the microscopic model were summarized as follows:

- Dendrite envelope motion

$$\frac{\partial}{\partial t}(f_S + f_d) = \frac{A_E}{v_E}\frac{D_L m_L(k-1)C_E}{\pi^2 \Gamma}[I^{-1}(\Omega_c)]^2 \qquad (18.34a)$$

- Solute balance of the solid phase

$$\frac{\partial(f_S\langle C_S\rangle^S)}{\partial t} = \bar{C}_{Sd}\frac{\partial f_S}{\partial t} + \frac{A_S}{v_S}\frac{D_S}{l_{Sd}}(\bar{C}_{Sd} - \langle C_S\rangle^S) \qquad (18.34b)$$

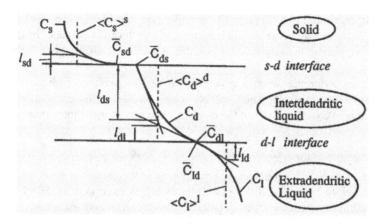

Fig. 18.9 Definition of interface compositions. (Wang and Beckermann 1993; with permission of Springer Science and Business Media)

- Solute balance of the interdendritic liquid

$$\frac{\partial(f_d\langle C_d\rangle^d)}{\partial t} = (C_E - \bar{C}_{dS})\frac{\partial f_S}{\partial t} + C_E\frac{\partial f_d}{\partial t} + \frac{A_S}{v_S}\frac{D_d}{l_{dS}}$$

$$(\bar{C}_{dS} - \langle C_d\rangle^d) + \frac{A_e}{v_e}\frac{D_d}{l_{dL}}(C_E - \langle C_d\rangle^d) \qquad (18.34c)$$

- Solute balance of the liquid phase

$$\frac{\partial(f_L\langle C_L\rangle)}{\partial t} = C_E\frac{\partial f_L}{\partial t} + \frac{A_L}{v_L}\frac{D_L}{l_{Ld}}(C_E - \langle C_L\rangle^L) \qquad (18.34d)$$

- Interfacial solute balance at the *S-d* interface

$$(\bar{C}_{dS} - \bar{C}_{Sd})\frac{\partial f_S}{\partial t} = \frac{A_S}{v_S}\frac{D_S}{l_{dS}}(\bar{C}_{dS} - \langle C_d\rangle^d) + \frac{A_S}{v_S}\frac{D_S}{l_{Sd}}(\bar{C}_{Sd} - \langle C_S\rangle^S) \quad (18.34e)$$

where the subscript *d* stands for the interdendritic liquid, double subscripts stand for interface quantity (e.g., *dS* means interdendritic S/L interface), *l* are the species diffusion lengths. $I^{-1}(\Omega_c)$ is the inverse Ivantsov function. The inverse Ivantsov function can be approximated with $I^{-1}(\Omega_c) = 0.4567(\Omega/(1-\Omega))^{1.195}$. The definitions of the interphase concentrations are summarized in Fig. 18.9. Other notations are as before. Note that the model for the dendrite tip velocity is essentially that used by Rappaz and Thévoz (1993).

When supplying the expressions for the diffusion lengths, thermodynamic conditions, and geometrical relations, these five equations have five unknowns and thus can be solved. They represent a complete model for solute diffusion, applicable to both equiaxed and columnar solidification. However, calculation of diffusion lengths and geometrical relations cannot be done without further simplifications.

For example, assuming a 1D plate-like dendrite arm geometry, it can be shown that $l_{dS} = d_S/6$, where d_S is the mean characteristic length or the diameter of the solid phase. A complete discussion is not possible within the constraints of the present text.

The main difference when applying this model to equiaxed (spherical grains) or columnar (cylindrical grains) solidification consists in the calculation of the interfacial area concentrations of the dendrite envelope. It was shown that they are:

for equiaxed grains: $\frac{A_E}{v_E} = \frac{1}{\chi_E}(36\pi)^{1/3} N^{1/3}(1 - f_L)^{2/3}$,

for columnar grains: $\frac{A_E}{v_E} = \frac{1}{\chi_E}(4\pi)^{1/2}\frac{1}{\lambda_1}(1 - f_L)^{1/2}$,

where $\lambda_1 = N^{-1/3}$ is the primary arm spacing of the columnar dendrites.

According to Ludwig and Wu (2002) the Wang–Beckermann model included a number of uncertainties such as detailed volumetric heat and mass transfer coefficients, and stereological formulations for interfacial area concentrations that are hard to quantify. Addressing some of these issues, they produced a two-phase (L, S) model for globular equiaxed solidification thus avoiding calculation of stereological equations. The exchange (source) terms take into account interactions between the melt and the solid, such as mass transfer (solidification and melting) friction and drag, solute redistribution, release of latent heat, and nucleation. In principle, they write the exchange of momentum, species, or enthalpy between the liquid and the solid as the sum of diffusional interaction (d) and phase change (p). For example, for species the source term is $S_{C-LS} = S_{C-LS}^d + S_{C-LS}^p$. The detailed formulation of the source terms is discussed in the paper. The conservation equations were solved numerically by using the fully implicit, computational fluid dynamics (CFD) code FLUENT (by Ansys).

To overcome the uncertainties discussed above, in later models (e.g., Wu and Ludwig 1954), spherical geometry was assumed for the equiaxed grains, and their growth velocity was calculated as:

$$V_{eq} = \frac{D_L}{r_{eq}} \frac{C_L^* - C_L}{C_L^* - C_{eq}^*} \tag{18.35}$$

where the subscript eq refers to the equiaxed grains. Similarly, the growth velocity of the dendrite trunks was derived as:

$$V_{col} = \frac{D_L}{r_{col}} \frac{C_L^* - C_L}{C_L^* - C_{col}^*} ln^{-1}\left(\frac{r_{max}}{r_{col}}\right) \tag{18.36}$$

where the subscript col refers to the columnar grains and r_{max} is the maximum radius available for cylindrical growth. This model was used to evaluate the effect of the grid size on calculated macrosegregation (Könözsy et al. 1991). For the case that was analyzed, the grid cells have been increased from 180 to 4300. It was found that the macrosegregation pattern did not change significantly above 2780 cells. This indicates that this, and other models, can predict grid-independent numerical solutions.

An important solidification event, in particular for steel and superalloy castings, is the CET. Research efforts in understanding the problem date at least as early as

1954 (Winegard and Chalmers 2000) with the first numerical model being proposed by Wang and Beckermann (1993b) as discussed above. Other CET numerical models followed. We note the model by Wu and Ludwig (1954) who extended their previous model to a three-phase model for mixed columnar-equiaxed solidification capable of predicting the CET. The three phases are the melt as the primary phase, and the solidifying columnar and equiaxed dendrites as two different secondary phases. The three phases are considered as spatially coupled and interpenetrating continua. The conservation equations are solved for all three phases, and an additional conservation equation for the number density of the equiaxed grains is defined and solved. The growth velocity of the globular equiaxed dendrites is calculated as:

$$V_{eq} = \frac{dr_{eq}}{dt} = \frac{D_L}{r_{eq}(1-k)}\left(1 - \frac{C_L}{C_L^*}\right) \tag{18.37}$$

The columnar dendrites are approximated as growing cylinders. The growth velocity of the tip is calculated with the Lipton–Glicksman–Kurz model (Eqs. 9.29 to 9.32). The growth velocity in the radial direction of the cylindrical trunk was calculated as:

$$V_{col} = \frac{dr_{col}}{dt} = \frac{D_L}{r_{col}}\frac{C_L^* - C_L}{C_L^* - C_S^*}\ln^{-1}\left(\frac{r_f}{r_{col}}\right) \tag{18.38}$$

where r_f is half of the primary arm spacing. The columnar tip front tracking (FT) ignores the growth anisotropy determined by crystallographic orientation. While the model shows promise, the authors recognize that to obtain realistic predictions, the model should consider more realistic approximations of the dendrite morphology and should include the effect of melt convection on the diffusion field ahead of the S/L interface. Also, mechanical interactions between moving equiaxed and stationary columnar dendrites, as well as dendrite fragmentation are phenomena that should be included in more advanced numerical models.

18.2.4 Microporosity Models

Kubo and Pehlke (1985) developed a two-dimensional (2D) model, with constant thermal properties. The fraction liquid in the source term was calculated based on the equilibrium or the Scheil model. The continuity equation was used for each volume element, equating the shrinkage (first left-hand term) to the interdendritic inflow of liquid metal (second and third left-hand term) and the growth of the pore (fourth left-hand-term):

$$(\rho_S/\rho_L - 1)\partial f_L/\partial t - (\partial f_L/\partial x)V_x - (\partial f_L/\partial y)V_y + \partial f_G/\partial t = 0 \tag{18.39}$$

The pressure drop through the mushy zone was calculated assuming Darcy type flow. The permeability was formulated based on the Blake–Kozeny model (Eq. 7.14) with $C_2 = 1/180$ and using λ_{II} rather than λ_I.

As long as no porosity has formed, P_{mush} is calculated from Eqs. 18.39 and 7.11. The gas pressure is then calculated from Eq. 12.1. The amount of porosity, f_G, is calculated from mass conservation assuming equilibrium, stating that the initial gas volume is:

$$v_i = v_S(1 - f_L) + v_L f_L + R_H P_G f_G / T \qquad (18.40)$$

where $R_H = 27,300/(f_L \rho_L + f_S \rho_S)$ is a unit transformation parameter. The first, second, and third right-hand term are the amount of hydrogen in solid, liquid, and porosity, respectively. v_S and v_L are obtained from Sievert's law.

Their calculations were in good qualitative agreement with the experiment. They predicted that the amount of porosity increases with the amount of gas dissolved but decreases with higher cooling rate. The basic concepts introduced by Kubo and Pehlke have been used by many other researchers that improved on some details of the calculation as well as on the database.

Poirier et al. (1987) assumed nucleation of cylindrical pores (i.e., r_1 and $r_2 = \infty$) between the primary dendrite arms. They argued that less excess pressure would be required for the gas phase existing between the primary arms than between the secondary arms since the former are larger. This resulted in a modified equation for the surface tension pressure, $P_\gamma = 4\gamma/(g_L \lambda_1)$, where λ_1 is the primary dendrite arm spacing. This model was able to show that increasing the cooling rate or the thermal gradient (which decreases λ_1), also decreases the amount of segregation.

Zhu and Ohnaka (2001) assumed that nucleation occurs when the hydrogen content in the residual liquid exceeds the solubility limit. Selecting an arbitrary number for this supersaturation (in this case $0.1 \, \mathrm{cm}^3/100 \, \mathrm{g}$), a supersaturation pressure for porosity nucleation, ΔP_N, can be calculated. Accordingly, Eq. 12.1 was rewritten as:

$$P_G - P_{local} - \Delta P_N > P_\gamma \qquad (18.41)$$

They further assumed that a pore having a radius 1/2 DAS nucleates, and that the pore grows as a cylinder of constant radius in the interdendritic space. The porosity fraction can be calculated from the mass balance equation $f_G = R (C_L - C_L^{max}) f_L \rho_L T/(100M P_G)$, where R is the gas constant, C_L^{max} is the gas solubility limit in the liquid, and M is the molecular weight of the gas.

In their experimental results, porosity was found between secondary dendrite arms. Only in the relatively gassy heat ($0.48 \, \mathrm{cm}^3/100 \, \mathrm{g}$) was any porosity found between primary arms. The model correctly showed the effect of increasing ambient pressure, decreasing initial hydrogen content, and decreasing solidification time.

Zou and Doherty (1992) modeled porosity occurring in upward directional solidification assuming again Darcy-type flow. They relaxed the steady-state flow assumption of Ganesan and Poirier (1990). This yielded an interdendritic flow equation:

$$\nabla P + \rho_L \mathbf{g} - \frac{\mu}{K} \cdot V = \rho_L \cdot \frac{\partial V}{\partial t}.$$

Assuming further that the pore volume is so small that it can be ignored in the mass balance, and constant solid and liquid densities, the continuity equation was simplified to:

$$\frac{\partial f_S}{\partial t}\left(\frac{\rho_S}{\rho_L}\right) - \frac{\partial f_G}{\partial f} + \nabla V = 0.$$

Solidification kinetics was also included in the model. The porosity versus distance from the chill results appears to have too little porosity, primarily because the authors did not include the effect of dissolved gas. However, in upward directional solidification it is entirely possible under some conditions that the gas pores be pushed ahead of the advancing solidification front.

Viswanathan et al. (2009) modified the equation for capillary pressure to $P_\gamma = (2\gamma/r)\sqrt{S(\theta)}$, where $S(\theta)$ is a factor that takes into account the effect of heterogeneous nucleation of a pore, and θ is the contact angle of the gas-melt interface with the pore. Knowing that the pore fraction must be greater than zero when pores form, and using the previous equation, they offered the following microporosity condition:

$$C_{H_2}^0 > (f_S k_H + f_L) K_L \sqrt{P_{mush} + \frac{2\gamma}{r}\sqrt{S(\theta)}} \tag{18.42}$$

where $C_{H_2}^0$ is the initial gas concentration within the liquid, k_H is the partition coefficient of hydrogen between solid and liquid, and K_L is the equilibrium constant in Sievert's law.

Sabau and Viswanathan (2002) used these concepts in their 3D model for porosity prediction in aluminum alloys. The model computes flow and pressure both in the liquid region and in the mushy zone and calculates shrinkage porosity when feeding flow is cut off. This resulted in numerical difficulties because the dynamic pressure in the liquid zone (typically < 1 Pa) is much lower than the pressure drop in the mushy zone (typically 1–10 KPa). When feeding in a region is cut off, they no longer solve for pressure or velocity in the region, but rather compute porosity such that it compensates for all the shrinkage occurring in that region.

Péquet et al. (2002) coupled a microporosity model based on Darcy flow and gas microsegregation with a macroporosity and shrinkage pipe prediction model in 3D. The governing equations of microporosity formation are solved in the mushy zone with boundary conditions imposed around this zone. The microporosity model is only applied in the mushy zone by superimposing a fine finite volume grid onto the coarser finite element mesh used for heat flow computations. Each liquid region of the casting is evaluated to find whether it is connected to a free surface, surrounded by solid or surrounded by a mushy zone. In the latter two cases, integral boundary conditions must be solved to determine the pressure boundary condition. Because of inaccuracies in this computation, the void fractions are adjusted to ensure global mass conservation.

Carlson et al. (2003) presented a one-domain multiphase model that predicts melt pressure, feeding flow, and porosity in steel castings during solidification and calculates microshrinkage, and closed and open shrinkage cavities throughout a shaped casting as it solidifies. The model assumes that each volume element in the casting is composed of some combination of solid metal (S), liquid metal (L), porosity (G), and air (a) such that:

$$g_S + g_L + g_G + g_a = 1 \qquad (18.43)$$

Mixture properties are obtained as the sum of the property values for each phase multiplied by their respective volume fractions. The mixture energy conservation assumed that the solid fraction is a known function of temperature and ignored the advection term due to shrinkage flow, which may be quite significant at riser necks.

By combining Darcy's law, which governs fluid flow in the mushy zone, with the equation for Stokes flow, which governs the motion of slow-flowing single-phase liquid, they derived a momentum equation that is valid everywhere in the solution domain:

$$\nabla^2 \mathbf{V} = \frac{g_L}{K}\mathbf{V} + \frac{g_L}{\mu_L}\nabla P - \frac{g_L}{\mu_L}\rho_{ref}\mathbf{g} \qquad (18.44)$$

where μ_L is the dynamic viscosity of the liquid (assumed to be constant), $\mathbf{g}$ is the gravity vector, and ρ_{ref} is the reference liquid density, taken as the melt density at the liquidus temperature. Note that this equation reduces to Stokes' equation in the single-phase liquid region, where K becomes very large. Also, in the mushy zone, the left-hand side (LHS) of this equation becomes very small relative to the permeability term, and the equation then reduces to Darcy's law.

The mixture continuity equation does not include the air phase and assumes that the solid metal and the porosity are stationary. Then:

$$\frac{\partial}{\partial t}(g_S\rho_S + g_L\rho_L + g_G\rho_G) + \nabla \cdot (\rho_L\mathbf{V}) = 0 \qquad (18.45)$$

where $\mathbf{V}$ denotes the superficial liquid velocity, $\mathbf{V} = g_L\mathbf{V}_L$. This equation shows that the divergence of the velocity field is a function of the solidification contraction, liquid density change, porosity evolution, and gradients in the liquid density.

By combining the momentum equation with the continuity equation, a pressure equation was derived. The concentration of each gas species dissolved in the melt is obtained from the mixture species conservation equation. The partial pressure of the gas in the pores is calculated from Sievert's law.

The governing equations are solved to give the melt pressure, gas pressure in the pore, and feeding velocity throughout the casting. The criterion for porosity nucleation is Eq. 12.1. However, in this model the capillary pressure was set to zero. When porosity forms, the melt pressure at that location is forced to $P_{mush} = P_G - 2\gamma/r$. This allows the continuity Eq. 18.45, to be solved for the pore fraction, g_G. Once the pore fraction is known, the liquid fraction is updated using Eq. 18.43.

To simulate the formation of a shrinkage cavity open to the atmosphere, the pressure is forced to atmospheric pressure in the volume elements that are emptying of liquid. Then, the continuity equation is solved for the liquid fraction, g_L, while keeping g_G constant (no pore formation when air is present). Finally, the air fraction is obtained from Eq. 18.43.

This multiphase model has been implemented in a general-purpose casting simulation code. The predicted porosity distributions were compared to radiographs of steel castings produced in sand molds.

A predefined pore density number was used in the Péquet et al. (2002) and Carlson et al. (2003) models. Although porosity fractions calculated with these models matched well the experimental measurements, the size and number of pores were dependent on the assumed pore density number.

Catalina et al. (2007) developed a model for porosity prediction that accounts for mass conservation of the gas and local gas/melt equilibrium. The local amount of porosity as well as the number and size of the gas bubbles can be predicted. The gas concentration in the alloy (liquid or solid) was calculated from Sievert's law:

$$C_G = \left(K_G/g_G^a\right)\sqrt{P_G/101325} \qquad (18.46)$$

where C_G is the concentration (wt.%) of the dissolved gas, g_g^a is the gas activity coefficient, P_G is the gas partial pressure in the gas phase (Pa), and K_G, a function of temperature, is the equilibrium coefficient.

Assuming a spherical shape of radius r_G for the gas bubbles, the relationship between P_G and local melt pressure, P_{local}, is given by $P_G = P_{local} + 2\gamma/r_G$. The local melt pressure is calculated from the momentum conservation equation

$$g_L\rho_L\frac{\partial}{\partial t}\left(\frac{\mathbf{V}}{g_L}\right) + \rho_L\mathbf{V}\cdot\nabla\left(\frac{\mathbf{V}}{g_L}\right) = -g_L\nabla P_L + \eta\nabla^2\mathbf{V} - g_L\frac{\eta}{K_{mush}}\mathbf{V} + g_L\rho_L\mathbf{g} \qquad (18.47)$$

and the continuity equation

$$\frac{1-g_G}{\bar{\rho}}\frac{\partial\bar{\rho}}{\partial t} - \frac{\partial g_G}{\partial t} + \nabla\cdot\mathbf{V} = 0 \qquad (18.48)$$

where the first term on the LHS represents the volume fraction of metal shrinkage during the time dt, the second term represents the volume fraction of gas bubbles generated during dt, and the third is the volume fraction of liquid supplied to the volume element. g_S, g_L, and g_G are the volume fractions of solid, liquid, and gas bubbles, respectively ($g_S + g_L + g_G = 1$), $\mathbf{V}$ is the superficial velocity vector, $\mathbf{g}$ is the gravity vector, and K_{mush} is the permeability of the mushy zone that can be calculated with an equation such as Eq. 7.15. To calculate the gas bubbles' evolution, it is assumed that the gas obeys the ideal gas law:

$$P_G v_G^1 = N_m RT/n_G \qquad (18.49)$$

where v_G^1 is the volume of one gas bubble, n_G is the number of bubbles, N_m is the total number of moles of gas in the gas phase of a volume element, and R is the gas constant. At the end of a time step N_m is composed of an initial number of moles, N_m^o, and the number of moles passing from solution into the gas phase, ΔN_m, during the time step ($N_m = N_m^o + \Delta N_m$). With gas concentration expressed in wt.% and assuming that only the gas dissolved in the liquid phase precipitates into bubbles, ΔN_m can be expressed as:

$$\Delta N_m = \frac{(1 - g_S - g_G)\rho_L v_{elem}}{100 M_G}(C_{LG}^o - C_{LG}^n) \qquad (18.50)$$

where v_{elem} is the volume of the volume element, M_G is the molar mass of gas, and C_{LG}^o and C_{LG}^n are the gas concentrations in the liquid before and after precipitation, respectively. Assuming that the bubble/liquid equilibrium is reached at the end of the time step:

$$C_{LG}^n = \left(\bar{K}_G/g_G^a\right)\sqrt{P_L + 2\gamma/r_G} = \left(\bar{K}_G/g_G^a\right)\sqrt{P_G} \qquad (18.51)$$

where the notation $\bar{K}_G = K_G/\sqrt{101325}$ has been used for simplicity. Further, by making use of Eqs. 18.49, 18.50, and 18.51, the following equation that relates r_G and n_G can be obtained:

$$\left(a_4 + a_3\sqrt{P_L + 2\gamma/r_G}\right) \cdot r_G^3 + a_2 r_G^2 + n_G^{-1}\left(a_1\sqrt{P_L + 2\gamma/r_G} + a_o\right) = 0 \qquad (18.52)$$

with the coefficients a_i ($i = 0$–4) given by:

$$a_o = -\left[\frac{(1 - g_S)\,\rho_L v_{elem} C_{LG}^o}{100 M_G} + N_m^o\right]; a_1 = \frac{(1 - g_S)\rho_L v_{elem}\bar{K}_G}{100 M_G}; a_2 = \frac{8\pi\gamma}{3RT};$$

$$a_3 = -\frac{4\pi}{3}\frac{\rho_L\bar{K}_G}{100 M_G}; a_4 = \frac{4\pi}{3}\left(\frac{P_L}{RT} + \frac{\rho_L C_{LG}^o}{100 M_G}\right).$$

The Newton–Raphson method can be used to solve Eq. 18.52 by first assuming a value for n_G and then solving it for r_G. First, a minimum radius of the bubble is calculated from the relationship $C_{LG}^o = \left(\bar{K}_G/g_G^a\right)\sqrt{P_L + 2\gamma/r_{min}}$ and then a maximum number of bubbles, n_{max}, of radius, r_{min}, that can fill the available volume $V_{available} = \left[\Delta t\left(1 - g_G^o\right)\bar{\rho}^{-1}\partial\bar{\rho}/\partial t + g_G^o\right]v_{elem}$. Then n_G is decreased iteratively as long as Eq. 18.52 has a real and positive solution for r_G. In the end, the liquid gas concentration is updated based on Eq. 18.51, and then the average concentration is updated with the relationship given by Eq. 18.28.

In poorly fed regions, it is possible that the computed liquid pressure drops below zero (cavitation). It was considered that gas porosity develops up to the point where pressure reaches the cavitation value after which the entire solidification shrinkage transforms into porosity. Application of this model to a Caterpillar's steel casting showed a good qualitative agreement with the actual defects (Catalina et al. 2007).

The concept of oxide entrapment during mold filling and heterogeneous nucleation on or in the entrapped oxide films suggested by Campbell (2003) was used by Sako et al. (2001) and Ohnaka (2002) to model microshrinkage formation in aluminum alloys. First, collision of free surfaces with each other and with the mold wall was evaluated. The free surface in a volume element was represented in Cartesian coordinates as shown in Fig. 18.10. A criterion for collision based on the filling

Fig. 18.10 Modeling of free surface of melt. (Sako et al. 2001)

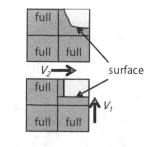

Fig. 18.11 Generation of markers. (Sako et al. 2001)

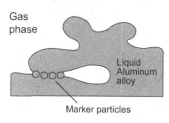

ratio of the element and the velocity on the surface of the element was formulated. Then marker particles were generated to symbolize entrapped oxides in the volume elements where the collision occurred (Fig. 18.11). The markers contain information of oxide number (N_{of}) and average surface area (S_{of}^{av}), which are calculated from the relative collision velocity ($|V_1 - V_2|$) and collision area with the following equations:

$$N_{of} = \alpha_1 |V_1 - V_2| + \alpha_2 \text{ and } A_{of}^{av} = \sum A_{of}/N_{of},$$

where α_1 and α_2 are parameters depending on alloy composition determined experimentally, and A_{of}^{av} and A_{of} are the average and individual surface area of the oxide films. Then, assuming that the markers move with the same velocity as the melt, their movement is calculated from the flow velocity of the melt. The final distribution of the oxide markers in an Al–7.5 % Si casting is compared with the observed shrinkage porosity in Fig. 18.12a and b.

Recognizing that the above method cannot evaluate macroporosity, Ohnaka et al. (2003) developed another model in which the temperature, the solid volume change, and pore growth are calculated for each element at each time step. Before the beginning of solidification the change in pore volume because of hydrogen diffusion is calculated under the assumption of quasi-steady-state diffusion and ideal gas. After the beginning of solidification the pressure field in the mushy zone is calculated through mass conservation and Darcy's law. To avoid solving a nonlinear equation the pore volume is estimated by assuming that all the hydrogen rejected from the solid moves to the pores and extrapolating the pore pressure from the previous pressure change. The error of this estimation is claimed to be less than 10 %. Another complication arises when calculating the pressure field after a mushy or liquid region is enclosed by a solid layer. It requires solving simultaneously for Darcy

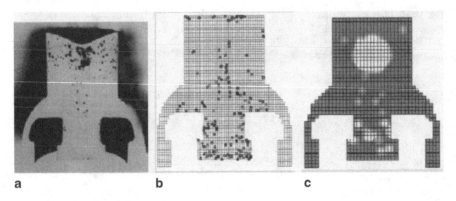

a   b   c

Fig. 18.12 Comparison of simulated and observed results for an Al–7.5 % Si casting. **a** Observed shrinkage. **b** Oxide distribution. **c** Porosity distribution. (Ohnaka 2002; copyright 2003 by The Minerals, Metals and Materials Society; reprinted with permission)

Fig. 18.13 Periodic distribution of pores (*left: dashed circles* indicate impingement of hydrogen depletion layers) and hydrogen profile (*right*). (Atwood et al. 2003; with permission from Elsevier)

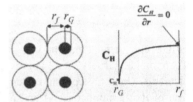

flow, pore growth, and solid layer deformation. In a simplified approach, the pressure change from a reference pressure and correction of the pore volume to satisfy mass balance are used. An example of calculation results is given in Fig. 18.12c.

Based on in situ experimental observations of the kinetic and pore nucleation and growth in Al–Cu alloys using an X-ray temperature gradient stage, Lee and Hunt (1997) concluded that, for small mushy zone, microporosity formation is controlled by gas diffusion. Accordingly, Atwood et al. (2003) developed a deterministic model to solve the hydrogen-diffusion-controlled growth. The pore grows in the liquid because of hydrogen diffusion which is enriched in the liquid because of rejection from the growing solid. Once the pore and the solid grain impinge, the pore grows between the grains assuming a regular closed packed spherical structure (Fig. 18.13). The change in curvature occurring during growth affects the pressure within the pore. The region around the pore was approximated as a sphere of radius r_f within which diffusion was governed by Fick's second law:

$$\frac{\partial C_G}{\partial t} = D_G \frac{1}{r^2} \frac{\partial}{\partial r} \left(r^2 \frac{\partial C_G}{\partial r} \right) + S_t \tag{18.53}$$

where S_t is a time-dependent source term representing hydrogen rejection as determined by the solidification rate. Impingement was calculated based on the maximum spherical size that could be incorporated in the octahedral interstitial site within a lattice made up of spherical grain envelopes. Although this model did not solve for fluid flow and thus ignored shrinkage drive growth of pores, the authors claim that the correlation with the in situ experimental observations was good for both final size of pores and growth of pores.

Atwood and Lee (2000) and then Lee et al. (1998) also developed a stochastic model that includes grain growth and pore growth sub-models. The descriptions of the nucleation algorithm for solid and pores are unclear. Growth of the solid is calculated through standard cellular automaton (CA) techniques. The pore radius is calculated from the combination of the following three equations:

$$P_G v_G = N_m RT \quad P_G = P_{st} + \frac{2\gamma}{r} \quad v_G = \frac{4\pi}{3} r_G^3.$$

The solubility of H_2 in Al is calculated as a function of temperature and the effect of dissolved alloying elements. The solubility of H_2 at equilibrium with the pore pressure was determined from Sievert's law. If v_G exceeded the total volume of the cell, the microshrinkage was allowed to expand into one of the neighboring cells as a complex agglomeration of spheres.

Statistical models based on regression analyses of empirical data have also been proposed. Tynelius et al. (1993) argued that the quantity and size of microporosity in a casting for which the thermal field of the casting/mold assembly is known can be predicted from a statistical model. Based on multiple regression analysis of experimental data on alloy A356, they expressed the percentage porosity as:

$$\%\text{porosity} = a[H]^2 + b[H]t_f + c[Sr][H] + d[H]V_S + e(m_{GR}),$$

where [H] is the hydrogen content, [Sr] is the strontium content, (m_{GR}) is the amount of grain refiner, and a, b, c, d, e are fitted coefficients.

18.3 Phase Field Models

The phase field (PF) method is an integrated simulation technique that typically solves two parabolic partial differential equations, where one of them governs the evolution of the PF variable, which describes the type of phase, solid or liquid, present in the system. The solution of the PF equation is affected by the solution of the second equation, either the solute or heat conservation equation, depending on the controlling process assumed for solidification.

PF models represent the S/L interface as a continuous transition layer with finite thickness between the two phases using an additional PF variable, which avoids explicit tracking of the phase boundaries. The form of the PF and conservation equations is derived from thermodynamics, assuming an expression for the entropy distribution in the system and assuring positive generation of entropy. The technique can simulate the kinetics of dendritic growth. In addition, several factors affecting dendritic growth can be incorporated, such as the crystallographic orientations, motion of the boundaries during impingement (Warren et al. 1995), and others.

However, the technique has significant limitations. The need to spread the change of the PF variable over several mesh points during the numerical solution of the equations demands massive computer resources. This limits the use of this technique on very small domains or for a few dendrites. Furthermore, the technique cannot readily incorporate the solution of both heat and solute conservation equations, implying that only simplified cases of limited practical interest are addressed, such as constant cooling rate or isothermal conditions. Moreover, some of the parameters do not have direct correlation with physical data, such as diffusion coefficients and mobility parameters, therefore functional forms are assumed.

They have been applied successfully to the solidification of pure metals (Karma and Rappel 1996) and alloys (Warren and Boettinger 1996; Loginova et al. 2001; Lan et al. 2003) producing very realistic dendritic growth patterns (Gránásy et al. 2004) and validating quantitatively that the growth kinetics of the dendrite tip agrees well with the results of microscopic solvability theory (Karma and Rappel 1998). However, the simulation results of PF models are significantly dependent on the prescribed interface thickness, which is required to be sufficiently small for accurate simulations. Because the solute-diffusion length is several orders of magnitude smaller than the heat-diffusion length, resulting in extremely thin solute boundary layers, sharp treatment of the interface is required for the simulations of dendritic growth in alloys. This imposes a very fine grid, leading to high computational times.

Karma (2001) proposed an anti-trapping current concept that allows elimination of nonequilibrium effects at the interface. Thus a relatively thick interface width can be used. Together with the efficient adaptive grid techniques, the computational speed of PF simulations has increased by several orders of magnitude. Recently, quantitative PF simulations of alloys for free dendrite growth were performed (Ramirez et al. 2004; Ramirez and Beckermann 2005). Greenwood et al. (2004) reported simulation of cellular and dendritic growth in directional solidification of dilute binary alloys using a PF model solved on an adaptive grid. The simulated spacing of primary branches as a function of pulling velocity for various thermal gradients and alloy compositions was found to have a maximum and agreed with experimental observations.

An example of dendrite growth simulation through the PF method is given in Fig. 18.14. At low nucleation rate there is enough space to develop a full dendritic morphology.

As early as 2006, Badillo and Beckermann (2006) used PF modeling to investigate the CET in two dimensions without an orientation field. The method relies on the solution of a solute conservation equation and an equation for the propagation of the PF on the scale of the developing microstructure. The nucleation sites were placed a priori on the crystal lattice. In more recent work (Gránásy et al. 2004), these simplifications were removed and quantitative computations were made for the CET in Al–Ti alloys. Older models were combined, allowing the use of thermodynamics in a quantitative model relying on the anti-trapping current. The thermodynamic properties were taken from a CALPHAD. Heterogeneous nucleation of the crystalline phase was approximated by the free-growth-limited model of Greer et al. (2000). The nuclei were assumed to follow a Gaussian size distribution. Selected results are displayed in Fig. 18.15. Apparently, the PF simulations were consistent with Hunt's model, as nucleation-controlled equiaxed structure appeared

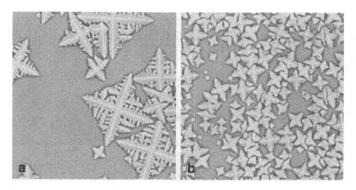

Fig. 18.14 Phase field simulation of two-dimensional anisotropic multigrain solidification as a function of composition and nucleation rate in the Cu–Ni system at 1574 K. **a** Low nucleation rate. **b** High nucleation rate. (Gránásy et al. 2004)

Fig. 18.15 PF modeling of CET: $G = 10^5$ K/m; the pulling velocity increases as $V = [4, 8, 16, \text{and} 32] \times 10^{-4}$ m/s from *right* to *left*. (Gránásy et al. 2014; with permission of Springer Science and Business Media)

for the calculations above Hunt's curve, whereas columnar structure was produced calculations under the curve.

Tiaden (1989) simulated the microstructure evolution during peritectic solidification of Fe–C alloys using a multiphase field approach. It was assumed that the process is nonequilibrium, diffusion controlled. Three phases, liquid, ferrite and austenite, were considered, and PFs were defined for each phase. The PF model was coupled with a solute diffusion model. An example of the calculation of the growth of four ferrite particles during constant cooling of an Fe–C alloy is shown in Fig. 18.16. Anisotropy was not considered. Below the peritectic temperature, single nuclei were placed on the four interfaces. It is seen that the peritectic γ grows around the primary δ by simultaneous consumption of both ferrite and liquid. The

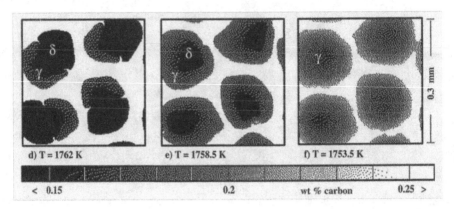

Fig. 18.16 Simulation of peritectic solidification at constant cooling. The carbon concentration is illustrated by the *gray scale*. (Tiaden 1989; with permission from Elsevier)

peritectic reaction is the fastest growth mechanism because at the peritectic temperature the carbon concentration in γ is higher than that in δ but smaller than that of the liquid. Thus the fastest growth will be where liquid and ferrite can directly react. Thus, the austenite grows around the ferrite.

Results of simulation of directional solidification under a constant thermal gradient are shown in Fig. 18.17. The δ-dendrite produced by morphological instability growth in the liquid until the peritectic temperature is reached. At this temperature, some random nucleation of γ was imposed on the system, and then austenite continued to grow on the undercooled dendrite consuming both the ferrite and the liquid. This model was later expanded to ternary systems (see Fig. 18.18) and incorporated in the commercial software MICRESS.

18.4 Stochastic Models

Deterministic continuum models for microstructure evolution have made tremendous progress rewarded by recognition and acceptance by the industry. However, their main shortcoming is their inability to provide a graphic description of the microstructure. Another problem is in the very mathematics that is used. Recent advances in electron microscopy make the examination of materials with near-atomic resolution possible. Such nanoscale structures cannot be described adequately through continuum approximations. As stated by Kirkaldy (1995), "Materials scientists who persist in ignoring the microscopic and mesoscopic physics in favor of mathematics of the continuum will ultimately be seen to have rejected an assured path towards a complete theoretical quantification of their discipline."

More recently, simulation of microstructure evolution has been approached through stochastic models based on the local description of the material combined

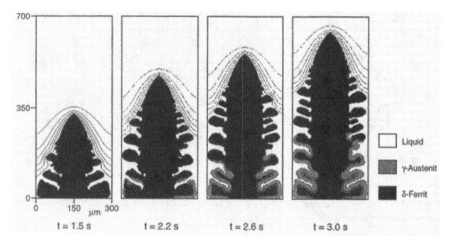

Fig. 18.17 Simulation of directional peritectic solidification. The morphology of the primary ferrite becomes unstable and a dendrite evolves. Under the peritectic temperature, austenite is formed and coats the dendrite. (Tiaden 1989; with permission from Elsevier)

Fig. 18.18 Simulation of peritectic solidification in an Fe–C–Mn system—L, δ, γ phases. (http://web.access.rwth-aachen.de/MICRESS/)

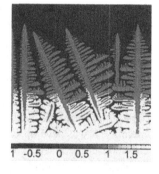

with some evolutionary rules rather than solving the differential and integral equations. Several techniques have been developed including the *Monte Carlo* (MC) technique, where the evolutionary rules are stochastic, and the *cellular automaton* (CA) technique, where algorithms or probabilistic rules control evolution.

Stochastic models approach solidification modeling at the mesoscale, microscale, nanoscale, or a combination of these scales. They may attempt to describe the growth of the dendrite grain envelope or of the eutectic grain (commonly referred to as *grain growth models*), or the development of the morphology of the dendrite grain (termed *dendrite growth* in this text). The two cases are described graphically in Fig. 18.19.

Grain growth models are in principle mesoscale models that ignore the growth morphology at the scale of the grain. The graphical output of these models is limited to showing the grain boundaries as they follow the dynamics of the grain envelope.

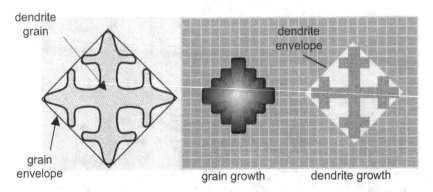

Fig. 18.19 Features described by stochastic growth models *(left)* and discretization of the computational space *(right)*

They are useful for eutectic solidification or for solid-state grain growth. However, they do not give a description of the interface dynamics within the dendrite envelope. The primary and higher-order dendrites arms, internal fraction of solid, and dendrite coherency are not modeled. However, these dendrite features influence fluid flow and thus the outcome of solidification.

18.4.1 Monte-Carlo Models

The MC technique is based on the minimization of the interface energy of a grain assembly. It was developed to study the kinetics of grain growth (Sahni et al. 1983; Anderson et al. 1984). The microstructure was first mapped on a discrete lattice as shown as an example in Fig. 18.20. Each lattice site was assigned an integer (grain index), I_j, from 1 to some number. Lattice sites having the same I_j belong to the same grain. The grain index indicates the local crystallographic orientation. A grain boundary segment lies between two sites of unlike orientation. The initial distribution of orientations is chosen at random, and the system evolves to reduce the number of nearest neighbor pairs of unlike crystallographic orientation. This is equivalent to minimizing the interfacial energy. The grain boundary energy is specified by defining an interaction between nearest neighbor lattice sites. When a site changes its index from I_j to that of its neighbor, I_k, the variation in energy can be calculated from the Hamiltonian[1] cycle describing the interaction between nearest neighbor lattice sites:

$$\Delta G = -\gamma \sum (\delta_{I_j I_k} - 1) \tag{18.54}$$

---

[1] A Hamiltonian cycle is a cycle that contains all the vertices of a graph.

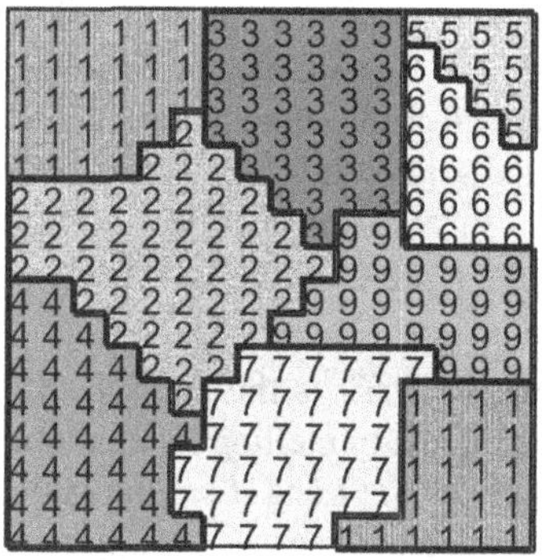

Fig. 18.20 Example of a microstructure mapped on a rectangular lattice; the *integers* denote orientation and the *lines* represent grain boundaries

where γ is the interface energy and $\delta_{I_j I_k}$ is the Kronecker delta. The sum is taken over all nearest neighbors. The Kronecker delta has the property that $\delta_{I_j I_k} = 0$ and $\delta_{I_j I_j} = 1$. Thus, unlike nearest pairs contribute γ to the system energy, while like pairs contribute zero. The change of the site index is then decided on the basis of the transition probability which is:

$$
W = \begin{cases} \exp\left(-\Delta G / k_B \cdot T\right) & \Delta G > 0 \\ 1 & \Delta G \le 0 \end{cases} \tag{18.55}
$$

where k_B is the Boltzmann constant. A change of the site index corresponds to grain boundary migration. Therefore, a segment of grain boundary moves with a velocity given by:

$$
V = C \left(1 - \exp\left(-\frac{\Delta G_i}{k_B T}\right) \right) \tag{18.56}
$$

C is a boundary mobility reflecting the symmetry of the mapped lattice, and ΔG_i is the local chemical potential difference. Note that this equation is similar to that derived from classical reaction rate theory.

Dendrite Envelope Growth Models—Mesoscale We use the term "dendrite envelope" rather than "dendrite grains" to emphasize that the object of this modeling is not to describe the morphology of the dendrite but rather the shape of the grain, equiaxed or columnar, at the end of solidification. Chronologically, Spittle and Brown (1989) seem to be the first to use the MC approach to model solidification. Their mesoscale model was able to qualitatively predict the grain structure of small castings. However, their study was based on a hypothetical material, and the correspondence between the MC time step used in the calculations and real time was not clear. Consequently, it was not possible to analyze quantitatively the effects of process variables and material parameters.

To correct this inadequacy of previous models, Zhu and Smith (2001) coupled the MC method with heat and solute transport equations. They accounted for heterogeneous nucleation by using Oldfield's nucleation model Eq. 18.1. The probability model of crystal growth was based on a lowest free energy change algorithm. ΔG in Eq. 18.55 was calculated as the difference between the free energy determined by the undercooling, ΔG_v, and the interface energy existing between L/S and solid/solid (S/S) grains, ΔG_γ:

$$\Delta G = \Delta G_v - \Delta G_\gamma = \frac{\Delta H_f \Delta T}{T_m} v_m + ld(n_{SL}\gamma_{SL} + n_{SL}\gamma_{SS}) \tag{18.57}$$

where ΔH_f is the latent heat of fusion; ΔT is the local undercooling; T_m is the equilibrium melting temperature; v_m is the volume of the cell associated with each lattice site; l and d are the lattice length and thickness, respectively; n_{SL} is the difference between the number of new L/S interfaces and overlapped L/S interfaces; n_{SS} is the number of S/S interfaces between grains with different crystallographic orientations; and γ_{LS} and γ_{SS} are the interface energies. The time elapsed (MC time step) was calculated using the macroscopic heat transfer method where the time elapsed corresponds to the amount of liquid that has solidified under given cooling conditions.

Some typical computational results of this mesoscale model were presented earlier in this text in Fig. 9.32. A clear transition from columnar to equiaxed solidification is seen when the superheating temperature, and therefore the undercooling, is increased.

Eutectic Solid–Liquid Interface Growth Models—Microscale A more fundamental approach was used by Xiao et al. (2006). They simulated microscopic solidification morphologies of binary systems using a probabilistic MC model that accounts for bulk diffusion, attachment and detachment kinetics, and surface diffusion. An isothermal two-component system contained in a volume element subdivided by a square grid was considered. Initially, the region was filled by a liquid that consists of particles A and B that occupy each grid point according to a preset concentration ratio. Diffusion in the undercooled liquid was modeled by random walks on the grid. Through variation of interaction energies and undercooling, a broad range of microstructures was obtained, including eutectic systems and layered and ionic compounds. It was shown that, depending on the interaction energies between atoms, the microscopic growth structure could range from complete mixing to complete segregation. For the same interaction energy, as undercooling increases, the phase spacing of lamellar eutectics decreases (Fig. 18.21).

Thus, continuum derived laws, such as $\lambda \cdot \Delta T = $ const., were recovered through a combination between a probabilistic model and physical laws. Further, the composition in the liquid ahead of the interface is very similar to that predicted from the classic Jackson–Hunt model for eutectics, as shown in Fig. 18.22. However, this model can only be used for qualitative predictions because the length scale of the lattice, a, is several orders of magnitude higher than the atomic size. Thus, this is not a nanoscale model, but rather a compromise between a nano- and a microscale model. The model is limited to microscale level calculations, and cannot be coupled to the macroscale because of computing time limitations. Nevertheless, the main merit of this work is that it demonstrated that physical laws can be successfully combined

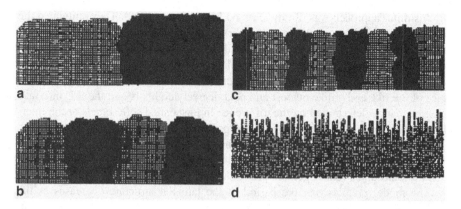

Fig. 18.21 Role of increased undercooling on microscopic growth morphology. **a** $\Delta\mu/k_BT = 0.1$. **b** $\Delta\mu/k_BT = 0.5$. **c** $\Delta\mu/k_BT = 5$. **d** $\Delta\mu/k_BT = \infty$. $\Delta\mu$ is the chemical potential. (Xiao et al. 2006; reprinted with permission from Phys. Rev.; copyright 1992 by the American Physical Soc.; www.aps.org)

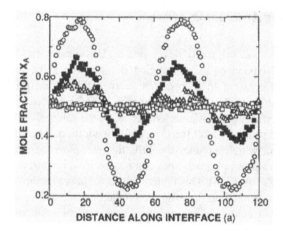

Fig. 18.22 Concentration in mole fraction parallel to the liquid/solid interface at $2a$ (o), $10a$ (■), $20a$ (△) and $50a$ (□) into the liquid; a is the lattice constant. (Xiao et al. 2006; reprinted with permission from Phys. Rev.; copyright 1992 by the American Physical Soc.; www.aps.org)

with probabilistic calculation for transformation kinetic modeling. Also, it proved that to obtain satisfactory results in terms of length scale of the microstructure, it is sufficient to model the diffusion of atom aggregates, perhaps as large as 1000 atoms per aggregate. These aggregates represent the statistically averaged trajectories of the atoms contained within.

A similar approach was use by Das and Mittemeijer (2000). They simulated the isothermal solidification structure of binary alloys taking into account the simultaneous diffusion of all liquid atoms, instead of random walk of single liquid atoms, one at a time, as considered in earlier simulations. The thermodynamic driving force is composed of a volume energy change during solidification. It includes bond energies of the like and unlike nearest and next nearest neighbors and the S/L interface energy change determined by the difference in unsatisfied solid bonds at the interface, before and after solidification. The probability of solidification or remelting of a particular atom is controlled by the change in Gibbs energy during phase transition. At the microscopic length scale, the energy minimization of the system is achieved through a rearrangement process at the S/L interface.

The model predicts that the period of the lamellar structure decreases as the undercooling increases and the S/L diffusivity decreases. As the growth temperature decreases the solidification velocity passes through a maximum, because of the increased undercooling and the decreased diffusivity in the liquid.

18.4.2 Cellular Automaton Models

The CA technique was originally developed by Hesselbarth and Göbel (1991). It is based on the division of the simulation domain into cells, which contain all the necessary information to represent a given solidification process, as shown schematically in Fig. 18.23. Each cell is assigned information regarding the state (solid, liquid, interface, grain orientation, etc.), and the value of the calculated fields (temperature, composition, solid fraction, etc.). In addition, a neighborhood configuration is selected, which includes the cells that can have a direct influence on a given cell. The definition of a neighbor can be modified by introducing a *weighted neighbor* (Dilthey et al. 1998). This means that its influence on the other cell is taken with some weight, depending for example on the distance that separates them. The fields of the cells are calculated by analytical or numerical solutions of the transport and transformation equations. The change of the cell states is calculated through transition rules, which can be analytical or probabilistic. When these rules are probabilistic, the technique is called *stochastic*. The important feature of the method is that all cells are considered at the same time to define the state of the system in the following time step. Thus, the computational time step can be directly related to the physical time step. This gives a certain advantage of the CA technique over the MC method where cells are chosen randomly.

Cellular Automaton Grain Growth Models—Mesoscale The CA method was perfected by Rappaz and Gandin (1987) and then Rappaz and co-workers (e.g., 1995, 1996). They coupled a probabilistic nucleation model with a CA growth model and a finite element heat flow model. The result is a mesoscale model (CAFE). First, the explicit temperature and the enthalpy variation were interpolated at a cell location. Then, nucleation and growth of grains were simulated using the CA algorithm.

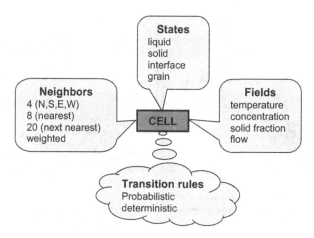

Fig. 18.23 Information attached to a cellular automaton (CA) cell

This algorithm accounts for the heterogeneous nucleation in the bulk and at the mold/casting interface, for the preferential growth of dendrites in certain directions, and for microsegregation. Since in their models dendritic tip growth law was used, and growth velocity was then averaged for a given micro-volume element, the final product of the simulation was dendritic grains, not dendritic crystals. In this respect, the method relies heavily on averaged quantities, just as continuum deterministic models do, but can display grain boundaries on the computer screen.

An example of the CA setup for a plate is provided in Fig. 18.24. The volume is divided into a square lattice of regular cells. The Von-Neumann neighborhood configuration was adopted in calculation (i.e., only the nearest neighbors of the cell are considered). The index that defines the state of the cell is zero when the cell is liquid and an integer when the cell is solid. The integer is associated with the crystallographic orientation of the grain and it is randomly generated.

For eutectic grains, the velocity is normal to the interface, while for dendritic grains it corresponds to the preferential growth direction of the primary and secondary arms. The extension of each grain is then calculated by integrating the growth rate over time. All the liquid cells captured during this process are given the same index as that of the parent nucleus. The computed structure can be either columnar or equiaxed.

One of the most interesting features of the CAFE models is their ability to predict the CET in the presence of fluid flow. As shown in Fig. 18.25, without convection the predicted grain structure is fully equiaxed. It does not reproduce the sedimentation cone and the columnar grain structure observed experimentally. Even such details as the deflection of columnar grains because of fluid flow parallel to the solidification interface can be modeled with the CA technique (Lee et al. 2004).

Cellular Automaton Dendrite Growth Models—Microscale Commonalities of CA models for the simulation of dendritic solidification include (Reuther and Rettenmayr 2014):

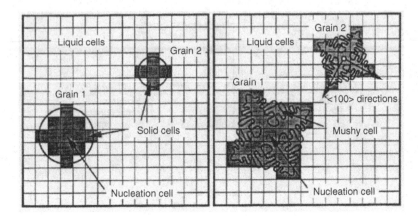

Fig. 18.24 Computational space of a 2D cellular automaton for dendrite envelope grain growth. (Rappaz et al. 1995)

- Representation of the simulation domain by a regular Cartesian grid.
- Definition of the state of a cell by three variables: local solute concentration, phase of the cell, and the volume fraction of the solid phase inside the cell.
- Integration of the diffusion equation in the bulk by finite differences with explicit time forward integration.
- Definition of three possible values for cell phase: solid, liquid, and interface. Only interface cells change their solid fraction, while solid cells have a solid fraction of unity, and liquid cells a solid fraction of zero.
- Separation of regions of solid cells from regions of liquid cells by a single layer of interface cells.

A wide divergence then exists between researchers in approaching these issues. In an early attempt to extend the CA technique to describe the crystallographic anisotropy of the grains and the branching mechanism of dendrite arms (Pang and Stefanescu 1996; Stefanescu and Pang 1974), stochastic modeling at a length scale of 10^{-6} m was coupled with deterministic modeling at a length scale of 10^{-4} m. Atoms attach much faster at the tips of the dendrite, which are growing in the $\langle 100 \rangle$ direction. The lowest rate is in the direction of $\theta = 45°$ with respect to the dendrite axes. Accordingly, the growth velocity of the dendrite liquid interface was expressed as $V_{dendr} = V_{max}f(\theta)$, where $f(\theta)$ is a function of the orientation of the cell with respect to the dendrite axes. This function must be 1 in the $\langle 100 \rangle$ directions and zero at $\theta = 45°$. V_{max} and arm thickening were calculated with deterministic laws derived from the dendrite tip velocity. However, the overall growth of dendrite arms was derived from CA probabilistic calculations. Branching of dendrites arm was allowed to occur based on the classic criterion for morphological instability. Thus, the dendrite morphology, rather than the grain structure, was simulated.

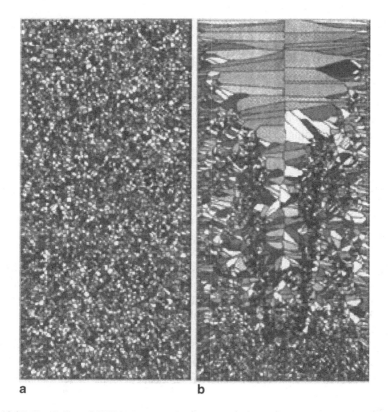

Fig. 18.25 Simulation of CET in a conventionally cast Al–7 % Si alloy. **a** Without grain movement. **b** With grain movement. (Gandin et al. 1998; copyright 1998 by The Minerals, Metals and Materials Society; reprinted with permission)

Expanding on the Dilthey et al. (1998) model, Nastac (1999) developed a stochastic model to simulate the evolution of dendritic crystals. The model includes time-dependent calculations for temperature distribution, solute redistribution in the liquid and solid phases, curvature, and growth anisotropy. Previously developed stochastic procedures for dendritic grains (Nastac and Stefanescu 1997) were used to control the nucleation and growth of dendrites. A numerical algorithm based on an Eulerian–Lagrangian approach was developed to explicitly track the sharp S/L interface on a fixed Cartesian grid. 2D calculations at the dendrite tip length scale were performed to simulate the evolution of columnar and equiaxed dendritic morphologies including the occurrence of the columnar-to-equiaxed transition (CET).

The classic transport equations in 2D and Cartesian coordinates were used, that is, Eqs. 17.1 and 17.2 for energy transport, and Eq. 5.2 assuming $\mathbf{V} = 0$ for solute transport. Eq. 5.2 was written for solid and then for liquid. The boundary conditions were:

- Solute conservation at the interface ($\vec{n}$ is the interface normal vector):

$$V_n(C_L^* - C_S^*) = D \left(\frac{\partial^2 C}{\partial x^2} + \frac{\partial^2 C}{\partial y^2} \right) \cdot \vec{n} \big|_-^+ \qquad (18.58)$$

- Local equilibrium at the interface: $C_S^* = kC_L^*$

Curvature was used in the formulation of both interface temperature and interface concentration according to the equations:

$$T^* = T_L + (C_L^* - C_o)m_L - \Gamma \bar{k} f(\varphi, \theta) \qquad (18.59)$$

$$C_L^* = C_o + (T^* - T_L^{EQ} + \Gamma \bar{k} f(\phi, \theta))m_L^{-1} \qquad (18.60)$$

where $\bar{k}$ is the mean curvature of the S/L interface, $f(\varphi, \theta)$ is a coefficient used to account for growth anisotropy, θ is the growth angle (i.e., the angle between the normal and the x-axis), and φ is the preferential crystallographic orientation angle.

The treatment of the interface curvature merits further discussion. A large number of models use the cell-counting technique, originally developed in a hydrodynamic model (Nichols et al. 1980). It was first applied to the simulation of dendritic growth by Brown and Spittle (1992) with the summation performed in the von Neumann neighborhood and then modified by Sasikumar and Sreenivisan (1994) to employ larger, circular neighborhoods. The cell-counting technique approximates the mean geometrical curvature and not the local geometrical curvature. Thus, the method produces a qualitative measure for the local curvature rather than a precise, quantitative one.

In the Nastac model, another modification of the cell-counting method was introduced. The average interface curvature for a cell with the solid fraction f_S was calculated as:

$$\bar{k} = \frac{1}{a} \left(1 - 2 \left(f_S + \sum_{k=1}^{N} f_S(k) \right) (N+1)^{-1} \right) \qquad (18.61)$$

where N is the number of neighboring cells taken as $N = 24$. This includes all the first- and second-order neighboring cells.

The anisotropy of the surface tension was calculated as per Dilthey and Pavlik (1997):

$$f(\phi, \theta) = 1 + \delta \cos(4(\phi - \theta)). \text{ with } \varphi = a \cos(V_x/[(V_x)^2 + (V_y)^2]^{1/2}), \quad (18.62)$$

where V_x and V_y are the growth velocities in the x and y directions, respectively, and δ accounts for the degree of anisotropy. For fourfold symmetry, $\delta = 0.04$.

Some typical examples of calculation results are given in Fig. 18.26. Note that all dendrites are oriented in the x–y direction.

Building on an earlier mesoscale CA model, Zhu and Hong (2003) developed a modified microscale CA model by incorporating the effect of the constitutional and the curvature undercoolings on the equilibrium interface temperature. The model

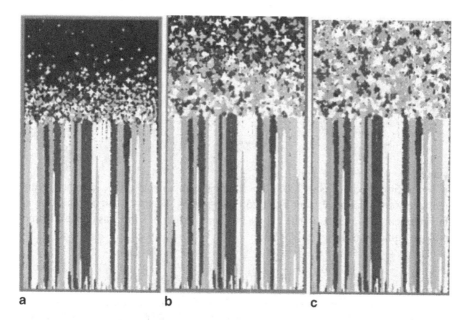

Fig. 18.26 Simulated microstructure (columnar cellular/equiaxed dendritic morphologies and CET formation) in unidirectional solidification of IN718–5 wt.% Nb alloy (initial melt temperature is 1400 °C, 10 × 20-mm domain, $a = 20$ μm, $h = 10^4$ W m^{-2} K^{-1}). (Nastac 1999; reprinted with permission from Elsevier)

has been applied to simulate 2D and 3D single- and multi-dendritic growth, non-dendritic and globular microstructure evolution in a semisolid process (Zhu et al. 2007), dendritic growth in the presence of melt convection, and microstructure formation in regular and irregular eutectics alloys (Zhu and Hong 1991). However, as the velocity of the S/L interface is calculated from the analytical theories of dendrite growth or by introducing a kinetics coefficient, the model leads to mostly qualitative graphical outputs.

Recognizing the problems of mesh-induced anisotropy in crystallographic orientation, Beltran-Sanchez and Stefanescu (BSS; 2003, 2004) developed a model for the simulation of solutal dendritic growth in the low Péclet number regime that does not use an analytical solution to determine the velocity of the S/L interface. The model adopted a methodology similar to that of Nastac (1999) by calculating growth kinetics from the complete solution of the solute and heat transport equations and by incorporating the boundary condition of solute conservation. The heat equation was solved using an Finite Difference (FD) method with an implicit scheme. For the case of the solute diffusion equation, a special scheme was used to overcome the problem of the discontinuity at the interface. The diffusion equation was solved in terms of a potential, which consists of an equivalent composition at every point in the domain. Thus, the whole domain is treated as a single phase for computational purposes. The potential is defined as:

$P = C_L$ for the liquid phase

$P = C_S/k$ for the solid phase

$P = C_L^*$ for the interface cells

Using these definitions the solute conservation equation becomes:

$$\frac{\partial \bar{C}}{\partial t} = D \left(\frac{\partial^2 P}{\partial x^2} \right) + D \left(\frac{\partial^2 P}{\partial y^2} \right) \tag{18.63}$$

where $\bar{C} = f_S C_S + (1 - f_S)C_L$.

In the first model, BSS (2003), the x and y components of the velocity at the interface were calculated from Eq. 18.58 using an FD scheme. The composition at the interface was calculated with Eq. 18.60. The fraction of solid was then evaluated with the equation (Dilthey et al. 1998):

$$\Delta f_s = \frac{\Delta t}{\Delta s} \left(V_x + V_y - V_x V_y \frac{\Delta t}{\Delta s} \right) \tag{18.64}$$

While the equations for V_x and V_y derived from the solute conservation at the interface Eq. 18.58 are correct from the mathematical point of view, they describe the motion of an interface that is planar in the two Cartesian directions, which cannot be true physically. The result is the evaluation of a solid fraction assuming that the interface moves in the x and y direction as a planar front. This produces an overestimation of the actual solid fraction. Since the solute balance is valid only at the limit of a vanishing S/L interface thickness in the normal direction, the only physically realistic velocity that can be calculated from this balance is the normal velocity and not the components. Accordingly, in their second model, BSS (2004) derived the following equation to evaluate the normal S/L interface velocity and solid fraction increments in 2D Cartesian coordinates:

$$V_n = \frac{1}{C_L^*(1 - k_o)} \left[-D_L \left(D_\mathbf{n} C_L \right)|_{\text{interface}} + D_S \left(D_\mathbf{n} C_S \right)|_{\text{interface}} \right],$$

where $D_\mathbf{n}$ is the directional derivative operator on the direction of the normal. The change in solid fraction was then computed as $\Delta f_S = V_n \Delta t / L_\varphi$, where L_φ is the length of the line along the normal passing through the cell center. The length L_φ represents the distance to be covered by a point on the S/L interface so that it could be considered solid. Note that this distance is normalized with the direction of motion of the S/L interface so as to minimize the effect of the artificial mesh anisotropy on the rate of advance of the interface. Thus, while the model is based on the CA concepts, it is using virtual tracking of the sharp S/L interface. Combined with the trapping rules for new interface cells, the mesh dependency of calculations was eliminated allowing the model to simulate dendrites growing at any crystallographic direction.

As the calculation of local curvature with a counting cell method used in previous models is mesh dependent, BSS introduced another method that converges to a

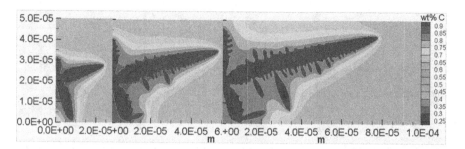

Fig. 18.27 Simulation of an Fe–0.6 wt.% C alloy dendrite with $\theta = 15°$. $\Delta s = 0.1\ \mu m$. From left to right: after 0.03, 0.05, and 0.07 s. (Beltran-Sanchez and Stefanescu 2004; with permission from Springer Science and Business Media)

finite value when the mesh is refined. It is based on the variation of the unit vector normal to the S/L interface along the direction of the interface. The final equation in vectorial form is $K = D_{\hat{T}}\hat{n} \cdot \hat{T}$, where $D_{\hat{T}}$ is the directional derivative operator in the direction of the tangent $\hat{T}$. The tangent vector is perpendicular to the unit normal $\hat{n} = -\nabla f_S / \|\nabla f_S\|$. In 2D Cartesian coordinates, the equation for curvature takes the following form:

$$K = \left[\left(\frac{\partial f_S}{\partial x} \right)^2 + \left(\frac{\partial f_S}{\partial y} \right)^2 \right]^{-3/2} \left[2\frac{\partial f_S}{\partial x}\frac{\partial f_S}{\partial y}\frac{\partial^2 f_S}{\partial x \partial y} - \left(\frac{\partial f_S}{\partial x} \right)^2 \frac{\partial^2 f_S}{\partial y^2} - \frac{\partial^2 f_S}{\partial x^2}\left(\frac{\partial f_S}{\partial y} \right)^2 \right].$$

This last equation was originally proposed by Kothe et al. (2009) using a definition of curvature as:

$$K = -\left(\nabla \cdot \hat{n} \right) = \frac{1}{|\vec{n}|}\left[\left(\frac{\vec{n}}{|\vec{n}|} \cdot \nabla \right)|\vec{n}| - \left(\nabla \cdot \vec{n} \right) \right].$$

Using this approach, the model does not need to use the concept of marginal stability and stability parameter to uniquely define the steady state velocity and radius of the dendrite tip. The model indeed contains an expression for the stability parameter but the process determines its value.

The simulated composition map of a single dendrite growing from one wall at a given orientation is presented in Fig. 18.27. The secondary and tertiary arms still grow perpendicular to the primary arms, which are not aligned with the x–y directions of the mesh. The effect of the upper wall prevents the side facing it to develop branching, while the lower side is fully branched. The dendrite tip approaches the parabolic shape. Another example of model output is shown in Fig. 18.28 for the case of multiple equiaxed dendrites.

The classic theories for dendritic growth predict that at low Péclet numbers the radius of the dendrite tip increases with lower growth velocity (Kurz et al. 1986) as shown in Fig. 18.29. The BSS model was used to simulate the influence of growth velocity on dendrite morphology. The growth velocity was changed by imposing

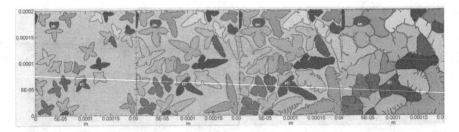

Fig. 18.28 Simulation of equiaxed solidification of Al–4 wt.% Cu alloy showing grain boundary formation. From left to right: after 0.04, 0.08, 0.16, and 0.2 s. (Beltran-Sanchez and Stefanescu 2004; with permission from Springer Science and Business Media)

Fig. 18.29 Schematic correlation between solidification velocity and dendrite tip radius

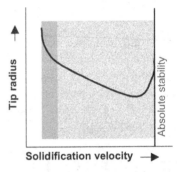

different cooling rates (Fig. 18.30). As the cooling rate decreases, the tip radius of the dendrite becomes larger. When the radius of curvature of the dendrite tip is of the same order of magnitude as that of the whole grain, the dendrite becomes a cell in directional solidification or globular in equiaxed solidification.

Zhu and Stefanescu (2004) developed a computationally efficient quantitative virtual front tracking (FT) model for the two-dimensional simulation of dendritic growth in the low Péclet number regime. The kinetics of dendritic growth is driven by the difference between the local equilibrium composition, calculated from the local temperature and curvature, and the local actual liquid composition, obtained by solving the solutal transport equation:

$$\partial C_i/\partial t = \nabla \cdot (D_i \nabla C_i) + C_i(1 - k)\partial f_S/\partial t \qquad (18.65)$$

where the subscript i indicates solid or liquid. The source term (second term on the right-hand side, RHS) is included only at the S/L interface and denotes the amount of solute rejected at the interface. For each interface cell with nonzero solid fraction, the compositions of liquid and solid are calculated and stored simultaneously. The solid composition is calculated with:

$$C_S = \sum_{n=1}^{N} k \Delta f_S(n) C_L(n) \Big/ \sum_{n=1}^{N} \Delta f_S(n),$$

Fig. 18.30 Composition maps of the transition from cellular to dendritic microstructure as function of the cooling rate (different heat transfer coefficients) for equiaxed grains of Fe–0.6 wt.% C alloy (Beltran-Sanchez and Stefanescu 2003; with permission from Springer Science and Business Media)

where N indicates the number of iterations (time step intervals) after the cell became an interface cell. The interface equilibrium temperature and composition were calculated with Eqs. 18.59 and 18.60, respectively. The calculated local interface equilibrium composition C_L^* from Eq. 18.60 is compared with the local actual liquid composition C_L, calculated with the solute transport Eq. 18.65. If $C_L^* - C_o > 0$, the solid fraction of the cell is increased. According to the interface equilibrium condition, during one time step interval, the increase in solid fraction of an interface cell can be evaluated through:

$$\Delta f_S = (C_L^* - C_L)/(C_L^*(1 - k)).$$

If at time t_N, the sum of the solid fraction in an interface cell equals one, i.e., $\sum_{n-1}^{n} \Delta f_S(n) = 1$, this interface cell has fully solidified. Then, the interface cell changes its state to solid. From the calculated increase in solid fraction at each time step, the normal growth velocity of the interface can be obtained with $V_n = \Delta f_S \, \Delta x / \Delta t$, where Δx is the cell size and Δt is the time step.

The model adopts previously proposed solutions for the evaluation of local curvature and interface-capturing rules with a virtual interface-tracking scheme, which make the model virtually mesh independent. The dynamics of dendrite growth from the initial unstable stage to the steady-state stage is accurately described. Side branching develops without the need to introduce local noise.

The Zhu–Stefanescu model was extended to 3D calculation (Pan and Zhu 2010). In the Gibbs–Thomson equation for anisotropic crystals, the capillarity undercooling was calculated based on the weighted mean curvature that is incorporated with the effect of anisotropic surface energy as suggested by Taylor (1984). Some model generated microstructures are presented in Fig. 18.31.

FT/CA models have found application in the simulation of dendritic and non-dendritic microstructure evolution in semisolid-processing solidification, the influence of fluid flow on dendrite morphology, and the dendritic-eutectic solidification (see review by Zhu et al. 2007).

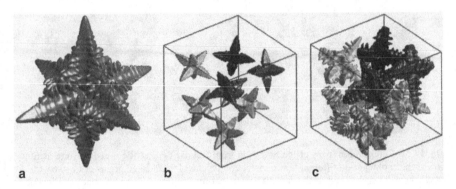

Fig. 18.31 Equiaxed dendrite generated through a CA code. **a** Simulated single-dendrite morphologies of an Fe–0.6 wt.% C alloy. **b** Simulated evolution of multi-equiaxed dendrites for an Al–3 % Cu alloy at $f_S = 0.05$. **c** Simulated evolution of multi-equiaxed dendrites for an Al–3 % Cu alloy at $f_S = 0.15$. (Pan and Zhu 2010; with permission from Elsevier)

With recent advances in computational power, multiscale finite differences/CA microstructure simulations were used to model the CET (Spittle and Brown 1995; Nastac 1999; Dong and Lee 2005). In spite of their success in producing realistic-looking dendritic growth patterns, some questions remain regarding the accuracy of CA models. For example, while it is generally accepted that dendritic growth of crystalline materials is greatly affected by the surface energy anisotropy, to the best of our knowledge, no CA model incorporates the effect of surface energy anisotropy.

For a recent extensive review of the advances in CA modeling the reader is referred to the review paper by Reuther and Rettenmayr (2014).

18.4.3 Lattice Boltzmann Models

The preceding discussion was based on the framework developed by Navier and Stokes whose differential formulation of the mechanics of incompressible flow accounts for the conservation of mass, momentum and energy, and the local conservation of these quantities. Applying these equation to problems related to the solidification of castings requires solving coupled sets of nonlinear partial differential equations through finite difference or finite element methods. However, the Navier–Stokes differential formulations do not apply theoretically and do not converge numerically under conditions which are characterized by large Knudsen numbers (the ratio between the mean free molecule path and the characteristic system length scale), that is for situations where the mean free path of the fluid molecules is similar to the geometrical system constraints. Typical engineering limitations include the simulation of multicomponent flows in the area of polymer and metal processing, liquid crystal processing, processing of metallic foams, and solidification in non-quiescent environments, to list just a few. In the range between the

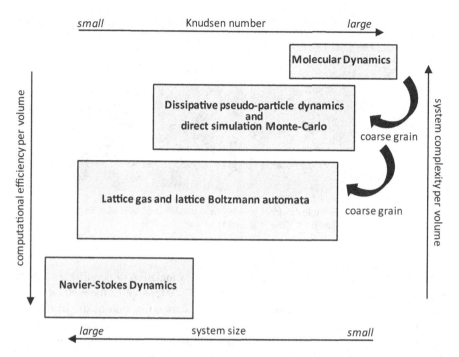

Fig. 18.32 Approaches to computational fluid dynamics and their preferred range of applicability. (After Raabe 2004)

atomic scale, appropriately simulated by molecular dynamics (MD) models, and the macroscale, amenable to continuum deterministic simulations, a new method is gaining traction—lattice Boltzmann automata.

As summarized by Raabe (2004), the lattice Boltzmann method (LBM) belongs to a broader group of pseudo-particle methods that can be grouped into lattice-based CA approaches (lattice gas method, LBM) and off-lattice approaches (dissipative particle dynamics method, direct simulation MC method). Their range of length scale interest is summarized in Fig. 18.32. Dissipative particle dynamics and direct simulation MC are off-lattice pseudo-particle methods in conjunction with Newtonian dynamics. The lattice gas method and the LBM treat flows in terms of coarse-grained fictive particles represented by a velocity distribution function. The particles reside on a mesh. System dynamics develops by the repeated application of local rules for the motion, collision, and redistribution of these particles. It is an ideal approach for mesoscale and multiscale problems such as flows under complicated geometrical boundary conditions, phase transformation in flows, and complex S/L interfaces, all typical for casting solidification. As a detailed coverage of this method is out of the scope of this text, the interested reader is referred to the paper by Raabe (2004).

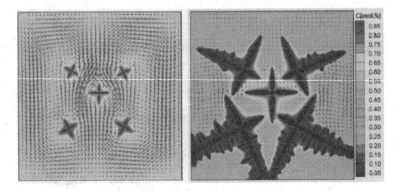

Fig. 18.33 Evolution of multiple dendrites growing in an undercooled melt with Rayleigh numbers of 5×10^3 at $f_S = 0.02$ and 0.18. (Zhu et al. 2014)

The LBM has been applied successfully to the simulation of dendrite growth in the presence of fluid convection. Miller and Succi (2002) applied the method to the effect of transition from laminar to turbulent flow as a function of the increasing roughness of a surface and of the viscosity of the fluid on the dendrite morphology. More recently, Zhu et al. (2014) developed a 2D LBM-CA model to investigate dendritic growth in binary alloys in the presence of natural convection. To numerically solve natural convection, thermal and solute transport simultaneously, three sets of distribution functions were employed. Based on the LBM-calculated local temperature and concentration at the S/L interface, the kinetics of dendritic growth was calculated based on a local solute equilibrium approach. They demonstrated that natural buoyancy flow, induced by thermal and solutal gradients, transports the heat and solute upward. Dendritic growth is thus accelerated in the downward direction, whereas it is inhibited in the upward direction, yielding asymmetrical dendrite patterns (Fig. 18.33). Increasing the Rayleigh number and undercooling will enhance and reduce, respectively, the influence of natural flow on the dendritic growth.

18.5 Molecular Dynamics Models

MD methods integrate Newton's equations of motion for a set of molecules on the basis of an intermolecular potential. The first applications of MD and MC methods to atomistic systems using classical potentials are now over 50 years old, and detailed descriptions of these techniques can be found in a number of recent textbooks (e.g., Rapaport 2004). Academic packages, such as the Vienna Ab Initio Simulation Package (VASP; Kresse and Furthmuller 1996), are available to researchers interested in MD. The term *ab initio* implies that the method is based on first principles, as opposed to empirical methods where experimental parameterization is required.

Because of the large computational load, the method cannot yet be used for bulk materials such as castings, but has found applications in the investigation of S/L interfacial energy, crystalline anisotropy, interface kinetic properties (e.g., Eq. 3.29), nucleation, and rapid solidification processes.

Xiong et al. (1992) studied the local atomic structure evolution in Al_2Au alloys during solidification from 2000 to 400 K by ab initio MD simulations. It was found that the icosahedral-like atomic clusters (with CN > 11), which are believed to be the most stable structure in many metallic liquids and metallic glasses, are negligible in the Al_2Au stable liquid and supercooled liquid states. The most abundant clusters are those having coordination numbers of 8, 9, and 10. These clusters are similar to the local atomic structures in the CaF_2-type Al_2Au crystal, revealing the existence of structure heredity between liquid and crystalline phase in the Al_2Au alloy. Ma et al. (2012) carried out a MD investigation of the rapid solidification of Zn. They found that amorphous structure was produced when the cooling rate exceeded 2.5×10^{12} K/s.

MD simulations were employed to investigate the microscopic or atomic scale mechanism of melting/solidification in relation to the macroscopic material thermo-mechanics using a simple model with 2D array of atoms and uniform temperature distribution (Inoue and Uehara 1999). The model was used to evaluate the melting point and some other material properties, and the mode of volume dilatation and generation of latent heat due to melting and hysteresis of temperature.

With the rapid progress in MD simulation, problems directly relevant to metal casting are coming within reach. In a recent study, the cavitation pressure of liquid Al was computed (Hoyt and Potter 2012). They started with an expression for the cavitation pressure derived from the application of classic nucleation theory to a bulk liquid that experiences an abrupt change in pressure:

$$P_c = -\left[\frac{16\pi\gamma^3}{3k_BT} \frac{1}{ln\left(\Gamma_o v\, t/ln2\right)}\right]^2 \text{ with } \Gamma_o \approx \left[\frac{4}{3}\pi\, r_{cr}^3\right]^{-1} \frac{k_BT}{h} \qquad (18.66)$$

where γ is the liquid surface tension, v is the volume of liquid, t is time, and h is Planck's constant. The term r_{cr} refers to the critical nucleus radius and is defined as $-2\sigma/P_c$. A number of 25,000–108,000 atoms was used in the MD simulation. An extrapolation, based on classical nucleation theory, of the cavitation pressure from MD length and time scales to those consistent with casting resulted in a cavitation pressure of $P_c \approx -670$ MPa. It was also found that the cavitation pressure is not affected by either the presence of S/L interfaces in the system (Fig. 18.34) or the addition of trace amounts of Mg (Fig. 18.35).

The cavitation pressure predicted by this MD model is much lower than the classic theory prediction for the cavitation strength of most liquid metals which lies in the range of 10 MPa–1 GPa (see Application 18.1). At the other extreme, cavitation pressures in the range of −2 to −500 kPa are required by hot-tearing continuum models to match experiments (Rappaz et al. 1999; Vernede et al. 1990). Campbell (2013) argued that the discrepancy between the high MD values and the low continuum models is caused by the existence of a dense population of macroscopic unbonded interfaces known as bifilms.

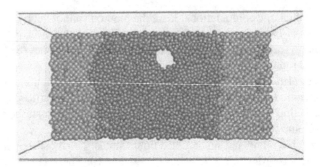

Fig. 18.34 MD simulation of crystallization of aluminum. Cavity forms in the bulk liquid and not at the solid–liquid boundaries. (Hoyt and Potter 2012; with permission of Springer Science and Business Media)

Fig. 18.35 A snapshot from a 108,000-atom-Al system containing one Mg atom. The cavity away from the single Mg atom indicates that the impurity does not act as a catalytic site for void nucleation. (Hoyt and Potter 2012; with permission of Springer Science and Business Media)

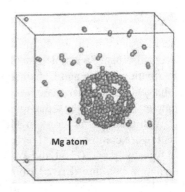

18.6 Applications

Application 18.1 Calculate the cavitation pressure (pressure required to form a void) in aluminum. Compare with the value calculated through the MD model of Hoyt and Potter (2012).

Answer The basic equation that describes the mechanical stability of a bubble is $\Delta P_c = 2\gamma/r_{cr}$. Following Campbell (2013), assuming $\gamma = 1000$ mJ/m^2, and assuming a bubble radius of one atom diameter 0.286 nm (to give an embryonic bubble consisting of approximately eight vacancies), the fracture pressure is calculated to be 7 GPa. In the Hoyt/Potter model the surface energy was taken to be 280 mJ/m^2. Using 1000 mJ/m^2, a cavitation pressure of 2.39 GPa is obtained.

References

Aizawa T (1978) Imono 50:33

Anderson MP, Srolovitz DJ, Grest GS, Sahni PS (1984) Acta Metall. 32:783

Atwood RC, Sridhar S, Zhang W, Lee PD (2000) Acta Mater. 48:405

Atwood RC, Lee PD (2003) Acta Materialia 51:5447

Avrami M (1939) J. Chem. Phys. 7:1103

Badillo A, Beckermann C (2006) Acta Mater. 54:2015–16

Basdogan MF, Kondic V, Bennett GHJ (1982) Trans. AFS 90:263

Beltran-Sanchez L, Stefanescu DM (2003) Metall. Mater. Trans. 34A:367

Beltran-Sanchez L, Stefanescu DM (2004) Metall Mater Trans 35:2471

Brown SGR, Spittle JA (1992) Scr. Metall. Mater. 27(11):1599–1603

Campbell J (2003) in: Stefanescu DM, Warren J, Jolly M, Krane M (eds) Modeling of Casting, Welding and Adv. Solidif. Processes X. TMS, Warrendale Pa p 209

Campbell J (2013) Metall. Mater. Trans. 44A:1158

Carlson KD, Lin Z, Hardin RA, Beckermann C, Mazurkevich G, Schneider MC (2003), in: Stefanescu DM, Warren J, Jolly M, Krane M (eds) Modeling of Casting, Welding and Adv. Solidif. Processes X. TMS, Warrendale Pa p 295

Catalina AV, Leon-Torres JF, Stefanescu DM, Johnson ML (2007) in: Proceedings of the 5th Decennial International Conference on Solidification Processing. Sheffield, UK pp 699–703

Chang S, Shangguan D, Stefanescu DM (1992) Metall Trans. 23A:1333

Das A, Mittemeijer EJ (2000) Metall. and Mater. Trans. 31A:2049

Dilthey U, Pavlik V, Reichel T (1997) in: Cerjak H (ed) Mathematical Modelling of Weld Phenomena. The Inst. of Materials, London p 85

Dilthey U, Pavlik V (1998) in: Thomas BG, Beckermann C (eds) Modeling of Casting, Welding and Advanced Solidification Processes-VIII. TMS, Warrendale Pa. p 589

Dong HB, Lee PD (2005) Acta Mater. 53:659

Dustin I, Kurz W (1986) Zeitschrift Metallkde 77:265

Fras E (1975) PhD Dissertation. Academy of Mining and Metallurgy, Cracow, Poland

Fredriksson H. and L. Svensson, 1985, in The Physical Metallurgy of Cast Iron, H. Fredriksson and M. Hillert eds., Elsevier pp.273–284

Fredriksson H, Svensson L (1988) in: Stefanescu DM, Abbaschian GJ, Bayuzick RJ (eds) Solidification Processing of Eutectic Alloys. TMS, Warrendale Pa. pp 153–162

Gandin CA (2000) Acta mater. 48:2483±2501

Gandin CA, Jalanti T, Rappaz M (1998) in: Thomas BG, Beckermann C (eds) Modeling of Casting, Welding and Advanced Solidification Processes-VIII. TMS, Warrendale Pa. p 363

Ganesan S, Poirier DR (1990) Met. Trans. 21B:173

Goettsch DD, Dantzig JA (1994) Metall. and Mat. Trans. 25A:1063

Gránásy L, Pusztai T, Warren JA (2004) J. Phys.: Condens Matter 16:R1205

Gránásy L, Ratkai L, Szallas A, Korbuly B, Toth G, Kornyei L, Pusztai T (2014) Metall. Mater. Trans. 45A:1694

Greenwood M, Haataja M, Provatas N (2004) Phys Rev Lett 93:246101

Greer AL, Brunn AM, Tronche A, Evans PV, Bristow DJ (2000) Acta Mater. 48:2823–35

Hesselbarth HW, Göbel IR (1991) Acta Metall. 39:2135

Hillert M (1955) Acta Metall. 3:653

Hoyt JJ, Potter AA (2012) Metall. Mater. Trans. 43A:3972

Hunt JD (1984) Mat. Sci. and Eng. 65:75

Inoue T, Uehara T (1999) in: IUTAM Symp. on Variations of Domain and Free-Boundary Problems in Solid Mechanics, Solid Mechanics and its Applications, 66:103–111

Jacot A, Rappaz M (2002) Acta Mater. 50:1909

Johnson WA, Mehl RF (1939) Trans. AIME 135:416

Juric D, Tryggvason G (1996) J Comput Phys. 123:127

Karma A (2001) Phys Rev Lett 87:115701

Karma A, Rappel WJ (1996) Phys Rev Lett 77:4050

Karma A, Rappel WJ (1998) Phys Rev E 57:4323

Kanetkar CS, Stefanescu DM (1988a) in: Giamei A, Abbaschian GJ (eds) Modeling of Casting and Welding Processes IV. TMS Warrendale, Pa. p 697

Kanetkar CS, Chen IG, Stefanescu DM, El-Kaddah N (1988b) Trans. Iron and Steel Inst. of Japan 28:860

Kirkaldy JS (1995) in: Chen LQ et al. (eds) Mathematics of Microstructure Evolution. TMS, Warrendale Pa. p 173

Kolmogorov AN (1937) Bulletin de L'Académie des Sciences de L'URSS (3):355–359

Kothe DB, Mjolsness RC, Torrey MD (1991) RIPPLE: A computer Program for Incompressible Flows with Free Surface, Los Alamos National Lab., LA-10612-MS, Los Alamos, NM

Könözsy L, Ishmurzin A, Mayer F, Grasser M, Wu M, Ludwig A (2009) Int. J Cast Metals Res. 22(1–4):175

Kresse G, Furthmuller J (1996) Phys. Rev. B, 54B (16): 11169–11186

Kubo K, Pehlke RD (1985) Met. Trans. 16B:359

Kurz W, Giovanola B, Trivedi R (1986) Acta Metallurgica 34(5):823–830

Lacaze J, Castro M, Lesoult G (1989) in: Exner HE, Schumacher V (eds) Advanced Materials and Processes, vol. 1. Informationsgesellschaft Verlag p 147

Lan CW, Chang YC, Shih CJ (2003) Acta Mater 51:1857

Lee PD, Hunt JD (1997) Acta Mater. 45:4155

Lee PD, Chirazi A, Atwood RC, Wang W (2004) Mat. Sci. Eng. A365:57

Lee SY, Lee SM, Hong CP (1998) in: Thomas BG, Beckermann C (eds) Modeling of Casting, Welding and Advanced Solidification Processes-VIII. TMS, Warrendale Pa pp 383–390

Lesoult G (1991) in: Rappaz M, Ozgu MR, Mahin KW (eds) Modeling of Casting, Welding and Advanced Solidification Processes-VI. TMS, Warrendale Pa p 363

Loginova I, Amberg G, Agren J (2001) Acta Mater 49:573

Ludwig A, Wu M (2002) Metall. and Mater. Trans. 33A:3673–3683

Ma RN, Qiu HC, Wu JJ (2012) Adv. Mater. Res. 383–390:7385–7389

Mampey F (1988) in: 55th International Foundry Congress, CIATF, Moscow paper 2I

Mampey F (1991) in: Rappaz M, Ozgu MR, Mahin KW (eds) Modeling of Casting, Welding and Advanced Solidification Processes-VI. TMS, Warrendale Pa. p 403

Maxwell I, Hellawell A (1975) Acta Metall. 23:229

Miller W, Succi S (2002) J. Stat. Phys. 107:173

Nastac L (1999) Acta mater. 47:4253

Nastac L (2004) Modeling and Simulation of Microstructure Evolution in Solidifying Alloys. Kluwer Academic Publishers, Boston

Nastac L, Stefanescu DM (1992) in: Beckermann C et al. (eds) Micro/Macro Scale Phenomena in Solidification. HTD-Vol. 218, AMD-Vol. 139, ASME, New York p 27

Nastac L, Stefanescu DM (1996) Metall. Trans. 27A:4061 and 4075

Nastac L, Stefanescu DM (1997) Modelling and Simulation in Mat. Sci. and Eng., Inst. of Physics Publishing 5(4):391

Ni J, Beckermann C (1991) Metall. Trans. 22B:349

Ni J, Incropera FP (1995) Int. J. Heat Mass Transf. 38:1271–1284 and 1285–1296

Nichols B, Hirt C, Hotchkiss R (1980) Sola-vof: A Solution Algorithm for Transient Fluid Flow with Multiple Free Boundaries, Tech. Rep., Los Alamos Scientific Laboratory

Ohnaka I, Iwane J, Sako Y, Yasuda H, Zhao H (2002) in: Proceedings of the 65th World Foundry Congress, Gyeogju, Korea p 639

Ohnaka I (2003) in: Stefanescu DM, Warren J, Jolly M, Krane M (eds) Modeling of Casting, Welding and Adv. Solidif. Processes X. TMS, Warrendale Pa p 403

Oldfield W (1966) ASM Trans. 59:945

Pan S, Zhu MF (2010) Acta Materialia 58(1):340–352

Pang H, Stefanescu DM (1996) in: Ohnaka I, Stefanescu DM (eds) Solidification Science and Processing. TMS, Warrendale, Pa p 149

Péquet Ch, Gremaud M, Rappaz M (2002) Metall. Mater. Trans. 33A:2095

Poirier R, Yeum K, Maples AL (1987) Met. Trans. 18A:1979

Raabe D (2004) Modelling Simul. Mater. Sci. Eng. 12:R13–R46

Ramirez JC, Beckermann C, Karma A, Diepers H-J (2004) Phys Rev A 69:051607

Ramirez JC, Beckermann C (2005) Acta Mater 53:1721

Rapaport DC (2004) The art of molecular dynamics simulation, Cambridge, Cambridge University Press

Rappaz M, Thévoz P (1987) Acta. Metall. 35:1487 and 2929

Rappaz M, Gandin CA (1993) Acta Metall. Mater. 41:345

Rappaz M, Gandin CA, Charbon C (1995) in: Solidification and Properties of Cast Alloys. Proc. Technical Forum, 61st World Foundry Congress, Beijing, Giesserei-Verlag p 49

Rappaz M, Gandin CA, Desbiolles JL, Thevoz P (1996) Metall. and Mat. Trans. 27A:695

Rappaz M, Drezet J-M, Gremaud M (1999) Metall. Mater. Trans. 30A:449

Reuther K, Rettenmayr M (2014) Comput. Mater. Sci. 95:213–220

Sabau AS, Viswanathan S (2002) Metall. Mater. Trans. 33B:243

Sahni PS, Grest GS, Anderson MP, Srolovitz DJ (1983) Phys. Rev. Lett. 50:263

Sako Y, Ohnaka I, Zhu JD, Yasuda H (2001) in: Proc. of the 7th Asian Foundry Congress, The Chinese Foundrymen's Assoc., Taipei p 363

Sasikumar R, Sreenivisan R (1994) Acta Metall. 42(7):2381

Spittle JA, Brown SGR (1989) Acta Metall. 37:1803

Spittle JA, Brown SGR (1995) J Mater Sci. 30:3989

Stefanescu DM (1973) PhD Dissertation, Inst. Politehnic Bucuresti, Romania

Stefanescu DM (1995) ISIJ International 35:637

Stefanescu DM (2001) in: Yu KO (ed) Modeling of Casting and Solidification Processing. Marcel Dekker Inc., New York p 123

Stefanescu DM, Trufinescu S (1974) Z. Metallkde. 65:610

Stefanescu DM, Kanetkar C (1985) in: Srolovitz DJ (ed) Computer Simulation of Microstructural Evolution. The Metallurgical Soc., Warrendale, Pa pp 171–188

Stefanescu DM, Kanetkar C (1986) in: Fredriksson H (ed) State of the Art of Computer Simulation of Casting and Solidification Processes. Les Edition de Physique, Les Ulis, France p 255

Stefanescu DM, Upadhya G, Bandyopadhyay D (1990) Metall. Trans. 21A:997

Stefanescu DM, Pang H (1998) Canadian Metallurgical Quarterly 37(3–4):229–239

Su KC, Ohnaka I, Yanauchi I, Fukusako T (1984) in: Fredriksson H, Hillert M (eds) The Physical Metallurgy of Cast Iron. North Holland, New York p 181

Sung PK, Poirer DR, Felicelli SD (2002) Mod. Sim. Mat. Sci. Eng. 10, 551

Taylor JE (1992) Acta Metall Mater 40(7):1475

Thévoz P, Desbioles JL, Rappaz M (1989) Metall. Trans. 20A:311

Tiaden J (1999) J. Crystal Growth, 198/199:1275–1280

Tian H (1992) Ph. D. Dissertation, University of Alabama, Tuscaloosa

Tian H, Stefanescu DM (1993) in: Piwonka TS, Voller V, Katgerman L (eds) Modeling of Casting, Welding and Advanced Solidification Processes-VI. TMS, Warrendale Pa p 639

Tynelius K, Major JF, Apelian D (1993) Trans. AFS 101:401

Udaykumar HS, Mittal R, Shyy Wei (1999) J Comput Phys. 153:535

Upadhya G, Banerjee DK, Stefanescu DM, Hill JL (1990) Trans. AFS 98:699

Vernede S, Dantzig JA, Rappaz M (2009) Acta Mater. 57:1554–69

Viswanathan S, Duncan AJ, Sabau AS, Han Q (1998) in: Thomas BG, Beckermann C (eds) Modeling of Casting, Welding and Advanced Solidification Processes-VIII. TMS, Warrendale Pa p 49

Wang CY, Beckermann C (1993a) Mater. Sci. and Eng. A171:2787

Wang CY, Beckermann C (1993b) Metall. Trans. 24A:2787

Wang CY, Beckermann C (1994) Metall. Trans. 25A:1081

Wang CY, Beckermann C (1996) Metall. Mater. Trans. 27A:2754

Warren JA, Boettinger WJ (1995) Acta Metall Mater 43:689

Warren JA, Kobayashi R, Carter WC (2000) in: Sahm PR, Hansen PN, Conley JG (eds) Modeling of Casting, Welding and Advanced Solidification Processes IX. Shaker Verlag, Aachen, Germany, p CII–CIX

Winegard WC, Chalmers B (1954) Trans ASM 46:1214

Wu M, Ludwig A (2006) Metall. Mater. Trans. 37A:1613–1631

Xiao R, Alexander JID, Rosenberg F (1992) Phys. Rev. A, 45:R571–R574

Xiong LH et al. (2014) Acta Materialia 68:1–8

Zhao P, Venere M, Heinrich JC, Poirier DR (2003) J Comput Phys 188:434

Zhu JD, Ohnaka I (1991) in: Rappaz M, Ozgu MR, Mahin K (eds) Modeling of Casting, Welding
   and Advanced Solidification Processes V. TMS, Warrendale, Pa p 435

Zhu MF, Hong CP (2001) ISIJ Int. 41:436

Zhu MF, Kim JM, Hong CP (2001) ISIJ Int. 41:992

Zhu MF, Hong CP (2004) Metall Mater Trans 35A:1555

Zhu MF, Stefanescu DM (2007) Acta Materialia 55:1741–1755

Zhu MF, Hong CP, Stefanescu DM, Chang YA (2007) Metall. Mater. Trans. 38B:517

Zhu MF, Sun DK, Pan SY, Zhang QY, Raabe D (2014) Modelling Simul. Mater. Sci. Eng. 22
   034006 (18pp)

Zhu P, Smith RW (1992) Acta metall. mater. 40:683 and 3369

Zou J, Doherty R (1993) in: Piwonka TS, Voller V, Katgerman L (eds) Modeling of Casting,
   Welding and Advanced Solidification Processes VI., TMS, Warrendale, Pa p 193

Chapter 19
Solidification of Some Casting Alloys of Commercial Significance

This chapter will cover the solidification of some alloys of commercial interest as well as selected numerical models developed to describe and predict the microstructure evolution and defect formation in these alloys. The research community and the commercial entities have produced and continue to produce quite an abundance of numerical models for the prediction of defect occurrence and microstructure development of alloys. For metal casting applications, the most commonly used commercial software are ProCAST and Magmasoft. General commercial codes, such as ABAQUS or Fluent, have also been used in conjunction with specialized routines. This chapter will attempt to illustrate the progress, success, and limitation of both research and commercial models in the prediction of microstructure and related features.

19.1 Steel

Steel is a Fe–C-based alloy that solidifies without eutectic. Plain carbon (unalloyed) steel is an alloy whose equilibrium phase diagram (the Fe–C diagram) exhibits partial solid solubility with a peritectic reaction (see Fig. 10.20). Consequently, the solidification microstructure of plain carbon and low-alloy steels is made of equiaxed or columnar α or γ dendrites.

19.1.1 Macrostructure

As carbon steel solidifies with a peritectic transformation, the primary grain structure of steel is hard to outline through optical metallography because of the post-solidification recrystallization. Sometimes the macrostructure can be observed in shrinkage cavities, which behave similar to interrupted solidification experiments. The typical dendritic structure of steel revealed by such a technique is presented in

© Springer International Publishing Switzerland 2015
D. M. Stefanescu, *Science and Engineering of Casting Solidification,*
DOI 10.1007/978-3-319-15693-4_19

Fig. 19.1 Scanning electron micrograph of the center of an as-cast low-carbon steel ingot showing dendrite primary and secondary arms. Unetched. 10 × (Bramfitt 1985). (Reprinted with permission of ASM International. All rights reserved. www.asminternational.org)

Fig. 19.2 Austenite grains (transformed to pearlite) outlined by ferrite network in the as-cast structure. 40 × (Briggs 1946)

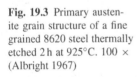

Fig. 19.3 Primary austenite grain structure of a fine grained 8620 steel thermally etched 2 h at 925°C. 100 × (Albright 1967)

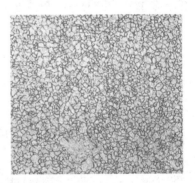

Fig. 19.1. Many times the as-cast grain structure of steel can be observed using classic metallographic etching because ferrite forms at the austenite grain boundaries during solid state transformation (Fig. 19.2).

In most cases more sophisticated metallographic techniques must be employed. Most of these techniques consist in outlining the microsegregation patterns (for details see Stefanescu and Ruxanda 2004). Thermal etching can be also used to outline the solidification of grain size of steel, as shown in Fig. 19.3.

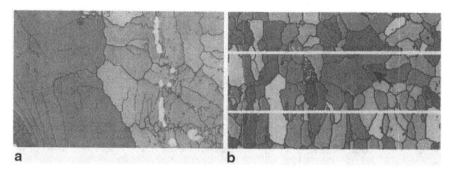

Fig. 19.4 Electron backscatter diffraction (EBSD) orientation micrographs on AISI 304 stainless steel strip. A ridged substrate (**b**) produced a higher density of randomly oriented grains at the chill surface, as compared with a smooth substrate (**a**) (Hunter and Ferry 2002). **a** As solidified surface cast on smooth substrate. **b** As solidified surface cast on ridged substrate. (With permission of Springer Science and Business Media)

Specimens of solidified steel less than 0.05-mm (0.002-in.) thick can be X-rayed. Electron backscatter diffraction (EBSD) is capable of evaluating the effect of casting conditions on the grain size of strip-cast steel, as shown in Fig. 19.4.

The macrostructure of a solidified steel ingot, casting, or strand-cast slab or bloom, consists of columnar or equiaxed grains, or a mixture of both, depending on solidification conditions. In most applications, either a columnar or an equiaxed structure is desired for the casting. Typically, castings in which a sudden columnar-to-equiaxed transition occurs are rejected because of the mechanical weakness of the transition zone.

19.1.2 Microstructure

As for all alloys, the solidification and as-cast microstructure of steel is a function of chemical composition and cooling rate. For plain carbon and low-alloy steels, the solidification structure consists of austenite grains. However, during cooling to room temperature after solidification, a peritectic and then a solid-state transformation occur that almost entirely conceal the original as-cast structure. The sequence of a peritectic transformation in a high-alloy steel is presented in Fig. 19.5.

For carbon steel, the austenite transforms into ferrite and pearlite if the composition is hypoeutectoid, and into cementite and pearlite for hypereutectoid steel. Depending on cooling rate and composition, low-alloy steels can have various microstructures consisting of different forms and combinations of pearlite, bainite, martensite, cementite, and ferrite. High-alloy steels may have an austenitic structure even after cooling to room temperature, as shown in Fig. 19.6 for austenitic 18Ni-26Cr steel.

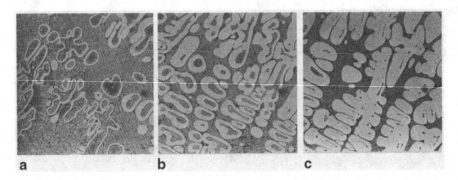

Fig. 19.5 Three stages of peritectic reaction in a DS high-speed steel (Fredriksson 1988). **a** *Dark gray* is austenite; *white* is ferrite. **b** Peritectic transformation of **a**. **c** Further transformation of **b**; *dark gray* in the *middle* of the ferrite is newly formed liquid. (With permission of ASM International. All rights reserved. www.asminternational.org)

Fig. 19.6 The as-cast microstructure of HK-40 steel: austenite grains surrounded by eutectic carbides; 250 × (Dillinger et al. quoted in Wieser 1980)

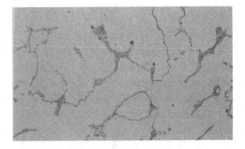

Table 19.1 Equilibrium partition coefficients for various elements in steel. (Fisher et al. 1976)

Element	P	Nb	Cr	Mn	Ni
k	0.14	0.23	0.33	0.71	0.83

During solidification significant segregation of the various elements occurs. As already discussed, the extent of the segregation is a function of the partition coefficient and the amount of element in steel. Some typical values for partition coefficients in steel are given in Table 19.1.

As segregation impacts negatively on properties, it is a cause of concern. One way to decrease segregation is to refine the microstructure, i.e., the dendrite arm spacing.

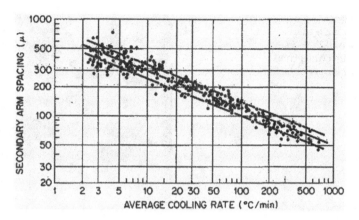

Fig. 19.7 Influence of cooling rate on the SDAS of commercial steels containing 0.1 to 0.9 % C (Suzuki et al. 1968)

19.1.3 Dendrite Arm Spacing

The primary and secondary arms spacing are directly related to the mechanical properties of cast steel. As discussed in Sect. 9.4, the primary arm spacing (PDAS) is mostly a function of cooling rate, while the secondary arm spacing (SDAS) depends on the cooling rate and on the local solidification time. For a 0.59 % C steel, Jacobi and Scherdtfeger (1976) suggested the following equations (spacing in μm):

$$\lambda_1 = 283 \, (\dot{T})^{-0.49} \tag{19.1}$$

$$\lambda_2 = 15.8 \, (t_f)^{-0.44}. \tag{19.2}$$

Slightly different number for SDAS were obtained by Suzuki et al. (1968) based on regression analysis of a large number of data for steel with 0.1–0.9 % C (Fig. 19.7):

$$\lambda_2 = 148 \, (\dot{T})^{-0.38}. \tag{19.3}$$

However, as shown by El-Bealy and Thomas (1966), while the cooling rate determines both the primary and secondary spacing, the carbon content is also an essential parameter. The following empirical equations for PDAS were developed from experimental data of several investigators:

$$\lambda_1 = 278.7 \, (\dot{T})^{-0.206} \, (C_o)^n \tag{19.4}$$

where C_o is the carbon content of the steel and n is an exponent that depends on the range of carbon content. For $0 \le C_o \le 0.15$, $n = -0.316 + 2.032 \, C_o$. For $0.15 \le C_o \le 1$, $n = -0.019 + 0.4916 \, C_o$. For steel with less than 0.53 % C an equation identical with Eq. 19.3 was suggested.

Fig. 19.8 Manganese sulfide
inclusions at grain boundaries in as-cast 0.25 % C steel
(Wieser 1980). (With permission of Steel Founders' Soc.
of America)

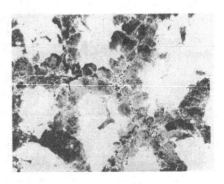

For the SDAS of steel with $0.53 \leq C_o \leq 1.5$, the following equation was proposed:

$$\lambda_2 = (21.53 - 9.4C_o)(t_f)^{(0.4+0.08C_o)} \tag{19.5}$$

where the solidification time is calculated with $t_f = (T_L - T_S)/\dot{T}$. All spacings are in μm and cooling rates are in K/s. The influence of the various carbon ranges are related to the delta and gamma primary solidification and the peritectic reaction. Indeed, at < 0.09 % C, the primary phase is δ, in the range of 0.09–0.15 % C the peritectic reaction produces $\delta + \gamma$, while in the range 0.15–0.53 % C we have $L + \delta \rightarrow L + \gamma$, and above 0.53 % C the primary phase is γ.

19.1.4 Nonmetallic Inclusions

Most nonmetallic inclusions are not affected by homogenization and therefore remain unchanged during phase transformations, thus retaining their original position and shape in the solidified casting. A notable exception is the manganese sulfides. Some nonmetallic inclusions form in the liquid before solidification, others, during solidification. Aluminates and silicates generally form before solidification, but sulfides form during solidification. Manganese sulfide inclusions frequently form in the interdendritic regions and at primary grain boundaries, where the last of the liquid solidifies (Fig. 19.8). These inclusions represent type II sulfides, which are commonly found in aluminum-killed steel ingots and castings (Sims 1959). By contrast, Fig. 19.9 illustrates how the inclusions formed from a higher melting point compound, in this instance, a silicate, are distributed as clusters in isolated regions in the cast product.

Lanthanides are added to liquid steel before solidification to change the morphology and composition of the sulfide inclusions. Recent technology involves injecting calcium into liquid steel to reduce sulfur levels and change inclusion morphology and composition. Figure 19.10 shows a typical calcium-containing inclusion, where calcium sulfide forms a shell around a core of calcium aluminate.

Fig. 19.9 Silicate inclusions (*dark gray* particles) randomly distributed in the ferrite matrix (*light*) of strand-cast 1017 steel that also contains pearlite (medium *gray*). Etchant: picral. 500 × (Kilpatrick quoted in Bramfitt 1985). (With permission of ASM International. All rights reserved. www.asminternational.org)

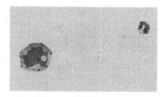

Fig. 19.10 Low-carbon steel showing inclusions of calcium sulfide outer rim (*light gray*) and calcium aluminate core (*dark gray*). As-polished. 500 × (Kilpatrick quoted in Bramfitt 1985). (With permission of ASM International. All rights reserved. www.asminternational.org)

19.1.5 Simulation of the Solidification of Steel

Macrosegregation, a common occurrence in shaped steel castings, as well as in steel ingots, has undesired effects on the properties of the cast material. In Chap. 5 we have discussed the analytical models for solute redistribution that leads to macrosegregation and in Sect. 17.4 the principles of numerical modeling of macrosegregation. Numerous macrosegregation models were developed for shaped and ingot casting of steel.

As early as 1986, Ohnaka presented a numerical macrosegregation model for steel ingots limited to a binary Fe–C alloy. Olsson et al. (1986) included the effects of settling of free crystals, but used approximations such as the estimation of flow velocity. Good agreement with experiments was achieved.

Binary alloy solidification models that couple mass, momentum, energy, and species conservation in all regions (solid, mush, and bulk liquid) were used by Roch et al. (1991) to model steel solidification considering only buoyancy-driven flow. It was assumed that the compositions in the mushy zone were governed by the lever rule (for carbon) or by the Scheil equation (for other elements). The solid and liquid compositions were then determined explicitly at the end of each time-step.

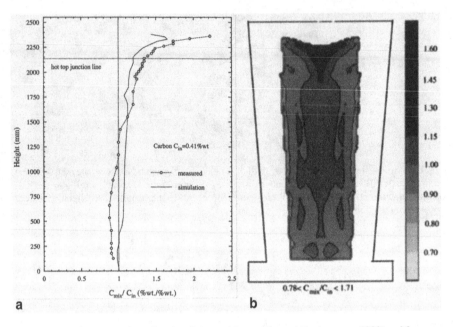

Fig. 19.11 Macrosegregation of carbon in a steel ingot (Gu and Beckermann 1999). **a** Measurements and calculations along the vertical centerline of the ingot. **b** Calculated macrosegregation pattern. (With permission of Springer Science and Business Media)

Capitalizing on an earlier model (Schneider and Beckermann 1995) for multicomponent steel solidification, Gu and Beckermann (1999) developed a single domain macrosegregation model for a steel ingot that considered 11 alloying elements, including their different segregation behaviors and effects on the liquid density. The model accounts for melt convection and involves the solution of fully coupled conservation equations for the transport in the liquid, mush, and solid. Melt convection and macrosegregation in a large steel ingot were simulated. While a general good agreement with measurements was claimed (Fig. 19.11), the model failed to predict the zone of negative macrosegregation observed in the lower part of the ingot because it did not consider the sedimentation of free equiaxed grains.

Macrosegregation development is closely related to the evolution of microstructure during solidification. For example, the permeability in the mushy zone, which determines the flow, is a function of the solid fraction that can be calculated at the macroscale, but is also a function of the dendrite arms spacing which requires microscale calculations. In addition, equiaxed grains or solid fragments that have separated from the mold wall or free surface, or melted off dendrites, can either float or settle depending on their density relative to the liquid. Thus, multiscale models that can solve the macro-transport equations and include microscale dendrite growth are necessary for a correct solution. Such models based on the continuum or volume averaged approach have been discussed in Sect. 18.2.2. Yet the future belongs to multiscale models that integrate continuum mechanics models with cellular automaton or phase-field models, and eventually with molecular dynamics ones.

A finite element model for simulation of dendritic solidification of multicomponent-alloy castings that calculates the pressure and redistribution of gas-forming elements during solidification and cooling, and the flow induced by solidification shrinkage was developed by Sung et al. (2002), and applied to the formation of microporosity in investment castings of AISI 8620 steel. The model solves the conservation equations of mass, momentum, energy, each alloy component, and gas-forming elements (i.e., hydrogen and nitrogen). By solving for the concentrations of H and N in the mushy zone and comparing the sum of their Sievert's pressures with the local pressure, the model predicts regions of possible formation of porosity. The thermal boundary conditions were extracted from a thermal calculation performed with the commercial code ProCAST. The critical contents of gas-forming elements for the porosity formation were predicted. It was also found that the effect of gasses is interrelated. For example, if the uncombined nitrogen in 8620 Steel is 100 ppm, then the hydrogen content must be 3 ppm or less to avoid the formation of porosity. Even if the uncombined nitrogen is taken to be zero, the simulations indicate that a hydrogen content of about 5.5 ppm or more can induce porosity. After adding titanium at a level of 0.087 %, the simulations predicted lower porosity.

A recent effort in this direction resulted in the development of a weakly integrated micro–macroscopic model including a cellular automata (CA) model to describe the growth of dendrites during solidification and a continuum macro-transport (MT) model to calculate macrosegregation during solidification (Liu 2013). The MT model was based on the classical mixture theory. The CA model generates the relationship between the fraction solid and the mean SDAS, which is then used to calculate the permeability in the mushy zone required by the MT model. While the CA model is rather primitive as it can only grow equiaxed dendrites in the 0, 45, and 90° of the square grid, it still demonstrated that calculations with SDAS as a function of the fraction solid yield significant differences compared to the case of fixed SDAS (Fig. 19.12).

19.2 Cast Iron

Cast iron is a multicomponent Fe–C-based alloy (Fe-C-Si-Mn-S-etc.) that solidifies with an eutectic (see phase diagram in Appendix C).

Primitive people worked with meteoric iron long before learning to extract iron from the iron ore. The Sumerian word AN.BAR, the oldest word designating iron, is made up of the pictogram "sky" and "fire". Similar terminology is found in Egypt ("metal from heaven") and with the Hittites ("black iron from sky"). In most ancient cultures, the metallurgist was believed to have a direct link to the divine, if not

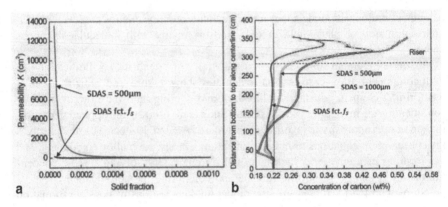

Fig. 19.12 Predicted effect of SDAS on permeability and macrosegregation (Liu 2013). **a** Effect of SDAS on permeability. **b** Carbon macrosegregation along the centerline of an 100 t ingot

Fig. 19.13 The cast iron bridge near Coalbrookdale in England. **a** General view. **b** Detailed view— the author with Prof. Campbell

of divine origin himself. However, the beginning of iron metallurgy on an industrial scale was not possible until the secret of smelting magnetite or hematite was discovered, followed by the art of hardening the metal through quenching (about 1200 – 1000 BC.) in the mountains of Armenia (Eliade 1978). The earliest dated iron casting is a lion produced in China in 502 BC. Introduction of cast iron in Europe did not occur until about 1200 – 1450 AD. Remarkable European cast iron artifacts include the sewer pipes in Versailles (1681) and the iron bridge near Coalbrookdale in England (1779) shown in Fig. 19.13.

Before the invention of microscope in 1860, only two types of u iron were known based on the appearance of their fracture: white and gray. In 1896, the first paper on cast iron to be published in the newly created Journal of the American Foundrymen's Association (1896) stated that "The physical properties of cast iron are shrinkage, strength, deflection, set, chill, grain, and hardness. Tensile test should not be used

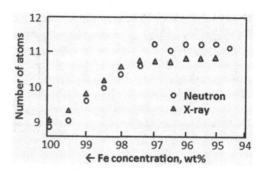

Fig. 19.14 Number of atoms in the first coordination sphere obtained by neutron diffraction and X-ray. (After Steeb and Maier 1974)

for cast iron, but should be confined to steel and other ductile materials. Compression test should be made, but is generally neglected, from the common erroneous impression that the resistance of a small cube or cylinder, which is enormous, is always in excess of loads which can be applied." It took another 50 years for ductile iron to be discovered (1938–1940 independently by Adey, Millis, and Morrogh). The major discoveries of cast iron ended in the 70s with the recognition of compacted graphite iron as a grade in its own merit. Today, cast iron remains the most important casting material accounting for about 75 % of the total world casting tonnage. The main reasons for cast iron longevity are the wide range of mechanical and physical properties associated with its competitive price.

19.2.1 The Structure of Liquid Cast Iron

X-ray and neutron wide-angle diffraction on molten Fe–C alloys with up to 5.5 wt. % carbon, in the temperature range of 1150–1600 °C, were performed by Steeb and Maier (1974). They found that for the pure iron the number of nearest neighbors (number of atoms in the first coordination sphere) is 9, and the nearest neighbor distance is 2.6×10^{-10} m. Up to 1 % C the packing density is increased, as the distance increases to 2.67×10^{-10} m, and the number of neighbors increases to 10.4 (see Fig. 19.14). Between 1.8 and 3 % C the nearest neighbor distance remains constant, but the number of neighbors increases to 11.2 atoms, which means that the packing density is further increased. At 3.5 % C the authors concluded that short-range ordered regions exist in the melt, but were unable to establish their structure. These clusters contain approximately 15 atoms (C_{15}) with a stability time interval of about 10^{-10} s. Note also that the melts containing short-range order regions show high values of viscosity (Fig. 19.15).

19.2.2 Graphite Shape

Cast iron is one of the most complex, if not the most complex, alloy used in industry, mostly because it can solidify with either a stable (austenite–graphite) or a metastable (austenite–Fe_3C) eutectic. For irons solidifying with stable eutectic, depending on composition and cooling rate, three main types of graphite shapes can

Fig. 19.15 Viscosity of Fe–C alloys as a function of carbon concentration. (After Krieger and Trenkler 1971)

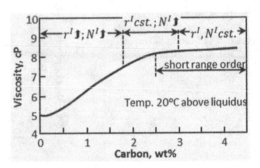

be obtained at the end of solidification, namely, lamellar (LGr), compacted (CGr), and spheroidal (SGr), as exemplified in Fig. 19.16. The LGr is further classified as a function of size and distribution. The most common shapes of lamellar graphite are type-A (coarse graphite with random distribution) and type-D (fine interdendritic lamellar graphite). This last graphite shape is typically obtained in irons of hypoeutectic composition at relatively high cooling rates, and is often referred to as undercooled graphite.

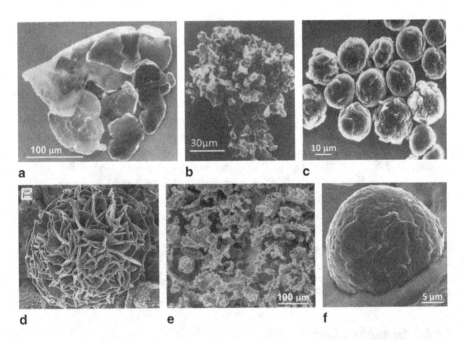

Fig. 19.16 Typical graphite shapes obtained in commercial cast iron: **a** extracted LGr aggregate (Fang 2000). **b** extracted CGr agregate (Bandyopadhyay et al. 1989). **c** extracted Gr spheroids (Fang 2000). **d** LGr aggregate after deep etching (Fras et al. 2007). **e** CGr after deep etching (Fang 2000). **f** SGr after deep etching (Amini and Abbaschian 2013)

Fig. 19.17 Superfine interdendritic graphite. (Aguado
et al. 2014)

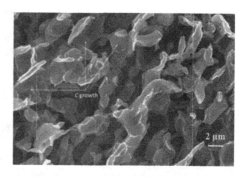

Other intermediate graphite shapes have been identified. Coral graphite is a highly branched fibrous type of graphite that is different from either interdendritic LGr or CGr. It has been obtained in pure Fe–C–Si alloys containing very small amounts of sulfur, typically around 0.001 %, at high cooling rates (Lux 1967) or under vacuum (Hatate et al. 1989). Superfine interdendritic graphite (SIGr) is short (10–20 μm) and stubby, exhibiting round edges similar to the coral graphite, as shown in Fig. 19.17. It is obtained in low-S and moderate-Ti additions (e.g., < 0.01 %S, ∼ 0.3 %Ti) irons (Aguado et al. 2014).

Irregular graphite shapes may result by degeneration of SGr. Chunky graphite forms by extensive branching of graphite spheroids (Liu et al. 1983; Itofuji and Uchikawa 1990). While chunky graphite is highly interconnected and does not consist of broken pieces of graphite spheroids (Fig. 19.18), some branches can be disturbed and fractured by the thermally induced turbulence of the melt during solidification. The tendency for formation of chunky graphite can be decreased by lowering the carbon equivalent and the silicon content and by increasing the cooling rate when possible.

When describing microstructure formation in cast iron during the liquid/solid transformation, we must explain two distinct stages: (I) solidification of the austenite dendrites and graphite crystallization from the liquid (before the beginning of eutectic solidification), and (II) solidification of the stable and metastable eutectics, including the various shapes of the carbon-rich phase (graphite or carbide).

19.2.3 Nucleation and Growth of Austenite Dendrites

The solidification of hypoeutectic Fe–C alloys starts with the nucleation and growth of austenite dendrites. Outlining the austenite on the metallographic structure is not trivial. Boeri and Sikora (2001) developed a macro-etching technique that allows visualization of the austenite dendrite grains. It consists in austempering the iron after solidification, without cooling to room temperature (Direct Austempering after Solidification—DAAS). A typical example of LGr iron macrostructures revealed by this technique is shown in Fig. 19.19a–c. As the carbon equivalent increases,

Fig. 19.18 Chunky graphite extracted from the matrix at two different magnifications. (Itofuji and Uchikawa 1990). (Copyright 1990 American Foundry Soc., used with permission)

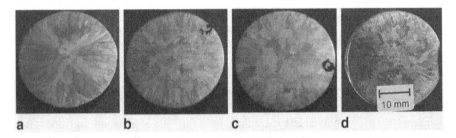

Fig. 19.19 Macrostructure of 30 mm diameter bars showing primary dendrites. Etching: DAAS + Picral 5%. (Boeri and Sikora 2001; Rivera et al. 2004). **a** Hypoeutectic LGr iron (3.94CE). **b** Eutectic LGr iron (4.27CE). **c** Hypereutectic LGr iron(4.64CE). **d** SGr iron. (**a** through **c** with permission of Elsevier)

the macrostructure changes from columnar to columnar-equiaxed, and then mostly equiaxed. Similar macrostructures are found for SGr iron (Fig. 19.19d). The size of the austenite grains is rather large. As it will be discussed later, these grains do not include only the primary austenite, but in most instances also include the eutectic austenite.

Nucleation of austenite in cast iron is still in need of research. Titanium is well known to be a deoxidizer and structure refiner in steel. For the case of cast iron, Basutkar et al. (1969) argued that titanium additions nucleate dendrites favoring the formation of small equiaxed dendrites. Yet, TiC did not appear to be a nucleation site for the primary austenite in low-sulfur irons as it was found mostly at the periphery of the secondary arms of the austenite, in the last region to solidify (Moumeni et al. 2013). Naro and Wallace (1969) and Wallace (1975) found that titanium additions refined the secondary arm spacing in both gray and ductile iron. Ruff and Wallace (1976) concluded that the number of austenite dendrites can be increased by reducing the carbon equivalent, adding elements, such as Ti and B,

Fig. 19.20 Austenite grain
density as a function of
quenching temperature and
cooling rate. (Tian and Stefanescu 1993)

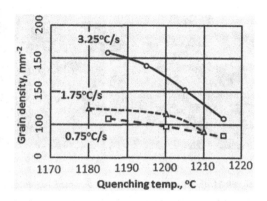

which increase the undercooling by reducing the nucleation potential for graphite
or restricting the growth of the eutectic grain, or by adding materials that serve as
substrates for austenite nucleation (nitrides, carbonitrides, and carbides of various
elements such as Ti and V). Okada and Miyake (1996) suggested that Ti additions
resulting in the formation of TiC produce low-carbon regions at the S/L interface,
favoring formation of type-D graphite. Using SEM/EDS analysis, Zeng et al. (2006)
identified the presence of different Ti compounds in hypoeutectic gray irons containing about 0.08 % S and up to 0.02 % Ti. The compounds included TiN (35 at
% N), (MnTi)S and TiC. Wilford and Wilson (1985) studied the influence of up to
0.4 % Ti in gray iron. They stated that first, Ti will react with N producing TiN or
Ti(CN) that affects the solidification of primary austenite. The excess Ti will then
react with S. Formation of TiS decreases the available S for MnS formation and
increases undercooling which is responsible for type-D graphite formation.

To quantify the dependency of area grain density of primary austenite on cooling rate and undercooling Tian and Stefanescu (1993) conducted liquid quenching
experiments on an iron having 2.98 % C and 1.65 % Si (Fig. 19.20). No significant
grain coalescence was observed. At a quenching temperature of 1180 °C, the correlation between the area grain density (N_γ in mm^{-2}) and cooling rate (instantaneous
nucleation) was found to be:

$$N_\gamma = 48.12 + 5.33\frac{dT}{dt} + 0.087 \, (dT/dt)^2. \tag{19.6}$$

The number of austenite grains in lamellar graphite (LGr) iron can also be expressed
as a function of undercooling (continuous nucleation) as:

$$N_\gamma = \mu_\gamma \, \Delta T^n. \tag{19.7}$$

Tian and Stefanescu (1993) proposed $\mu_\gamma = 2.45$ and $n = 0.93$ for N_γ in mm^{-2},
based on the experimental work presented previously. Fras et al. (1993) suggested
$\mu_\gamma = 500$ and $n = 2$ for N_γ in mm^{-3}.

To evaluate the austenite dendrites' tip radius and spacing, Tian and Stefanescu
(1992) have conducted DS experiments combined with a liquid metal decanting
technique. This method allowed separation of the solidification front from the liquid

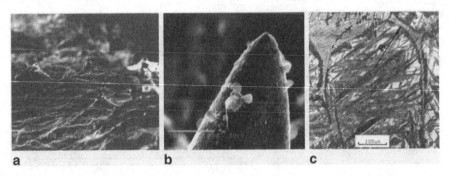

Fig. 19.21 Interface morphology at decanted L/S interface in a Fe− 3.08 % C− 2.01 %Si alloy ($G = 50$ K/cm) (Tian and Stefanescu 1992). **a** Array of austenite dendrites. **b** Paraboloid-shaped austenite dendrite tip. **c** Austenite cell. (With permission of Springer Science and Business Media)

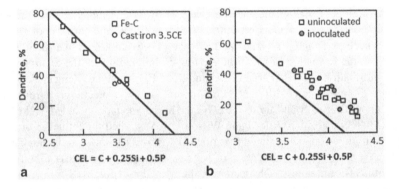

Fig. 19.22 Experimental and equilibrium (lever rule shown as *solid line*) dendrite volume in iron-base alloys; CEL is the carbon equivalent liquidus. **a** *White* cast iron. **b** *Gray* cast iron. (After Mampey 2001)

by means of gravitational and additional centrifugal forces. The typical morphology of dendrites in a Fe− 3.08 % C− 2.01 % Si alloy is shown in the SEM micrographs in Fig. 19.21. It is apparent that the dendrite tip has the form of a paraboloid, which is consistent with observations on other systems. The experimental results plotted on Fig. 9.21, as well as metallographic observations, indicated that at a growth velocity of 0.65 μm/s a cellular-to-dendritic transition occurred.

Mampey (2001) compared the amount of austenite calculated with the lever rule from the Fe–C equilibrium diagram with that measured on 100 mm diameter bars cast in green sand. As shown in Fig. 19.22, for white irons the calculated equilibrium value is very close to the experimental one. This means that metastable (white) solidification is not influenced by the austenite dendrites. There is no measurable growth of the eutectic austenite on the austenite dendrites.

For gray irons, the measured volume of austenite is much larger than predicted by the lever rule. There seem to be no difference between inoculated and uninoculated irons. Thus, this difference cannot be attributed to the difference in undercooling. It can be explained through the difference in their eutectic-coupled zone. As shown by Jones and Kurz (1980), the white eutectic remains in the coupled zone regardless of the undercooling, while for the gray eutectic the coupled zone is asymmetric, and below an undercooling of 7.3 °C austenite and eutectic will grow together. Thus, more austenite than predicted by equilibrium will form. Another explanation is that the austenite dendrite in LGr iron include both primary and some eutectic austenite.

19.2.4 Nucleation of Graphite

The analysis of the vast literature on this subject suggests that the various mechanisms for graphite nucleation in the liquid belong to one of these two categories:

1. One-stage nucleation by single-type inclusions: direct nucleation by one type of inclusion such as an oxide or carbide, or on carbon molecules
2. Multi-stage nucleation by multi-particles: sequential nucleation consisting of the catalytic nucleation of an inclusion on a preexisting inclusion, followed by graphite nucleation on the new inclusion

Nucleation of Lamellar Graphite Iron A large body of inoculation experiments indicates that graphite is a potent one-stage nucleant in LGr iron. Direct experimental evidence is missing because, even with electron microscopy, it is not easy to distinguish the graphite nucleus from the graphite that has grown on it. However, if we accept the idea that inoculated cast iron is an iron-saturated colloidal system with graphite molecules of 10^{-9} to 10^{-8} m size (Vertman and Samarin 1969), or that clusters containing about 15 atoms of carbon exist in Fe–C liquids with higher than 3.5 % C (Steeb and Maier 1974), then we can also admit that these molecules can serve as homogeneous nuclei for graphite. Indeed, early solidification experiments on iron with low sulfur (Alonso et al. 2015) revealed growth of graphite lamellae at the γ/L interface in the absence of any MnS or other inclusions, as illustrated in Fig. 19.23.

The salt-like carbides nucleation theory advanced by Lux (1968) is another example of one-stage nucleation model. Experiments demonstrate that a number of pure metals such as Li, Ca, Ba (Lux and Tannenberger 1962), Sr, and Na (Stefanescu 1972) can be effective in promoting graphite nucleation in cast iron. Lux suggested that these and all elements from groups I, II, and III from the periodic table, when introduced in molten iron, form salt-like carbides that develop epitaxial planes with the graphite, and thus constitute nuclei for graphite. Yet, as all these metals are strong oxide and sulfide formers, some other explanations are possible.

Examples of TiC or Ti(CN) acting as nuclei for LGr have been presented by Sun and Loper (1983). Yet, they argued that while the atomic structure of TiC is such as to act as a nucleus for graphite, the degree of mismatch between the lattices would minimize the nucleation effectiveness of Ti nitrides and carbonitrides. Indeed, the shortest distance between the $\langle 111 \rangle$ planes occupied by C atoms in CaC_2

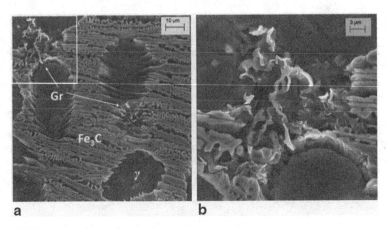

Fig. 19.23 SEM images of early solidification ($f_S = 0.49$) for a 3.5 % carbon equivalent iron with 0.58 %Mn, 0.011 %S, and 0.18 %Ti. **a** Graphite lamellae at the γ/L (Fe$_3$C) interface. **b** Detail of image in (**a**) (Alonso et al. 2015)

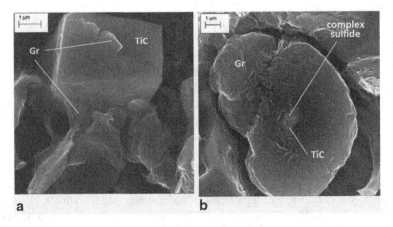

Fig. 19.24 Nucleation of graphite by TiC (Alonso, Stefanescu, Suarez private communication). **a** Lamellar graphite growing on TiC. **b** Spheroidal graphite growing on TiC

(a good nucleant) is $a_{CaC_2} = 3.41 \times 10^{-10}$ m, while that between the layers of graphene is $a_{Gr} = 3.35 \times 10^{-10}$ m, which amounts to a disregistry of 1.76 %. For TiC, $a_{TiC} = 3.74 \times 10^{-10}$ m, with a disregistry of 11.6 %. The situation improves when interatomic distances within the $\langle 111 \rangle$ graphite planes are considered, i.e., $a_{TiC} = 3.06 \times 10^{-10}$ m, and $a_{Gr} = 2.84 \times 10^{-10}$ m. The disregistry is then 7.7 %. Further, one must consider that the TiC found in ferrous alloys is more complex, as it dissolves nitrogen. Nucleation of both lamellar and spheroidal graphite on TiC was found in recent quenching experiments (Fig. 19.24).

The first multistage nucleation mechanism seems to have been proposed by Weiss (1974), who argued that nucleation of lamellar graphite occurs on SiO$_2$ oxides

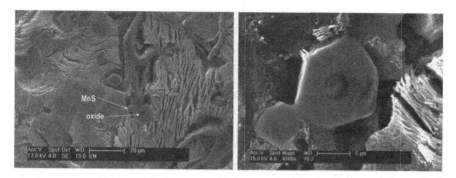

Fig. 19.25 Graphite nucleated on a MnS sulfide which in turn nucleated on an Al, Mg,Si, Ca oxide (Moumeni et al. 2013). (With permission of Springer Science and Business Media)

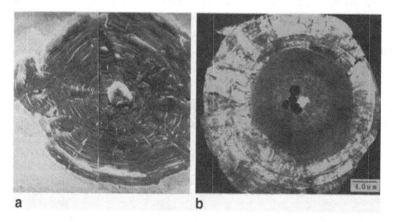

Fig. 19.26 Nuclei in the center of a graphite spheroids. **a** Ionic etching (Liu et al. 1983). **b** TEM micrograph (Fang 2000). (**a** Copyright 1983 American Foundry Soc., used with permission)

formed by heterogeneous catalysis of CaO, Al_2O_3, and oxides of other alkaline metals. After Wallace (1975) revealed the role of MnS in graphite nucleation, a consensus was reached that graphite flakes nucleate on MnS or complex (MnX)S compounds, which have low crystallographic misfit with graphite (De and Xiang 1991; Chisamera et al. 1998; Skaland et al. 1993). When small amounts of Mg were added to the melt, the manganese sulfides contained additional elements such as Al (as Al_2O_3), or Mg as (Mn, Mg)S (Sommerfeld and Ton 2009). These type of information and their own research moved Riposan et al. (2003) to suggest that LGr nucleation occurs on complex (Mn, X)S sulfides, which in turn grow on complex oxides of Al, Si, Zr, Mg, and Ti. Similar findings were reported by Moumeni et al. (2013), as illustrated in Fig. 19.25.

Nucleation of Spheroidal Graphite Nuclei for spheroidal graphite have been observed on optical micrographs as well as on ion-etched SEM and TEM micrographs, as shown in Fig. 19.26. Recent research has done much to elucidate the nature of these nuclei.

A two-stage nucleation theory of double-layered (cored) nucleation was proposed as early as 1974 by Jacobs et al. for spheroidal graphite. Using the results of SEM analysis, they contended that SGr nucleates on duplex sulfide–oxide inclusions (1 μm dia.); the core is made of Ca–Mg or Ca–Mg–Sr sulfides, while the outer shell is made of complex Mg–Al–Si–Ti oxides. This idea was further developed by Skaland et al. (1993). They argued that SGr nuclei are sulfides (MgS, CaS) covered by Mg silicates (e.g., MgO · SiO$_2$) or oxides that have low potency (large disregistry). After inoculation with FeSi that contains another metal (Me) such as Al, Ca, Sr, or Ba, hexagonal silicates (MeO · SiO$_2$ or MeO · Al$_2$O$_3$ · 2SiO$_2$) form at the surface of the oxides, with coherent/semi-coherent low-energy interfaces between substrate and graphite.

A two-stage nucleation is probable also for the example in Fig. 19.24b, where the sequence of nucleation seems to be a complex oxisulfide–TiC–graphite.

Nucleation of SGr iron has been described through empirical relationships by a good number of investigators (see Sect. 18.1). One of the most complete was developed by Skaland et al. (1993) who also included the effect of fading time (number of nuclei in mm^{-3}):

$$N = c \, ln((1.33 + 0.6 \, t_{end})/(1.33 + 0.6 \, t_{start})) \tag{19.8}$$

where c is a kinetic constant to be evaluated experimentally, and t_{start} and t_{end} are the time intervals between inoculation and start and end of solidification, respectively.

19.2.5 Growth of Graphite from the Liquid

Before approaching the difficult subject of explaining the complexities of graphite growth, let us discuss the main factors affecting it. The two main influences responsible for the changes in graphite morphology are the impurities in the melt (type and level) and the cooling rate of the alloy. It is well established that higher cooling rates favor the transition LGr-to-CGr-to-SGr. While the impurities affecting graphite growth are many, they can be divided in two categories:

a. Reactive impurities favoring the transition from LGr to SGr, such as Mg, Ca, Y and Lanthanides (Ce, La); we will call them compacting or spheroidizing
b. Surface-active impurities favoring the transition from SGr to LGr, such as S, O, Al, Ti, As, Bi, Te, Pb, Sb; we will call them anti-compacting or anti-spheroidizing

All elements decrease the surface energy of liquid Fe–C alloys. However, while Ni, Cu, and Si slightly reduce the surface energy, Ca, Mg, Ce, S, Se, and Te have a much stronger effect (Fig. 19.27). For example, sulfur decreases the surface energy from 1.38 J/m^2 at 0.01 %S, to 0.92 J/m^2 at 0.07 %S (Keverian and Taylor 1957).

There are extensive chemical reactions between the impurities in the two categories. Mg reacts with both oxygen and sulfur as can be inferred from Fig. 19.28a, which shows that at the same Mg level in the melt, the oxygen activity is higher at

Fig. 19.27 Influence of some elements on the surface energy of liquid iron at 1400 °C. (After Cosneanu 1966)

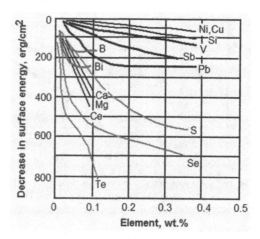

higher sulfur content because some of the Mg is used to produce MgS. The relation for oxygen activity in the presence of magnesium is given by:

$$\log a_o = -25751\frac{1}{T} + 6.28 - \log a_{Mg} + \log a_{MgO} \qquad (19.9)$$

where T is in OK, and a are the activities of Mg and MgO (Mampaey et al. 2008). Another equation was proposed by Hummer (1991):

$$\log\{[O][Mg]\} = -24973\frac{1}{T} + 7.36 \qquad (19.10)$$

where [O] and [Mg] are the mass% of the elements. These relationships and Fig. 19.28a, b show that oxygen activity deceases with higher Mg and lower temperature. The correlation between the oxygen and nodularity summarized by Fig. 19.28c indicates that the decreasing oxygen favors the LGr-to-SGr transition, but also that oxygen is not the only factor affecting it (see the large data spread).

In general, reactive impurities such as Mg and Ce remove surface-active impurities generating high interface energy (e.g., 1.45 J/m² measured by Washchenko and Rudoy 1962). There is however a difference between the surface energy of the melt and the L/Gr interface energy as summarized in Table 19.2. The data in the table show that in Fe–Mg alloys $\gamma Fe/Gr_{prism} > \gamma Fe/Gr_{basal}$. This led McSwain and Bates to conclude that graphite grows from the melt normal to the plane with the lowest interfacial energy, which is the C-direction for the Fe–C–Mg alloy and the A-direction for the Fe–C–S alloy.

The graphite shape is the result of the interaction between these two categories of impurities. Typical limits of selected impurities associated with the various graphite shapes are summarized in Table 19.3. These limits should be considered only as guidelines, as cooling rate and the other impurities will alter the ranges. A careful look at this table suggests that the interplay between the reactive compacting element Mg and the surface-active anti-compacting elements O and S dictates the final

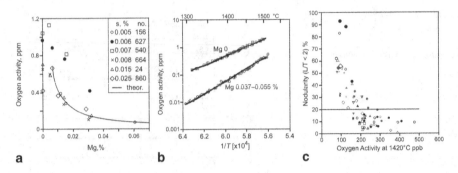

Fig. 19.28 Oxygen activity in cast iron (Mampaey and Beghyn 2006; Mampaey et al. 2008). **a** As a function of residual Mg for different S content. The curve is the theoretical relation as given by Eq. 19.9). **b** As a function of magnesium and temperature. **c** As a function of nodularity

Table 19.2 Surface properties of Fe–C–Si alloys on graphite in the absence of oxygen. (McSwain and Bates 1975)

Alloy	Graphite	Surface tension of iron, J/m^2	Fe/Gr interfacial energy[a], J/m^2
Fe-3.7 C-2.8Si-0.037Mg	Basal	1.128	1.460
	Polycrystalline	1.167	1.621
	Prism	1.147	1.721
Fe-3.7 C-2.4Si-0.05 S	Basal	1.057	1.270
	Polycrystalline	1.017	951
	Prism	1.153	846

[a]calculated from contact angle, surface energy of graphite, and surface tension of iron

Table 19.3 Graphite shape as a function of the typical impurity levels for small and medium size castings

Graphite shape	S (%)	O (ppm) at 1420° C	Mg (%)
Lamellar type-A (LGrA)	> 0.03	> 0.75	< 0.01
Lamellar interdendritic type-D (LGrD)	> 0.012	> 0.75	< 0.01
Superfine interdendritic (SIGr)	< 0.012	0.5–1.5	< 0.01
Compacted (CGr)	< 0.014	0.17–0.5	~ 0.02
Spheroidal (SGr)	< 0.01	0.09–0.15	> 0.035

graphite shape. In principle the LGrA-to-SGr transition is favored by higher cooling rate, decreasing amounts of anti-compacting impurities, and increasing amounts of compacting impurities.

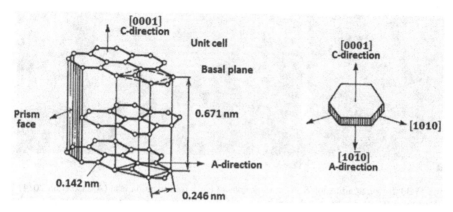

Fig. 19.29 Crystal lattice of graphite

The practical procedure to trigger the LGr-to-SGr transition is the addition of small amounts of Mg or lanthanides to a low-sulfur iron. Alternatively, similar structural changes can be achieved by holding molten LGr iron in vacuum for long time prior to casting to remove impurities such as O, S, and P, leaving a "clean" melt (Oldfield et al. 1967, Sadocha and Gruzleski 1975, Dhindaw and Verhoeven 1980). High cooling rates helped.

Crystal Lattice of Graphite The crystal lattice of graphite is hexagonal, with A-B-A-B stacking of semi-infinite hexagonal monolayers (Fig. 19.29). The bonding is strong covalent in the plane of the layers and weak van der Walls between them. Each monolayer is a two-dimensional polymeric sheet, sometimes called "graphene" (Dresselhaus et al. 1988), to which atoms can add easily in the A-directions, but with a much lower probability of attachment in the C-direction, normal to a monolayer. This results in a hexagonal unit cell with dimensions $c = 0.671$ nm and $a = 0.246$ nm and four atoms per cell (Wyckoff 1963). The layers A and B are displaced by half the crystallographic c-axis spacing, i.e., 0.335 nm.

Due to the difference in the bonding forces between the A- and C-directions, it is reasonable to assume that the preferred (normal) growth habit for graphite is in the A-direction, producing a sheet. While a large number of intermediate graphite forms exist, the two basic forms of graphite are lamellar (sheets) and spheroidal (spheres).

Microscopy Information on Graphite Shape Examination of early solidification microstructure obtained through interrupted experiments provided insight in the various transitions of the graphite morphology. In gray iron with low sulfur (< 0.01 %), the lamellar graphite nucleates at the γ/L interface and grows in contact with the liquid (Fig. 19.30), as parallel platelets (Fig. 19.31). The platelets, which are the building blocks for the Gr lamellae, are very thin and have micrometer-size width. Upon further solidification they develop into lettuce-leaf plates (Fig. 19.16a), typical for LGr.

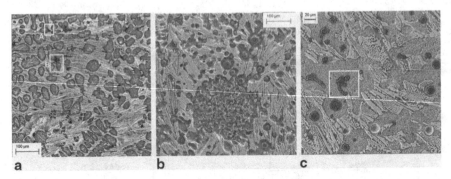

Fig. 19.30 Low magnification SEM micrographs of early solidification iron (Alonso et al. 2014). **a** No Mg; austenite dendrites and graphite nucleation at the γ/L interface. **b** 0.013 % Mg iron; SGr and curly graphite. **c** 0.02 % Mg iron; SGr and TPGr (in the square)

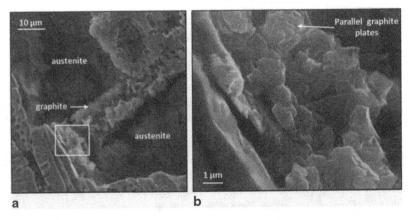

Fig. 19.31 SEM images of early solidification ($f_S = 0.64$) for a 3.6 % carbon equivalent iron with 0.57 %Mn and 0.011 %S (Alonso, Stefanescu, Suarez private communication). **a** General view. **b** Magnification of the region marked by the rectangle on **a**

In sulfur-containing irons the sulfur appears to accumulate in the graphite (Fig. 19.32). Yet, Auger analyses show concentrations of O and S in Fe, but not in the Gr, adjacent to the Fe/lamellar graphite interfaces, of about 20at% O and 5at% S in some two or three atomic layers (Johnson and Smartt 1979). Scanning Auger microscopy found that both S and Te segregate to the Fe/Gr interface (Park and Verhoeven 1996). In the S-doped alloys, type-A lamellae were generally covered with a monolayer of S with patches of O in the form of iron oxide having a thickness of about 2 nm.

In Mg-treated iron, neither Mg, nor O or S, were detected on the Gr surface, but appeared isolated in combined form as Mg–S–P compounds. This seems to imply that Mg does not act directly on the Gr, but rather that it acts as a scavenger of the

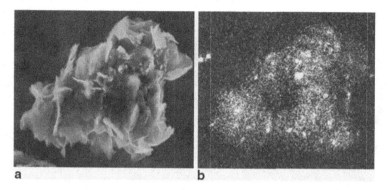

Fig. 19.32 SEM images of lamellar graphite extracted from a high-sulfur cast iron, and sulfur distribution. (Fang 2000)

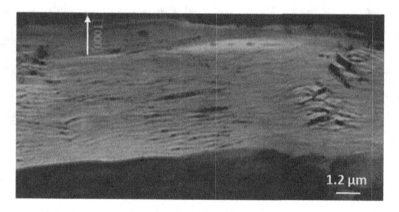

Fig. 19.33 TEM micrograph of a graphite lamella. (Fang 2000)

impurities that stabilize lamellar graphite. However, as good SGr cannot be obtained by simply reducing the S and O content to zero, and as Mg-containing irons produce good Gr spheroids, while Ce- or Ca- containing irons only produce quasi-spheroidal graphite, it is reasonable to conclude that the reactive impurities also play a direct role in graphite growth.

TEM observations reveal that a graphite lamella is made of a large number of thin plates and incorporates numerous defects (Fig. 19.33). Higher magnification unveil that the lamellae are imperfectly crystalline on a local scale and may contain amorphous regions (Purdy and Audier 1984).

Magnesium additions dramatically change graphite morphology. Early solidification experiments on iron undertreated with low Mg exhibit graphite spheroids surrounded by liquid or austenite shells (Fig. 19.30b). The morphology of these spheroids depends on the amount of Mg, as exemplified in Fig. 19.34. At very low Mg contents (0.013 %) the spheroids are made of radially growing aggregates. At > 0.02 % Mg, full spheroids made of platelets with cabbage-leaf orientations

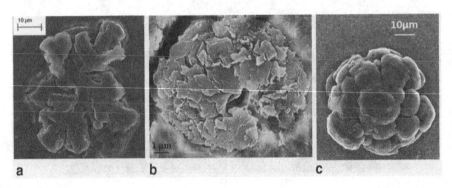

Fig. 19.34 Graphite spheroids precipitated at f_S < 0.4 in undertreated irons (Alonso et al. 2015). **a** 0.013 % Mg—radial growth. **b** 0.02 % Mg—cabbage-leaf platelets. **c** 0.02 % Mg—conical sectors

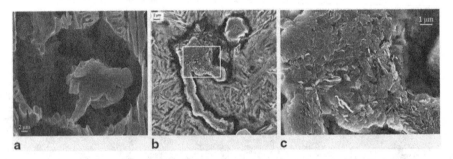

Fig. 19.35 SEM images of tadpole graphite in 0.02 % Mg iron (Alonso et al. 2014). **a** TPGr growing in contact with the liquid. **b** Magnification of tadpole graphite in Fig. 19.30c. **c** Magnification of tadpole graphite in **b**

(Fig. 19.34c), or graphite with conical radial sectors (Fig. 19.34c) were found. The building blocks of initial spheroids are thin platelets 1–2 μm wide that grow mostly in A-direction, through twinning, branching, twisting, and curving. The occurrence of conical sectors will be discussed later.

As the fraction solid increases to about 0.4, some of the graphite spheroids develop tails (*tadpole graphite*—TPGr) as illustrated in Fig. 19.30c. In irons with low Mg (0.013 %), further solidification results in spherical γ/Gr eutectic grains with *curly graphite* (Fig. 19.30b), while in higher Mg irons (0.02 %) the tadpoles develop into compacted graphite. The tadpole is partially enveloped in austenite at the time it emerges from the spheroid, but continues to grow in contact with the liquid (Fig. 19.35a).

The graphite morphology of the TPGr includes a large number of graphite platelets of diverse orientations (Fig. 19.35b, c). This was also confirmed through TEM work (Hara et al. 2014). The platelets consolidate into 3D graphite aggregates with polygonal plates developed in the A-direction [1010], but with significant

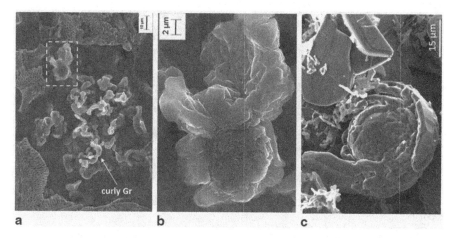

Fig. 19.36 SEM images of curly and tadpole graphite in 0.013 % Mg irons. **a** Curly graphite (Alonso, Stefanescu, Suarez private communication). **b** Tadpole graphite (Alonso, Stefanescu, Suarez private communication). **c** Tadpole graphite (Amini and Abbaschian 2013). (**c** with permission from Elsevier.)

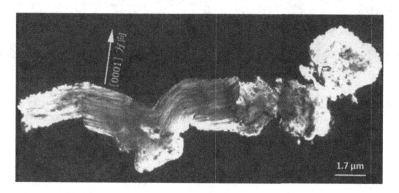

Fig. 19.37 TEM image of compacted graphite. (Fang 2000)

thickness in the *C*-direction [0001] (Fig. 19.36). This configuration is true for curly as well as for compacted graphite, as demonstrated from TEM work on CGr (Fig. 19.37).

If enough Mg has been added to the iron (typically > 0.035 %) fully developed graphite spheroids will crystallize. They may have three concentric layers as suggested by Fig. 19.26b. Recent TEM work (Theuwissen et al. 2012) found a microcrystalline structure at the center of the spheroid (small areas with different orientations) while another TEM report (Hara et al. 2014) identified an amorphous central region, surrounded by annular rings of a layered intermediate region, and

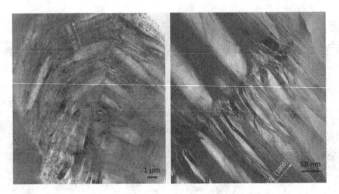

Fig. 19.38 TEM images of the misfit at the junctions between radial sectors in Gr spheroids. (Fang 2000)

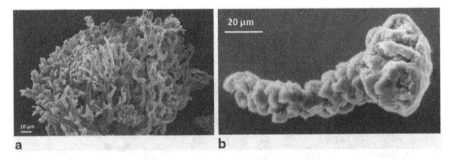

Fig. 19.39 SEM images of extracted chunky graphite. **a** SEM images of extracted chunky graphite colony (courtesy of Theuwissen from Lacaze et al. 2013). **b** SEM images of extracted segment of chunky graphite. (Fang 2000)

then an outer region made up of large polygonal crystalline platelets in a mosaic-like structure. TEM image of fractured graphite found amorphous graphite at the growth steps at the γ/SGr interface (Purdy and Audier 1984). This would be consistent with graphite growth through carbon diffusion through the austenite shell and deposition of amorphous carbon at the γ/Gr interface, followed by crystallization of the amorphous carbon at the growth steps.

Many times, but not always, the graphite spheroids may exhibit radial sectors joining in the center (Fig. 19.26a). The angle between the [0001] directions of Gr in the two adjacent radial sectors has been measured to be 31° (Theuwissen et al. 2012). Yet, other angles seem also possible (Fig. 19.38). The misfit at the joining of radial sectors seems to preclude the continuity of Gr plates across the joining.

As the solidification time increases, SGr tends to degenerate into exploded or into chunky graphite. The chunky graphite appears to be a highly branched Gr aggregate (Fig. 19.39a), with the branches being the result of helical growth (Fig. 19.39b).

Fig. 19.40 Plan of graphene sheet with growth in the A-direction inhibited by adsorption of oxygen or sulfur. (After Double and Hellawell 1995)

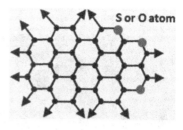

Fig. 19.41 Growth of lamellar graphite

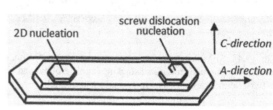

Growth Mechanism of Graphite in the Liquid It is now widely accepted that the growth of LGr starts with the formation of graphene sheets that can grow easily in the A-direction. To produce a Gr lamella (plate), the sheets will have to also grow in the C-direction and generate multilayer sheets. It was suggested that as the surface-active impurities (S, O, N) are adsorbed at the unsaturated edges of a sheet (Fig. 19.40), growth in the A-directions is partially inhibited, and growth in the C-direction becomes more probable. Thus, in gray cast iron that is relatively rich in O and S, the 3D Gr lamellae grow in the crystallographic A-directions with the {0001} basal planes parallel to the plane of the lamellae and thicken in the C-direction.

As found from the previously discussed early solidification experiments, graphite platelets nucleate on a variety of inclusions (oxides, sulfides, carbides, etc.) or on carbon molecules at the austenite/liquid interface. These polygonal platelets 1–2 μm wide that grow in the A-direction are the building blocks (structural building units) of the Gr aggregates. Thickening of the platelets occurs through spiral growth at screw dislocation steps or by two-dimensional nucleation of the sheet in the C-direction (Fig. 19.41). There is microscopy evidence for both (see for example Fig. 9 in Zakhartchenko et al. 1979, Fig. 12 in Roviglione and Hermida 2004).

Graphite lamellae exhibit two types of defects: twin boundaries, which tilt the flake out of the basal plane, and twist boundaries (stacking faults or coincident boundaries) that lie on the basal planes (Fig. 19.42a, b). Twin boundaries defects may result in Gr branching through splitting along its basal plane while growing in the A-direction (Fig. 19.43). Twist boundaries cause a rotation about the C-axis of the Gr. Based on experiments on Ni–C alloys, it was suggested that successive layers are stacked together in one of the three ways such that they are related by rotations of about 13°, 22°, or 28° about the C-axis that produce coincidence boundaries (Fig. 19.42c). Thus, graphite lamellae are composed of layers of fault free crystal some 10^{-4} mm thick.

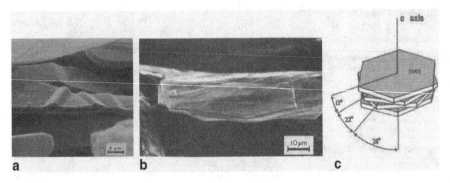

Fig. 19.42 SEM micrographs of defects in graphite (Bollman and Lux 1985), and schematic representation of C-axis rotational stacking faults (Double and Hellawell 1995). **a** Twinning of plates. **b** Twist boundaries. **c** C-axis rotational stacking faults

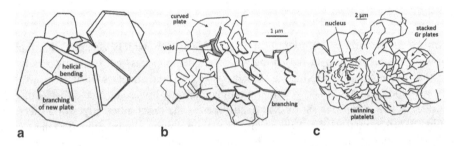

Fig. 19.43 Drawings of details of graphite growth based on SEM micrographs. **a** Helical growth of a graphite plate in chunky graphite (after Fig. 14 in Liu et al. 1983); the original plate bends upon itself and branches to produce a new plate. **b** Cabbage-leaf growth of graphite platelets in SGr. **c** Tadpole graphite—Gr spheroid made of twinning platelets and beginning of CGr made of stacked plates

While growth of lamellar Gr is fairly well understood, that of spheroidal Gr is still a subject of much debate. Many theories have been proposed over the years to describe the mechanisms of formation of various graphite shapes, and reviews of these theories have been periodically written (e.g., Minkoff 1983; Elliott 1988; Stefanescu 1998). The subject is still one in need of research.

The role of the reactive (Mg, Ce, La) and surface-active elements (S, O, Pb, Sb, Ti) has been recognized early in a theory (Herfurth 1965) that postulates that the change from lamellar to spheroidal graphite occurs because of the change in the ratio between the growth velocity on the $[10\bar{1}0]$ face (A-direction) and growth on the $[0001]$ face of graphite (C-direction). For equilibrium conditions, the Gibbs–Curie–Wulf law states that the crystalline phase with the higher interface energy grows more rapidly in the normal direction. For graphite $\gamma_{[10\bar{1}0]} = 7.7 \cdot \gamma_{[0001]}$. Thus $V_{[10\bar{1}0]} > V_{[0001]}$. Bravais' rule stipulates that the growth rate normal to a plane is inversely proportional to the density of atoms located on the plane. Thus, again, the preferred growth direction should be the A-direction. However, according to Herfurth, under the nonequilibrium conditions that prevail during the solidification of cast iron, kinetic considerations become important. Assuming growth by 2D nucleation, the highest rate of growth will be experienced by the face with the higher density of atoms, where the probability for nucleation is higher. Indeed, experimental evidence shows that in directional solidification of pure metals the highest growth velocity is normal to the most densely occupied face (Hellawell and Herbert 1962; Rosenberg and Tiller 1957). Therefore, in a pure environment (pure Fe–C–Si alloy), the highest growth rate will be in the [0001] direction of the graphite crystal, that is in the C-direction. This will result in the formation of single crystals (coral graphite). In a contaminated environment, surface-active elements are adsorbed on the high-energy planes [10$\bar{1}$0], which have fewer satisfied bonds. Subsequently, the [10$\bar{1}$0] plane face achieves a lower surface energy and higher atomic density than the [0001] face. Growth becomes predominant in the A-direction and lamellar graphite results. If reactive impurities (e.g., Mg, Ce, and La) are available, they scavenge the melt of surface-active elements, after which they also block growth on the [10$\bar{1}$0] prism face. The preferred growth direction becomes the C-direction, and a polycrystalline spheroidal graphite results. However, there is little if any experimental evidence of graphite growing from the liquid preferentially in the C-direction.

The significant change in surface energy caused by the addition of reactive elements prompted a number of investigators to conclude that the higher surface energy promotes spheroidal graphite as the system attempts to decrease its energy. According to Buttner et al. (1951), there is a critical Gr/L interface energy above which polycrystalline SGr is favored over single-crystal LGr. This theory was supported by many investigators (e.g., De. Sy 1949; Keverian 1953; Marincek 1953; Milman 1958; Geilenberg 1964; McSwain and Bates 1975). However, if a crystal becomes larger than 1 µm, the change in free energy because of departure from equilibrium becomes small compared with the supersaturation necessary for crystal growth (Frank 1958). Furthermore, these theories cannot explain the occurrence of conical radial sectors in SGr (*e.g.* Fig. 19.34c) as well as their cabbage-leaf morphology (*e.g.* Fig. 19.16c).

It can be assumed that magnesium, cerium, or lanthanide additions will scavenge the oxygen (MgO) and sulfur (MgS), thus promoting growth and wrapping of the graphene sheets in the A-direction, resulting in spheroidal growth. The hexagonal ring structure nucleated from the liquid grows into a monolayer sheet, and then into multilayer sheets (lamellar graphite). According to Double and Hellawell (1969, 1975, 1995), the ability of a graphene sheet to bend in steps of 20°45′ about three $\langle 1\bar{1}00 \rangle$ axes mutually inclined at 120° makes it possible for a lamellar crystal to roll upon itself as conical helices. There is SEM evidence for helical growth (see Figure 19.39c and Figure 19.43a), although not for complete conical helices. The close packed conical helices radiating from a common center may grow into a spheroid (Fig. 19.44a). Indeed, the TEM work previously discussed indicates that the Gr planes lie with their normal in the radial direction which is consistent with the model in Fig. 19.44a.

Growth of a Gr spheroid may also occur through repeated bending of the Gr sheets, as suggested by Sadocha and Gruzleski (1975) (Fig. 19.44b). A large number

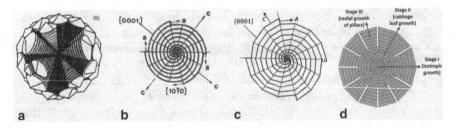

Fig. 19.44 Models for spheroidal graphite growth. **a** Growth by helical bending—conical helices radiating from a common center (Double and Hellawell 1975). **b** Circumferential growth of graphite (Sadocha/Gruzleski 1975). **c** Circumferential growth with boundary tilt through twinning. **d** Three stages growth (Amini and Abbaschian 2013). (**d** with permission from Elsevier.)

of steps on the surface of the spheroid are assumed to grow in the A-direction by curved crystal growth (see for example Fig. 19.43b), with the low-energy basal plane of graphite exposed to the liquid. The growing steps run into one another forming boundaries on the surface, from which new steps can develop, producing a cabbage-leaf effect. In the presence of surface-active impurities that decrease the surface tension, the spherical shape deteriorates into a lamellar one. The major role of spheroidizing elements is to act as scavengers, and to produce a high-interface surface tension, which helps the curved growth of graphite.

While there is ample evidence of curved growth of graphite, the Sadocha/ Gruzleski model cannot explain the occurrence of the radial sectors in Fig. 19.38. This can be corrected by assuming that while growing in the A-direction the Gr sheets tilt periodically through twinning to minimize surface energy (Fig. 19.44c). Yet, not even the modified model can explain the formation of neighboring detached radial sectors as seen in Fig. 19.34a or 19.34c. Note that all the models in Fig. 19.44 are postulating the A-direction as the main growth direction, albeit through different mechanisms.

The role of the growth velocity ratio in the A- and C-direction in determining the graphite shape has also been advanced by Amini and Abbaschian (2013), although not based on surface energy arguments, but on kinetic ones. Attempting to explain the occurrence of SGr in hypereutectic Ni− 3 % C alloys solidified at very high cooling rate (quenching), they argued that a roughening transition from faceted to diffuse Gr/L interfaces is responsible for the LGr to SGr transition. They demonstrated that the growth in length (A-direction) of LGr is diffusion controlled, while the thickening (C-direction) is surface controlled through 2D poly-nucleation growth (a large number of 2D nuclei form on the entire surface and also on the already growing layers before the layer spreading is complete). As shown in Fig. 19.45, the calculated values, which are based on the assumption that the prism plane is atomically rough while the basal plane is atomically flat, agree well with the measured ones.

Further, they assume that at small solidification rates, the graphite crystals with basal and prismatic planes are faceted. As the interface velocity increases, the supersaturation increases and the faceted interface becomes gradually rough (Fig. 19.46).

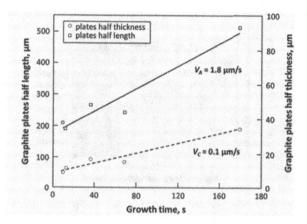

Fig. 19.45 Graphite plates half-length and thickness as a function of growth time; the lines are calculated; their slopes are the growth rates. (After Amini and Abbaschian 2013)

The prismatic face roughening transition occurs at a smaller driving force than the basal plane. Thus, at intermediate rates the prismatic interfaces become rough and grow faster, while the basal plane remain faceted leading to the formation of LGr. At high growth rates ($> \Delta C_{L \to S}$) both interfaces grow with similar velocities, resulting in "bulky" spherical morphology. The model then assumes that the graphite spheres grow in three sequential stages: I) basal and prismatic planes' isotropic growth; II) circumferential growth of graphite tiles; III) radial growth of sectors (Fig. 19.44d). While the Amini/Abbaschian model claims to explain both the occurrence of conical sectors and cabbage-leaf growth, it seems to contradict microscopic evidence that shows the conical sectors extending to the core of the Gr spheroid, as well as their own microscopic evidence that shows cabbage-leaf growth on the surface of the spheroid (Fig. 19.16c).

The three concentric layers sometimes observed in Gr spheroids (see example in Fig. 19.26b) can be understood as resulting from growth in three different stages: in the liquid in direct contact with the liquid, in the liquid through the austenite shell, and in the solid during cooling to the eutectoid temperature.

From this analysis, it follows that there is a wide range of folding and wrapping possibilities for the graphene sheets which could describe the many graphite morphologies encountered in industrial cast irons. The lamellar and spheroidal forms of graphite may be regarded as two extremes. Other intermediate graphite shapes, such as compacted graphite, are the result of a mixed growth mechanism.

To summarize, at low surface energy (melt with S, O, N) platelets grow initially parallel to one another when nucleated at the austenite/liquid interface. Extensive twinning, branching, and growth in 2D, followed by thickening through spiral growth at screw dislocations steps, or by 2D nucleation, and then stacking of platelets produce lamellar graphite. The graphite plates arrangement in Fig. 19.47a has been termed foliated dendrites (Saratovkin quoted by Roviglione and Hermida

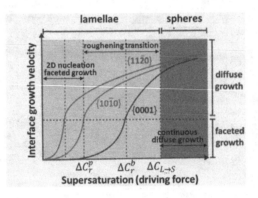

Fig. 19.46 Qualitative interface growth rate curves of basal and prismatic interfaces of graphite: ΔC_r^p and ΔC_r^b are the necessary supersaturations for kinetically roughening transition of basal and prismatic interfaces, respectively, $\Delta C_{L\to S}$ is the necessary supersaturation for lamellar to spherical morphology change. (Amini and Abbaschian 2013, with permission from Elsevier)

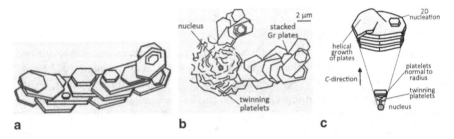

Fig. 19.47 Growth mechanisms of various graphite shapes. **a** Growth of curly graphite. **b** Growth of compacted graphite from a tadpole graphite. **c** Growth of conical sector in spheroidal graphite

2004). Flemings considered them to be the results of morphological instability of faceted crystals.

As Mg is added to the melt in amounts roughly under 0.015 %, a mixed spheroidal/lamellar structure appears in the early stages of solidification. Curved graphite growth, driven by minimization of surface energy, produces highly curved lamellae (curly graphite Fig. 19.47a). Initial spheroids develop through circumferential growth or stacking of platelets. The graphite planes in the initial spheroids start splitting through twinning or twisting. The surface of the graphite becomes rough with planes protruding and growing in different directions. The spheroids produce tails (tadpole graphite) in which platelets have random orientations (Fig. 19.47b).

At intermediate surface energy (e.g., 0.018–0.022 %Mg), the tails of the graphite tadpoles thicken and stack in the [0001] direction producing compacted graphite. In addition to the twining faults, rotational stacking faults occur. A large number of thin or thick plates of various orientations are generated (see Figs. 19.35c and 19.47b).

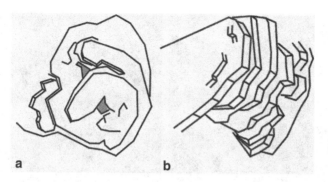

Fig. 19.48 Drawing of chunky graphite after SEM pictures. **a** Helical growth of tip (after Fig. 19.39b). **b** stacking of parallel graphite plates (after Fig. 7 in Liu et al. 1983)

At high surface energy (pure metal or reactive elements Mg, Ce), the situation is less clear. It is necessary to explain the growth of conical sectors first. Initial platelets are nucleated perpendicular to the bounding facets of the nucleus. Curved graphite growth becomes dominant because of the high interface energy. The platelets that initially grow with random orientation will curve, forming curved plates normal to the radius. These plates will grow in the C-direction through helical growth and repeated bending (Figs. 19.43a and 19.47c), or through 2D nucleation of new plates. The result is a conical sector made of stacked graphite plates. The conical sectors may occupy the whole volume of the sphere forming a graphite spheroid, or only part of it like in exploded graphite. Chunky graphite can be the product of a succession of conical sectors growing on top of one another (Fig. 19.39b) through helical growth (Fig. 19.48a). Stacking of graphite plates was also observed (Fig. 19.48b) in chunky graphite.

It is most certain that in strongly hypereutectic irons, both LGr and SGr grow directly from the liquid. For hypoeutectic irons, solidification starts by the formation of austenite dendrites. The graphite then forms in the interdendritic liquid by an eutectic reaction, with major differences between the growth of LGr and SGr.

19.2.6 Eutectic Solidification of Cast Iron

Two different solidification processes must be considered: i) directional solidification where microstructure formation is controlled by the G/V ratio, and ii) continuous cooling solidification where the controlling factor is the $G \cdot V$ product (the cooling rate). Also, the discussion must include the two equilibria, stable and metastable. Accordingly, we will discuss the solidification of the stable austenite–graphite eutectic (gray) and that of the metastable austenite–iron carbide eutectic (white).

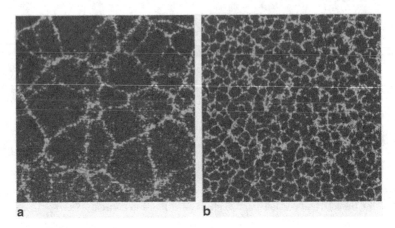

Fig. 19.49 Room temperature eutectic grain structure in LGr iron. Etching: 2.5 h in Stead's reagent. Magnification: 14 X. (Moore 1973). **a** coarse grains. **b** fine grains. (Reprinted with permission of ASM International. All rights reserved. www.asminternational.org)

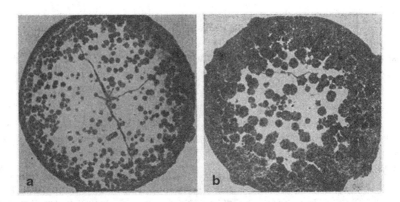

Fig. 19.50 Macrostructure of interrupted solidification of rods showing spherical eutectic grain growth at two different times during solidification. (Oldfield 1966)

Stable Solidification of the Austenite-Graphite Eutectic—Continuous cooling The macrostructure of gray iron can be visualized by using selective etching to outline the phosphide eutectic which is the last to solidify, and therefore delineates the eutectic grain boundaries (Fig. 19.49). In the old or industrial vernacular the eutectic grains are also termed "eutectic cells" or "eutectic colonies". This terminology should be avoided as it creates confusion with cellular solidification of single phase alloys (see for example Fig. 19.14). The length scale (size) of the eutectic grains depends on cooling rate and inoculation, with higher cooling rate and inoculation promoting higher number of grains. An alternative visualization technique is the use of interrupted solidification experiments (Fig. 19.50). Direct evidence of the quasi-spherical shape of the LGr eutectic grains was also provided by Motz (1975) who was able to decant the grains from the solidifying liquid.

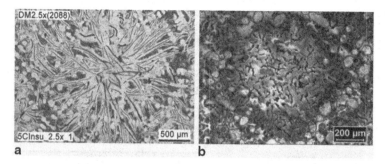

Fig. 19.51 Eutectic grains in cast iron (courtesy of A. Dioszegi.). **a** Lamellar graphite eutectic grain. **b** Compacted graphite eutectic grain

Fig. 19.52 Calculated eutectic coupled zone for pure Fe-C alloys (after Jones and Kurz 1980)

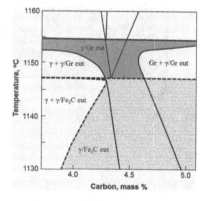

Eutectic lamellar and compacted Gr irons solidify with spherical eutectic grains in which Gr grows radially from a nucleus together with the cooperatively growing eutectic austenite (Fig. 19.51).

More recently Rivera et al. (2008) advanced the hypothesis that solidification of LGr iron, whether hypo- or hyper-eutectic, starts with independent nucleation of austenite and graphite from the melt. As the austenite dendrites grow they interact with the graphite and then grow cooperatively forming "units" or "eutectic cells" or "eutectic colonies", but not eutectic grains?! They argue that this sequence is supported by the large austenite dendrites evidenced through the DAAS technique shown earlier in Figure 19.19a, b, c. Yet, the need of having primary austenite in the solidification structure of LGr iron contradicts experimental evidence such as shown in Fig. 19.50. Furthermore, experiments by Jones and Kurz (1980) demonstrate that a coupled zone exists where the γ/Gr eutectic solidifies without the presence of primary austenite even for pure Fe-C alloys. Primary austenite only appeared at growth velocities in excess of 10^{-7} μm/s.

The following general picture of the solidification of LGr iron emerges. In alloys solidifying within the coupled zone the eutectic grain will nucleate in

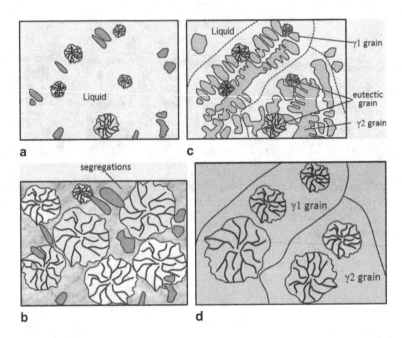

Fig. 19.53 Solidification of the eutectic in LGr iron. **a, b** eutectic iron; **c, d** hypoeutectic iron or high cooling rate; **a, c** early solidification; **b** late solidification; **d** end of solidification

the liquid mostly on heterogeneous nuclei such as (MnX)S or other compounds (Figure 19.53a). Cooperative growth between the austenite and graphite will produce spherical eutectic grains. Some limited amount of primary austenite may solidify away from the eutectic grain, depending on the cooling rate. If the iron solidifies outside of the coupled zone, even for eutectic compositions, solidification will start with the formation of primary austenite (Figure 19.53c). Carbon rejection by the solidifying austenite produces carbon rich envelopes around the dendrite in which the eutectic graphite can nucleate on homogeneous graphite nuclei, as seen for example in Fig. 19.23. Conditions for the existence of these nuclei seem also supported by X-ray experiments on liquid Fe-C alloys discussed earlier. Then cooperative austenite/graphite growth of the eutectic grain will occur. As the eutectic austenite starts growing together with the graphite at the primary austenite/liquid interface, the eutectic austenite will grow on the preexisting primary austenite and will have the same grain orientation with the primary austenite ($\gamma 1$ or $\gamma 2$ on Figure 19.53c). One primary austenite dendrite can be the site of several graphite (eutectic grains) nucleation events. Thus, at the end of solidification a large austenite grain may include several eutectic grains. This explains the large austenite grains seen in LGr iron of all compositions in Figure 19.19. Similar behavior has been observed for the Al-Si eutectic (Dinnis et al. 2005).

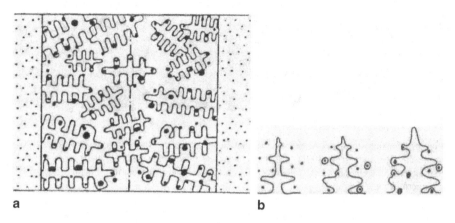

Fig. 19.54 Models for the solidification morphology of near-eutectic SGr iron. **a** after Ellerbrok and Engler (1981); **b** after Stefanescu and Bandyopadhyay (1989)

The γ-SGr eutectic is a divorced eutectic (see section 10.1.5 for the analytical model) that starts with the solidification of austenite. Indeed, very early after the discovery of SGr iron, in 1949, Patterson and Scheil used experimental findings to state that SGr forms in the melt and is later encapsulated in an austenite (γ) shell. Measurements of the radii of the graphite and the γ shell led to the conclusion that they develop such as to conserve a constant ratio between the radii of the γ shell and the SGr ($r_\gamma/r_{Gr} = 2.3$) throughout the microstructure (Scheil and Hütter 1953). This ratio was confirmed theoretically (Wetterfall *et al.* 1972) through calculations for the steady-state diffusion-controlled growth of graphite.

Ellerbrok and Engler (1981), quoted in Rickert and Engler 1985, described as early as 1981 the two important features of the eutectic solidification of SGr iron: i) the dendritic nature of the eutectic with several graphite spheroids within each austenite dendrite, and ii) the mushy type of solidification (Figure 19.54a). Later it was also suggested that some austenitic shell may develop around the graphite spheroids before any contact with the primary austenite (Figure 19.54b).

This solidification scenario is supported by microstructures found in the microshrinkage of thin SGr iron plates presented in Figure 19.55. Microshrinkage can be viewed as a region of interrupted solidification. Two types of austenite exhibiting non-similar morphologies can be identified: *primary austenite dendrites and eutectic austenite containing SGr*. The morphology of the primary austenite dendrites is typical for dendrites in metallic alloys. They exhibit clear primary and secondary arms (Figure 19.55a). The eutectic austenite is made of several austenite shells surrounding the graphite (Figure 19.55b). While displaying branching, there is no clear distinction between primary and secondary arms. It appears that the γ/SGr eutectic aggregates are made of several graphite nodules surrounded by quasi-spherical austenite envelopes. Thus it is apparent that an eutectic grain cannot be clearly defined in SGr iron. The main solidification unit is the primary austenite

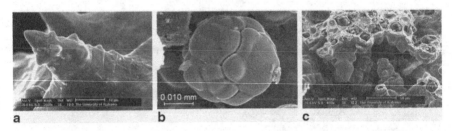

Fig. 19.55 Microstructures of SGr iron found in the same microshrinkage cavity from a SGr iron plate (Ruxanda et al. 2001). **a** Primary austenite dendrite. **b** Eutectic austenite dendrite—SGr aggregate. **c** Overall view of microshrinkage. (Copyright 2001 American Foundry Soc., used with permission)

dendrite within which graphite nodules grow through a divorced eutectic reaction together with the eutectic austenite.

The solidification sequence of eutectic SGr iron may be summarized as follows. At the beginning of eutectic solidification, graphite nodules and austenite crystals are nucleating independently in the liquid. Natural convection will move both phases. Once nucleated, the austenite dendrites will grow freely in the liquid. This is possible even though the composition is eutectic, as cast irons solidify following an asymmetric phase diagram. While the austenite dendrites grow, carbon is continuously rejected at the g/L interface, increasing the probability of further SGr nucleation. As seen in Figure 19.55c, multiple nodules may grow within the same austenite dendrite.

Once incorporated in austenite, the graphite particles grow isotropically through solid-state diffusion of carbon from the liquid through the austenite shell to the graphite particle. The first graphite particles to be incorporated will have the longest time to grow and will become the largest nodules. Smaller graphite spheroids will be found in the austenite closer to the interface.

The primary austenite growing into the liquid will tend to grow anisotropically in its preferred crystallographic orientation. However, restrictions imposed by isotropic diffusion growth will impose an increased isotropy on the system. Consequently, the dendritic shape of the austenite will be altered and the γ-L interface will exhibit only small protuberances instead of clear secondary arms. This process is dominant in the last regions to solidify.

This interpretation is consistent with phase-field modeling of dendritic growth. Karma and Rappel (1998) have compared 3D growth of dendrites with cubic symmetry for the extreme cases of high anisotropy, Fig. 19.56a, and no anisotropy, Fig. 19.56b. It is seen that a metal with a cubic lattice will grow in a "cauliflower" shape in the absence of anisotropy. The SEM investigation showed remarkable similarities between the simulated and real austenite morphology for the case of isotropic growth (compare Fig. 19.56b with Fig. 19.56c).

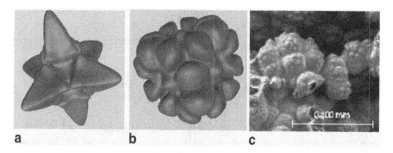

Fig. 19.56 Simulated growth of the dendrite morphology (Karma and Rappel 1998) a) and b), and real dendrite morphology (Ruxanda et al. 2001) c). **a** Simulated growth with high anisotropy. (Reprinted with permission from Phys. Rev. Copyright 1998 by the American Physical Soc. www.aps.org). **b** Simulated growth without anisotropy. (Reprinted with permission from Phys. Rev. Copyright 1998 by the American Physical Soc. www.aps.org). **c** SGr eutectic dendrites where the loss of anisotropy lead to an effective modification of the dendrite morphology

Fig. 19.57 Schematic illustration of solidification mechanisms of continuously cooled lamellar and spheroidal graphite cast iron

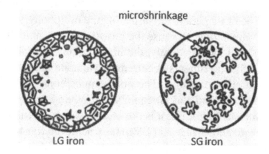

As shown in Fig. 19.57, the solidification mechanisms of lamellar and spheroidal graphite cast iron during continuous cooling are quite different. LGr iron solidifies with skin formation while SGr iron is characterized by mushy solidification.

There has been some debate over the uni-nodular or multi-nodular morphology of the eutectic grain in SGr iron. The uni-nodular concept permeated from the modeling community that was using the picture of one nodule growing through carbon diffusion through a surrounding austenite shell. Color etching then demonstrated that several Gr spheroids are typically surrounded by a highly segregated last-to-freeze region, and this aggregate was then defined as "multi-nodular eutectic grains" (Rivera et al. 1976). Further research with the DAAS technique and Electron Back Scattering Diffraction (EBSD) (Boeri and Sikora 2001, Rivera et al. 2004) revealed that the solidification microstructure includes large austenite grains with numerous graphite nodules.

After this discussion we hope that it is clear to the reader that an eutectic grain cannot be defined in SGr iron. Several nodules grow in contact with the primary austenite resulting in additional eutectic austenite that cannot be distinguished from the primary austenite, with which it shares the same crystallographic orientation.

Fig. 19.58 Influence of *G/V* ratios and percent cerium on structural transitions in cast iron (Bandyopadhyay et al. 1989). FG stands for flake graphite, i.e., lamellar graphite

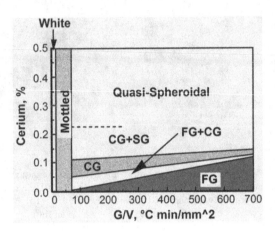

Stable Solidification of the Austenite-Graphite Eutectic—Directional Solidification
The basic parameters affecting the morphology of the eutectic are the *G/V* ratio and composition. Because the growth velocity and the temperature gradient can be controlled independently, the information obtainable through directional solidification (DS) experiments is extremely valuable in understanding the intricacies of solidification of cast iron. The first composition-*G/V* diagram for cast iron was produced by Lakeland and Hogan (1968). When varying *G/V* and/or the level of impurities, such as Mg or Ce, it is possible to obtain a variety of graphite and matrix structures. Argo and Gruzleski (1986) have achieved transition from a SGr through CGr to LGr structure by directional solidification of Mg containing SGr iron. The cooling rates used in their work varied from 0.0015 to 0.013 °C/s, which is relatively low. Consequently, complete structural transition, to include the stable/metastable one, was not obtained.

The complete structural transition from metastable to stable, and for different graphite morphologies, has been documented for cast irons of hypoeutectic composition, as a function of growth velocity and temperature gradients at the S/L interface, and cerium concentration (Fig. 19.58). Note that because Ce was used as a graphite shape modifier, the graphite was not fully spheroidal. It was found that while the metastable-to-stable (white-to-gray) transition depends mostly on the *G/V* ratio, the transition between different graphite shapes (lamellar-to-compacted-to spheroidal) depends mostly on the cerium concentration.

A sequence of changes in the eutectic morphology of directionally solidified cast iron was proposed, as shown in Figure 19.59a. As the *G/V* ratio decreases, or the composition *Co* (e.g., Mg or Ce) increases, the S/L interface changes from planar, to cellular, and then to equiaxed, while graphite remains lamellar. Cooperative growth of austenite and graphite occurs. Further change of *G/V* or of *Co* brings about formation of an irregular interface, with austenite dendrites protruding in the liquid. Graphite becomes compacted and then spheroidal. Eutectic growth is divorced. From these structures, those that have practical importance are the first three that is the irregular SG, the irregular CG, and the eutectic grains of lamellar graphite.

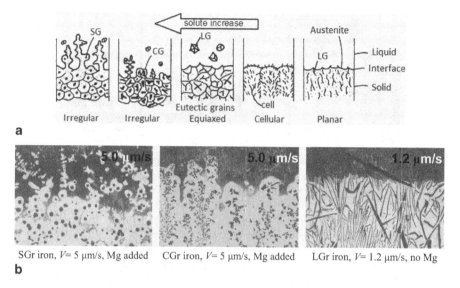

SGr iron, V = 5 μm/s, Mg added   CGr iron, V = 5 μm/s, Mg added   LGr iron, V = 1.2 μm/s, no Mg

b

Fig. 19.59 Influence of composition and solidification velocity on the morphology of the S/L interface. **a** Schematic representation (Stefanescu and Bandyopadhyay 1989). **b** Interrupted directional solidification experiments (Li et al. 1990, copyright 1990 American Foundry Soc., used with permission)

Interrupted directional solidification (DS) experiments confirmed the proposed sequence From Fig. 19.59b, it is seen that the S/L interfaces of SGr and CGr irons are coarse, while that of LGr iron is closer to planar. SGr iron solidifies with an austenite–dendritic interface that incorporates graphite nodules. CGr iron shows an austenite–cellular interface that includes the CGr. In both cases the leading phase during solidification is the austenite. LGr iron solidifies with a rather flat interface where graphite and austenite grow cooperatively. Graphite is the leading phase in this case.

For solidification of regular eutectics, a number of relationships have been established between process and material parameters based on the extremum criterion (see Sect. 10.1.3 for analytical models). Eutectic gray iron can solidify with a planar interface. The interlamellar spacing of LGr is a function of cooling rate and composition. On Fig. 19.60, summarizing some DS experiments, it is seen that as the cooling rate (growth velocity) decreases the lamellar spacing increases. Sulfur additions clearly increase the spacing, even in amounts as low as 0.001 %. For the sulfur containing irons, a sudden decrease in spacing is observed at growth velocities around 10^{-5} m/s. It can be attributed to the transition from type-A to type-D graphite. For irons containing more than 0.006 % S, the graphite tip protrudes into the melt (Nakae and Shin 1999).

Many other DS data were generated for LGr iron and summarized as λ–V relationships: Lakeland (1964) $\lambda = 3.8 \cdot 10^{-5} \, V^{-0.5}$ cm; Nieswaag and Zuithoff (1975)

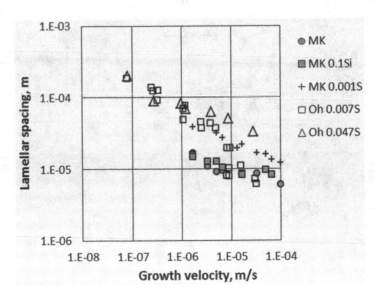

Fig. 19.60 Effect of growth velocity and composition on the lamellar spacing of LGr. Data from Magnin and Kurz 1988 (MK) and Ohira et al. 1974 (Oh). MK—Fe-C eutectic; MK 0.1Si—Fe-C– 0.1 % Si; Oh 0.0075 S—Fe-C– 0.01 %Si– 0.007 %S; Oh 0.0475 S—Fe-C– 0.01 %Si– 0.047 %S

$\lambda = 0.56 \cdot 10^{-5} \ V^{-0.78}$ cm (for 0.004 %S) and $\lambda = 7.1 \cdot 10^{-5} \ V^{-0.57}$ cm (for S > 0.02 %). All these data are positioned above the theoretical Jackson–Hunt model $\lambda = 1.15 \cdot 10^{-5} \ V^{-0.5}$ cm. As discussed earlier (Sect. 10.1.4), lamellar graphite iron is an irregular eutectic, and both an extremum and branching spacing can be defined. The average spacing should be higher than the extremum one predicted form Jackson–Hunt theory.

Based on DS experimental information the sequence of changes in the eutectic morphology of directionally solidified gray cast iron is summarized in Fig. 19.61. As the cooling rate (GV) increases, the interface of LGr iron changes from planar to cellular, and then to equiaxed. For relative higher cooling rates, the cellular interface may breakdown into a dendritic-equiaxed mushy zone. Cooperative growth of austenite and graphite occurs. Before the breakdown of the planar interface, the graphite becomes much finer and twisted, as seen in Fig. 19.60. An increase in the cooling rate may result in higher nodularity for the CGr iron, and finer structure for SGr iron. Increasing the amount of reactive impurities, or decreasing the content of surface-active impurities, brings about formation of an irregular interface, with austenite dendrites protruding in the liquid, and changes graphite shape from LGr to CGr and then to SGr. Eutectic growth is divorced.

Metastable Solidification of the Austenite-Fe Carbide Eutectic Metastable solidification of cast iron produces what is commercially called a white iron whose microstructure consists of austenite dendrites + austenite—iron carbide (Fe_3C)

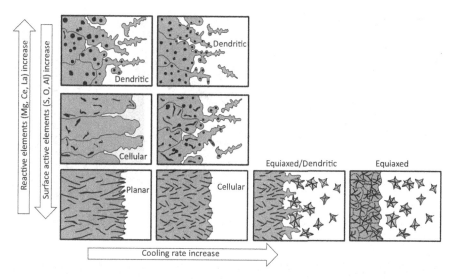

Fig. 19.61 Solid/liquid interface morphology and graphite shape in directionally solidified austenite–graphite eutectics

eutectic (ledeburite). The white eutectic consists of iron carbide plates or rods in an austenitic matrix that becomes pearlite at room temperature (Fig. 19.62). Growth of ledeburite starts with the development of a cementite plate on which an austenite dendrite nucleates and grows. This destabilizes the Fe_3C which then grows through the austenite. As a result, two types of eutectic structure develop, a lamellar eutectic with Fe_3C the leading phase in the edgewise direction, *a*, and a rod-like eutectic in the sidewise direction, *c* (Fig. 19.63). Under specific directional solidification conditions ledeburite behaves like a regular eutectic, as shown in Fig. 19.64.

It was suggested that as the cooling rate increases, the lamellar part of the original parallelepipedic eutectic grain becomes larger at the expense of the rod eutectic, the grain starts bending inward, and eventually a spherulitic eutectic grain results (Fig. 19.65).

Unalloyed white irons do not have significant practical applications. On the contrary, medium and high-alloyed irons are extensively used for their abrasion resistance. The alloying elements change the composition and morphology of the carbides, as well as the microstructure of the matrix. For example, irons having relatively high carbon, 3–5 % Ni, and 1.4 to 4 % Cr solidify with a martensitic matrix (Fig. 19.66a). Chromium promotes white solidification with a continuous network of alloyed iron carbides, $(Fe, Cr)_3C$. When the chromium content increases from 7 to 11 %, the composition and morphology of the carbides change to discontinuous Cr_7C_3 eutectic carbides (Fig. 19.66b).

The amount of Cr_7C_3 carbides can be increased by increasing the amount of chromium to 15–25 %. For the high-chromium irons in Fig. 19.67, as the carbon content increases the amount of carbides increases. At low carbon, the carbides

Fig. 19.62 Microstructure of the iron–iron carbide eutectic (ledeburite) (Stefanescu and Ruxanda 2004). (Reprinted with permission of ASM International. All rights reserved. www.asminternational.org)

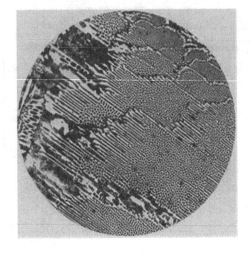

Fig. 19.63 Schematic representation of the growth of the ledeburite eutectic. (Hillert and Steinhauser 1960)

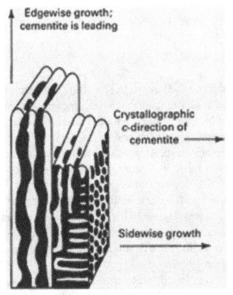

form a continuous network around the austenite dendrites. When the carbon content exceeds the eutectic carbon content, large hexagonal carbide rods precipitate from the melt as primary phases.

19.2.7 The Gray-to-White Structural Transition

Of particular interest to the cast iron manufacturer is the stable-to-metastable microstructure transition (also known as the gray-to-white transition, GWT), as it

Fig. 19.64 Longitudinal section of DS white cast iron (Magnin and Kurz 1988). (Reprinted with permission of ASM International. All rights reserved. www.asminternational.org)

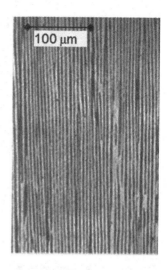

signals the occurrence of unwanted iron carbides in the gray iron. In a binary Fe–C alloy, the difference between the stable (T_{st}) and metastable (T_{met}) eutectic temperatures is only 3K. Thus, during cooling of a Fe-C casting the temperature of the melt may become smaller than T_{met} before any stable structure nucleates and grows. This may happen even at very low cooling rates. In the Fe–C–Si system the $T_{st} - T_{met}$ interval is much larger, and stable solidification may occur before the temperature reaches T_{met}.

The first rationalization of the GWT was based on the influence of cooling rate on the stable (T_{st}) and metastable (T_{met}) eutectic temperatures. As the cooling rate increases, both temperatures decrease (Fig. 19.68). However, since the slope of T_{st} is steeper than that of T_{met}, the two intersect at a cooling rate which is the critical cooling rate, $(dT/dt)_{cr}$, for the GWT. At cooling rates smaller than $(dT/dt)_{cr}$ the iron solidifies gray, while at higher cooling rates it solidifies white.

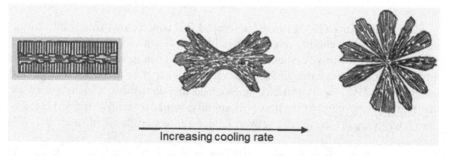

Increasing cooling rate

Fig. 19.65 Schematic illustration of the change in the morphology of the austenite-Fe$_3$C eutectic grain at increasing cooling rate. (Bunin et al. 1969)

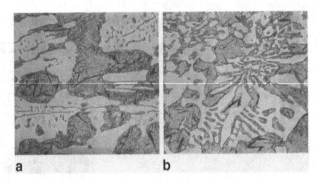

Fig. 19.66 Microstructures of nickel–chromium abrasion resistant white irons. Magnification: 340 X. (Gundlach 1988). **a** 3– 3.6 % C, 3.3– 5 % Ni, 1.4– 4 % Cr. **b** 2.5–3.6 % C, 5–7 % Ni, 7–11 % Cr. (Reprinted with permission of ASM International. All rights reserved. www.asminternational.org)

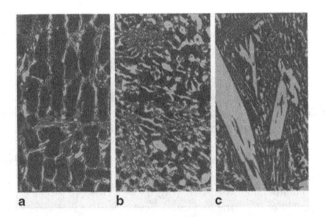

Fig. 19.67 Microstructures of high-chromium white irons. Magnification: 75 ×. (Gundlach 1988). **a** Low carbon (hypoeutectic). **b** Eutectic. **c** High carbon (hypereutectic). (Reprinted with permission of ASM International. All rights reserved. www.asminternational.org)

Magnin and Kurz (1985) further developed this concept by using solidification velocity rather than cooling rate as a variable. As shown in Fig. 19.69, the curves describing the growth velocities of the two eutectics intersect at V_{cr}. Since above this velocity the metastable growth velocity is higher than the stable one at any undercooling ($V_{met} > V_{st}$), it can be assumed that above this velocity (or under this temperature) only white iron will solidify, while under V_{cr} the structure is completely gray. Thus, if only growth considerations are invoked, a clear GWT exists.

These arguments have ignored nucleation. It is well accepted that nucleation of white iron is more difficult than that of gray iron. Consequently, additional undercooling, in excess of that predicted from growth velocity considerations, is required for a complete GWT. This is shown in the figure as ΔT_n^{met}, that is the nucleation temperature of the metastable cementite, displaces the critical velocity to the right,

Fig. 19.68 Critical cooling rate for the GTW transition

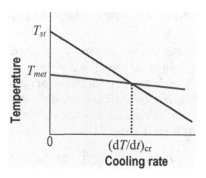

Fig. 19.69 Structural transitions in cast iron as function of undercooling and growth velocity. (After Magnin and Kurz 1985)

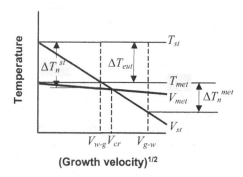

from V_{cr} to V_{g-w}. Thus, a complete white iron is obtained only at growth velocities larger than V_{g-w}. A similar argument holds for the white-to-gray transition, when a complete transition cannot occur unless the undercooling is smaller than ΔT_n^{st}, which is the nucleation temperature of the stable gray eutectic. Thus, a region of mixed structure, gray and white, will exist at growth velocities between V_{w-g} and V_{g-w}. This is the mottled region.

Based on this discussion it is clear that a model that predicts the GWT must describe nucleation and growth of both the gray and white eutectic. An analytical model was proposed by Fras and Lopez (1993). For eutectic iron the *chilling equivalent* is:

$$E = \frac{T_{st}^{1.08}}{1.9\Delta T_{pour}^{0.5}}\left(\frac{1}{\mu_1\mu_2{}^3c^5\left(\Delta T_{eut} + \Delta T_n{}^{met}\right)^{10}}\right)^{1/6} \qquad (19.11)$$

where ΔT_{pour} is the superheating above the eutectic temperature, μ_1 and μ_2 are nucleation and growth coefficients, respectively, and c is the specific heat. It is seen that the chilling tendency (chilling equivalent) decreases as the number of eutectic grains and their growth rate increase (μ_1 and μ_2 increase), and as the eutectic interval, ΔT_{eut}, and the pouring temperature increase.

19.2.8 Thermal Analysis of Cast Iron

One of the first applications of thermal analysis (TA) in cast iron was for the estimation of the chemical composition, or more precise of the carbon equivalent (*CE*). *CE* includes the contribution of carbon and of the other important elements (Si, Mn, P, S, etc.), allowing the multicomponent iron to be treated like a binary Fe–C alloy. It can be calculated from equilibrium thermodynamics as (see Stefanescu and Katz 2008 for details):

$$CE = \%C + 0.31 \cdot \%Si + 0.33 \cdot \%P - 0.027 \cdot \%Mn + 0.4 \cdot \%S. \qquad (19.12)$$

However, because of the discrepancy between the theoretical and actual liquidus line, Humphreys (1991) has introduced the Carbon Equivalent Liquidus (*CEL*), which is different than *CE,* based on experiments with standardized test sand cups:

$$TL = 1669 - 124 \cdot CEL \quad \text{where} \quad CEL = \%C + \%Si/4 + \%P/2. \qquad (19.13)$$

Further work (Donald and Moore 1973) using two cups, a standard one and one that had some tellurium addition, resulted in the ability to determine both C and Si. The tellurium addition has the effect of promoting metastable (white) solidification which because of the high growth rate of the white eutectic (ledeburite) eliminates the eutectic undercooling. The eutectic arrest, TE_{white}, becomes flat. The equations proposed by Heraeus Electro-Nite are:

$$CEL = 14.45 - 0.0089 \cdot TL$$
$$\%C = -6.51 - 0.0084 \cdot TL + 0.0178 \cdot TE_{white}$$
$$\%Si = 78.411 - 4.28087 \cdot Si\text{-adj.} - 0.06831 \cdot TE_{white} \qquad (19.14)$$

where Si-adj. is a correction factor, mainly depending on the phosphorus content of the iron. For SGr iron, the same equation as for LGr iron can be used to calculate CEL by substituting *TL* with (*TL* + 5).

For other applications such as the prediction of the general eutectic structure (i.e., gray or white), of the degree of inoculation, of graphite shape, of shrinkage propensity, of melt oxidation, and even of mechanical properties using one, two, or three thermocouples, the reader is referred to the review paper by Stefanescu (2015).

A successful industrial process based on the two-thermocouple DTA method is the SinterCast process for production of CGr iron (Dawson and Popelar 2013). It uses a steel sheet cup with two thermocouples. The walls of the cup are coated with a reactive coating that consumes active magnesium in order to simulate the fading of magnesium in the ladle. This allows for the simultaneous measurement of the solidification behavior at the start of casting (through the center thermocouple) and also after a predetermined loss of magnesium (through the bottom thermocouple). Computer-aided cooling curve analysis allows then the prediction of the solidification microstructure of the iron and, more importantly, the corrections necessary to achieve the desired CGr microstructure.

19.2.9 Simulation of Solidification of Cast Iron

The age of computational modeling of cast iron was started by Oldfield (1966) who developed a computer model that could calculate the cooling curves of LGr iron. This appears to be the first attempt to predict solidification microstructure through computational modeling, and the first attempt to validate such a model against cooling curves. The nucleation process was approximated to a time independent model, in which all available nuclei within the undercooling range are assumed to grow simultaneously (see Sect. 18.1 and Table 18.1). The growth of spherical eutectic grains was calculated with the parabolic law $V = \mu \Delta T^2$. For the first time, it was demonstrated that the growth constant is significantly affected by the sulfur content of the iron. Values for this constant were evaluated by fitting the calculated and experimental undercooling. Correction for grain impingement against one another and against the wall, and a computer model for heat flow across a cylinder similar to FDM were also proposed. Other models using analytical solution for energy transport and time-stepping procedures were developed in the 70s and 80s (see review by Stefanescu 2007).

Simulation of Spheroidal Graphite Iron Solidification The first coupled FDM energy transport–solidification kinetics model for SGr iron was proposed by Su et al. (1985). They used Oldfield's nucleation model, carbon diffusion-controlled growth through the γ shell, and performed some validation against experiment. This model was improved by Rappaz et al. (1989) who accounted for the volume change during solidification of the SGr eutectic. Equation 10.27 was modified to:

$$\frac{dr_\gamma}{dt} = \left[D_C^\gamma \frac{r_G \left(C^{G/\gamma} - C^{L/\gamma} \right)}{(r_G - r_\gamma) r_\gamma} + \frac{\rho_L}{\rho_\gamma} \frac{dC^{\gamma/L}}{dt} \frac{r_o^3 - r_\gamma^3}{3 r_\gamma^2} \right] \left(C^{\gamma/L} - C^{L/\gamma} \right)$$

(19.15)

where r_o is the equivalent radius of the eutectic domain ($r_o = \sqrt{3} \cdot 0.5(1/N)1/3$).

According to Chen et al. (1995), this model can experience problems because of the assumption of complete mixing of C in the liquid, and because the second term in the first parenthesis can become very large in the early stages of the eutectic reaction (small r_γ value), which results in a bouncing effect. A transient carbon solute boundary layer, which allowed for the carbon balance to be maintained during the eutectic solidification, was introduced.

Mampey (1997) included fluid flow in the transport calculations, compared filling simulation with experiment, and demonstrated the influence of mold filling on the final distribution of nodule count. He also illustrated the shifting of thermal center and the reduction of radial temperature differences when flow was included.

Onsøien et al. (1999) used the internal state variable approach to model the multiple phase changes occurring during solidification and subsequent cooling of near-eutectic SGr iron. The microstructure evolution was captured mathematically in terms of differential variation of the primary state variables with time for each of the relevant mechanisms. Separate response equations were developed to convert the current values of the state variables into equivalent volume fractions of constituent phases utilizing the constraints provided by the phase diagram. The results may conveniently be represented in the form of C-curves and process diagrams to illuminate how changes in alloy composition, graphite nucleation potential, and thermal program affect the microstructure evolution at various stages of the process. The model can be implemented in a dedicated numerical code for the thermal field.

Recently, Escobar et al. (2014) compared solidification of SGr iron for two growth models. In the first model, dubbed the multinodular model, it is assumed that austenite and graphite nucleate and grow independently in the liquid phase until the nodule reaches 6 μm in size, when the Gr is enveloped by the austenite shell. Subsequent growth occurs by carbon diffusion trough the austenite shell. The second model, dubbed the uninodular model, assumes new graphite spheroids are immediately surrounded by austenite and that all austenite formed during the eutectic reaction grows as an envelope around the graphite nodules.

For the multinodular model, the austenite grows by the lever rule and the growth of graphite in contact with the liquid is expressed by Zener's equation for the growth of an isolated spherical particle in a matrix with low supersaturation:

$$\frac{dr_G}{dt} = \frac{1}{2r_G} \frac{C^{L/\gamma} - C^{L/G}}{C^G - C^{L/G}} \frac{\rho_L}{\rho_G} D_C^L. \tag{19.16}$$

Once the graphite nodules are enveloped by austenite their growth rate is calculated as:

$$\frac{dr_G}{dt} = \frac{0.9 r_\gamma / r_G}{r_G(r_\gamma / r_G - 1)} \frac{C^{\gamma/L} - C^{\gamma/G}}{C^G - C^{\gamma/G}} \frac{\rho_\gamma}{\rho_G} D_C^\gamma (1 - f_S)^{2/3}. \tag{19.17}$$

For the uninodular model, the growth of graphite and austenite are given by the equations:

$$\frac{dr_G}{dt} = \frac{r_\gamma}{r_G(r_\gamma - r_G)} \frac{C^{\gamma/L} - C^{\gamma/G}}{C^G - C^{\gamma/G}} \frac{\rho_\gamma}{\rho_G} D_C^\gamma \quad and$$

$$\frac{dr_\gamma}{dt} = \frac{r_G}{r_\gamma(r_\gamma - r_G)} \frac{C^{\gamma/L} - C^{\gamma/G}}{C^G - C^{\gamma/G}} \left[1 + \left(\frac{\rho_\gamma}{\rho_G} - 1 \right) \frac{C^{L/\gamma} - C^{\gamma/L}}{C^G - C^{\gamma/G}} \right] D_C^\gamma. \tag{19.18}$$

Note that the eqaution for dr_G/dt is essentially the same as Eq. 10.27, with the added influence of the densities of the phases. No significant differences were found between predictions with the two models.

An example of the use of a commercial code coupled to user-developed subroutines is presented by Meijer et al. (1999). They developed a mathematical heat-transfer/microstructural model based on the commercial FE code ABAQUS. Specialized routines based on relationships describing nucleation and growth of equiaxed primary austenite, gray iron eutectic, and white iron eutectic have been incorporated into ABAQUS through user-specified subroutines. The phase distribution predicted with the model captured the correct trend, but exhibited a general error of 10–20 %.

Fig. 19.70 Virtual solidification microstructures of SGr cast iron. (Beltran-Sanchez 2003)

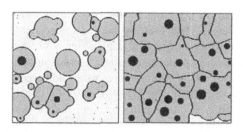

The transformation of the computer into a dynamic microscope that propelled cast iron in the category of virtual materials was pioneered by Rappaz and his collaborators with their application of the cellular automaton (CA) technique to microstructure evolution modeling. The first application of the CA technique to cast iron is due to Charbon and Rappaz (1997) who used the classic uninodular model for diffusion-controlled graphite growth through the austenite shell to describe SGr iron solidification. Beltran-Sanchez and Stefanescu (Ruxanda et al. 2001; Beltran-Sanchez 2003) improved on the previous model by including solidification of primary austenite grains and by initiating graphite growth once graphite nuclei came in contact with the austenite grains. After contact graphite was allowed to grow through the diffusion-controlled growth mechanism. Using Object Oriented Programming in C++, a grain was defined as an object that had several properties: type (primary austenite grain, graphite, eutectic austenite), index (e.g., color, to distinguish two grains of same type of austenite for instance), growth law (one for primary austenite, another for the graphite, and coupled for the eutectic austenite), location (x, y, z-coordinates), and state (liquid or solid only, no mushy zone or transforming cells). Random nucleation was used for Gr and γ. All grains grow like spheres using deterministic laws. All cells within that sphere are transformed into solid. The primary and eutectic austenite with all the graphite nuclei being touched by the austenite became at the end, for display purposes only, a single grain. Since each grain is an object, each one of them operated more or less independently of each other, following its own local properties at the grain level (Fig. 19.70).

In their CA models, Burbelko et al. (2011, 2012) account for the continuous nucleation of austenite and graphite from liquid, and the subsequent growth of the γ-Gr eutectic under nonsteady-state temperature field. The interface migration rate was considered to be a linear function of the local kinetic undercooling, which requires a kinetic growth coefficient for computation. The model, which does not include the effect of density difference between graphite and iron, was used to investigate the effect of the growth of graphite spheroids on the inhomogeneity of the carbon concentration field and on the growth of the austenite dendrite.

Zhao et al. (2011) also proposed a multiphase CA model to simulate the solidification of the divorced eutectic microstructures of SGr cast iron. In the model, the growth kinetics of both graphite nodules and austenite dendrites were determined using a local composition equilibrium approach which allows calculation of the growth velocities for the two phases without the need of introducing kinetic coefficients. However, as the average concentration was used to calculate the growth velocity of graphite, and as the γ-Gr density difference was ignored, the simulated graphite fraction and average nodule size were lower than the experimental data. These shortcomings were addressed by the authors in a following paper (Zhu et al. 2015). The effect of the γ-Gr density difference was included in the calculation of the graphite fraction, and the actual local carbon concentration was used to evaluate the growth kinetics of graphite nodules. The computation domain was divided into uniform square cells (200×200 mesh with a mesh size of $\Delta x = 2$ μm) characterized by several variables, such as composition, temperature, phase fractions (liquid, austenite, and graphite), crystallographic orientation (for austenite dendrite), index (for graphite nodule), and states including liquid cell ($f_L = 1$), graphite cell ($f_{Gr} = 1$), austenite cell ($f_\gamma = 1$), the graphite/liquid (Gr/L) interface cell ($0 < f_{Gr} < 1$, and $f_\gamma = 0$), the austenite/liquid (γ/L) interface cell ($0 < f_\gamma < 1$, and $f_{Gr} = 0$), the graphite/austenite (Gr/γ) interface cell ($f_{Gr} > 0$, $f_\gamma > 0$, and $f_{Gr} + f_\gamma = 1$), and the graphite/austenite/liquid (Gr/γ/L) interface cell ($f_{Gr} > 0$, $f_\gamma > 0$, and $f_{Gr} + f_\gamma < 1$), where f_L, f_{Gr} and f_γ are the fractions of liquid, graphite, and austenite, respectively. The cells that belong to the same graphite nodule share the index, radius, and volume of the nodule.

The governing equation for solute diffusion in the entire domain is given by:

$$\partial C_i / \partial t = \nabla \cdot (D_i \nabla C_i) + C_L^*(1 - k_\gamma) \partial f_\gamma / \partial t + (C_{\text{int}}^* - C_{Gr}) \partial f_{Gr} / \partial t \quad (19.19)$$

where t is the time, C_i is the composition, D_i is the diffusivity, and the subscript i indicates the γ, the Gr or the L. The second term on the right-hand side, $C_L^*(1 - k_\gamma) \partial f_\gamma / \partial t$, is calculated only at the γ/L and Gr/γ/L interfaces, and denotes the amount of solute carbon rejected at the γ/L and Gr/γ/L interfaces due to the growth of austenite phase. The third term on the right-hand side, $(C_{\text{int}}^* - C_{Gr}) \partial f_{Gr} / \partial t$, is calculated only at the Gr/L, Gr/γ and Gr/γ/L interfaces, and indicates the amount of carbon absorbed by the growing graphite at those interfaces. C_{int}^* is the composition of austenite at the Gr/γ interface, or the composition of liquid at the Gr/L and Gr/γ/L interfaces when the graphite grows from austenite or liquid, respectively.

For each γ/L interface cell with $0 < f_\gamma < 1$, the compositions of liquid and austenite are calculated simultaneously. The austenite composition is calculated with:

$$C_\gamma = \frac{\sum_{n=1}^{N} \Delta f_\gamma(n) k_\gamma C_L(n)}{\sum_{n=1}^{N} \Delta f_\gamma(n)} \quad (19.20)$$

where N indicates the number of iterations after the cell becomes a γ/L interface cell, i.e., when the cell is captured as a γ/L interface cell, $N = 1$.

To handle the problem of discontinuous properties at the interfaces when solving Eq. 19.19, an equivalent composition C_e and equivalent diffusion coefficients, D_e, are defined. Then, the entire domain can be treated as a single phase for solute transport, which was solved with an explicit FD scheme with second-order accuracy.

Since the thermal diffusivity of Fe–C alloys is about three orders of magnitude larger than the solute diffusivity in the liquid, it can be assumed that the temperature field in a microscale domain is uniform and the temperature decreases with a time dependent cooling rate that can be evaluated with:

$$\frac{\partial T}{\partial t} = -\dot{T} + \frac{\Delta H_\gamma}{\rho c_L} \frac{\partial f_\gamma}{\partial t} + \frac{\Delta H_{Gr}}{\rho c_L} \frac{\partial f_{Gr}}{\partial t} \qquad (19.21)$$

where $\dot{T}$ is an imposed heat extraction, ρ is the density of liquid, c_L is the specific heat of liquid, ΔH_γ and ΔH_{Gr} are the volumetric latent heat of austenite and graphite, respectively.

Cooling rate-dependent instantaneous nucleation laws were used for both austenite and graphite. The driving force for graphite nodule growth is a function of the difference between the local equilibrium composition and the local actual composition at the interfaces. For the case of a graphite nodule growing in the liquid, the equilibrium composition at the Gr/L interface is:

$$C_{Gr/L}^{eq} = C_o + \frac{T^* - T_{Gr/L}^{eq}}{m_{L,Gr}} + \frac{\Gamma_{Gr} K_{Gr/L}}{m_{L,Gr}} \qquad (19.22)$$

where T^* is the local temperature at the interface, $T_{Gr/L}^{eq}$ is the equilibrium liquidus temperature at the initial composition C_0, $m_{L,Gr}$ is the liquidus slope of graphite, and $K_{Gr/L}$ is the curvature of the graphite nodule at the Gr/L interface. Γ_{Gr} is the Gibbs–Thomson coefficient of graphite. In the present 2D model, the average interface curvature of each graphite nodule is calculated as $K_{Gr/L} = 1/\bar{R}$ with $\bar{R} = (V_C \sum f_{Gr}/\pi)$, where $\bar{R}$ is the average radius of a graphite nodule, V_C is the volume of each cell, and $\sum f_{Gr}$ is the sum of graphite fractions of all cells in the same graphite nodule.

During one-time step interval, the increment in graphite fraction of a Gr/L interface cell is evaluated with

$$\Delta f_{Gr} = G_{Gr} \cdot \frac{(C_{Gr/L}^* - C_{Gr/L}^{eq})}{(C_{Gr} - C_{Gr/L}^{eq})} \qquad (19.23)$$

where $C_{Gr/L}^*$ is the local actual liquid composition at the Gr/L interface obtained with Eq. 19.19, and G_{Gr} is a geometrical factor used to eliminate the artificial anisotropy caused by the CA square grid.

The equilibrium composition at the Gr/γ interface for graphite growing in the liquid is:

$$C_{Gr/\gamma}^{eq} = C_o + \frac{T^* - T_{Gr/\gamma}^{eq}}{m_{S,Gr}}. \qquad (19.24)$$

During one-time step interval, the increase in graphite fraction of a Gr/γ interface cell is calculated with an equation similar to Eq. 19.23, where the subscript Gr/L is substituted with Gr/γ. If at time $t = t_n$, the sum of the graphite fraction in a Gr/L or a Gr/γ interface cell equals one, i.e., $\sum_n f_{Gr}(n) = 1$, this cell transforms its state from a Gr/L or Gr/γ interface to graphite, and gets the same graphite nodule index

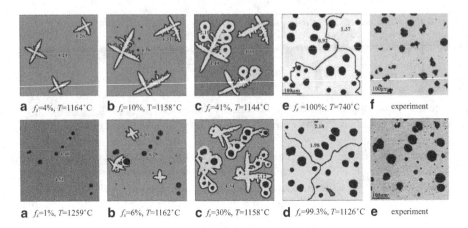

a f_s=4%, T=1164°C   **b** f_s=10%, T=1158°C   **c** f_s=41%, T=1144°C   **e** f_s=100%; T=740°C   **f** experiment

a f_s=1%, T=1259°C   **b** f_s=6%, T=1162°C   **c** f_s=30%, T=1158°C   **d** f_s=99.3%, T=1126°C   **e** experiment

Fig. 19.71 Simulated and experimental morphologies for SGr cast iron: *upper row*—hypoeutectic with $C_0 = 4.14\%$; *lower row* -hypereutectic $C_0 = 4.7\%$. C_0 is the carbon equivalent. Numbers on the figures show the local carbon concentration, and f_s is the total solid fraction (Zhu et al. 2015). (With permission of Elsevier).

as its graphite neighbor. This newly formed graphite cell in turn captures a set of its liquid or austenite neighbors to be the new *Gr/L* or *Gr/γ* interface cells. The graphite nodule growth will thus continue in the next time step.

Similar equations were used for the growth of austenite based on the difference between the local equilibrium liquid composition and local actual liquid composition. The degree of anisotropy of the surface energy was accounted for.

Selected model predictions are presented in Fig. 19.71 for hypo- and hypereutectic SGr irons. Note the reasonable prediction of the experimental microstructure, including the formation of a double graphite spheroid in Fig. 19.71d and 19.71e.

An important phenomenon in the production of iron casting is the graphite expansion during solidification, which decreases the need for risers and even allows for riserless castings. However, calculation of the amount and timing of the graphite expansion with respect to the solidification time is essential. Gurgul et al. (2014) attempted prediction of SGr iron density changes during solidification through a cellular automaton-finite differences (CA-FD) model. Their model takes into account the independent nucleation of graphite and austenite as a function of undercooling, the migration rate of the grain boundaries, nonuniform distribution of temperature and concentration in the calculation domain, and diffusion of the carbon in the liquid and the austenite. The FD component solved the heat flow and carbon diffusion in the computational domain. The interface migration rate was assumed as a linear function of the local kinetic undercooling, $V = \mu \, \Delta T_\mu$, where μ is the kinetic growth coefficient, different for austenite and graphite. The CA model has a set of six states of cells. At the beginning of simulation all cells have the same state—liquid. When the temperature inside a cell decreases below the equilibrium

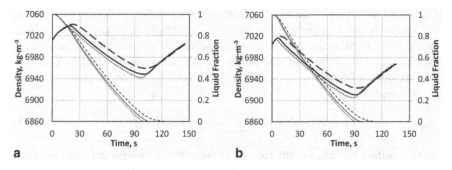

Fig. 19.72 The density changes of SGr iron (*bold lines*) and the liquid fraction (*thin lines*) during solidification; the graphite nucleation coefficient N_{max}: 10^9 m^{-2}– *dashed lines*, 10^{10} m^{-2}– *solid lines*, 10^{11} m^{-2}...... *dotted lines* (Gurgul et al. 2014). **a** Hypoeutectic alloy. **b** Eutectic alloy

temperature, the cell may change its state from liquid to transitional state (inside this cell a phase transformation can happen).

There are three types of the transitional states: $L{\rightarrow}\gamma$, $L{\rightarrow}$Gr, and, when a graphite nodule grows in an austenite shell, $\gamma{\rightarrow}$Gr. An increment Δf of a new phase inside each transitional cell is calculated by the fallowing equation:

$$\Delta f = \frac{V\Delta\tau}{a(|\cos\theta| + |\sin\theta|)} \qquad (19.25)$$

where $\Delta\tau$ is the time step, a is the side length of the CA square cell, and θ is the angle between the x-axis and the normal direction to the front. When a transitional cell finishes the transformation it changes its state into a solid state. There are two types of solid states. If the transformation is $L{\rightarrow}$Gr or $\gamma\rightarrow$Gr, the cell acquires the graphite state. For the transformation $L\rightarrow\gamma$ the cell is changed into the austenite state.

The densities of specific phases were calculated with the Thermo-CALC® software. The following regression equations were obtained (ρ in kg/m^3):

$$\rho_L = 8192.2 - 0.5402 \cdot T - 9805.3 \cdot C \qquad (19.26a)$$

$$\rho_\gamma = 8238.7 - 0.48684 \cdot T - 3876 \cdot C - 5982 \cdot C^2 \qquad (19.26b)$$

$$\rho_{Gr} = 2292.9 - 0.067442 \cdot T. \qquad (19.26c)$$

The calculated density changes for SGr iron, together with the changes of liquid fraction, are presented in Fig. 19.72. There are three stages for the density changes: pre-eutectic shrinkage, eutectic expansion, and final shrinkage. They have been proven experimentally through the use of Combined Liquid Displacement and Cooling Curve Analysis (Stefanescu et al. 2012; Alonso et al. 2014).

Simulation of Lamellar Graphite Iron Solidification To describe the solidification of eutectic LGr grains, Catalina and Stefanescu (1996) developed a spherical symmetry model assuming carbon diffusion-controlled growth and isothermal S/L interface. While the model was not accurate for the beginning and end of solidification, it produced very good agreement with the experiment for 75 % of solidification. The instantaneous average lamellar spacing was calculated as $\lambda_{Gr} = 2\pi r^*/N^*$, where r* is the instantaneous grain radius and N* is the instantaneous number of graphite lamellae on the grain circumference. At the end of the solidification process, an average lamellar spacing defined at the level of the whole grain, λ_{Gr}^{av}, can be calculated as:

$$\lambda_{Gr}^{av} = (1/r_t)\sum_{i=1}^{n} [\lambda_{Gr}(dr^*/dt)dt]_i \qquad (19.27)$$

where r_t is the total radius of the eutectic grain, dr^*/dt is the growth rate of the grain during the time-step i, Δt is the duration of the time-step, and n is the total number of time-steps required for eutectic solidification in the considered volume element. Also, the average thickness of graphite lamellae at the end of eutectic transformation, $\overline{S}_{Gr}$, can be calculated with the relationship $\overline{S}_{Gr} = f_{Gr}\lambda_{Gr}^{av}$. The model was implemented in Caterpillar's proprietary solidification software.

The first attempt to use a numerical model to predict the gray-to-white transition (GWT) in cast iron appears to belong to Stefanescu and Kanetkar (1987), who developed an axisymmetric implicit FDM heat transport model coupled with the description of the solidification kinetics of the stable and metastable eutectics. They validated model predictions against cast pin tests. Later, Nastac and Stefanescu (1995) produced a complete FDM model for the prediction of the GWT, which was incorporated in ProCAST. The model included the nucleation and growth of the stable and metastable phases and accounted for microsegregation. The model demonstrated such phenomena as the influence of Si segregation on the $T_{st}-T_{met}$ interval for gray and white irons (Fig. 19.73), and the influence of cooling rate and amount of Si on the gray-to-white and white-to-gray transitions (Fig. 19.74). Similar work was also reported later by Zhao and Liu (2003).

Leube et al. (1998) derived a branching criterion from the Magnin–Kurz model for irregular eutectics to calculate the maximum spacing in LGr iron. They calculated the minimum spacing as per Jackson–Hunt. Their model allows calculation of the average length of lamellae for all grains in one volume element and of the average of the longest graphite lamella. An interesting innovation was the simulation of the eutectic grain structure for 3D Johnson-Mehl-tesselation (Leube and Arnberg 1999) through a Monte-Carlo simulation based on a pixel gradual calculation of the structure. As shown in Fig. 19.75, low nucleation potential produced a nonuniform grain structure because of late nucleation during the undercooling at the end of solidification. The calculated final grain structure matched well with the experimental one.

Currently, a number of commercial software such as Magmasoft have the capability of simulating solidification of lamellar and spheroidal graphite irons, including output of the microstructure features.

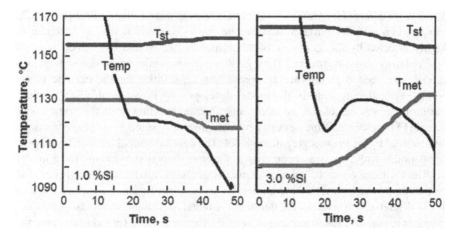

Fig. 19.73 Calculated influence of Si on the equilibrium temperatures and on the cooling curve for a volume element solidifying at an initial cooling rate of 9.3 °C/s (Nastac and Stefanescu 1995). (Copyright 1995 American Foundry Soc., used with permission)

Fig. 19.74 The influence of Si and initial cooling rate on structural transition in a 3.6 % C, 0.5 %Mn, 0.05 %P, 0.025 %S cast iron (Nastac and Stefanescu 1995)

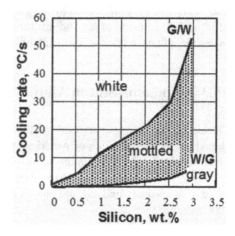

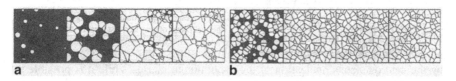

Fig. 19.75 Calculated development of eutectic grain structure in LGr iron (Leube and Arnberg 1999). **a** Low nucleation potential. **b** High nucleation potential

Multiphase Microsegregation As discussed, during eutectic solidification of SGr iron, carbon diffuses through an austenite shell. The carbon flux in austenite is highly affected by the gradients of other species like Si and Cu, which is often referred to as "cross-diffusion." Thus, multicomponent multiphase microsegregation models are needed to improve the predicting capabilities for the eutectic transformation of ductile iron. In the model developed by Pustal et al. (2009), solute partitioning was calculated assuming interface equilibrium with the commercial CALPHAD software. Time-sensitive microsegregation patterns and phase fractions were solved by the microsegregation model. The development of fraction liquid over temperature with time was compared to Gulliver–Scheil simulations for a model with and without cross-diffusion. The microsegregation model was directly coupled to the temperature solver of a commercial process simulation tool, which uses an FVM discretization. Due to the enormous amount of volume elements, it was not possible to simulate the precipitation kinetics at each element for industrial castings within reasonable time.

Microsegregation simulations for higher heat extraction rates when species diffuse only in their own gradient (normal diffusion) found that solidification may even be slower than predicted by Gulliver–Sheil. This is due to higher partition coefficients for Ni, Si, and Cu leading to negative gradients at the interfaces restraining the growth. This was not the case when cross-diffusion effects were accounted for. C and Ni are intensely influenced by the gradients of other species leading to the powerful up-hill diffusion of C. The resulting graphite fraction is considerably higher under the influence of cross-diffusion effects.

19.3 Aluminum–Silicon Alloys

Out of the about 238 compositions for casting aluminum alloys registered with the Aluminum Association of America, 46 % consists of aluminum–silicon alloys. However, 90 % of all aluminum alloy-shaped castings are made of Al–Si alloys (Granger and Elliott 1988). Most of these alloys are hypoeutectic. The length scale of the primary and eutectic phases is controlled through grain refinement and modification.

19.3.1 Nucleation and Growth of Primary Aluminum Dendrites

Most properties of practical interest depend on the length scale of the primary aluminum dendrites (grain size) of the aluminum alloy. Grain refinement is obtained by the heterogeneous nucleation of the α-aluminum phase. While it is possible to achieve this goal by dynamic nucleation, in metal casting practice this is mostly done by addition of chemicals. An efficient grain refiner must provide stable nuclei at the liquidus temperature of the alloy. The most common grain refiner is a master alloy containing titanium and boron. When adding more than 0.15 % titanium to the aluminum alloy, $TiAl_3$ forms, as can be seen on the phase diagram in Fig. 19.76. The

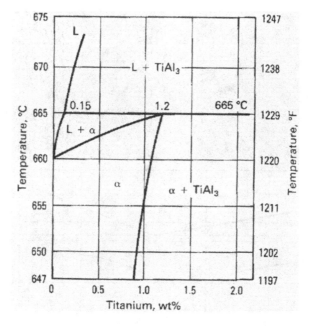

Fig. 19.76 The aluminum–titanium phase diagram (Granger and Elliott 1964). (Reprinted with permission of ASM International. All rights reserved. www.asminternational.org)

TiAl$_3$ is solid at a temperature higher than the melting point of the Al–Si alloy, and serves as heterogeneous nuclei. If boron is also added, complex borides (Al,Ti)B$_2$ serve as nuclei.

19.3.2 Eutectic Solidification of Al–Si Alloys

Typical microstructures resulting during the solidification of *Al–Si* eutectic alloys are shown in Fig. 19.77. A rod-angular morphology of the silicon phase is typical for the unmodified eutectic (type D). This type of structure is very brittle. Modification is used to change the shape of the silicon phase to fibrous.

Just as for cast iron, two types of modifications are possible:

- Through rapid solidification (quench modification)
- Impurity modification (chemical modification): elements in groups I and IIa of the periodic table and lanthanides can be used, but the most efficient are Na and Sr.

The effect of growth velocity and composition on microstructure is also shown in Fig. 19.77. Typical morphologies occurring in the coupled zone as a function of growth velocity are shown in Fig. 19.78. Impurity modification produces similar structures.

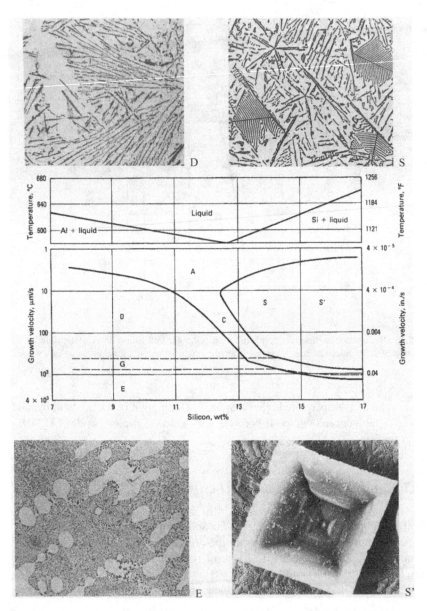

Fig. 19.77 Coupled zone diagram for Al–Si alloys obtained through directional solidification under a thermal gradient of 125 °C/cm; D: unmodified structure of flake eutectic Si and Al dendrites, 100 × ; S: complex regular and star-like primary Si and flake eutectic Si, 100 × ; E: chill modified structure showing fibrous eutectic Si and Al dendrites, 100 × ; S': SEM showing a (100) section through an octahedral primary Si particle revealing four {111} planes, 1500 × (Granger and Elliott 1988). (Reprinted with permission of ASM International. All rights reserved. www.asminternational.org)

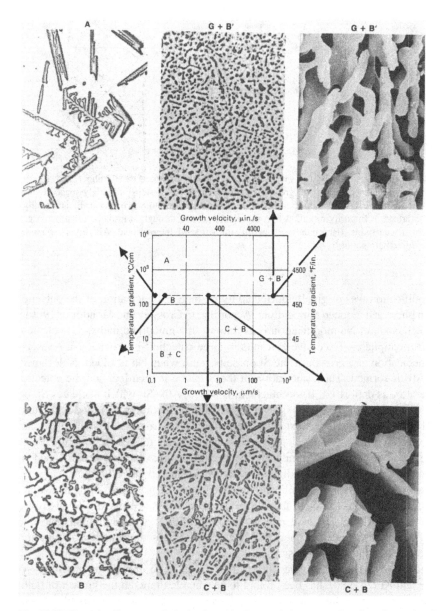

Fig. 19.78 Eutectic silicon morphologies found in the coupled zone as a function of growth velocity and temperature gradient in the liquid at the solid/liquid interface; *A*: massive faceted silicon and Al dendrites (100 ×); *B*: rod eutectic silicon (100 ×); *B + C*: angular and some flake eutectic silicon; *C + B*: flake and some angular eutectic silicon (100 × and 1500 × ; *G + B'*: quench modified fibrous silicon (100 × and 1500 ×). (Hellawell 1973)

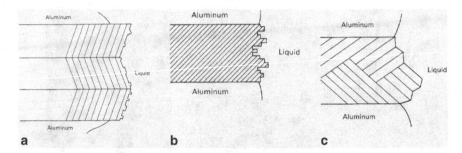

Fig. 19.79 Interface morphology of the eutectic liquid/solid interface (Granger and Elliott 1988): **a** Unmodified flake silicon, growth through propagation of steps on widely spaced twins across the interface; **b** Quenched-modified fibrous silicon, continuous growth on the atomically rough interface; **c** Impurity-modified fibrous silicon, growth through twin-plane reentrant edge of finely spaced twins. (Reprinted with permission of ASM International. All rights reserved. www.asminternational.org)

Modification was originally attributed to the repeated nucleation of the eutectic silicon phase at a reduced temperature. According to Crossley and Mondolfo (1966), the mechanism of Na modification is as follows. In unmodified melts, Si nucleates on AlP precipitates. Excessive P results in large eutectic Si particles. Na restricts both nucleation and growth of the Si phase because when Na is added, NaP rather than AlP is formed. Thus, nucleation of the Si phase is retarded, and the eutectic temperature is depressed. If too much Na is added, AlNaSi, which nucleates Si, is formed (Sigworth 1983). This is over modification. According to Flood and Hunt (1981), the effect of Na is to smooth the solidification front and thus increase the front velocity, which is equivalent to a chilling effect that refines the structure.

However, because of the similarities between structures obtained through quench and impurity modification, modification is now considered the result of a change in the growth mechanism of the silicon phase from faceted growth to a more isotropic growth (Fig. 19.79). Although the eutectic temperature is depressed, silicon grows continuously without repeated nucleation (Elliott 1983). The aluminum phase is not affected. There is experimental evidence that both sodium and strontium are concentrated in the silicon phase.

The Jackson–Hunt parabolic law of irregular eutectic growth, as modified by Fisher and Kurz (1980), has been found to be suitable to model the growth of both modified and unmodified equiaxed eutectic grains of irregular eutectics like Al–Si (Degand et al. 1996). The modifiers restrain the growth of Si in favorable directions increasing the twin density, and thus the Si kinetic undercooling. Including the kinetic undercooling of the Si phase and using weighted averages for the two phases lead to:

$$\overline{\Delta T} = \frac{m_{Si}\overline{\Delta T^{Al}} + |m_{Al}|\overline{\Delta T^{Si}}}{|m_{Al}| + m_{Si}} = \Delta T_c + \Delta T_r + \Delta T_k \qquad (19.28)$$

with $\Delta T_k = |m_{Al}| \overline{\Delta T_k^{Si}} / (|m_{Al}| + m_{Si})$, where m_{Al} and m_{Si} are the slopes of the liquidus in the Al–Si phase diagram. To obtain an equation uniquely relating ΔT to the growth velocity V, one needs to know ΔT_k^{Si} as a function of V. Magnin and Trivedi (1991) have derived some values of ΔT_k^{Si} from experimental data. When ΔT_k^{Si} was plotted versus V, it was noticed that a parabolic law fits better the experimental values than the exponential law suggested by Tiller (1969). This can be written as:

$$V = \mu_{Si} \left(\Delta T_k^{Si} \right)^2 . \tag{19.29}$$

This means that a parabolic dependency of V on ΔT is still valid in the case of modified irregular eutectics with a different growth parameter:

$$\mu = \left[\sqrt{K_1 K_2} \left(\phi + \frac{1}{\phi} \right) + \frac{|m_{Al}|}{(|m_{Al}| + m_{Si})} \mu_{Si}^{-1/2} \right]^{-2} = \left[\mu_{ex}^{-1/2} + A\mu_{Si}^{-1/2} \right]^{-2} \tag{19.30}$$

where ϕ is a coefficient relating the lamellar spacing of regular and irregular eutectics: $\lambda_{ir} = \phi \lambda_{ex}$. For Al–Si alloys, ϕ has values between 2.2 for rod morphology and 3.2 for lamellar morphology. This law has been verified experimentally by several authors (e.g., Magnin and Trivedi 1991) by using directional solidification experiments.

The growth parameter was evaluated experimentally (Degand et al. 1996). It was found that it decreases as the degree of modification increases (Fig. 19.80). During solidification of sodium-modified Al–Si alloys, the grain morphology changes from the center to the outside. As shown in Fig. 19.81, a coarse microstructure surrounds the nucleus while a finer fibrous structure is observed at the periphery of the grain. This suggests that a variable growth parameter should be used to simulate the eutectic grain growth. μ is therefore supposed to be initially large and then decreases because of sodium rejection by the solidification interface. Such a growth parameter will impose less undercooling and recalescence on the cooling curve.

This microlevel analysis can be included in a macro-model by using an Avrami-type equation for the growth parameter:

$$\mu = \mu_{rod} + (\mu_{lam} - \mu_{rod}) \exp\left(-V t^{1/2} / D^{1/2}\right) \tag{19.31}$$

where μ_{lam} is the initial growth parameter (lamellar eutectic $\approx 5 \cdot 10^{-5}$ m s^{-1} K^{-2}) and μ_{rod} the growth parameter of the fibrous eutectic ($\approx 5 \cdot 10^{-7}$ m s^{-1} K^{-2}). The factor $V/D^{1/2}$ has been used to account for the opposite effects of interface velocity and diffusion in the liquid on the sodium segregation ahead of the liquid/solid interface.

The spherical morphology of the eutectic grain of modified Al-Si alloy in Figure 19.81 is very similar to that of the γ-LGr or γ-CGr eutectic grain in Figure 19.51. This morphology was confirmed for Sr-modified alloys by Dahle et al. (2001). By comparing the orientation of the aluminum in the eutectic to that of the surrounding primary aluminum dendrites by EBSD they further noted that that the eutectic nucleation and subsequent growth mode are strongly dependent on the Sr level. At

Fig. 19.80 Influence of modification on the average value of the growth parameter (modification rating: 2—low; 5—high) (Degand et al. 1996). (Copyright 1996 by The Minerals, Metals & Materials Society. Reprinted with permission)

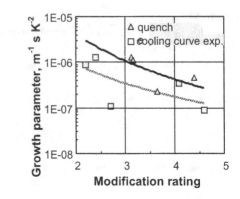

Fig. 19.81 Microstructure of an Al— 11 %Si eutectic grain core in a sodium-modified casting (Degand et al. 1996). (Copyright 1996 by The Minerals, Metals & Materials Society. Reprinted with permission)

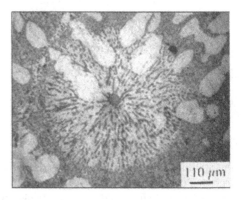

very low Sr, the eutectic nucleated and grew from the primary Al dendrites. Increasing the Sr level to 70–110 ppm resulted in nucleation of independent eutectic grains with no relation to the primary dendrites. At a Sr level of 500 ppm, the eutectic again nucleated on and grew from the primary phase while a well-modified eutectic structure was still present.

19.3.3 Effect of Oxides

The concept of an entrained double oxide film (also styled bifilm) and its damaging effects on the properties of aluminum castings has been advocated and demonstrated by Campbell (2003). As shown in Fig. 19.82, when the surface of the liquid aluminum folds upon itself during pouring or mold filling, the surface oxide film is entrained in the bulk liquid. Double oxide films that trap air between the two surfaces will form, possibly resulting in cracks in the solidified casting. It was also postulated that they can open and expand because of a variety of driving forces, such as hydrogen diffusion into the atmosphere of the double oxide film defect, and the shrinkage of the liquid metal during solidification. The oxide films can separate entire regions of the casting (Fig. 19.83) producing nonuniform microstructures and thus variability in the local mechanical properties.

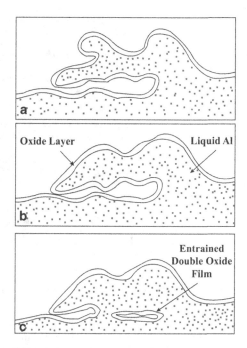

Fig. 19.82 Sequence of bifilm generation: **a** Surface turbulence. **b** Formation of bifilm having two unwetted sides of oxide films contacting with each other. **c** Double oxide film (bifilm) submerged in the bulk liquid. (Campbell 2003)

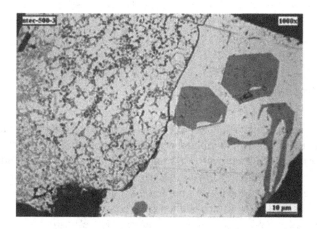

Fig. 19.83 The silicon phase forms on the oxide barrier as a preferential substrate. (Walker 1959)

Experiments by Nayebi and Divandari (2012) seem to indicate that in pure aluminum and Al-7Si-0.4Mg alloy folds of 100–500 nm thickness occur before solidification due to the melt flow stress. Some wrinkles form before solidification because of the contraction stress, while others are produced during solidification on dendrite tips (Fig. 19.84).

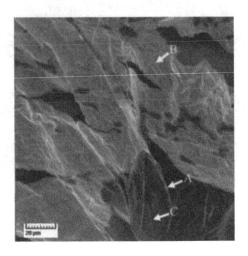

Fig. 19.84 Oxide films in Al alloys: A—fold, B—wrinkle formed before solidification, C—wrinkle formed during solidification. (Nayebi and Divandari 2012)

It is now generally accepted that bifilms act as cracks to initiate failures in reactive metals such as Al and Mg and, thereby, control the properties of casting. To quote Campbell (2003, 2006), "bifilms appear to be the source of most of our familiar casting defects; in Al alloys they can exert major control of the cast microstructure, including (a) grain size, (b) DAS, and (c) the modification of the eutectic silicon." While the brilliant work of Campbell in this field is significant for the understanding of many phenomena, a certain caution must be exercised in overemphasizing the role of bifilms. For example, Wang et al. (2006) show that while fatigue life decreases with the defect area, pores are more detrimental to fatigue than oxide films in precision sand castings (Fig. 19.85).

19.3.4 Ultrasonic Processing

High-intensity ultrasonic vibrations in liquids are accompanied by cavitation, i.e., by the formation of gas-filled bubbles which pulsate in a highly nonlinear manner. During ultrasonic melt treatment (UST) above the cavitation threshold, a cavitation bubble will most likely form in the proximity of a gaseous phase where the cavitation strength is the lowest. As numerous nonmetallic particles are present in aluminum melts, and their surface is often covered with cracks filled with a gaseous phase, the bubbles tend to form on these cracks. Once formed, the bubbles will start to grow and pulsate. The energy stored during the growth of the bubble in the rarefaction phase of the acoustic field is released when the bubble collapses in the positive phase of the acoustic field. This effect is accompanied by acoustic noise, shock waves, and light emission. At the moment of the collapse, hot spots will be produced with equivalent temperature of roughly 5000 K, pressure of ~ 1000 atm., and heating and cooling rates above 1010 K/s (Eskin 1998). The explosion is also

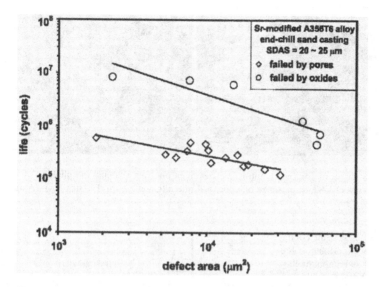

Fig. 19.85 The effect of the type and size of defect on the fatigue life in an A356-T6 casting (Wang et al. 2006). (With permission of Springer Science and Business Media)

accompanied by the formation of cumulative jets. The impingement of this jet and related shock waves can create localized erosion, improve mass transport, and cause fragmentation of particles or grains. They may also result in sonocapillary effect (filling of the crack with the melt and subsequent solidification of this melt). Consequently, the amount of solidification sites might be increased and, as a result, the grain size will be reduced. In addition, movement of dislocations associated with the propagation of ultrasound waves in media will occur.

Relevant applications include structural refinements of solidifying metals with potential applications in die-casting, semisolid metal processing, and forging of high fractions of fine spherical grains.

Atamanenko et al. (2009) found that ultrasonic melt treatment applied continuously in the liquid and semisolid state had different influences on solidification of aluminum and its alloys with different amounts of eutectics. In all cases studied, it resulted in significant grain structure refinement as exemplified in Fig. 19.86. Yet, the mechanism of refinement appears to be different. In the case of Al and the Al–4Cu alloy, cavitation-aided grain refinement was produced by simultaneous action of the energy dissipation during UST and the shock waves produced by cavitation. During solidification of an Al–11Cu alloy, ultrasound mostly affected the primary nucleation.

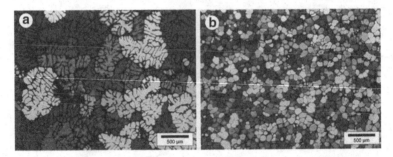

Fig. 19.86 Effect of continuous ultrasonic treatment on grain size and morphology of Al–11Cu alloy: **a** Without UST in presence of immersed idle ultrasonic horn; **b** After UST (Atamanenko et al. 2009)

Fig. 19.87 Schematic cooling curve showing primary solidification of an aluminum alloy and role of grain refinement

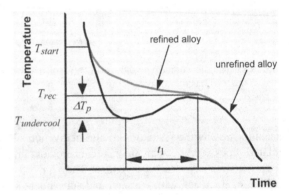

19.3.5 Thermal Analysis of Aluminum Alloys

The evaluation of grain refinement efficiency can be performed through cooling curve analysis. As shown in Fig. 19.87, grain refinement decreases the amount of undercooling observed in the initial stages of the solidification. However, the undercooling alone is not enough to evaluate the degree of grain refinement. Indeed, from Fig. 19.88, it is seen that both very fine and very coarse grain structures may have minimal undercooling. Thus, other features of the cooling curve must be used, such as the time difference between the maximum undercooling and the recalescence or elements of the first derivative curve, t_1.

Bäckerud et al. (1990) demonstrated that the maximum of the first derivative in the region of primary solidification of aluminum alloys and the area under it can be used to evaluate the efficiency of grain refinement. As seen from Fig. 19.89, grain refinement reduces the recalescence. The recalescence area (shaded area in the figures) and the recalescence time directly correlate to the efficiency of grain refinement.

Fig. 19.88 Correlation between grain size and time, and temperature of undercooling in hypoeutectic aluminum alloys. (After Charbonier 1984)

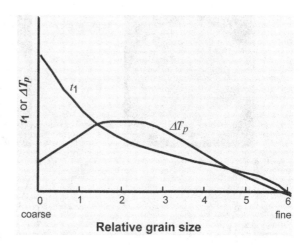

Cooling curve analysis can also be used for evaluation of modification efficiency. Modification results in higher undercooling (Fig. 19.90). This is similar to the results of Mg-treatment of cast iron, which is modification, but opposed to the FeSi treatment of lamellar graphite iron, which decreases undercooling because it increases the number of eutectic grains. However, just as in the case of grain refinement, evaluating only the undercooling is not enough to decide whether the required degree of modification has been achieved. Indeed, as shown in Fig. 19.91, the highest undercooling is obtained for the under-modified eutectic having an acicular lamellar structure.

A modified differential thermal analysis (DTA), in which the reference sample was not a neutral body, but a second thermocouple placed in the same cup close to the wall was used to detect solidification events in aluminum alloys (Bäckerud et al. 1990). Two cooling curves were thus recorded for the same alloy (Fig. 19.92a). The temperature difference between the wall and center thermocouples that describe the temperature gradient across the sample can be used to determine the occurrence of solidification events such as nucleation at wall, dendrite coherency, and formation of various phases such as Mg_2Si (Fig. 19.92b).

19.3.6 Simulation of the Solidification of Aluminum-Based Alloys

Equiaxed dendritic solidification in the presence of melt convection and solid-phase transport was investigated by Wang and Beckermann (1996). They developed a multiphase model to predict composition and structure evolution in alloys solidifying with equiaxed morphology. The model accounts for the transport phenomena occurring at the macroscopic scale, as well as the grain nucleation and growth mechanisms occurring over various microscopic length scales. When applying the model to study the solidification of an Al–4Cu alloy, some interesting conclusions were

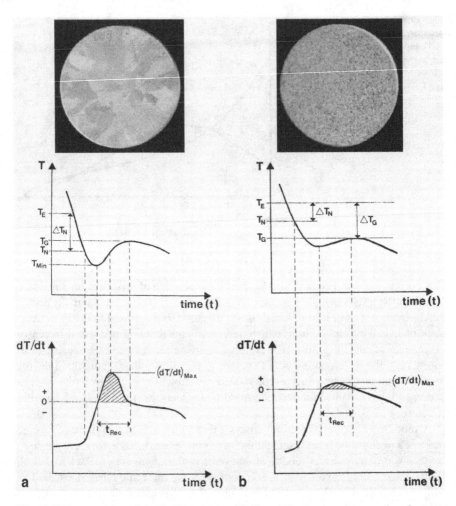

Fig. 19.89 Interpretation of the cooling curve and its first derivative to estimate grain refinement in aluminum alloys. (After Bäckerud et al. 1990). **a** No grain refinement. **b** Grain refinement

drawn. Channel segregates were produced because of thermosolutal convection within a stationary mushy zone. The channels are oriented downward, as observed in laboratory experiments. Solid movement reduces macrosegregation. No direct correspondence was found between the nucleation rate and the local grain size whenever solid motion was present. Thus, the final grain size obtained from micrographs of solidified materials cannot be directly used to infer the nucleation rate, because the grain density distribution may be altered by flow during solidification. However, the knowledge of the measured final grain size combined with the prediction of solid transport would enable the development of a realistic nucleation model.

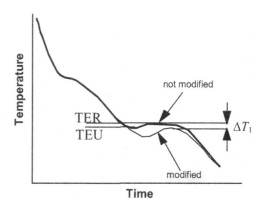

Fig. 19.90 Schematic diagram of cooling curves for non-modified and modified hypoeutectic Al–Si alloys

A particular issue in the solidification of Al alloys is the formation and effects of oxide films. They play a significant role in microporosity formation. Modeling of microporosity because of oxide films has been addressed by a number of researchers and discussed in Sect. 18.2.3. Dai et al. (2012) developed an oxide film entrainment-tracking algorithm method to simulate the movement of oxide films on the liquid metal surface. Patel and Nastac (2014) produced a mathematical model based on the Niyama criterion, which considers the stochastic behavior of pore nucleation during casting solidification and provided adequate validation against experiments.

Another important issue in aluminum casting alloys is the effect of ultrasonic treatment. Nastac (2014) performed a numerical analysis to determine the fluid flow

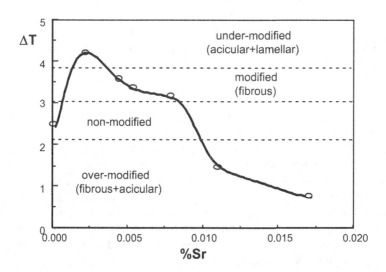

Fig. 19.91 Influence of Sr content on $\Delta T = $ TER—TEU for an Al– 7 % Si alloy (based on data from Argyropoulos et al. 1983)

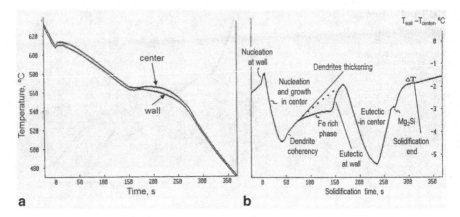

Fig. 19.92 DTA with two thermocouples in the same sample (after Bäckerud et al. 1990). **a** Cooling curves. **b** Temperature difference between wall and center

characteristics and cavitation phenomena during molten A356 alloys processed via high-power UST as well as the resulting solidification microstructure of the ultra-sonically processed A356 alloys. The modeling approach was based on two main models: (i) a previously developed cavitation model (implemented into Ansys's Fluent—CFD software) applied to simulate acoustic streaming, cavitation, and bubble dynamics during molten alloy processing and (ii) a stochastic mesoscopic model to predict microstructure evolution during alloy solidification. The first model can predict the amount of the cavitated phase that is used to determine the number of potential nuclei formed due to the UST processing.

In the ultrasonic cavitation model, a two-phase cavitation model that consists of the standard viscous flow equations governing the transport of phases (Eulerian multiphase) and the k-ε turbulence model was used. In cavitation, the liquid–bubble mass transfer is governed by the cavity (bubble) transport equation (Nastac 2011):

$$\frac{\partial}{\partial t}(f_b \rho_b) + \nabla \cdot (f_b \rho_b \vec{V}_b) = R_G - R_C \qquad (19.32)$$

where b subscript denotes the cavitation bubble phase, f_b is the bubble volume fraction, ρ_b is the bubble density, $\vec{V}_b$ is the bubble phase velocity, and R_G and R_C are the mass transfer source terms related to the growth and collapse of the cavitation bubbles, respectively. In this equation, the interphase mass transfer rates per unit volume account for the liquid and bubble phases in cavitation. They are calculated using the growth of a single bubble based on the Rayleigh–Pleasset model. The model assumes no barrier for nucleation; thus, the bubble dynamics can be obtained from the general Rayleigh–Plesset equation:

$$r_b \frac{d^2 r_b}{dt^2} + \frac{3}{2}\left(\frac{dr_b}{dt}\right)^2 = \frac{p_b - p}{\rho_L} - \frac{2\sigma_L}{\rho_L r_b} - \frac{4\nu_L}{r_b}\frac{dr_b}{dt} \qquad (19.33)$$

where r_b is the bubble radius, σ_L is the surface tension coefficient of the liquid phase, ρ_L is the liquid density, ν_L is the kinematic viscosity of the liquid phase, p_b is the bubble surface pressure, and p is the local far-field pressure.

The mesoscopic model developed previously (Nastac and Stefanescu 1997; Nastac 1999) was adapted to include the effect of UST on the grain refinement of alloys. It includes computations of the grain size and columnar-to-equiaxed transition (CET) as well as of segregation. It also includes computations at the dendrite tip length scale (mesoscopic scale) for prediction of columnar and equiaxed dendritic morphologies and of microsegregation patterns.

The model was used to simulate the flow characteristics during ultrasonic processing of molten metallic alloys under two different gravity conditions: (i) gravitational acceleration oriented downward and (ii) gravitational acceleration oriented upward. It was found that the latter case has significantly stronger bulk stirring and a larger cavitation zone that is more favorable for ultrasonic stirring and cavitation processing of alloys and metal-matrix-nanocomposites.

19.4 Superalloys

Superalloys are alloys developed for elevated temperature service, requiring resistance to severe mechanical stressing and to creep, as well as high surface stability, and corrosion and oxidation resistance. The driving forces for the development of these alloys include the quest for higher temperature capability, higher fuel efficiency and higher thrust/weight ratio. As the turbine entry temperature increases, considerable gains in the specific thrust are obtained. Since their emergence in 1940, the high temperature capabilities of superalloys have increased considerably (Fig. 19.93).

Superalloys include a large number of elements that contribute to attaining the desired properties (Table 19.4). The basic types are: Ni-base (10–20 % Cr, 5–10 % Co, < 8 %Al + Ti), Co-base (20–30 % Cr, 10–22 %Ni, 7–14 %W), and Ni–Fe-base (25–60 %Ni, 15–60 %Fe). For Ni-based superalloys, the metallic elements can be divided into three types:

a. Nickel and other transition metals: Co, Fe, Cr, Mn
b. Heavy, high-melting temperature metals: W, Ta, Mo, Nb, Hf, and Re. They provide high temperature solution hardening
c. Elements that improve mechanical strength by the formation of γ' phase precipitates: Al and Ti

The first metal to solidify tends to be rich in the high-melting elements, leaving a lower density liquid. Subsequent convection can lead to the lighter liquid rising and breaking dendrite arms, which leads to the formation of misaligned grains.

Typical applications of superalloys include aerospace parts such as turbine blades, turbine disks, turbine rear frame, and automotive parts such as turbochargers (Mazda turbocharger achieved a world speed record—238.445 mph).

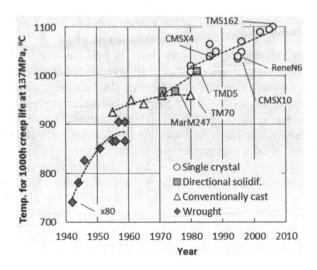

Fig. 19.93 Evolution of high-tempearture properties of selected superalloys. (After Reed 2006)

Table 19.4 Elements found in superalloys and their purpose

Purpose	Cr	Al	Co	Fe	Mo	W	Ti	Ta	Nb	Hf	C	B	Zr	Mg
Matrix strengtheners	•		•	•	•									
γ' formers		•					•	•	•					
Carbide formers	•				•	•	•	•	•	•				
Oxide-scale formers	•	•												
Grain-boundary strengtheners										•	•	•	•	•

19.4.1 Microstructure of Superalloys

The typical matrix of superalloys is an austenitic face-centered cubic crystal structure. The base-alloying elements are usually Ni, Co, or Ni–Fe. The microstructure of
a typical superalloy includes the following phases: γ-solid solution, γ' precipitates,
carbides, and topologically close-packed (TCP) phases. In an Ni-based superalloy
the γ phase is a continuous matrix of Ni-based austenite strengthened by the addition
of solid solution elements. The atomic diameters of these elements differ from Ni by
1 (for Co) to 12 % (for Mo). The γ' phase is a long-range order phase obtained by
precipitation hardening. It is of the FCC A_3B-type compound, typically of the form
$(Ni, Co)_3(Al, Ti)$. Since the degree of order increases with temperature, the alloys

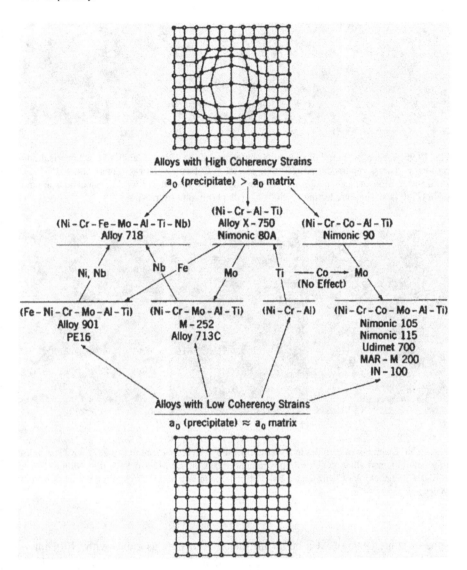

Fig. 19.94 Classification of Ni-base superalloys according to the degree of mismatch. (Decker and Mihalisin 1969)

with a high-volume fraction of γ' show higher strength with increasing temperature up to about 800 °C. γ' contributes to strengthening by dislocation interaction forcing the dislocations to bypass or to cut the particles.

Depending on the alloying elements various degree of mismatch between the γ/γ' phases can be obtained (Fig. 19.94). The γ/γ' mismatch determines the γ' particle morphology. At a small mismatch of $\sim 0.05\%$ the morphology is

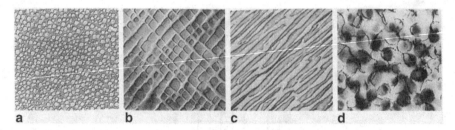

Fig. 19.95 Various γ' morphologies in Ni-base superalloys (Sims et al. 1987). **a** Spherical and cooling γ' (stress rupture tested U500.5). 450×. **b** Cubical γ' (stress rupture tested IN-100). 13,625×. **c** Elongated γ' in alloy 713 C (stress rupture testing at 815 °C). **d** Dislocations around γ' in U500 (stress rupture testing at 790 °C) (With permission from Wiley)

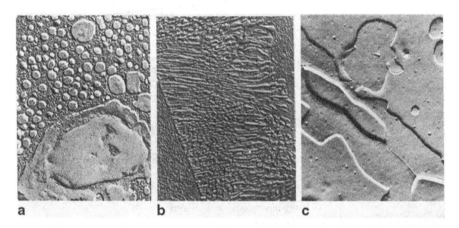

Fig. 19.96 Examples of carbides in Ni-base superalloys. 4900 × (Sims et al. 1987). **a** MC particles in γ'-strengthened alloy; degeneration commenced. **b** Cellular M23C6 formed in Nimonic 80 A at 650 °C. **c** Blocky M6C surrounded by γ' at grain boundary of AF-1753. (With permission from Wiley)

spheres (Fig. 19.95a). At a larger mismatch γ' occurs as cubes with {100} interfaces (Fig. 19.95b). At a mismatch larger than 1.25 % semicoherent plates form (Fig. 19.95c). In some instances a γ'' phase, Ni_3Nb, may be found.

The carbon content of superalloys varies between 0.02 and 0.6 %. Various morphologies and types of carbides can be found as shown in Fig. 19.96, including MC (where M stands for Ti, Ta, Nb or W), M_6C (where M stands for Mo or W), and $Cr_{23}C_6$. MC carbides are unstable at elevated temperatures:

$$MC + \gamma = M_{23}C_6 + \gamma'$$

The product of this reaction can be blocky or cellular. Blocky carbides inhibit grain boundary sliding, and the γ' resulting from this reaction envelops the carbides

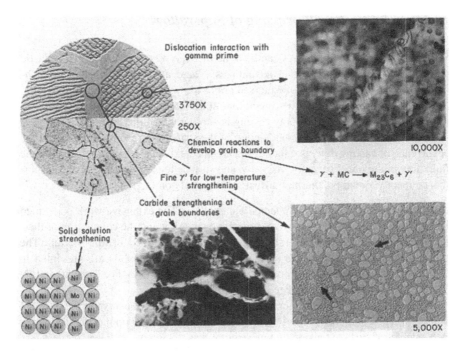

Fig. 19.97 Major features for achieving useful strength in nickel-base alloys (Sims et al. 1987). (With permission from Wiley)

and the grain boundary in a relatively ductile, creep-resistant layer. If a cellular structure is formed, ductility is reduced sharply.

The topologically close-packed (TCP) phases that include σ, μ, and Laves (χ) can form at the end of solidification, during heat treatment, or during service. These phases that usually form as thin plates parallel to the {111}γ can lead to low rupture strengths and loss in rupture ductility. They also may remove solid solution strengthening elements.

A summary of the contribution of the microstructural features to the properties of superalloys is illustrated in Fig. 19.97.

Modern superalloys were developed in the 1980s with the advent of single crystal solidification through Bridgman-type techniques that allowed production of castings without grain boundaries, and thus of alloys with considerable higher creep resistance. Carbides became unnecessary as grain boundary strengtheners, and the volume fraction of the γ' precipitates was increased to 50–70 % through additions of Al, Ti, Ta, and Nb. Examples of this first generation superalloys include: PWA 1480, René N4, and SRR99. The 2nd (e.g., CMSX-4 and René N5) and 3rd (e.g., CMSX-10, and René N6) generation superalloys introduced about 3–6 % rhenium. Fourth, 5th, and even 6th generation superalloys have been developed which incorporate ruthenium additions (Reed 2006).

19.4.2 Solidification Processing of Superalloys

While processing of superalloys can be done through powder metallurgy or plastic deformation, casting has certain advantages. There are three main casting techniques: (i) investment casting of equiaxed grains alloys, (ii) directional solidification of columnar grains alloys, (iii) directional solidification of single-crystal alloys. Rapid solidification through spray casting is another possibility.

For analytical or numerical calculation of the scale of the microstructure and the occurrence of casting defects reliable data are needed for the properties affecting the fluid flow of the metal (density, viscosity, and surface tension) and the heat transport (heat capacity, enthalpy, thermal diffusivity, thermal conductivity).

Thermophysical Properties of Selected Superalloys The extensive work performed by Quested et al. (2009) on the experimental evaluation and the calculation of thermophysical properties of several superalloys is summarized in this section. The chemical compositions and the surface tensions of the elements are presented in Table 19.5. Thermophysical properties are given in Table 19.6. For Rene 80 and IN 738 the specific heat was estimated.

Surface tension and its temperature dependency are very sensitive to the levels of soluble oxygen and sulfur. For IN 738 alloy samples with 10 ppm S, the following dependency of surface tension on temperature was obtained:

$$\gamma = 1842 - 0.11(T - 2000) \tag{19.34}$$

with γ in mN m^{-1}, and T in K.

Numerical analysis of literature data by Mills et al. (2006) for solidus and liquidus temperatures produced the following relationships:

$$T_L = 956.9 + 8.2\text{Ni} + 7.6\text{Co} + 2.2\text{Al} + 3.4\text{Ti} - 3.7\text{Ta} + 3.9\text{Cr} + 7.2\text{Mo}$$
$$+ 11.5\text{W} + 21.1\text{Re} + 12.1\text{Ru} + 7.4\text{Fe} - 3.3(\text{Nb} + \text{Hf}) \tag{19.35a}$$

$$T_S = -120 + 17.7\text{Ni} + 16\text{Co} + 15.2\text{Al} + 2.1\text{Ti} - 13.4\text{Ta} + 192\text{Cr} + 5\text{Mo}$$
$$+ 20.4\text{W} + 34.3\text{Re} + 23.1\text{Ru} + 15.4\text{Fe} - 6.4(\text{Nb} + \text{Hf}). \tag{19.35b}$$

The elements are expressed in % mass. Most calculated values for T_S and T_L are within ± 10 and ± 5K, respectively.

The viscosities of liquid Ni-based superalloys can be estimated using the following equation (Quested et al. (2009):

$$log_{10}\eta_{alloy} = \frac{2570}{T} - 0.8224 + 1.75 \cdot 10^{-3}\text{Cr} + 1.1\text{Fe} \tag{19.36}$$

$$+ 10.2 \cdot 10^{-3} \cdot \sum (W + Re + Hf + Nb + Ta + Mo) \tag{19.37}$$

where the viscosity is in m Pa s, and the elements are in mass%.

Table 19.5 Chemical compositions of alloys (mass%) and surface tension of elements at the melting point. (Quested et al. 2009)

Alloy	Ni	Al	C	Co	Cr	Mo	Nb	Ta	Ti	W
CMSX-4	Balance	5.6	0.06	9	6.5	0.6	–	6.5	1.0	6.4
CMSX-10	Balance	5.78	0.1	3.4	2.3	0.41	–	8.3	0.32	5.6
IN 738	52.5	3.53	0.1	8.26	15.85	1.73	0.62	1.73	3.49	2.57
Rene 80	Balance	2.95	0.17	9.6	3.9	4.05	–	–	4.95	3.9
γ_{Tmelt}, N m^{-1}	1.8	0.89		1.9	1.7	2.25	1.937	2.05	1.65	2.32

Table 19.6 Experimental and estimated thermophysical properties of selected superalloys. (Quested et al. 2009)

Alloy	Property	Value	Property	Temperature						
				298	600	1000	1400	T_S	T_L	1700
CSMX-4	T_S	1593	c_p	397	448	540	925	630	630	630
	T_L	1653	ρ	8700	8559	8374	8193	8107	7754	7710
	ΔH_f	240[a]	α		3.1	4.35	5.1		4.9	4.9
			k		12	19.9	24.4			25
CSMX-10	T_S	1635	c_p	395	446	536	805	880	630	630
	T_L	1683	ρ	9046	8945	8793	8600	8442	8133	8128
	ΔH_f	207[a]	α	2.43	3.30	4.47			5.1–6.2	
			k	8.7	13.2	21.1				
IN 738	T_S	1512	c_p	450	511	631			700	700
	T_L	1628	ρ	8177	8082	7934	7702	7600	7324	7280
	ΔH_f	282[a]	α	2.65	3.53	4.67	4.87			
			k	9.8	14.6	23.4				
Rene 80	T_S	1485	c_p	457	519	644	1100	750	830	830
	T_L	1611	ρ	8130	8037	7891	7666	7600	7359	7300
	ΔH_f	209[a]	α	2.77	3.56	4.60	4.87		4.46	
			k	10.3	14.8	23.4				

Units: temperature in K, specific heat c_p in J K^{-1}kg^{-1}, heat of fusion ΔH_f in kJ kg^{-1}, density ρ in kg m^{-3}, thermal diffusivity α in $\times 10^6$ m^{2}s^{-1}, thermal conductivity k in W m^{-1}K^{-1}

[a]values for heating curves

Investment Casting of Equiaxed Grains Superalloys The casting process has two main advantages over other processes. The first one is material conservation. Indeed, for example, for a finished 10 kg disk we need 200 kg ingot to produce a wrought disk compared to 40 kg for a cast disk. The second major advantage is the possibility of integral casting of wheels and rotors, which translates in significant time savings.

There are, however, some disadvantages such as increased inclusions and porosity, and sometimes poor microstructural control (large and nonuniform grains). Melting of superalloys through vacuum induction melting (VIM) generates inclusions from the ceramic crucible. Newer processes such as electrode slag remelting (ESR), electron beam copper hearth remelting (EBCHR), and use of filtration significantly reduce the amount of inclusions. Porosity can be corrected through hot isotactic pressing (HIP). Finally, grain control can be achieved through a number of processes:

1. Mechanical vibration during solidification which breaks up the dendrites
2. Thermal control:

   - Heat sinks and insulation around the mold to control temperature gradients
   - Reduction of melt superheat (time and temp.) to retain the carbides in the melt which serve as nuclei
   - Reduction of the pouring temperature to increase cooling rate in the mold

3. Inoculation of melt or in-mold

Directional Solidification of Columnar Structure and Single-Crystals Superalloys Directional solidification (DS), such as the Bridgman technique, can produce superalloys with columnar or single-crystal structure that have clear advantages over the equiaxed structure, as illustrated in Fig. 19.98 for creep. The advantages of columnar structure over equiaxed structure include improved stress-rupture strength in the grain growth direction, and 45 % lower elastic modulus than conventional equiaxed castings, which is beneficial in reducing thermally induced stress in the blades. The advantages of single crystals include greater temperature capability than columnar structure, higher creep strength, higher thermal fatigue resistance, and higher corrosion resistance. Single-crystal superalloy technology has combined the design of alloys for exclusive use in single-crystal form with advances in directional solidification to produce single-crystal castings with complex internal cooling passages.

A schematic drawing of an industrial setup for the DS of superalloys has been presented in Fig. 6.3b. The melt produced by VIM or EBM is poured into the ceramic mold. The temperature gradient during solidification is controlled by the heat input to the mold from the multizone furnace, by the heat extraction through a water-cooled cooper chill under the mold, through thermal baffles at the base of the hot zone, and through the mold translation mechanism for mold withdrawal from the furnace.

The mechanism of heat extraction from the ceramic mold involves: (a) conduction from the liquid to the solid to the chill in the extraction direction, (b) conduction from the solid to the mold perpendicular to the extraction direction, (c)

Fig. 19.98 Evolution of creep strain with time (206 MPa load at 982° C) for the super-alloy Mar-M200 cast by different processes. (After Kear and Piearcey 1967)

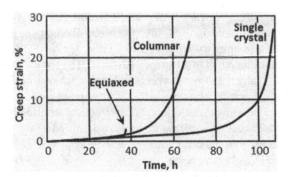

radiation from the mold to the furnace atmosphere perpendicular to the extraction direction. For calculations, three heat transfer coefficients are required: solid-chill (h_{Sch}), solid-mold (h_{Sm}), and mold-atmosphere radiation (h_{rmat}). Thus, the heat flux through convection (Newton's law) can be written as:

$$q = (h_{Sch} + h_{Sm} + h_{rmat})(T - T_o). \qquad (19.38)$$

For additional discussion on heat transfer coefficients see also Sect. 6.4.3. Some typical solid-mold coefficients are given in Table 19.7. The casting geometry and the composition of the molding aggregate are the main factors affecting the value of the temperature-dependent heat transfer coefficient.

At steady state a thermal balance exist between the rate of heat loss by conduction along the extraction direction, and the rate of heat loss by radiation perpendicular to the extraction direction. However, at the beginning of solidification most heat is lost through the conduction path liquid—solid casting—chill. As the thickness of the solid increases this contribution diminishes in favor of heat loss by radiation.

During the DS of superalloys, texture develops along the direction of withdrawal by a process of competitive growth. To produce single crystals, a grain selector is used (Fig. 19.99). Competitive growth of the many initial grains results in only one grain growing out of the selector.

The solidification conditions affect the length scale of the microstructure. Indeed the primary dendrite arm spacing (PDAS) has been derived to be a direct function of the $G \cdot V$ product (see Eq. 9.53), which is the cooling rate, or of the $G^{-1/2} \cdot V^{-1/4}$ product (see Eq. 9.54). However, as seen in Fig. 19.100, the proportionality constant (the slope) in these equations is different for various superalloys.

Table 19.7 Solid-mold heat transfer coefficients for some investment cast superalloys

Alloy	Casting geometry	h_{Sm} W m^{-2} K^{-1}	Reference
IN718	Cylindrical	300	Sahai and Overfelt 1995
	Plate	50-5000	
IN738LC	Plate	300–660	Konrad et al. 2011

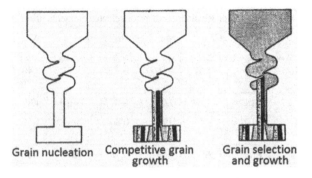

Grain nucleation   Competitive grain   Grain selection
                         growth   and growth

Fig. 19.99 Grain selector for the DS of single crystals

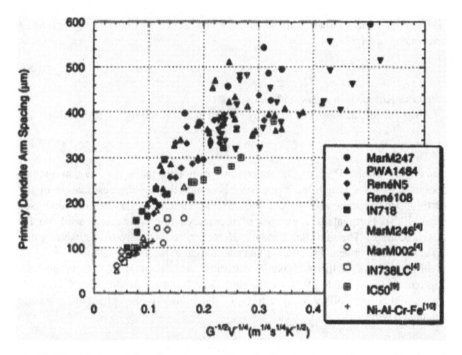

Fig. 19.100 Variation of the primary arm spacing of selected DS superalloys as a function of the product $G^{-1/2} \times V^{-1/4}$ (Whitesell et al. 2000). (With permission of Springer Science and Business Media)

Ma et al. (1993) established the following experimental relationship for columnar CMSX-6 (G in K/mm, V in mm/min, and λ_1 in mm):

$$\lambda_1 = 1.08 \cdot G^{-1/2} V^{-1/4}. \tag{19.39}$$

Calculated results of the constant for a CSMX-4 alloy are given in Application 19.3.

Table 19.8 Pseudo-binary alloy phase diagram data for selected elements in superalloys

Quantity/ Element	Al	Co	Cr	Mo	Ta	Ti	W	Source
m_i	− 4.1	0.40	− 1.90	− 0.001	− 2.0	− 11.3	2.40	Murray et al. 1986
	− 2.27	0.18	− 1.64	0.65	− 1.94	− 6.3	4.87	Pandat
k_i		1.10		0.70	0.80		1.60	Kurz et al. 1986
	0.81	1.01	0.92	1.03	0.78	0.76	1.60	Pandat

Experimental equations have been derived by a number of researchers. McLean (1983) found the following correlation for the SDAS of directionally solidified investment cast IN738LC alloy:

$$\lambda_2 = 47.85 \cdot (G\ V)^{-0.367} \tag{19.40}$$

with G in K/mm, V in mm/s, and λ_2 in μm.

A similar relationship was found by Franke et al. (2011) for the same alloy:

$$\lambda_2 = 46 \cdot (G\ V)^{-1/3} \tag{19.41}$$

Calculation of the DAS of DS superalloys is not trivial as they are multicomponent alloys, for which the main issue becomes the accurate estimation of the partition coefficients and slopes. Limited experimental data are available (see Table 19.8). Fortunately, a number of thermodynamic software is available for this task (e.g., ThermoCalc, Pandat). However, as seen from the table, often disagreements between experiment and calculation exist.

In the absence of thermodynamic software, for fast analytical calculations one approach is to use the Eq. 9.49 for the equivalent k and m introduced in Sect. 9.3.6. The liquidus and solidus temperatures in the pseudo-binary phase diagram are calculated with:

$$T_L = T_f + \overline{m}_L \overline{C}_L \quad T_S = T_f + \overline{m}_L \overline{C}_L / \overline{k} \tag{19.42}$$

where T_f is the melting temperature of the base metal, in this case Ni.

Using this method, Kermanpur et al. (2000a) estimated that the SDAS of an IN738LC alloy can be calculated with the following equation:

$$\lambda_2 = 55.41 \cdot (G\ V)^{-0.56} \tag{19.43}$$

with G in K/mm, V in mm/s, and λ_2 in μm.

However, the use of the Eq. 9.49 to determine the pseudo-binary phase diagram data was questioned by Ma (2001), who pointed out that the calculated T_L and T_S

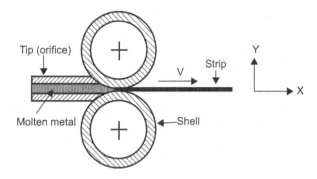

Fig. 19.101 Schematic diagram of a horizontal twin roll caster

were quite different from the measured ones (see Application 19.1). He offered a different alternative for the calculation of the equivalent slope and partition coefficient based on the measured liquidus and solidus temperatures of the alloy:

$$\overline{m} = \frac{T_L - T_{Ni}}{\overline{C}_L} \quad \overline{k} = \frac{T_{Ni} - T_L}{T_{Ni} - T_S}. \tag{19.44}$$

Better results from calculation are obtained when Eq. 9.59 is used for μ_0 in $\lambda_2 = (\mu_0 t_f)^{1/3}$. As demonstrated in Application 19.2 for a CSMX-4 superalloy, SDAS equations similar to the experimental ones for the IN738LC alloy are obtained:

$$\lambda_2 = 11.21(t_f)^{1/3} \quad \text{and} \quad \lambda_2 = 55.31(GV)^{-1/3}. \tag{19.45}$$

Some post-processing of the single-crystal castings is required. Typically, it includes heat treatment to correct undesirable solidification structure features, such as dendritic segregation, coarse γ/γ' eutectic, coarse secondary γ'. Also, coating with Al_2O_3 and elements that retard spallation of Al_2O_3 may be performed.

Strip Casting Strip casting is a continuous casting process based on continuous melt extraction and solidification to near-net shape products. It involves the flow of metal through a distributor (tundish) into the casting mold (typically one or two water-cooled rolls) followed by continuous solidification and withdrawal of the casting (strip) from the mold (Fig. 19.101). Such a process has been used to produce 0.2–0.8 mm thick strips of the Ni-based Inconel 606 (Yukumoto and Yamane 1995). The dendritic structure closed to the chill surface was fine columnar with SDAS = 2 to 3.5 μm, which gave an estimated cooling rate of 1 to 3 $\cdot$ 10^3 K/s. Equiaxed crystals were observed in the centerline of the strip.

19.4.3 Simulation of the Solidification of Superalloys

The objectives of computational modeling of the solidification of superalloys include simulation of the length scale of the microstructure (PDAS and SDAS), of the general grain structure, and of the defects such as segregation and freckles.

Grain Structure Probably the most widely used software for modeling solidification of superalloys is the ProCAST finite elements macro-transport code. As early as 1996, Nastac and Stefanescu incorporated their continuum macro/micromodel for dendritic growth into ProCAST to simulate the solidification of Inconel 718 superalloy. Regression analysis of the experimental correlation between the grain radius evaluated on metallographic samples cut of the experimental castings and the cooling rate in the middle of the plate was used to calculate the nucleation law in Table B4, Appendix B. The partition coefficient and the slope of the liquidus line were obtained from the pseudobinary phase diagram of the 718 as $k = 0.993 - 0.118\,\mathrm{Nb}_L^* + 0.0046(\mathrm{Nb}_L^*)^2$ and $m = -0.743 \cdot 10^{0.0873\,\mathrm{Nb}_L^*}$, where Nb_L^* is the niobium concentration in the liquid at the interface. Good validation against cooling curves was demonstrated.

Due to the small thermal effect associated with the solidification of Laves phases, it is difficult to evaluate the amount of Laves phases from the cooling curve analysis. However, it can be readily obtained by following the evolution of niobium concentration, as shown in Fig. 19.102. When the concentration of niobium reached 19.1 %, it was considered that the remaining liquid solidifies as eutectic ($\gamma +$ Laves phase). The amounts of Laves phases calculated based on this assumption were compared with the experimental ones for different cooling rates in Fig. 19.103. Note that Scheil's model predicted 7.16 % Laves irrespective of cooling rate. It is seen that a critical cooling rate exists at which a maximum amount of Laves phases are formed.

A major step forward in solidification simulation was achieved when a mesoscale 3D CAFE cellular automaton model, originally developed by Gandin et al. (1994, 1999), was coupled to ProCAST. It allows the output of microstructure of equiaxed, columnar, and single-crystal superalloys. However, there are a number of input parameters such as heat transfer coefficients, nucleation, and dendrite growth velocities that must be calibrated if correct predictions are to be obtained.

Szeliga et al. (2013) used the ProCAST–CAFE model to study the directional solidification of rod castings made of CMSX-4 nickel superalloy. The emphasis was on the verification of the boundary conditions and on the values assumed for nucleation and grain growth rate coefficients. They used the nucleation law given by Eq. 18.3 and a growth law controlled by the undercooling: $V = a_2 \Delta T^2 + a_3 \Delta T^3$. The coefficients in the growth law were calculated using the CAFE module and found to be: $a_2 = 3.149 \cdot 10^{-7}$ m s^{-1} K^{-2} and $a_3 = 4.257 \cdot 10^{-7}$ m s^{-1}K^{-3}. Other quantities used in calculations included capillary length $\Gamma = 3.659 \cdot 10^{-7}$ Km, and liquid diffusivity $D_L = 3.69 \cdot 10^{-9}$ m^2 s^{-1} (Nastac 2004). The maximum consistency of simulation results with the experimental ones was obtained for the nuclei quantity of $n_{max} = 10^8$ m^{-2} and $\Delta T_N = 0.5$ K, $\Delta T\sigma = 0.3$ K.

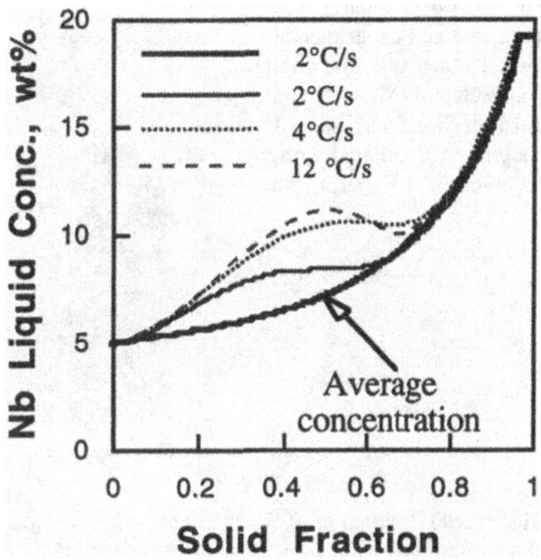

Fig. 19.102 Calculated variation of liquid Nb concentration at the S/L interface (*thinner lines*) and intrinsic volume average concentration for various cooling rates (Nastac and Stefanescu 1996). (With permission of Springer Science and Business Media)

Fig. 19.103 Calculated and experimental Laves phase content as a function of the calculated cooling rate immediately above the liquidus temperature (Nastac and Stefanescu 1996). (With permission of Springer Science and Business Media)

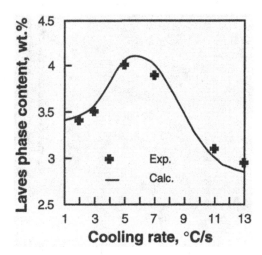

Similar experiments and modeling approach was used by Matache et al (2015) to simulate single crystal development in a CMSX-4 Ni-based superalloy produced through the Bridgman method. The experimental cylindrical bars were produced in an industrial vacuum melting and casting furnace. After numerical experimentation, the following heat transfer coefficients were used: a temperature-dependent casting/shell coefficient included in the ProCAST software, 1000 W m^{-2} K^{-1} casting/plate coefficient, and 50 W m^{-2} K^{-1} shell/plate coefficient. For the range of parameters used for simulation, the experimental measurements of grain density in the grain selector correlated best with a nucleation potential of 10^7 m^{-2}, as shown qualitatively in Fig. 19.104.

Fig. 19.104 Comparison of simulated and experimental surface macrostructure on the grain selector (Matache et al. 2015). **a** Simulated with 10^9 nucl/m². **b** Simulated with 10^7 nucl/m². **c** Experimental

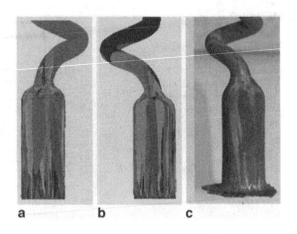

a   b   c

Fig. 19.105 Prediction of single grain selection over the distance from the cooling plate for two nucleation potential. (Matache et al. 2015)

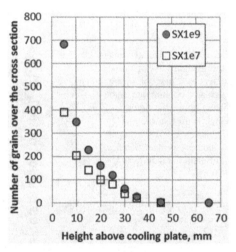

Good agreement between the simulation and the experiment was found for the height from the cooling plate at which a single grain is selected, and for the grain deviation from de direction <001>. Again nucleation potential between 10^{-7} and 10^{-9} made little difference (Figs. 19.105 and 19.106).

Important parameters of single-crystal superalloy behavior at high temperature, such as the PDAS and the SDAS, were measured and calculated from simulation results. It was found that both PDAS and SDAS fit equations of proportionality with the cooling rates $(GV)^{-0.34}$ and $(GV)^{-0.37}$, respectively. The ProCAST calculated values for PDAS and SDAS based on temperature gradient and solidification velocity produced a reasonable match to the experimental data.

The ProCAST–CAFE software has also been used for optimization of the grain selector when producing single-crystal superalloys. Simulation results demonstrated

Fig. 19.106 Simulated grain
orientation deviation with
respect to the growth direction
as a function of the distance
from the cooling plate. (Matache et al. 2015)

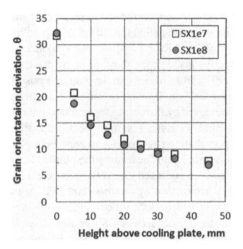

that the spiral grain selector becomes more efficient when decreasing the spiral diameter and increasing its overall (helical) diameter (Dai et al. 2009).

One of the limitations of the Bridgman DS method is that as the mold is lowered in the furnace, radiative heat transfer on the lateral surface of the mold becomes the dominant cooling mode. This makes it difficult to maintain a steep thermal gradient. An alternative method is the liquid metal cooling (LMC) process (e.g., Giamei and Tschinkel 1976; Shalin. and Pankratov 1992) in which the mold containing the melt is withdrawn from the hot zone through a baffle opening and directly immersed into a liquid metal bath (e.g., aluminum). In the case of large-size turbine blades, defects such as stray grains, freckles, slivers, microporosity, stable oxides, and carbides may occur. To optimize the LMC process renewed efforts were directed to simulation.

The LMC process has been simulated by Hugo et al. (1999) in two dimensions using the ANSYS program. A constant heat-transfer coefficient has been assumed for the heat exchange between the casting and the liquid metal.

Kermanpur et al. (2000b) simulated the production of a cored turbine blade in 3D using ProCAST. The heat exchange between the liquid metal and the casting was replaced by a temperature- and time-dependent heat-transfer coefficient. The value of this coefficient was first determined on the basis of an axisymmetric dummy-blade simulation performed with a proprietary software (CALCOMOS Calcom SA, Lausanne, Switzerland). Simulation and experimental work demonstrated that the most important parameters are the baffle dimensions, the level of the liquid bath, the configuration of the starter and gate, and the thermal conductivity of the mold material. A flatter liquidus front and increased vertical thermal gradient were produced with a new design at a withdrawal rate of 5.5 mm/min. This produced the desired ⟨001⟩ texture in the airfoil and eliminated the stray grains at the top of the blade. However, the higher withdrawal velocity increased the missorientation of the grain boundaries.

Freckles Prediction The occurrence of macrosegregation with its worst scenario, channel segregation (freckles), in superalloy has been identified as early as 1970 by Giamei and Kear. Experimental evidence suggests that freckles are initiated by convective instabilities above or in the mushy zone. As summarized by Schneider et al. (1997), instabilities can develop in a positive thermal gradient when the segregated interdendritic liquid is less dense than the bulk liquid. Solutal plumes of low-density liquid, fed by flow of segregated liquid through the surrounding mush, emanate from the mushy zone. This flow delays dendrite growth or remelts dendrites producing open channels in the mushy zone below each plume. The channel provides a self-sustaining path for feeding the plume. The channels eventually freeze and appear as chains of equiaxed grains (i.e., freckles) in the solidified superalloy. Casting geometry seems to play an important role on freckles formation. According to Ma and Bührig-Polaczek (2014), freckles could not only occur on vertical surfaces but also on sloping surfaces of DS samples. The component portions having inward sloping surfaces are very freckle prone because a wider supporting reservoir for freckling flow is provided from below. Those with outward sloping surface are mostly freckle free due to the geometrical lack of the convection reservoir from below. Therefore, the geometrical feature of the components can more effectively affect the freckle formation than the local thermal conditions.

Copley et al. (1970) suggested that in steady columnar growth, there exists a critical cooling rate below which freckles are likely to form (if the thermal gradient is also below a certain critical value). Another relationship proposed by Pollock and Murphy (1996), where the λ_1 is proportional to the product $G^{-1/2} \cdot V^{-1/4}$, produced a similar or slightly better criterion for freckle formation. However, accurate predictions require the description of solute partitioning during solidification of superalloys and its effect on liquid density, as well as of the influence of dendrite morphology on melt flow in the mushy zone.

A model developed by Schneider et al. (1997) linked a previously developed thermodynamic phase equilibrium subroutine with an existing code for simultaneously solving the macroscopic mass, momentum, energy, and species conservation equations for solidification of a multicomponent alloy. The phase equilibrium subroutine developed by Boettinger et al. (1995) based on the CALPHAD method relies on the use of thermodynamic free energy functions for the relevant phases in the alloy. The subroutine accepts the liquid concentrations as input and calculates the corresponding equilibrium liquidus temperature, the interfacial solid concentrations, and the liquidus slopes. The phase equilibrium subroutine was coupled with the solidification model and used to calculate the two-phase equilibrium between the liquid and the primary γ-solid phases only. Because three-phase equilibrium calculations were not done, solidification was assumed to end via an isothermal eutectic reaction.

Simulation results indicated that for a given alloy composition, the onset of convection and freckle formation occurs at a critical primary dendrite arm spacing of ~ 320 μm for a classic CMSX2 alloy, as predicted by Pollock and Murphy (1996), and hence depends on the imposed cooling rate. The predicted number and shape of the freckle chains in the unstable cases was much lower. It was also demonstrated how the onset and nature of convection and macrosegregation vary with alloy composition.

19.5 Applications

Application 19.1

Consider a superalloy with the following composition: 15.9 % Cr, 8.1 % Co, 3.4 %Al, 3.3 %Ti, 1.9 %Mo, 3 %W, 1.6 %Ta, and balance Ni. The measured liquidus and solidus temperature are $T_L = 1332°C$ and $T_L = 1212°C$ (Kermanpur et al. 2000a). Calculate the equivalent slope and partition coefficient of the multicomponent alloy and the liquidus and solidus temperature. Compare with the experimental data.

Answer The Pandat calculated data in Table 19.8 will be used. The liquid composition is $\overline{C}_L = 37.2$. Using Eq. 9.49, we calculate that $\bar{m}_L = -1.08$ and $\bar{k} = 0.55$. Then through Eq. 19.41, we find $T_L = 1415°C$ and $T_S = 1382°C$. Both these temperatures are considerably higher than the measured ones.

Using the approach suggested by Ma (2001), that is Eq. 19.43, we obtain $\bar{m}_L = -3.31$ and $\bar{k} = 0.51$. Note that while the partition coefficients calculated by the two methods are close, the slopes are quite different.

Application 19.2

Consider a CMSX-4 superalloy having the composition 5.6 %Al, 9 % Co, 6.5 % Cr, 0.1 %Hf, 0.6 %Mo, 3 %Re, 1.6 %Ta, 3.3 %Ti, 3 %W, and 61.7 %Ni. The measured liquidus and solidus temperature (Heckel et al. 2010) are 1390 and 1360°C, respectively. Using the Pandat calculated slopes and partition coefficients for the binary phase diagrams, calculate the material constant μ_0 in the SDAS equation $\lambda_2 = (\mu_0 t_f)^{1/3}$.

Answer Panda calculations of the m_L, k, and eutectic composition at $f_S = 89.3\%$ (eutectic start) are given in Table 19.1. Using $\Gamma = 2.7 \cdot 10^{-7}$ and $D = 3.2 \cdot 10^{-9}$ and Eq. 9.59, it is obtained that $\mu_0 = 1.41 \cdot 10^{-15}$. This is equivalent to the following equations for which SDAS is in μm: $\lambda_2 = 11.21(t_f)^{1/3}$ and $\lambda_2 = 55.31(GV)^{-1/3}$. These equations are quite reasonable in the sense that they are in line with the experimental equations for the IN738LC alloy.

Application 19.3

For the CSMX-4 alloy in Application 19.2, calculate the material constant in the primary dendrite arm spacing equation (Eq. 9.54).

Answer Using the approach suggested by Ma (2001), the equivalent partition coefficient for the pseud-binary phase diagram is calculated with Eq. 19.43 as $\bar{k} = (1455 - 1390)/(1455 - 1360) = 0.68$. Then, $\mu_{\lambda 1} = 4.3(\Delta T_o D_L \Gamma / \bar{k})^{1/4} = 4.3(30 \cdot 3.2 \cdot 10^{-9} \cdot 2.7 \cdot 10^{-7}/0.68)^{1/4} = 1897.5 \mu m$. This value compares well with the experimental one for a CSMX-6 alloy given in Eq. 19.38.

References

Aguado E, Stefanescu DM, Sertucha J, Larrañaga P, Suárez R (2014) AFS Trans. 122: 249–258

Albright DL (1967) in: Transformation and Hardenability in Steels. Climax Molybdenum Co. of Michigan, p 14

Alonso G., Stefanescu DM, Larrañaga P, Suárez (2014) in: Boeri R, Massone J, Rivera G (eds) Proc. 10th Science and Processing of Cast Iron. Mar del Plata, Argentina

Alonso G, Stefanescu DM, Suarez R, Loizaga A, Zarrabeitia G (2014) Int. J. Cast Met. Res. 27:87–100

Alonso G, Stefanescu DM, Larrañaga P, De la Fuente E, Aguado E, Suarez R (2015) in: Nastac L et al. (eds) Advances in Science and Engineering of Casting Solidification. TMS Wiley, Hoboken NJ p 347–354

Amini S, Abbaschian R (2013) Carbon 51:110–123

Argo D, Gruzleski JE (1986) Mat. Sci. Tech. 10(2):1019

Argyropoulos S, Closset B, Gruzleski JE, Oger H (1983) Trans. AFS 91:351

Atamanenko TV, Eskin DG, Katgerman L (2009) Int. J. Cast Metals Res. 22(1–4):26

Bandyopadhyay DK, Stefanescu DM, Minkoff I, Biswal SK (1989) in: Ohira G, Kusakawa T, Niyama E (eds) Physical Metallurgy of Cast Iron IV. Tokyo, Mat. Res. Soc. Proc., Pittsburgh, Pa p 27

Basutkar P, Yew S, Loper C Trans. (1969) Trans. AFS 77:321–28

Bäckerud L, Chai GC, Tamminen J (1990) Solidification Characteristics of Aluminum Alloys. AFS/Skanaluminium, Stockholm

Beltran-Sanchez L (2003) Ph. D. Dissertation. University of Alabama, Tuscaloosa Al

Boeri RE, Sikora JA (2001) Int. J. Cast Metals Res. 13(5):307–313

Boettinger WJ, Kattner UR, Coriell SR, Chang YA, Mueller BA (1995) in: Cross M, Campbell J (eds) Modeling of Casting, Welding and Advanced Solidification Process VII. TMS, Warrendale, Pa pp 649

Bollman W, Lux B (1975) in: Lux B, Minkoff I, Mollard F (eds) The Metallurgy of Cast Iron. Georgi Publishing, St. Saphorin, Switzerland, p 462–70

Bramfitt BL (1985) in: Metals Handbook 9th Edition Metallography and Microstructures, vol. 8. American Soc. Metals, p 623–628

Briggs CW (1946) The Metallurgy of Steel Castings. McGraw-Hill, NY

Bunin KP, Malinotchka IN, Taran IN (1969) Osnovi Metallographyia Tchuguna. Metallurghyia, Moscow

Burbelko A, Fras E, Gurgul D, Kapturkiewicz W, Sikora J (2011) Key Eng Mater 457:330

Burbelko A, Gurgul D, Kapturkiewic W, Górny M (2012) IOP Conf. Series: Mater Sci Eng 33:012083

Buttner FH, Taylor HF, Wulff J (1951) American Foundryman 20(4):49

Campbell J (2003) Castings, 2nd ed. Butterworth-Heinemann, Oxford, U.K. p 17–69

Campbell J (2006) Metall. Mater. Trans. B 37B (Dec.):857

Catalina A, Stefanescu DM (1996) Metall. and Mater. Trans. 27A:4205

Charbon Ch, Rappaz M (1997) in: LesoultG, Lacaze J (eds) Physical Metallurgy of Cast Iron V. Scitec Publications, Switzerland pp 453–460

Charbonier J (1984) Trans. AFS 92:907

Chen Q, Langer EW, Hansen PN (1995) in: Cross M, Campbell J (eds) Modeling of Casting, Welding and Advanced Solidification Processes VII. TMS, Warrendale, Pa p 633

Chisamera M, Riposan I, Barstow M (1998) in: AFS International Inoculation Conference. Rosemont, Il, USA paper 3

Copley SM, Giamei AF, Johnson SM, Hornbecker MF (1970) Metall. Trans. 1:2193-2204

Cosneanu C (1966) Metalurgia (Romania) no. 10:563

Crossley PB, Mondolfo LF (1966) Modern Casting 49:53

Dahle AK, Nogita K, Zindel JW, McDonald SD, Hogan LM (2001) Metall. and Mater. Trans. 32A:949

Dai HJ, Gebelin J-C, D'Souza N, Brown PD, Dong HB (2009) Int. J. Cast Metals Res. 22(1–4):55

Dai X, Jolly M, Yang X, Campbell J (2012) Mater. Sci. Eng. 33:1–10

Dawson S, Popelar P (2013) in: 2013 Keith Millis Symposium on Ductile Iron. AFS, Schaumburg. IL p 59

De LR, Xiang YJ (1991) Trans. AFS 99:707–712

De Sy A (1949) Met. Progr. 55:138

Decker RF, Mihalisin JR (1969) Trans. ASM 62:481

Degand C, Stefanescu DM, Laslaz G (1996) in: Ohnaka I, Stefanescu DM (eds) Solidification Science and Processing. TMS, Warrendale Pa p 55

Dhindaw B, Verhoeven JD (1980) Metall. Trans. A 11A:1049–57

Dinnis CM, Dahle AK, Taylor JA (2005) Mat. Sci. Eng A 392:440–448

Donald W, Moore A (1973) BCIRA Report no. 1128

Double DD, Hellawell A (1969) Acta Metall., 17:1071–83

Double DD, Hellawell A (1975) in: Lux B, Minkoff I, Mollard F (eds) The Metallurgy of Cast Iron. Georgi Publishing, St. Saphorin, Switzerland, p 509

Double DD, Hellawell A (1995) Acta metall. mater. 43(6):2435

Dresselhaus M S, Dresselhaus G, Surihara K, Spain IL, Goldberg HA (1988) in: Graphite Fibers and Filaments. Springer, Berlin

Eliade M (1978) The Forge and the Crucible. The University of Chicago Press

Ellerbrok R, Engler S (1981) in: Erstarrung metallischer Schmelzen. Proc. Symp. DGM, Aachen pp 249–260

Elliott R (1983) Eutectic Solidification Processing. Butterworth, London

Elliott R (1988) Cast Iron Technology. Butterworth, London

El-Bealy M, Thomas BG (1966) Metall. Mater Trans. B 27B(August):689

Escobar A, Celentano D, Cruchaga M,. Lacaze L, Schulz B, Dardati P, Parada A (2014) Int. J. Cast Metals Res. 27(3):176

Eskin GI (1998) Ultrasonic treatment of light alloy melts. Amsterdam, Gordon and Breach Science Publishers

Fang KM (2000) Atlas of the Morphology and Microstructure of the Graphite in Cast Iron. Science Publ. Co. of China

Fisher RM, Speich GR, Cuddy LJ, Hu H (1976) Proc. Darken Conf. "Physical Chemistry in Metallurgy," US Steel, Monroeville, Pa pp 463–88

Fisher DJ, Kurz W (1980) Acta Metallurgica 28:777–94

Flood SC, Hunt JD (1981) Met. Sci. 15 (7):287–92

Frank FC (1958) in: Doremus, Roberts, Turnbull (eds) Growth and Perfection of Crystals. John Wiley N.Y.

Franke MM, Hilbinger RM, Konrad CH, Glatzel U, Singer RF (2011) Metall. Mater. Trans. A 42A(July):1847

Fras E, Lopez HF (1993) Trans. AFS 101:355

Fraś E, Górny M, Lopez H (2007) Metall. Mater. Trans. A 38A:385–395

Fredriksson H (1988) in: Stefanescu DM (ed) Metals Handbook 9th Edition vol. 15 Casting. ASM Int., Metals Park, OH p 129

Gandin CA, Rappaz M (1994) Acta Metall. Mater. 42(7):2233–2246

Gandin CA, Desbiolles JL, Rappaz M, Thevoz P (1999), Metall. Trans. A 30(12):3153–3165

Geilenberg H (1964) Giesserei Techn.-Wiss. Beih. 16:35
Giamei AF, Tschinkel JG (1976) Metall. Trans. A 7A: 1427–34
Giamei AF, Kear BH (1970) Metall. Trans. 1:2185–92
Granger D, Elliott H (1988) in: Stefanescu DM (ed) ASM Handbook 9th Edition Vol. 15 Casting. ASM International, Metals Park, Oh p 159
Gu JP, Beckermann C (1999) Metall. Mater. Trans. A 30A:1357–1366
Gundlach RB (1988) in: Stefanescu DM (ed) ASM Handbook 9th Edition Vol. 15 Casting. ASM International, Metals Park, Oh p 678–685
Gurgul D, Burbelko A, Guzik E, Kopycioski D, Krlikowski M (2014) in: Boeri R, Massone J, Rivera G (eds) 10th Int. Symp. on the Science and Processing of Cast Iron. Mar del Plata, Argentina, CD Proceedings paper 51
Hara T, Kitagawa T, Kuroki K, Saikawa S, Terayama K, Ikeno S, Matsuda K (2014) Mater. Trans. 55:1500–1505
Hatate M, Nakamura K, Sumimoto H (1989) in: Ohira G, Kusakawa T, Niyama E (eds) Physical Metallurgy of Cast Iron IV. Tokyo, Mat. Res. Soc. Proc., Pittsburgh, Pa pp 149–156
Heckel A, Rettig R, Singer RF (2010) Metal. Mater. Trans. A 41A(January):202
Hellawell A, Herbert PM (1962) Proc. Royal Soc. A269:560
Hellawell A (1973) Progr. Mater. Sci. 15:1
Herfurth K (1965) Freiberg Forschungs 105:267
Hillert M, Steinhauser H (1960) Jernkontorets Annaler 144:520
Hummer R (1991) 58th World Foundry Congress, Krakow, Poland, paper 2
Hunter A, Ferry M (2002) Metall. Mater. Trans. A. 33A:3747–54
Hugo F, Betz U, Ren J, Huang SC, Bondarenko JA, Gerasimov V (1999) in: Int. Symp. on Liquid Metal Processing and Cast-ings, Santa Fe, NM, Vacuum Metallurgy Division of American Vacuum Soc., New York pp 16–30
Humphreys JG (1961) BCIRA J. 9:609–621
Itofuji H, Uchikawa H (1990) AFS Trans. 98:429–448
Jacobi H, Scherdtfeger K (1976) Metall. Trans. A. 33A:3747–3754
Jacobs MM, Law TJ, Melford DA, Stowell MJ (1974) Metals Technology, 1(Part II Nov.):490
Johnson WC, Smartt HB (1979) in: Solidification and Casting of Metals. The Metal Society, Book No. 192 p 129
Jones H, Kurz W (1980) Metall. Trans. 11A:1265
Journal of the American Foundrymen's Assoc. (1896) 1
Karma A, Rappel W-J (1998) Physical Revue E 57(4):4323
Kear BH, Piearcey BJ (1967) Trans. Metall. Soc. AIME 239:1209-1218
Kermanpur A, VarahramN, Engilehei E, Mohammadzadeh M, Davami P (2000a) Mater. Sci. Technol. 16:579–86
Kermanpur A, Varahram N, Davami P, Rappaz M (2000b) Metall. Mater. Trans. B 31B:1293
Keverian J, Taylor HF, Wulff J (1953) American Foundryman 23:85
Keverian J, Taylor HF (1957) AFS Trans. 65:212-221
Konrad CH, Brunner M, Kyrgyzbaev K, Völkl R, Glatzel U (2011) J. Mater. Proc. Technology 211:181–186
Krieger W, Trenkler H (1971) Arch. Eisenhuttenwesen 42(3):175
Kurz W, Giovanola B, Trivedi R (1986) Acta Metall. 34:823–30
Lakeland KD (1964) BCIRA J. 12:634
Lakeland KD, Hogan LM (1968) in: Merchant HD (ed) Recent Research on Cast Iron. Gordon and Breach, NY pp 417–448
Lacaze J, Magnusson Aberg L, Sertucha J (2013) in: Keith Millis Symposium on Ductile Iron, AFS
Leube B, Arnberg L, Mai R (1998) in: Thomas BG, Beckerman C (eds) Modeling of Casting, Welding and Advanced Solidification Processes VIII. TMS, Warrendale Pa p 463
Leube B, Arnberg L (1999) Int. J. Cast Metals Res. 11:507–514
Li YX, Liu BC, Loper CR (1990) AFS Trans. 98:483–488
Liu DR (2013) Int. J. Cast Metals Res. 26(3):143
Liu PC, Li CL, Wu DH, Loper CR (1983) Trans. AFS 91:119–126

Lux B, Tannenberger H (1962) Modern Casting (March): 57

Lux B (1967) Giesserei Forschung 19:141

Lux B (1968) in: Merchant HD (ed) Recent Research on Cast Iron. Gordon and Breach, London p 241

Ma D (2001) Metall. Mater. Trans. B 32B:1212–13

Ma D, Meyer Ter Vehn M, Busse P, Sahm PR (1993) J. de Physique IV, 3(November):339

Ma D, Bührig-Polaczek A (2014) Metall. Mater. Trans. A 45A:1435

Magnin P, Kurz W (1985) in: Fredriksson H, Hillert M (eds) The Physical Metallurgy of Cast Iron. North Holland, New York p 263

Magnin P, Kurz W (1988) Metall. Trans. A 19A:1955–1963

Magnin P, Trivedi R (1991) Acta metal.mater. 39:453

Mampey F (1997) in: Lesoult G, Lacaze J (eds) Physical Metallurgy of Cast Iron V. Scitec Publications, Switzerland p 73–88

Mampey F (2001) in: Proceedings of Cast Iron Division. AFS 105th Casting Congress, American Foundry Soc., Des Plaines IL, p 51

Mampaey F, Beghyn K (2006) Trans. AFS 114:paper 06–115

Mampaey F, Habets D, Plessers J, Seutens F (2008) Int. Foundry Research/Giessereiforschung 60(1):2

Marincek B et al. (1953) Giesserei Techn.-Wiss. Beih. 12:587

Matache G, Stefanescu DM, Puscasu C, Alexandrescu E, Bührig-Polaczek A (2015) Int. J. Cast Metals Res. in print

McLean M (1983) Directionally Solidified Materials for High Temperature Service. The Metal Soc. London pp 35–37

McSwain RH, Bates CE (1975) in: Lux B, Minkoff I, Mollard F (eds) The Metallurgy of Cast Iron. Georgi Publishing, St. Saphorin, Switzerland p 423

Meijer D, Cockroft SL, Patt W (1999) Metall. Mater. Trans. A 30A:2147

Mills KC, Youssef Y, Li Z, Su Y (2006) ISIJ Int. 46:623–632

Milman BS (1958) Litejnoe Proizvodstvo no.6:11–17

Minkoff I (1983) The Physical Metallurgy of Cast Iron. John Wiley & Sons, New York

Moore JC (1973) in: Metals Handbook Eight Edition vol. 8 Metallography. ASM Int., Metals Park, Oh p 93

Motz J (1975) in: Lux B, Minkoff I, Mollard F (eds) The Metallurgy of Cast Iron. Georgi Publishing, St. Saphorin, Switzerland pp 356-369

Moumeni E, Stefanescu DM, Tiedje NS, Larrañaga P, Hattel JH (2013) Metall. Mater. Trans. 44A:5134

Murray JL, Bennet LH, Baker H (1986) Binary Alloys Phase Diagrams, ASM, Metals Park, Oh 1-2

Nakae H, Shin H (1999) Int. J. Cast Metals Res. 11:345–349

Naro R, Wallace JF (1969) Trans. AFS 77:311–320

Nastac L (1999) Acta Mater. 47(17):4253

Nastac L (2004) Modeling and Simulation of Microstructure Evolution in Solidifying Alloys. Kluwer Academic Publishers, Boston

Nastac L (2011) Metall. Mater. Trans. B 24B:1297

Nastac L (2014) ISIJ International 54(8):1830–1835

Nastac L, Stefanescu DM (1995) AFS Trans. 103:329–337

Nastac L, Stefanescu DM (1996) Metall. Mater. Trans. A 27A:4075

Nastac L, Stefanescu DM (1997) Model. Simulat. Mater. Sci. Eng. 5(4):391

Nayebi B, Divandari M (2012) Int. J. Cast Metals Res. 25(5):270

Nieswaag H, Zuithoff AJ (1975) in: Lux B, Minkoff I, Mollard F (eds) The Metallurgy of Cast Iron. Georgi Publishing, St. Saphorin, Switzerland p 327

Ohira G, Sato T, Sayama Y (1974) in: Lux B, Minkoff I, Mollard F (eds) The Metallurgy of Cast Iron. Georgi Publishing, St. Saphorin, Switzerland p 295

Ohnaka I (1986) in: Fredriksson H (ed) State of the Art of Computer Simulation of Casting and Solidification Processes. E-MRS, Mat. Res. Soc., Pittsburgh, Pa pp 211–23

Okada A, Miyake H (1996) The Unknown World of Cast Iron (in Japanese). KANSAI Univ. Press, p 141

Oldfield WA (1966) Trans. AFS 59:945–960

Oldfield WA, Geering GT, Tiller WA (1967) in: The Solidification of Metals. Iron and Steel Inst., London, Publ. No. 110:256

Olsson A, West R, Fredriksson H (1986) Scand. J. Met. 15:104

Onsøien MI, Grong Ø, Gundersen Ø, Skaland T (1999) Metall. Mater. Trans. 30A:1053

Patel A, Nastac L (2014) Int. J. Metalcasting 8(1):21

Park JS, Verhoeven JD (1996) Metall. and Mater. Trans. A 27A:2740

Pollock TM, Murphy WH (1996) Metall. Mater. Trans. A 27A:1081–94

Purdy GR, Audier M (1984) in: Fredriksson H, Hillert M (eds) The Physical Metallurgy of Cast Iron. Mater. Res. Soc. Proc., North-Holland N.Y. 34:13–23

Pustal B, Berger R, Subasic E, Laschet G, Schäfer W, Bührig-Polaczek A (2009) Int. J. Cast Metals Res. 22(14):252

Quested PN, Brooks RF, Chapman L, Morrell R, Youssef Y, Mills KC (2009) Mater. Sci. Techn. 25(2):154

Rappaz M, Richoz JD, Thevoz P (1989) in: Exner HE (ed) European Conf. Advanced Materials Proc. Aachen p 135

Reed RC (2006) The Superalloys. Fundamentals and Applications. Cambridge eBooks Online

Rickert A, Engler S (1985) in: Fredriksson H, Hillert M (eds) The Physical Metallurgy of Cast Iron. Mater. Res. Soc. Proc., North-Holland, NY 34:165

Riposan I, Chisamera M, Stan S, Skaland T (2003) Int. J. Cast Metals Res 16(1-3):105–111

Rivera GL, Boeri RE, Sikora JA (1997) Advanced Materials Research 4–5:169

Rivera GL, Boeri RE, Sikora JA (2004) Scripta Materialia 50:331–335

Rivera G, Calvillo PR, Boeri R, Houbaert Y, Sikora J (2008) Mater. Characterization 59:1342–1348

Roch F, Combeau H, Chevrier JC, Lesoult G (1991) in: Modeling of Casting, Welding and Adv. Solidif. Processes V. TMS, Warrendale, PA, p 789–95

Rosenberg B, Tiller WA (1957) Acta Metall. 5:565

Roviglione AN, Hermida JD (2004) Metall. Mater. Trans. 35B:313

Ruff G, Wallace JF (1976) Trans. AFS 84:705–728

Ruxanda R, Beltran-Sanchez I, Massone J, Stefanescu DM (2001) Trans. AFS 109:1037

Sadocha JP, Gruzleski JE (1975) in: Lux B, Minkoff I, Mollard F (eds) The Metallurgy of Cast Iron. Georgi Publishing, St. Saphorin, Switzerland, p 443

Sahai V, Overfelt RA (1995) Trans. American Foundrymen's Soc. 103:627632

Shalin RE, Pankratov VA (1992) Metall. Sci. Technol. 10:3-9

Scheil E, Hütter L (1953) Arch. Eisenhuttenwesen 4:24

Schneider MC, Beckermann C (1995) Metall. Mater. Trans. A 26A:2373–88

Schneider MC, Gu JP, Beckermann C, Boettinger WJ, Kattner UR (1997) Metall. Mater. Trans. A 28A:1517

Sigworth GK (1983) Trans. AFS 91:7

Sims CE (1959) Trans. Met. Soc. AIME 215:367–393

Sims C, Stoloff N, Hagel W (1987) Superalloys II: High Temperature Materials for Aerospace and Industrial Power. John Wiley & Sons

Skaland T, Grong F, Grong T (1993) Metall. Trans. 24A:2321 and 2347

Sommerfeld A, Tonn B (2009) in: Stefanescu DM, Voigt RC, Prucha T (eds) The Carl Loper Cast Iron Symposium. AFS CD Proceedings, p 168–178

Steeb S, Maier U (1974) in: Lux B, Minkoff I, Mollard F (eds) The Metallurgy of Cast Iron. Georgi Publishing, St. Saphorin, Switzerland, p 1–11

Stefanescu DM (1972) Giesserei-Praxis 24:430

Stefanescu DM (1988) in: Stefanescu DM (ed) ASM Handbook Vol. 15 Casting. ASM International, Metals Park, Ohio, p 168

Stefanescu DM (2007) Metall. and Mater. Trans. A 38A:1433–1447

Stefanescu DM (2015) Int. J. Metalcasting 9(1):7–22

Stefanescu DM, Kanetkar CS (1987) AFS Trans. 95:139–144

Stefanescu DM, Bandyopadhyay DK (1989) in: Ohira G, Kusakawa T, Niyama E (eds) Physical Metallurgy of Cast Iron IV. Tokyo, Mat. Res. Soc. Proc., Pittsburgh, Pa pp 15–26

Stefanescu DM, Ruxanda R (2004) in: ASM Handbook vol. 9, Metallography and Microstructures. ASM International p 71-131

Stefanescu DM, Katz S (2008) in: ASM Handbook vol. 15 Casting. ASM Int., Metals Park, Oh pp 41–55

Stefanescu DM, Moran M, Boonmee S, Guesser WL (2012) AFS Proceedings, American Foundry Society, Schaumburg, IL. USA Paper 12–045

Su KC, Ohnaka I, Yamauchhi I, Fukusako T (1985), in: Fredriksson H, Hillert M (eds) The Physical Metallurgy of Cast Iron. Mater. Res. Soc. Proc., North-Holland, NY 34:181–189

Sun GX, Loper CR (1983) Trans. AFS 91:639–646

Sung PK, Poirier DR, Felicelli SD (2002) Modell. Simul. Mater. Sci. Eng. 10: 551–56

Suzuki A, Suzuki T, Nagaoka Y, Iawata Y (1968) Nippon Kingaku Gakkai Shuho 32

Szeliga D, Kubiak K, Burbelko A, Motyka M, Sieniawski J (2013) J. Materials Eng. and Performance, published online

Theuwissen K, Lafont MC, Laffont L, Viguier B, Lacaze L (2012) Trans. Indian Inst. Met., published on line TP2602

Tian H, Stefanescu DM (1992) Metall. Trans. 23A:681

Tian H, Stefanescu DM (1993) in: Piwonka TS, Voller V, Katgerman L (eds) Modeling of Casting, Welding and Advanced Solidification Processes-VI. TMS, Warrendale Pa, p 639

Tiller WA (1969) in: Solidification. American Society for Metals p 84

Vertman AA, Samarin AM (1969) Svoista Rasplavov Jeleza Ed. Nauka, Moscow

Walker JL (1959) in: St. Pierre GR (ed) The Physical Chemistry of Process Metallurgy. Interscience, New York, NY, part 2, pp 845–47

Wallace JF (1975) Trans. AFS 83:363-378

Wang CY, Beckermann C (1996) Metall. Mater. Trans. A 27A:27542764 and 2765–2783

Wang QG, Davidson CJ, Griffiths JR, Creapeau PN (2006) Metall. Mater. Trans. B 37B(6):887

Washchenko KI, Rudoy AP (1962) AFS Trans. 70:855

Weis W (1974) in: Lux B, Minkoff I, Mollard F (eds) The Metallurgy of Cast Iron. Georgi Publishing, St. Saphorin, Switzerland, p 69–79

Wetterfall SE, Fredriksson H, Hillert M (1972) J. Iron and Steel Inst. 323

Whitesell HS, Li L, Overfelt RA (2000) Metall. Trans. 31B:546–551

Wieser PF (ed) (1980) Steel Castings Handbook Fifth Edition. Steel Founders Soc. of America, Rocky River, OH, p 19–6

Wilford KB, Wilson FG (1985) British Foundryman 78:301 and 364

Wyckoff RWG (1963) Crystal Structures, Vol. 1, Interscience, New York

Yukumoto M, Yamane H (1995) ISIJ International 35(6):778

Zakhartchenko EV, Akimov EP, Loper CR (1979) AFS Trans. 87:471

Zeng, D, Zhang Y, Liu J, He H, Hong X (2006) in: Li Y et al. (eds) Science and Processing of Cast Iron VIII. Tsinghua Univ. Press, p 86–91

Zhao H, Liu BC (2003) Int. J. of Cast Metals Research 16:281

Zhao HL, Zhu MF, Stefanescu DM (2011) Key Eng Mater 457:324

Zhu MF, Zhang L, Zhao HL, Stefanescu DM (2015) Acta Mater. 84:413–425

Appendix

Appendix A: Some Solutions of the Diffusion Equations

PDE	IC and BC	Solution
$\dfrac{\partial^2 \phi}{\partial x^2} = 0$	$\phi(0) = \phi_1$ and $\phi(L) = \phi_2$	Becomes ordinary differential equation $\phi(x) = C_1 x + C_2$ with $C_1 = (\phi_2 - \phi_1)/L$ and $C_2 = \phi_1$
$\dfrac{\partial^2 \phi}{\partial x^2} = 0$	$\phi(x,0) = f(x)$ $\dfrac{d\phi}{dx}(0) = 0$ and $\dfrac{d\phi}{dx}(L) = 0$	Becomes ordinary differential equation, initial boundary problem: $\phi(x) = C_1 x + C_2$ with $C_1 = 0$ and $C_2 = \phi(x) = \dfrac{1}{L}\int_0^L f(x)dx$ from $\rho c \int_0^L C_2 dx = \rho c \int_0^L f(x)dx$
$\dfrac{\partial \phi}{\partial t} = \Gamma \dfrac{\partial^2 \phi}{\partial x^2}$	$\phi(x,0) = f(x)$ $\phi(0,t) = 0$ and $\phi(L,t) = 0$	Separation of variables, Fourier series $\phi(x,t) = \sum_{n=1}^{\infty} B_n \sin\dfrac{n\pi x}{L} \exp\left[-\left(\dfrac{n\pi}{L}\right)^2 Dt\right]$ with $B_n = \dfrac{2}{L}\int_0^L f(x)\sin\dfrac{n\pi x}{L}dx$
$\dfrac{\partial \phi}{\partial t} = \Gamma \dfrac{\partial^2 \phi}{\partial x^2}$	$\phi(x,0) = f(x)$ $\phi(0,t) = \phi_1$ and $\phi(L,t) = \phi_2$	1) Nonhomogeneous problem $\phi(x,t) = \phi_E(x) + \sum_{n=1}^{\infty} a_n \sin\dfrac{n\pi x}{L} \exp\left[-\left(\dfrac{n\pi}{L}\right)^2 \alpha t\right]$ with $\phi_E(x) = \phi_1 + \dfrac{\phi_2 - \phi_1}{L}x$ and $a_n = \dfrac{2}{L}\int_0^L \left[f(x) - \phi_E(x)\right]\sin\dfrac{n\pi x}{L}dx$ 2) $\phi(x,t) = A + B erf\dfrac{x}{2\sqrt{\alpha t}}$
$\dfrac{\partial \phi}{\partial t} = \Gamma \dfrac{\partial^2 \phi}{\partial x^2}$	$\phi(x,0) = f(x)$ $\dfrac{d\phi}{dx}(0,t) = 0$ and $\dfrac{d\phi}{dx}(L,t) = 0$	Separation of variables, Fourier series $\phi(x,t) = \sum_{n=0}^{\infty} A_n \cos\dfrac{n\pi x}{L} \exp\left[-\left(\dfrac{n\pi}{L}\right)^2 \alpha t\right]$ with $A_0 = \dfrac{1}{L}\int_0^L f(x)dx$ and $A_n = \dfrac{2}{L}\int_0^L f(x)\cos\dfrac{n\pi x}{L}dx$

PDE	IC and BC	Solution
Laplace's equation $$\frac{\partial^2 \phi}{\partial x^2} + \frac{\partial^2 \phi}{\partial y^2} = 0$$	$\phi(x,0) = f_1(x)$ $\phi(x, H) = f_2(x)$ $\phi(0, y) = g_1(y)$ $\phi(L, y) = g_2(y)$	For $g_2 = f_1 = f_2 = 0$ $$\phi(x, y) = \sum_{n=1}^{\infty} A_n \sin \frac{n\pi y}{H} \sinh \frac{n\pi (x - L)}{H} \text{ with}$$ $$A_n = \frac{2}{H \sin (-n\pi L/H)} \int_0^H g_1(y) \sin \frac{n\pi y}{H} dy.$$ The original solution is obtained by adding together four such solutions
Laplace's equation for a circular disk $$\frac{1}{r} \frac{\partial}{\partial r} \left(r \frac{\partial \phi}{\partial r} \right) + \frac{1}{r^2} \frac{\partial^2 \phi}{\partial \theta^2} = 0$$	$\phi(a, \theta) = f(\theta)$ where a is the radius	$$\phi(r, \theta) = \sum_{n=0}^{\infty} A_n r^n \cos n\theta + \sum_{n=1}^{\infty} B_n r^n \sin n\theta$$ with $A_0 = \frac{1}{2\pi} \int_{-\pi}^{\pi} f(\theta) d\theta$ $$A_n a^n = \frac{1}{\pi} \int_{-\pi}^{\pi} f(\theta) \cos n\theta d\theta$$ $$B_n a^n = \frac{1}{\pi} \int_{-\pi}^{\pi} f(\theta) \sin n\theta d\theta$$
$$\rho c \frac{\partial T}{\partial t} = \frac{\partial}{\partial x} \left(k \frac{\partial T}{\partial x} \right) + \dot{Q}_{gen}$$ for $\dot{Q}_{gen} = \alpha_1 T$		Sturm–Liouville eigenvalue problem
Laplace's equation in a circular cylinder		Separation of variables, Fourier–Bessel functions

PDE partial differential equation; *IC* initial condition; *BC* boundary conditions

Appendix B: Properties of Selected Materials

Selected Constants

Boltzman constant	$k_B = 1.38 \cdot 10^{-23}$ J K^{-1}
Avogadro constant	$N_A = 6.023 \cdot 10^{23}$ mol^{-1}
Gas constant	$R = 8.31$ J·mol^{-1} K^{-1}
Stefan–Boltzman constant	$\sigma = 5.67051 \cdot 10^{-8}$ W m^{-2} K^{-4}
Planck constant	$h = 6 \cdot 63 \cdot 10^{-34}$ J s

Table B.1 Properties of selected metals and alloys near their melting point

Property	Units	Ni	IN-718[a]	CSMX-4	Cu	Pb	Sn–Pb
C_o	wt%		5Nb	39			–
C_E	wt%		γ+Laves: 19.1	43.9			38.1
$C_{\alpha m}$	wt%						2.5
$C_{\beta m}$	wt%						81
T_f, T_L	°C	1455	1336	1390	1085	328	–
ΔT_o	°C			50			–
T_E	°C			1340			183
m_α	K wt%$^{-1}$		$-0.74 \cdot 10^{(0.09\text{Nb})}$	-1.7			-0.83
m_β	K wt%$^{-1}$						2.43
f_α	–						0.63
f_β	–						0.37
k	–		$0.99 - 0.12\text{Nb} + 0.0046\text{Nb}^2$				
ΔH_f	J kg^{-1}	$2.91 \cdot 10^5$	$2.95 \cdot 10^5$	$2.4 \cdot 10^5$	$2.05 \cdot 10^5$		
ΔS_f	J m^{-3} K^{-1}			$1.19 \cdot 10^6$	$1.2 \cdot 10^6$		
k_L	W m^{-1} K^{-1}		30.5	0.68	166		
k_S	W m^{-1} K^{-1}	80			342	35	
c_L	J kg^{-1} K^{-1}				495		

Table B.1 (continued)

Property	Units	Ni	IN-718[a]	CSMX-4	Cu	Pb	Sn-Pb
c_S	J kg⁻¹ K⁻¹ → $\mathrm{J\,kg^{-1}\,K^{-1}}$	670			473		
ρ_L	$\mathrm{kg\,m^{-3}}$				8000		
$\rho_{S\alpha}$	$\mathrm{kg\,m^{-3}}$	7850	7620	8107	7670	10300	7300
$\rho_{S\beta}$	$\mathrm{kg\,m^{-3}}$						10300
D_L	$\mathrm{m^2\,s^{-1}}$	$3\cdot10^{-9}$	$3\cdot10^{-9}$	$3.2\cdot10^{-9}$			$1.1\cdot10^{-9}$
D_S	$\mathrm{m^2\,s^{-1}}$		$56\cdot10^{-5}\cdot\exp(-2.8\cdot10^{5}/RT)$				
α_L	$\mathrm{m^2\,s^{-1}}$				$42\cdot10^{-6}$		
α_S	$\mathrm{m^2\,s^{-1}}$				$67\cdot10^{-6}$		
v_m^S	$\mathrm{m^3\,mol^{-1}}$	$7.1\cdot10^{-6}$			$8.3\cdot10^{-6}$	$18.7\cdot10^{-6}$	
γ_{LV}	$\mathrm{J\,m^{-2}}$	1.782			1.286	0.460	
γ_{LS}	$\mathrm{J\,m^{-2}}$	0.322			0.177	0.040	
Γ_α	m K		3.6510^{-7}	$2.7\cdot10^{-7}$	$1.5\cdot10^{-7}$	$0.48\cdot10^{-7}$	$0.79\cdot10^{-7}$
Γ_β	m K						$0.48\cdot10^{-7}$
θ_α	°						65
θ_β	°						35
μ_c	$\mathrm{K\,s\,m^{-2}}$						$5.93\cdot10^{9}$
μ_r	m K						$0.207\cdot10^{-6}$
μ_v	$\mathrm{K^{-2}\,s^{-1}\,m}$						$2.04\cdot10^{-4}$
φ	–						1
$\Delta v/v$	–				4.2×10^{-2}		

[a] 18.5Fe, 52.5Ni, 19Cr, 3Mo, 5Nb

Table B.2 Properties of aluminum and of selected aluminum-base alloys near their melting point

Property	Units	Al	Al–Cu	Al–Cu	Al–Si
C_o	wt%	–	4.5	32.7	–
C_E	wt%	–	32.7	32.7	12.6
$C_{\alpha m}$	wt%	–	5.65	5.65	1.64
$C_{\beta m}$	wt%	–		52.5	99.98
$T_f - T_L$	°C	661	650	–	–
ΔT_o	°C	–	102	–	–
T_E	°C	–	548.2	548.2	577.2
m_α	K wt%$^{-1}$	–	-3.6	-4.6	-7.5
m_β	K wt%$^{-1}$	–		3.8	17.5
f_α	–	–	–	0.54	0.879
f_β	–	–	–	0.46	0.121
k	–	–	–	0.14	
ΔH_f	J kg^{-1}	$3.91 \cdot 10^5$			
ΔS_f	J m^{-3} K^{-1}	$1.02 \cdot 10^6$			
k	W m^{-1} K^{-1}	93			
k_S	W m^{-1} K^{-1}	213			
c_L	J kg^{-1} K^{-1}	1070			
c_S	J kg^{-1} K^{-1}	1170			
ρ_L	kg m^{-3}	2390			2350
ρ_E	kg m^{-3}				$2602 - 0.29T$
$\rho_{S\alpha}$	kg m^{-3}		$2564 - 0.29 \cdot T$	$2564 - 0.29 \cdot T$	2500
$\rho_{S\beta}$	kg m^{-3}	–		4000	2150
D_L	m^2/s	–	$2.8 \cdot 10^{-9}$	$2.8 \cdot 10^{-9}$	$5.4 \cdot 10^{-9}$
D_S	m^2/s	–	$5.54 \cdot 10^{-13}$		
α_L	m^2 s^{-1}	$37 \cdot 10^{-6}$			
α_S	m^2 s^{-1}	$70 \cdot 10^{-6}$			
v_m^S	m^3 mol^{-1}	$11 \cdot 10^{-6}$			
γ_{LV}	J m^{-2}	0.865			
γ_{LS}	J m^{-2}	0.093			
Γ_α	m K	$0.9 \cdot 10^{-7}$	$2.4 \cdot 10^{-7}$	$2.4 \cdot 10^{-7}$	$1.96 \cdot 10^{-7}$
Γ_β	m K	–	–	$0.55 \cdot 10^{-7}$	$1.7 \cdot 10^{-7}$
θ_α	°	–	–	65	30
θ_β	°	–	–	55	65

Table B.2 (continued)

Property	Units	Al	Al–Cu	Al–Cu	Al–Si
μ_c	K s m^{-2}	–	–	$4.62 \cdot 10^9$	$8.3 \cdot 10^9$
μ_r	m K	–	–	$0.47 \cdot 10^{-6}$	$0.94 \cdot 10^{-6}$
μ_V	K^{-2} s^{-1} m		–	$1.89 \cdot 10^{-3}$	$1.07 \cdot 10^{-5}$
φ	–		–	1	3.2
$\Delta v/v$	–	6.5×10^{-2}			

Table B.3 Properties of iron and of selected iron-base alloys in the vicinity of their melting point

Property	Units	Fe	δFe–C		γFe–Gr	γFe–Fe$_3$C
C_o	wt%		0.09	0.6	4.26	4.3
C_E	wt%		Perit. 0.53	4.26	4.26	4.30
$C_{\alpha m}$	wt%		–		2.08	2.11
$C_{\beta m}$	wt%		–		99.9	6.67
T_f	°C	1538	1531	1490	–	–
ΔT_o	°C		36	72	–	–
T_E	°C		Perit. 1493	1154.5	1154.5	1147.1
m_α	K wt%$^{-1}$		– 81	– 80	– 140	– 140
m_β	K wt%$^{-1}$		–		400	60
f_α	–		–		0.924	0.515
f_β	–		–		0.076	0.485
k_α	–		0.17	0.35	0.49	
k_β					0.001	
ΔH	J kg^{-1}		$2.72 \cdot 10^5$		$2.6 \cdot 10^5$	$1.5 \cdot 10^5$
ΔS	J m^{-3} K^{-1}		$1.07 \cdot 10^6$			
k_L	W m^{-1} K^{-1}	35			27.5	
k_S	W m^{-1} K^{-1}	40				
c_L	J kg^{-1} K^{-1}	820				
c_S	J kg^{-1} K^{-1}	794			880	
ρ_L	kg m^{-3}	7000			7050	7050
$\rho_{S\alpha}$	kg m^{-3}	7210			7000	7400
$\rho_{S\beta}$	kg m^{-3}		–		2110	7200
D_L	m^2/s		2.10^{-8}		$4.5 \cdot 10^{-9}$	$4.7 \cdot 10^{-9}$
D_S	m^2/s		$6 \cdot 10^{-9}$	$1 \cdot 10^{-9}$	$5.54 \cdot 10^{-13}$	
α_L	m^2 s^{-1}		$6.1 \cdot 10^{-6}$			

Table B.3 (continued)

Property	Units	Fe	δFe–C		γFe–Gr	γFe–Fe$_3$C
α_S	m^2 s^{-1}		$5.8 \cdot 10^{-6}$			
v_s	m^3 mol^{-1}	$7.7 \cdot 10^{-6}$				
γ_{LV}	J m^{-2}	1.806				
γ_{LS}	J m^{-2}	0.204				
Γ_α	m K	$1.9 \cdot 10^{-7}$			$1.9 \cdot 10^{-7}$	$1.9 \cdot 10^{-7}$
Γ_γ	m K		–	–	$7.0 \cdot 10^{-7}$	$2.4 \cdot 10^{-7}$
θ_α	°		–	–	25	50
θ_β	°		–	–	85	55
μ_c	K s m^{-2}		–	–	$1.51 \cdot 10^{9}$	$6.03 \cdot 10^{9}$
μ_r	m K		–	–	$2.36 \cdot 10^{-6}$	$0.752 \cdot 10^{-6}$
μ_V	K^{-2} s^{-1} m			–	$9.18 \cdot 10^{-8}$	$3.96 \cdot 10^{-5}$
φ	–			–	5.4	1.8
$\Delta v/v$	–		$3.6 \cdot 10^{-2}$			

For additional data on the thermodynamics of Fe–C and Fe–C-multicomponent systems, the reader is referred to Stefanescu and Katz (2008)

Table B.4 Data for solidification kinetics modeling of selected alloys

Quantity	Units	Cast iron	Al–Si	IN-718
T_L	°C	$1521 - 44 \cdot C - 9.9 \cdot C^2$		
T_E	°C	St: $1154 + 4 \cdot Si - 2 \cdot Mn$		
	°C	Met: $1148 - 15 \cdot Si + 3 \cdot Mn$		
$C_{equivalent}$	%	$C + 0.31 \cdot Si + 0.33 \cdot P -$ $0.027 \cdot Mn + 0.4 \cdot S$		
μ_V	K^{-2} s^{-1} m	St: $(3–9.5) \cdot 10^{-8}$	$10^{-6} – 10^{-7}$	
	K^{-2} s^{-1} m	Met: $2.4 \cdot 10^{-6}$		
N_{eut}	m^{-2}	St: $1 \cdot 10^{5} + 3.36 \cdot 10^{4}$ (dT/dt)		
	m^{-2}	Met: $5 \cdot 10^{5} + 1 \cdot 10^{5}$ (dT/dt)		
$N_{primary}$	m^{-2}	$4.81 \cdot 10^{7} + 5.33 \cdot 10^{6}$ (dT/dt) $+ 8.7 \cdot 10^{4}$ (dT/dt)^2		$2.39 \cdot 10^{9}$ (2.31 − 0.17 (dT/dt))^2
$N_{Gr\ SGI}$	m^{-3}	$1.12 \cdot 10^{11} + 9.55 \cdot 10^{11}$ (dT/dt)		

Note that the number of grains per unit area can be transformed into grains per unit volume with the following relationship:

$$N_V = 0.87 \, (N_A)^{1.5} \quad N_V \ (\text{m}^{-3}) \text{ and } N_A \ (\text{m}^{-2})$$

Table B.5 Properties of some molding materials

Material	k_m, W m^{-1} K^{-1}	ρ_m, kg m^{-3}	c_m, J kg^{-1} K^{-1}
Silica sand	0.52	1600	1170
Magnesite	4.15		
Mullite	0.38	1600	750
Zircon sand	1.04	2720	840
Plaster	0.35	1120	840

Table B.6 Heat-transfer coefficients in the gap at the metal/mold/air interfaces

Process	h, W m^{-2} K^{-1}
Splat cooling	10^5–10^6
Powder atomization	42–420
Steel in cast iron or steel mold	
Before gap formation	400–1020
After gap formation	400
Ductile iron in cast iron mold (carbon coated)	1700
Cast iron in sand mold	315
Aluminum in copper mold	1700–2550
Aluminum die casting	
Before gap formation	2500–5000
After gap formation	400
Sand-mold—air	6.9

Variations in the heat transfer coefficient before gap formation result from imperfect mold/casting thermal contact because of oxide skin, surface tension, mold coatings, etc

Appendix C: Selected Phase Diagrams

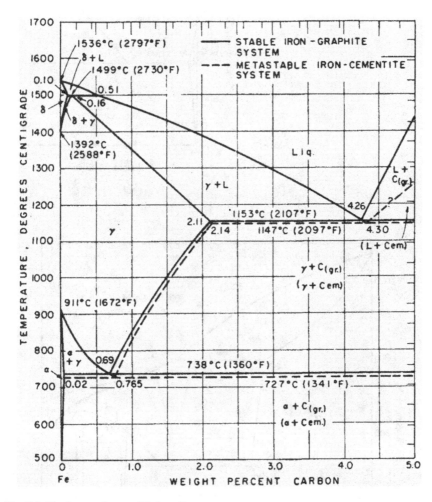

Fig. C.1 The iron–carbon equilibrium diagram

Fig. C.2 The aluminum–
silicon phase diagram

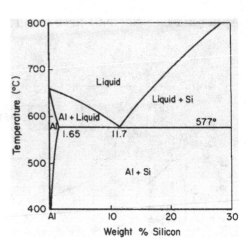

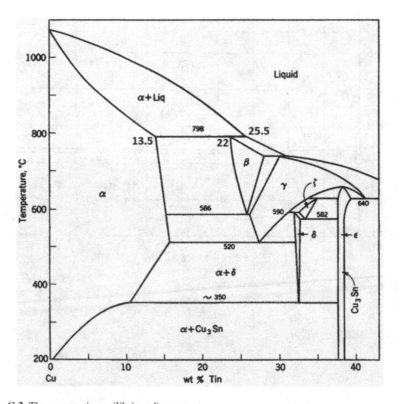

Fig. C.3 The copper–tin equilibrium diagram

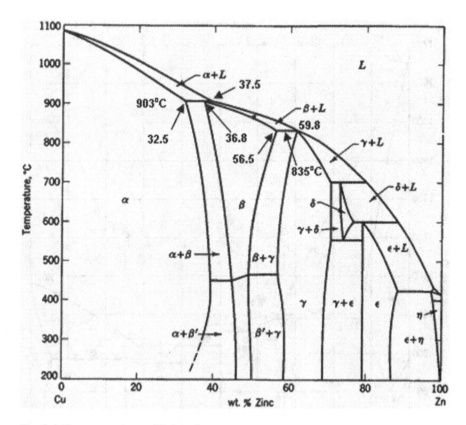

Fig. C.4 The copper–zinc equilibrium diagram

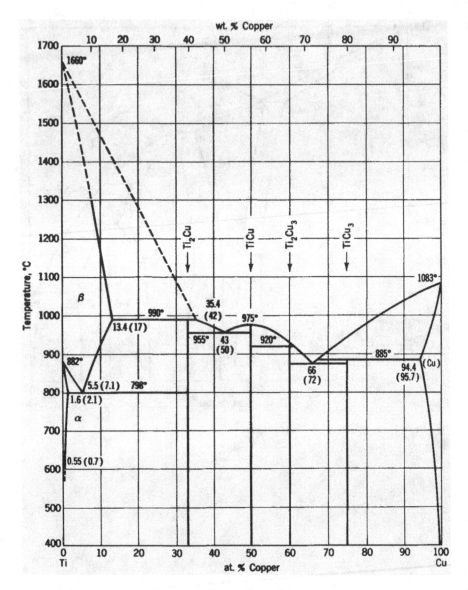

Fig. C.5 The titanium–copper equilibrium diagram

Fig. C.6 Section through the vanadium–carbon phase diagram

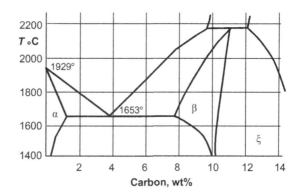

References

Chen C, Thompson RG, Davis DW (1991) in: Loria EA (ed) Superalloys 718, 625 and Various Derivatives. TMS, Warrendale, Pa p 81

Flemings MC (1974) Solidification Processing, McGraw Hill, NY

Holman JP (1986) Heat Transfer, 6th edition, McGraw Hill, New York

Kanetkar CS, Chen IG, Stefanescu DM, El-Kaddah N (1988) J. Iron Steel Inst Jap. 28:860

Kurz W, Fisher DJ (1989) Fundamentals of Solidification, 3rd ed. Trans Tech Publ., Switzerland

Ma D (2001) Metall. Mater. Trans. B 32B:1212–13

Magnin P, Trivedi R (1991) Acta metall. mater. 39(4):453

Magnusson T, Arnberg L (2001) Metall. Mater. Trans. 32A:2605

Nastac L, Stefanescu DM (1995) AFS Trans. 103:329

Nastac L, Stefanescu DM (1996) Metall. Mater. Trans. A 27A:4075

Poirier DR, Poirier EJ (1993) Heat Transfer Fundamentals for Metal Casting, 2nd ed. TMS, Warrendale Pa

Stefanescu DM, Katz S (2008) Thermodynamic Properties of Iron-Base Alloys. In: ASM Handbook vol. 15 — Casting. ASM International pp 41–55

Index

A

Aluminum-silicon alloys, 197, 220
Amorphous alloys, 283, 285, 286, 288,
  290–292
Arm spacing, 173
Atomization, 100, 284, 287
Austenite dendrites, 199, *see also* Dendrites
  contribution of, 216

B

Bifilms, 133, 276, 277, 429
  oxide, 404, 405
Blake-Kozeny model, 124
Boundary conditions (BC), 91, 92, 94, 97, 104,
  349, 351
Boussinesq approximation, 121, 357
Burst feeding, 130

C

Cast iron, 1, 49, 199
  hypoeutectic, 220
Cellular-automaton models, 416
Chemical potential, 7, 11
  gradients, 254
  in global equilibrium, 21
  in liquids and solids, 23
  stable and metastable local equilibrium, 22
Chunky graphite, 469
Clapeyron equation, 18, 19
Coherent interface, 40, 41
Columnar-to-equiaxed transition (CET), 182,
  184, 185, 187, 188
Compacted graphite, 460
Constitutional undercooling, *see* Undercooling
Constrained growth, 150
Contact angle, 37, 41, 199, 205, 235, 237, 328,
  336, 401

Continuous
  cooling solidification, 469
  growth, 25, 48, 51, 52, 498
Continuum modes, 63, 64, 390, 410
Control volume formulation, 354
Cooperative growth, 197
Coupling MT and TK models, 388
Criteria functions, 133
Crystal lattice of graphite, 457
Curly graphite, 460, 468
Curvature undercooling, *see* Undercooling

D

Dendrite arm spacing (DAS), 23, 123, 173, 440
Dendrites, 21, 47, 76, 123, 124, 130, 145, 148,
  219, 271, 287, 308, 389, 487, 526
  and cells
    analytical tip velocity models for, 157
  dynamic coarsening of, 175, 184
  equiad/plate, 139
  equiaxed/plate, 165, 188
  models describing growth of, 389, 443
  types of, 390, 399
Departure from equilibrium, 21, 283, 465
Differential thermal analysis (DTA), 106, 108,
  110, 112, 113
Diffusion boundary layer, 17, 67, 199
Diffusive energy transport, 89
  governing equation for, 89, 91
Diffusivity, 488
  calculations, 65
  carbon, 214, 230
  chemical diffusivity matrix
    assessment of, 255
  effective, 292
  infinite, 142
  interfacial, 26

© Springer International Publishing Switzerland 2015
D. M. Stefanescu, *Science and Engineering of Casting Solidification,*
DOI 10.1007/978-3-319-15693-4

liquid, 26, 34, 39, 278, 522
liquid and solid, 69
solid, 84, 85, 135, 143
specific phase quantities and, 61
thermal, 89, 121, 270
of mold, 364
Directional solidification, 22, 26, 68–70, 75,
    81, 82, 93, 127, 165, 219, 235, 247, 251,
    307, 325
experiments, 80
of rod castings, 522
stable solidification of the austenite-graphite
    eutectic, 476
Discretization, of governing equations, 348
Disregistry, 41, 44, 452
planar, 44, 45
Divorced growth, 197
Dynamic nucleation, *see* Nucleation

E

Enthalpy, 10, 11, 62, 90, 398
coupling of the macro- and
    micro-calculations using, 388
definition of, 346
enthalpy method, 346, 352, 373
Entropy, 9
of fusion, 45, 198
Equilibrium, 7
coefficient, 403
criterion, 8
departure from, 21
diagrams, 216
diffusion models, 192
global, 21
lever rule, 71, 144
local, 22, 23, 31
    interface, 22
melting temperature, 414
partition coefficient, 15
    of gas, 278
peritectic temperature, 227, 232
phase diagram, 8, 286, 435
reaction of gas (see also Sievert's law), 273
solidification, 26, 70, 72, 140, 278, 391
temperature, 10, 12, 19, 193
    distribution, 90
thermodynamic(s), 484
    calculations, 253
velocity, 312, 315
Error function, 96
Eutectic grain, 47, 220, 396, 417
Eutectic solidification
classification of, 197, 198

cooperative, 199, 201
divorced, 213–215
equiaxed eutectic solidification, 220
interface stability, 216–218, 220
irregular eutectic growth, 207–209, 211, 212
of cast iron, 469
regular eutectic growth, 201, 203–206

F

Finite difference method, 348
explicit formulation, 348, 353
Fluid flow effects, 261
Flux laws, 64, 65
Force(s)
drag force, 314, 315
gravity force, 315
interface force, 312, 314
lift force, 315
on a particle, 311, 314
Freckles, 124, 355, 525, 526
segregation and, 522
Free energy, 7, 199
calculation of, 10, 18, 30, 33
Gibbs, 7, 8, 251
Helmholtz, 9, 42
of solid, 12
partial derivatives of, 7
phase transformation, 9
use in CALPHAD method, 526
Front tracking models, 389

G

Gas pressure infiltration, 335
Gibbs free energy , *see* Free energy
Grain refining, 44, 180, 221, 494
cavitation-aided, 503
efficiency of, 504
of alloys, 509
Grain size, 42, 181, 301, 358, 384
and columnar-to-equiaxed transition (CET),
    509
of the aluminum alloy, 494
Graphite shape, 455, 464, 484
Gray-to-white transition (GWT), 480
Growth of graphite, 454

H

Hagen-Poiseuille model, 123
Heat of fusion, 10, 52, 91, 99, 146, 361, 414,
    514
Heat transfer coefficient, 92, 103, 104, 116,
    518
Helical growth, 469
Helmholtz free energy, *see* Free energy

Heterogeneous nucleation, *see* Nucleation
Homogeneous nucleation, *see* Nucleation

I

Inoculation, *see also* Grain refining
  degree of, 484
  experiments, 451
  role of, 274
Interdendritic feeding, 130
Interface
  faceted, 49, 466
  non-faceted, 198
Interface energy, 12, 30, 40, 169, 254, 255, 291
  liquid-solid particle, 45
Interface non-equilibrium, 22, 23, 283
Inverse segregation, 124
Investment casting, 92, 94, 443, 514
  of equiaxed grains superalloys, 517

J

Jackson Hunt (JH) model, 202

K

Kinetic undercooling, *see* Undercooling

L

Lamellar graphite, 124, 221, 459, 465, 467
Lamellar spacing, 199, 201, 259
  dependence of, 240
Lateral growth, 49, 51
  kinetics of, 53
Lattice Boltzman models, 427, 428
Limited diffusion in solid, 75, 77–79
Liquid
  feeding, 130
  shrinkage, 128
Local equilibrium, *see* Equilibrium
Local interface equilibrium, *see* Equilibrium

M

Macro-scale, *see also* Macrostructure, 369
  calculations, 414
  defect, 133
  local chemical differences (See also
    Macrosegregations), 355
  microsegregation models
    quantitative evaluation of, 135
  modeling of solidification, 345
  phenomena, 2
  segregation, 126
  use in solving transport equations, 61
Macrosegregation, 2, 80, 81, 124–126, 128,
    355, 359, 398, 441, 442, 526
  physics of, 125, 126

Macroshrinkage, 133
  metal shrinkage and feeding, 128, 130, 132
  modeling, 360
  shrinkage defects, 133
  thermal models, 360, 361
    fluid flow models, 363, 364
    volume calculation models, 362
Magnin-Kurz model, 207
Marangoni convection, 122, 240
Melt spinning, 22, 284
Meso-scale, 2
  3-D CAFE cellular automaton model, 522
  CA model, 420
  cellular automaton grain growth models, 416
  concept of, 3
  dendrite envelope growth models, 413
  models, 411
  problems, 427
  solidification at, 3
  stochastic models approach solidification
    modeling at, 411
Metal matrix composites, solidification of
  ex-situ, processing of
    reinforcements, infiltration of, 335
    spray casting, 336
    stir casting, 332, 334
    ultrasonic cavitation, 336
  freely moving particles, presence of, 307
    cellular/dendritic interface, 325, 326
    kinetic models, 312–316, 318, 319, 321
    material properties models, 311
    microstructure visualization models, 322
    planar interface, 308, 311
    S/L interface, 323, 324
  in-situ, processing of, 338, 339
  infiltration pressure
    microstructure effects, 331
    surface energy considerations, 327, 328
    transport phenomena considerations, 329,
      331
Metallic glasses, 288, 289, 291, 293
Micro-scale, *see also* Microstructure
  CA model, 420
  cellular automaton dendrite growth models,
    417
  dendrite growth, 442
  domain, 489
  eutectic S/L interface growth models, 414
  level, 135
    calculations, 414
    nonfaceted and faceted interfaces at, 49
  modeling of solidification, 380, 411
  solidification, 197

Micro-solute redistribution, 136
Microporosity models, 400, 402
Microsegregation models, 135, 136, 139, 140, 142, 143
Microshrinkage, 133
  defect size and shape, 263, 267
  shrinkage porosity formation, physics of, 267, 268
    gas evolution, liquid, 273
    gas pressure, pore, 273
    mushy zone, pressure, 270, 272
    pore growth, mushy zone, 278, 280
    pore nucleation, 274, 275, 277
Mixture-theory model, 63, 64, 355, 365
Modification, 221
Molecular dynamics models, 427–429
Momentum mass transport
  fluid flow
    controlled segregation, 124–126
    solute diffusion controlled segregation, 126, 128
  macroshrinkage
    metal shrinkage and feeding, 128, 130, 131
    shrinkage defects, 133
  Marangoni convection, 122
  mushy zone, 123
    Blake-Kozeny model, 124
    Hagen-Poiseuille model, 123
  natural convection, 119, 121, 122
  shrinkage flow, 119
Monotectic solidification
  classification of, 235
  mechanism of, 235, 236, 238, 240
Monte-Carlo models, 413, 414, 416
Multi-component alloys
  multi-component alloys, 255
  thermodynamics of, 251, 253
  thermophysical properties
    interface energy, 255
    microstructure, 255, 256, 258, 260, 261
    multi-component diffusion, 254, 255

N
Nano-scale
  model, 414
  phenomena, 2
  solidification solid-liquid (S/L) interface, 2
  structures, 410
Non-steady-state solidification, 94
Nonmetallic inclusions, 440
Nonvariant eutectics, 260
Nucleation, 400, 485
  and fragmentation, 185, 221
  and growth kinetics, 4, 29
  and growth of austenite dendrites, 447, 448
  and growth of primary aluminum dendrites, 494
  and growth of the (crystals, 229
  and growth of the Si phase, 498
  and growth rate, 289, 360, 380
  classic nucleation theory, 29
  classical theory of nucleation, 42
  dynamic, 45
  heterogeneous, 380, 383, 413
    nucleation models, 35, 37, 38, 40, 45, 380, 382, 385
  homogeneous, 30, 31
    theories for, 37
  instantaneous
    equation, 40
    nucleation model, 380, 382, 383
  multi-stage, 452
  of graphite, 22, 451
  of lamellar graphite iron, 451
  of spheroidal graphite, 453
  on bifilms, 276
  pore, 272, 274, 279
  random, 487
  reactions, 10
  sites
    volumetric density estimation of, 47
  time-dependent
    concept of, 43
  two-dimensional, 53, 467
  two-stage nucleation theory, 454
  velocity, 38
    calculations, 32–34

O
Oxide film, 500, 502, 507
Oxides, 37, 525

P
Particle
  engulfment, 238, 307, 324, 325
  pushing, 314, 322
Partition coefficient, 7, 16, 24, 25, 69, 172, 287
  derivation of, 83
Patternmaker shrinkage/t See Solid shrinkage, 129
Peritectic reaction, 229
Peritectic solidification
  classification of, 221
  mechanism of, 228–232, 234
  peritectic microstructures and phase selection, 223–225, 228
Peritectic transformation, 229, 230, 232

Phase field models, 407, 408, 410
Plate solidification, 131, 132
Pressure undercooling, *see* Undercooling

R

Rapid solidification, 283, 285, 286, 288, 290, 291, 294, 514
Resistance
 at mold/metal interface, 116
 in the mold, 97, 99, 100
 in the solid, 104

S

Scheil model, 80, 144, 193, 221, 399
Segregation, 79, 81
 comparison of, 85, 86
 negative, 358
 positive, 358
 type of, 124
Semi-solid processing, 295
 material models/systems, 301, 302
 phenomenology, 295, 297
 semi-solid slurry
  forming of, 301
  processing, 299, 300
Semi-solid slurry
 forming of, 301
 processing, 299, 300
Shrinkage porosity, 369, 370
Sievert's law, 273
Simulation method, 5
Simulation of solidification
 cast iron of, 485
Single crystal
 advantages of, 517
 behavior
  effect of temperature on, 524
 castings, 81
 development in a CMSX-4 Ni-based
  superalloy, 523
 formation of, 465
 growth processes, 93
 macrosegregation effects on, 355
 production of, 93
 solidification
  Bridgman-type techniques, 513
Single-phase alloys, solidification of, 47
 applications, 189, 191, 194
 cells and dendrites
  dendrite tip velocity, 170, 171
  extremum criterion, 162, 163
  interface anisotropy, 169
  marginal stability criterion, 163, 164, 166, 168

microsolvability theory, 168, 169
multicomponent alloys, 171, 172
solute diffusion controlled growth, 157, 158, 160
thermal diffusion controlled growth, 161
CET, 182–184, 186, 187
DAS and coarsening
 dendrite coherency, 181
 primary arm spacing, 173, 174
 secondary arm spacing, 175, 177–181
interface stability
 combined effects, 150–153
 convection, influence of, 153, 154
 solutal instability, 147, 149
 thermal instability, 145, 146
 primary phases, morphology of, 154, 156
Skin forming alloys, 132
Solid shrinkage, 129
Solidification
 contraction, 119, 124, 358
 shrinkage, 128
Solutal undercooling, *see* Undercooling
Solute redistribution, 69, 72, 73, 126, 135, 139, 356, 358, 359, 398, 441
Specific heat method, 347
Spheroidal graphite (SG), 45, 214
 nucleation of, 453
Spheroidal graphite iron, 485, 487, 489–491
Spray casting, 336
Squeeze casting infiltration, 335
Stable solidification, 471, 481
Steady-state solidification, 93, 174
Steel, 106, 156, 271, 435
 low-carbon, 364
 microstructure of, 437
 radiographs of, 403
Stir casting, 332, 334
Stochastic models, 410
Strip casting, 284, 521
Superalloys, 156, 263, 509
 applications of, 509
 columnar structure and single-crystals
  directional solidification of, 517, 518
 databases for, 251
 equiaxed grains
  investment casting of, 517
 melting of, 517
 microstructure of, 510
 modern, 513
 properties of, 513
 single-crystal, 522, 524
 solidification, 94
  processing of, 514

simulation of, 522
thermophysical properties of, 514
types of, 509
Surface energy, 13, 35, 151, 327

T
Tadpole graphite, 468
Thermal analysis (TA), 105, 106
  cast iron of, 484
  differential, 106, 109, 111, 113
Thermal undercooling, *see* Undercooling
Thermophysical properties, 254
  time-dependent, 354
Thixotropy, 297
Transport equations, 22, 329, 355, 358, 365,
      421
  and Cartesian coordinates, 419
  general conservation, 61
  macroscopic, 63, 64

U
Ultrasonic
  cavitation, 336
  processing, 502, 509
Unconstrained growth, 150
Undercooling, 9, 16, 17
  and disregistry

parabolic relationship between, 45
constitutional, 1, 15, 36, 148, 149, 152, 153,
      156, 183, 189, 224
  equations, 164
  critical, 34
  curvature of, 12, 14
  definition of, 11, 14
  for homogeneous nucleation, 31, 36
  for solidification, 21
  kinetic, 14, 19–21, 26, 52, 54, 145, 167
  local, 22
  nucleation, 227, 232
  pressure, 18, 23, 319
  solutal, 156, 165, 166, 189, 384, 391, 392
  thermal, 14, 15
Univariant eutectics, 256, 258–260

V
Volume-averaged models, 63, 385–387, 389,
      391, 396, 398
  continuum and, 385, 387

W
Wedge solidification, 131, 132

Z
Zone melting, 81, 83, 93

Printed in the United States
By Bookmasters